Integrated Life-Cycle and Risk Assessment for Industrial Processes and Products

Second Edition

Integrated Life-Cycle and Risk Assessment for Industrial Processes and Products

Second Edition

Edited by

Guido Sonnemann

Michael Tsang

Marta Schuhmacher

CRC Press
Taylor & Francis Group
Boca Raton London New York

CRC Press is an imprint of the
Taylor & Francis Group, an **informa** business

CRC Press
Taylor & Francis Group
6000 Broken Sound Parkway NW, Suite 300
Boca Raton, FL 33487-2742

First issued in paperback 2020

ISBN 13: 978-0-367-57088-0 (pbk)
ISBN 13: 978-1-4987-8069-8 (hbk)

Library of Congress Cataloging-in-Publication Data

Names: Sonnemann, Guido, editor. | Tsang, Michael, editor. | Schuhmacher, Marta, editor.
Title: Integrated life-cycle and risk assessment for industrial processes and products / [edited by] Guido Sonnemann, Michael Tsang, and Marta Schuhmacher.
Description: Second edition. | Boca Raton : Taylor & Francis, 2019. | Revised edition of: Integrated life-cycle and risk assessment for industrial processes / Guido Sonnemann, Francesc Castells, Marta Schuhmacher. c2004. | Includes bibliographical references.
Identifiers: LCCN 2018025095 | ISBN 9781498780698 (hardback : alk. paper)
Subjects: LCSH: New products--Environmental aspects. | Product life cycle--Environmental aspects. | Manufacturing processes--Environmental aspects. | Environmental risk assessment.
Classification: LCC TS170.5 .I58 2019 | DDC 658.5/75--dc23
LC record available at https://lccn.loc.gov/2018025095

Visit the Taylor & Francis Web site at
http://www.taylorandfrancis.com

and the CRC Press Web site at
http://www.crcpress.com

Dedication

For Marc and Julia Sonnemann Riba
and
Amara and Aymen Tsang-Alkharusi
and the next generations

Contents

Foreword to the Second Edition

To tackle climate change, increase energy security, and foster overall social and environmental sustainability, the transition to a low-carbon modern society is a central priority. Recent studies have identified potential limitations in economic access to a number of minerals and metals that may become bottlenecks in the supply chains necessary for wide-scale implementation of new energy technologies. These raw materials have been labeled as "critical" by academic authors and governmental agencies.

Whereas, risk assessment looks at events without focusing on the specific systems leading to the event, criticality assessment evaluates failure of specific elements within a larger system. Analyzing risks is important to prepare for, to mitigate against, and to reduce the likelihood of negative impacts of events. Establishing criticality assessment in common practice draws attention and points to the need for resources to be better allocated to ensure reliability of critical systems, products and components.

I am pleased to see that our recent work on integrating criticality assessment of raw materials into life-cycle sustainability assessment has become part of the new version of this book on *Integrated Life-Cycle and Risk Assessment for Industrial Processes and Products*. In this volume, Guido Sonnemann, Michael Tsang and Marta Schuhmacher deliver a practical introduction to life-cycle assessment, to risk assessment—and now also to criticality assessment. The main focus of the book, like the first edition, remains on the integrated assessment of risks to human health. It now covers not only chemical substances, but also new engineered nanomaterials in the life-cycle assessment sections of the new edition.

This textbook continues to be an important reference on how to integrate life-cycle assessment and risk assessment—both are fundamental tools in the modern sustainability toolbox. The book advocates for integration at the methodological level, rather than after analysis is done.

I wish all users of this new edition every success in their learning of these approaches, and in their contributions to sustainable development.

Steven B. Young
University of Waterloo, Canada

Foreword to the First Edition

Much of the twentieth century's focus has been on economic progress, toward which humankind has made giant steps. Increasingly, unwanted side effects such as acid rain, climate change and various forms of environmentally induced toxic pollution are becoming more manifest and demanding increased attention. With the need for equal development opportunities for all, the issue of sustainability is becoming increasingly important; in my opinion, it will be one of the key issues for the twenty-first century. As such, sustainable development is about the welfare of human beings and a natural environment that does not reduce the possibilities of future generations, without losing sight of economic continuity of the current generation.

We are beginning to recognize that the path toward sustainability requires a life-cycle approach. The internal logic of life-cycle thinking extends the traditional focus of environmental engineering on production facilities to all stages of the value chain, which are relevant from an environmental point of view, including the production, consumption, use and waste management phases. This implies a holistic, system-analytical point of view and the cooperation between the different stakeholders throughout the life of the product.

A largely used instrument in the assessment of environmental impacts is *risk assessment* for chemical substances applied, for instance, to accident forecasting and regulatory monitoring of industrial facilities. There is a clear necessity to link this approach with the existing environmental analysis tool that applies a life-cycle perspective, *life-cycle assessment*, that is alive and well in the U.S.

This is why I enthusiastically welcome *Integrated Life-Cycle and Risk Assessment for Industrial Processes*. In this book, Guido Sonnemann, Francesc Castells, and Marta Schuhmacher provide not only an updated introduction to life-cycle assessment, but also to risk assessment. While they demonstrate the potential for a further integration of these two approaches, they also show the limits and constraints. I certainly agree with them on the usefulness and need for a toolbox comprised of varied environmental system analysis methods and approaches.

There are quite a lot of endeavors to be undertaken in environmental system analysis. We are still at the beginning. I look forward to our continued journey together.

Mary Ann Curran

Preface

The simplistic perspective that emission source control is *the* way to improve the environmental behavior of an industrial process has to be replaced by a more systematic approach that integrates all of potential environmental impacts that can be assigned to a product, process or service. This book is a direct response to this challenge, developing environmental impact analysis with a global systems approach.

Life-cycle assessment is a methodology used to evaluate the environmental impacts of a product, process, or service during the span of its life cycle, and *risk assessment* is a tool to evaluate potential hazards to human health and the environment introduced by pollutant emissions. This textbook does not focus on these specific analytical tools in isolation of one another, but it looks instead at their complementary use, embedding them in a toolbox for environmental impact assessments.

Such a convergence of tools is necessary due to potential contradictory results when applying each tool separately. In recent years, life-cycle impact assessment has evolved, particularly from incorporating elements of risk assessment associated with chemical products, engineered nano-materials, and critical raw materials. It has also been increasingly embedded in environmental management practices. However, until now, these developments have not been reflected in the literature for graduate students, environmental professionals, and a host of other interested stakeholders. This book aims to fill in that demand. Before issuing the first addition of this book, we had identified a clear need for information provided in the form of a textbook with exercises designed for checking the understanding of the content, and a manual with flowcharts and examples for guidance. Since then, there have been many new trends in the marketplace and emerging industries that have challenged how existing impact assessment models can adequately address conventional environmental questions. Thus, the second issue of this book looks at some of these particular instances and draws out of them ways in which the complementary and integrated use of life-cycle assessment and risk assessment can be fostered.

The book covers life-cycle assessment, risk assessment, and a combined framework of both for environmental damages estimations in industrial process chains and site-dependent impact assessment. Methodological explanations accompany the description of practical applications of environmental impact analysis for industrial products, processes, and services.

The first four chapters of the book give a general overview of environmental management strategies, describe life-cycle assessment and risk assessment and place them in the so-called environmental management toolbox. The fifth chapter provides information on the recently developed approaches for the criticality assessment of raw materials. Chapter 6 focuses on the combination and integration of life-cycle assessment and risk assessment, followed by Chapter 7 that focuses on life-cycle toxicity assessment. The eighth chapter provides insights into a newly developed approach to integrated life-cycle and criticality assessment. Chapter 9 supplies additional information on techniques for data analysis that are commonly used in the analysis of environmental impacts. The tenth and eleventh chapters show interfaces between life-cycle assessment and risk assessment and provide ways of integrating the two beyond life-cycle toxicity assessment. In the final and twelfth chapter of the book, several case studies of life-cycle, risk assessment, and integrated life-cycle and risk assessments are presented.

The parallel presentation of both tools, life-cycle assessment and risk assessment, is a unique feature of this book that explains its success, which has led to the second edition. Only recently have both tools been further developed in a way that allows assessing engineered nanomaterials and critical materials in an integrated way. Therefore, the new sections and case studies of the book dealing with these challenges can be considered as pilots. Now, not only does the case study on the Municipal Solid Waste Incinerator of Tarragona (Spain), from the first edition, develop from one chapter to another, but also there is an additional case study following an engineered nanomaterial.

In this updated series of case studies, the combined application of life-cycle assessment and risk assessment casts new light on the controversy of the recourse efficiency of engineered nanomaterials on one hand and concerns over their human health effects on the other.

Some specialized knowledge in environmental science and technology is necessary to best understand the book in its entirety. Mathematical parts of the book are expressed by clearly explained equations. Figures systematically illustrate the content. Tables provide basic data and exemplary results of the case study and the exercises.

Apart from being a reference book for graduate and postgraduate students in the field of environmental management, the book intends to catalyze communication between life-cycle assessment and risk assessment experts and scientists from academia, consultancies, industry, and governmental agencies. In this way, the book is a manual for analyzing situations that are relevant for decision-making. The reader profits directly from the practical format of the book, including flowcharts, examples, exercises, and concrete applications, making the book a useful guide and facilitating understanding of the content.

The UN's Sustainable Development Goals adopted in 2015 calls for, among other objectives, responsible consumption and production by decoupling environmental resource use and environmental impacts from economic growth and human well-being. Life-cycle assessment and risk assessment are both analytical systems approaches that allow science-based knowledge creation according to the current state of understanding of environmental mechanisms. This book can still be considered as one of the first attempts to illustrate the existing interfaces between life-cycle assessment and risk assessment and to indicate options for the further integration of both tools.

Finally, the authors would like to take this opportunity to thank everyone who has contributed to this book both directly and indirectly, in particular, the researchers in the field of life-cycle assessment and risk assessment, who have supported our work through numerous information exchanges and who are not specifically mentioned in the acknowledgments.

Guido Sonnemann

Michael Tsang

Marta Schuhmacher

Acknowledgments

The previous edition and the core of the updated second edition of this book are based on a Ph.D. thesis elaborated at the Chemical Engineering Department of the Universitat Rovira i Virgili in 2002 by Guido Sonnemann, with the supervision of Prof. Dr. Francesc Castells and Prof. Dr. Marta Schuhmacher. The entire Environmental Management and Analysis Group (AGA) proved to be an excellent group with a rich intellectual stimulus. In particular, we would like to mention former AGA group members Ralph Harthan, Dr. Israel Herrera, Dr. Luiz Kulay, Dr. Montse Meneses, Yolanda Pla, Julio Rodrigo, and Dr. Haydée Yrigoyen, who provided valuable contributions to Chapters 4, 8, 10, and 12 of this book. Dr. Andreas Ciroth (GreenDelta TC, Berlin) and Marcel Hagelüken shared with us the modular model for municipal solid waste incinerators. Juan Carlos Alonso and Christoph Schäfer prepared and reviewed the life-cycle assessment case study of the SIRUSA waste incinerator. The ideas of Dr. Karl-Michael Nigge (Simon–Kucher & Partners, Bonn) were essential in the development of site-dependent impact assessment factors in Chapter 10. Many thanks to all! Furthermore, at the Chemical Engineering School, our special gratitude goes to Prof. Dr. Josep Maria Mateo for resolving several doubts in statistics and to Dr. Miguel Angel Santos for doing the same in mathematics.

Most additions to the second edition come from the Ph.D. thesis "Life-cycle Assessment of 3rd-Generation Organic Photovoltaic Systems: Developing a Framework for Studying the Benefits and Risks of Emerging Technologies" completed at the University of Bordeaux in 2016 by Dr. Michael Tsang with the supervision of Prof. Dr. Guido Sonnemann and Dr. Dario Bassani, both from the Institute of Molecular Sciences (ISM). Based on this Ph.D., Dr. Michael Tsang took the lead in revising the book. The contributions from those parts of the book could not have happened without the indispensable collaboration and support from Dr. Danail Hristozov and Dr. Antonio Marcomini and their colleagues at Ca' Foscari University in Venice, Italy, as well as from Dr. Sangwon Suh and Dr. Arturo Keller and their colleagues at the University of California, Santa Barbara, U.S.A.

Moreover, Prof. Dr. Guido Sonnemann and Prof. Dr. Marta Schuhmacher, supported by Dr. Dieuwertje Schrijvers from the Sustainable Chemistry and Life Cycle Group (CyVi) at the University of Bordeaux and Dr. Montserrat Mari from the AGA group at the the Universitat Rovira i Virgili, revised Chapters 1, 2, 3, and 4. Dr. Ioannidou Dimitra and Dr. Eskinder Gemechu, current and former Post-Docs at the CyVi group, contributed to Chapters 5 and 8 with the integration of criticality assessment of raw materials. Finally, Francisco Sanchez from the AGA group provided a new case study in Chapter 12. Again many thanks to all! We wish to also thank Ramon Nadal (SIRUSA) for the data from the municipal solid waste incinerator (MSWI) of Tarragona, Spain. We would like to express our gratitude to the Meteorological Service of Catalonia (SMC), especially to Jordi Cunillera, for providing meteorological data. We specially acknowledge the financial support provided by SIRUSA (Municipal Waste Incinerator Plant) and the General Directorate of Environmental Quality (Ministry of Territory and Sustainability, Generalitat de Catalunya, Spain); without it, the transformation of the Ph.D. thesis into this textbook would not have been possible.

Special gratitude goes to Ahmed Al Bualy for his support of the final editing and submission process and to Irma Britton and Claudia Kisielewicz from the CRC Taylor & Francis Group for providing us the chance to prepare a second edition, and for their patience while preparing for this, as well as to our families who have also supplied a gracious amount of patience, understanding, and support to us while preparing this book.

Editors

Guido Sonnemann is a full professor at the University of Bordeaux, France, where he holds a chair of excellence on Life Cycle Assessment and heads the Life Cycle Group CyVi at the Institute of Molecular Sciences. The group does research on life cycle assessment and sustainable chemistry. Dr. Sonnemann's main research interests are in the fields of life-cycle assessment and sustainable materials management; he is especially interested in the linkages of life-cycle approaches to risk assessment, including organic chemicals, engineered nanomaterials, and critical raw materials.

Until 2012, Prof. Dr. Sonnemann was UNEP Programme Officer for Sustainable Innovation and Life Cycle Management. In this function, he served as the Science Focal Point for the UNEP's Resource Efficiency sub-programme. He worked in the Sustainable Consumption and Production (SCP) Branch, Division of Technology Industry and Economics (DTIE), in Paris. He was co-initiator of the International Resource Panel and was responsible for the working groups on global metal flows and water efficiency. He set up and oversaw the Secretariat of the UNEP/SETAC Life Cycle Initiative throughout the 10 years in UNEP. In addition, he was involved in the development of several other innovative projects.

Before joining UNEP in 2002, Prof. Dr. Sonnemann worked as consultant with the Chemical Technology Centre of Tarragona/Spain. He holds a Ph.D. in Chemical Engineering from the University Rovira & Virgili in Tarragona, Spain (2002), M.Sc. in water biology and chemistry from the Engineering School of Poitiers, France (1996), and graduated as environmental engineer from the Technical University of Berlin, Germany (1995), with a specialization in waste management and pollution prevention. He is an author of a number of books and UN reports, as well as more than 100 scientific papers and book chapters. He is co-chair of the Executive Committee of the Forum for Sustainability through Life Cycle Innovation (FSLCI) and a member of the Life Cycle Initiative and the Society of Environmental Toxicology and Chemistry (SETAC).

Michael Tsang, Ph.D., is the owner and principal consultant at Three Pillars Consulting, an environmental and sustainability services company based in the Sultanate of Oman. He has been working for many years in the fields of environmental sciences and public health and how they related to industries and economic activity. Specifically, Dr. Tsang has focused on the risks associated with and the benefits offered by emerging technologies to society. Such technologies include renewable energies such as solar photovoltaics and biofuels, as well as engineered nanomaterials used in agriculture, textiles, and electronics. He has presented at major international conferences on the subject and has published the results of his work in highly ranked international journals such as Nature Nanotechnology.

Prior to starting Three Pillars Consulting, Dr. Tsang was a contract-consultant to the U.S. Army Corps of Engineers Engineer Research and Development Center, in the department of Risk and Decision Sciences. There, he worked on risk assessments, life-cycle assessments, and multi-criteria decision analysis. Dr. Tsang was also a research fellow at the U.S. Environmental Protection Agency in Washington, D.C., within the Office of Research and Development's Chemical Safety for Sustainability program. There, he worked on building novel datasets and environmental models for chemicals.

Dr. Tsang holds a Ph.D. on the topic of Eco-Design of Third Generation Photovoltaic Solar Cells from the University of Bordeaux. He graduated with a bachelor's degree in biology from the University of California, San Diego, and with a master's degree in public health, with a focus on environmental sciences and toxicology, from the University of California, Los Angeles.

Marta Schuhmacher is a full professor in the School of Chemical Engineering at the Rovira i Virgili University (Tarragona, Spain). She received her B.Sc. in chemistry from the University of Zaragoza (1976), her B.Sc. in oceanography chemistry from UNED (1991), her Ph.D. in analytical chemistry from the University of Zaragoza (1991), and her master's degree in environmental engineering and management from the School of Industrial Organization (EOI) (1995).

She has been invited as scientific expert to different research groups, such as the Department of Environmental Health Sciences, School of Hygiene and Public Health, Johns Hopkins University, Baltimore, MD; Nelson Institute of Environmental Medicine, New York University; Departement de Medicine et Hygiene du Travail et de l'Environnement (MDTE), the University Catholique de Louvain; Environmental Organic Chemistry and Ecotoxicology Research Group, Environment Science, University of Lancaster.

Prof. Dr. Schuhmacher co-managed the master environmental engineering and management program and the master occupational health and safety program at Rovira i Virgili University (Tarragona). Her main research interests are environmental impact pathway analysis, risk assessment, environmental toxicology and health, occupational exposure, multimedia model, life-cycle assessment, substance flow analysis, environmental indicators, and Monte Carlo analysis. Prof. Dr. Schuhmacher has published more than 300 scientific papers and book chapters. She has served in international symposia as a co-chairman and an invited lecturer. She is a member of the NEHA (National Environmental Health Association) and the SETAC (Society of Environmental Toxicology and Chemistry).

Contributors

Francesc Castells
Universitat Rovira i Virgili (URV)
Tarragona, Spain

Eskinder Demisse Gemechu
University of Alberta
Edmonton, Alberta, Canada

Ralph O. Harthan
Oeko-Institut
Freiburg, Germany

Dimitra Ioannidou
University of California in Santa Barbara
(UCSB)
Santa Barbara, California

Luiz Kulay
Universidade de Sao Paulo
Sao Paulo, Brazil

Vikas Kumar
Universitat Rovira i Virgili (URV)
Tarragona, Spain

Montserrat Mari
Universitat Rovira i Virgili (URV)
Tarragona, Spain

Montserrat Meneses
Universitat Autònoma de Barcelona
Barcelona, Spain

Karl-Michael Nigge
Volume Graphics
Heidelberg, Germany

Yolanda Pla
Tecnicas Reunidas, Spain

Julio Rodrigo
La Fundació Centre de Difusió Tecnológica
Fusta I Moble de Catalunya - CENFIM
Tarragona, Spain

Francisco Sánchez-Soberón
Universitat Rovira i Virgili (URV)
Tarragona, Spain

Dieuwertje Schrijvers
Université de Bordeaux
Bordeaux, France

Marta Schuhmacher
Universitat Rovira i Virgili (URV)
Tarragona, Spain

Guido Sonnemann
Université de Bordeaux
Bordeaux, France

Michael Tsang
Three Pillars Consulting
Muscat, Oman

Haydée Yrigoyen
Danone
Cuajimalpa de Morelos, Mexico

General Abbreviations, Symbols, and Indices

ABBREVIATIONS

a-Si	Amorphous silicon
AA	Air acidification
ADI	Acceptable daily intake
AE	Aquatic eco-toxicity
AER	Air exchange rate
AETP	Aquatic eco-toxicity potential
AGA	Environmental Management and Analysis Group
AGTS	Advanced acid gas treatment system
AoP	Area of protection
AP	Acidification potential
APEA	Air pathway exposure assessment
APME	Association of Plastic Manufacturers in Europe
AT	Air toxicity
ATSDR	Agency for Toxic Substances and Disease Registry
b_m	Baseline mortality
BAL	Bronchi-alveolar lavage
BAT	Best available technologies
BMC	Benchmark concentration
BMD10	Benchmark dose
BMR	Benchmark response
BOD	Biological oxygen demand
BOS	Balance of system
BUWAL	Swiss Federal Office of Environment, Forests and Landscape
C_{60}	Fullerenes
CalTOX	Californian EPA Multimedia Total Exposure Model for Hazardous-Waste Sites
CBA	Cost-benefit analysis
CCE	Cumulative carbon equivalents
CDI	Context-dependent integration
CEA	Cost-effectiveness analysis
CED	Cumulative energy demand
CF	Characterization factor
CHAINET	European network on chain analysis for environmental decisions support
CMA	Chemicals Manufacturers Association
CML	Centre of Environmental Science at Leiden University
COD	Chemical oxygen demand
COI	Costs of illness
CP	Cleaner production
CPBT	Carbon payback time
CTU	Comparative toxic unit

CV	Coefficient of variation
CVM	Contingent valuation method
CyVi	Sustainable chemistry and life cycle group, University of Bordeaux
DALY	Disability-adjusted life years
DDE	Dynamic data exchange
DEAM	Data for environmental analysis and management
DeNOx	Control device to reduce No_x emissions
DfE	Design for environment
DI	Dustiness index of particulate matter
DM	Dematerialization
DNRR	Depletion of nonrenewable resources
DOL	Depletion of ozone layer
D–R	Dose–response
DT	Detoxification
DW	Disability weight
DWD	Deutscher Wetterdienst
EB	Ecobalance
Eco-ind.	Eco-indicator
EcoSense	Integrated impact assessment model for impact pathway analyses
ED	Energy depletion
ED$_{10}$	Effect dose, including a 10% response over background
EDIP	Environmental design of industrial products
EEC	External environmental cost
EF	Effect factor in life-cycle impact assessment
E$_g$	Energy generated over during the use of a photovoltaic panel
EIA	Environmental impact assessment
EIME	Environmental information and management explorer
EL	Environmental load
ELG	Eco-labeling
ELV	End-of-life vehicles
EMA	Environmental management and audit
EMAS	Environmental management and audit scheme
EMPA	Swiss Federal Laboratories for Materials Testing and Research
EMS	Environmental management system
ENM	Engineered nanomaterial or nanoparticle
EPA	U.S. Environmental Protection Agency
EPBT	Energy payback time
EPD	Environmental product declaration
EPR	Extended producer responsibility
EPS	Environmental priority strategies
Equiv.	Equivalent
ER	Excess risk
E–R	Exposure–response
ERA	Environmental risk assessment
ERV	Emergency room visit
ES	Exposure scenario
ETH	Eidgenössische Technische Hochschule

ETL	Electron transport layer
EUSES	European Union System for the Evaluation of Substances
FDA	U.S. Food and Drug Administration
FP7	The Seventh Framework Programme (2007–2013) of the European Commission
FTO	Fluorine-doped tin oxide
FU	Functional unit
GE	Greenhouse effect
GenCat	Generalitat de Catalunya
GI	Global integration
GIS	Geographic information system
GLC	Ground level concentration
GM	Geometric mean
GSD	Geometric standard deviation
GW	Global warming
GWP	Global warming potential
HHRA	Human health risk assessment
HQ	Hazard quotient
HT	Human toxicity
HTF	Human toxicity factor
HTL	Hole transport layer
HTP	Human toxicity potential
HU	Hazard unit
HWP	Hazardous waste production
IA	Inventory analysis
IE	Industrial ecology
IEA	International Energy Agency
iF	Intake fraction in life-cycle impact assessment
IOA	Input–output analysis
IP	Intermediate product
IPA	Impact pathway analysis
IRIS	Integrated risk information system
ISC	Industrial source complex
ISCST	Industrial source complex short term model
ISO	International standard organization
ISM	Institute of Molecular Sciences, University of Bordeaux
I-TEQ	International toxic equivalence
ITO	Indium tin oxide
LCA	Life-cycle assessment
LCI	Life-cycle inventory
LCIA	Life-cycle impact assessment
LCM	Life-cycle management
LO(A)EL	Lowest observable (adverse) effect level
m-Si	Multi-crystalline silicon
MC	Monte Carlo
MCE	Multi-criteria evaluation
MEI	Maximal exposed individual
MFA	Material flow accounting

MHSW	Mixed household solid waste
MIPS	Material intensity per service unit
MRL	Minimum risk level
MSWI	Municipal solid waste incinerator
mUS$	10^{-3} US$
NA	Not available
NF	Near-field workroom volume
NIOSH	U.S. National Institute of Occupational Safety and Health
NMVOC	Non-methane volatile organic compounds
NO(A)EL	No observed (adverse) effect level
NOEC	No effect concentration
NP	Nitrification potential
OD	Ozone depletion
ODP	Ozone depletion potential
OPV	Organic photovoltaics
P3HT	Poly(3-hexyl)thiophene
PAF	Potentially affected fraction
PAH	Polyaromatic hydrocarbons
PBPK	Physiologically based pharmacokinetic
PCBM	[6,6]-Phenyl C61 butyric acid methyl ester
PCDD/Fs	Dioxins and furans
PCS	Phagocytizing cells
PE	Population exposure
PEC	Predicted environment concentration
PEDOT:PSS	Poly(3,4-ethylenedioxythiophene) polystyrene sulfonate
PET	Polyethylene terephthalate
PM	Particulate matter (PM_{10}: particle with an aerodynamic diameter of ≤ 10)
PNEC	Predicted no effect concentration
POC	Photochemical ozone creation
POCP	Photochemical ozone creation potential
POF	Photochemical oxidant formation
POP	Workplace population
PP	Pollution prevention
PR	Process
PV	Photovoltaic
PWMI	European Center for Plastics in the Environment
QSAR	Quantitative structure activity relationship
R*Y	Reserve size times remaining years
RA	Risk assessment
RCW	Relative concentration weighted
RDW	Relative deposition weighted
RE	Receptor exposure
REL	Recommended exposure limit
REW	Relative exceedance weighted
RfC	Reference concentration
RfD	Reference dose

RM	Raw material
RiF	Retained-intake fraction for use in life-cycle impact assessment
RIVM	Netherland's National Institute for Public Health and the Environment
RMD	Raw material depletion
RME	Reasonable maximum exposure
SAR	Structure activity relationship
SETAC	Society of Environmental Toxicology and Chemistry
SF	Slope factor
SFA	Substance flow accounting
SIRUSA	Societat d'Incineració de Residus Urbans, S.A.
SMC	Servei Meteorologic de Catalunya
SPI	Sustainable process index
SPOLD	Society for the Promotion of LCA Development
SSI	Site-specific integration
ST	Short term
TC	Technical committee
TD	Tumor dose
TE	Terrestrial eco-toxicity
TEQ	Toxicity equivalent
TEAM	Tool for environmental analysis and management
TOC	Total organic carbon
TQM	Total quality management
TSP	Total suspended particulate matter
UF	Uncertainty factor
UNEP	United Nations Environment Programme
US-EPA	United States Environmental Protection Agency
UWM	Uniform world model
VLYL	Value of year lost
VOC	Volatile organic carbon
VSL	Value of statistical life
WBCSD	World Business Council of Sustainable Development
WD	Water depletion
WE	Water eutrophication
WHO	World Health Organization
WISARD	Waste integrated systems assessment for recovery and disposal
Wp	Watt-peak
WT	Water toxicity
WTA	Willingness to accept
WTM	Windrose trajectory model
WTP	Willingness to pay
XF	Exposure factor in life-cycle impact assessment
YLD	Years of life disabled
YOLL	Years of life lost
Yr	Year

SYMBOLS

ρ	Receptor density
$\tilde{e}$	Weighted eco-vector
φ	Angle
η	Efficiency
ξ	Standard deviation (of the Gaussian variable)
Δ	Increment
μ	Ordinary mean
λ	Specific Weighting Factor
θ	Standard deviation
σ	Standard deviation (of log-normal distribution)
μ_g	Geometric mean
σ_g	Geometric standard deviation
A	Area
τ	Residence time
c	Concentration
C	Cost
D	Damage
E	Effect (factor)
E	Energy flow
E_M	Eco-technology matrix
e_v	Eco-vector
F	Fate and exposure (factor)
H	Variation insensitivity in the human population
h	Height
I	Incremental receptor exposure per mass of pollutant emitted
k	Removal velocity
M	Mass
P	Quantity of pollutants
Q	Emission rate
r	Radius or discount rate
R	Outer boundary of the modeling area
T	Duration
tkm	Transport unit equivalent to a mass of 1 t (1000 kg) transported along 1 km
u	Mean wind speed
v	Vector
V	Volume
W	Waste
WE_M	Weighted eco-matrix or damage matrix
W_M	Weighting or damage-assigning matrix
X_o	Damage today
X_t	Damage through the years
z_s	Height above mean sea level

INDICES

1	Primary pollutant
2	Secondary pollutant
CR	Concentration response
e	Energy or ecological
e, eco	Ecological
EE	Ecosystem exceeded
eff	Effective
Env	Environment
far	Long-range contribution
g	Geometric
i	Emission situation
m	Mass or target medium
M	Matrix
n	Initial medium
near	Short-range contribution
p	Pollutant
P	Product
Prod	Production
r	Receptor
RE	Relative exceedance
s	Stack
sed	Sediment
uni	Uniform
y	Lateral
z	Vertical

1 Basic Principles in Sustainability Management

Guido Sonnemann, Francesc Castells, and Marta Schuhmacher

CONTENTS

1.1 PROBLEM SETTING

Now and in the coming years, industry must play a paramount role with respect to the sustainability, not only as one of the main sources of environmental impact but also as one of the main actors in the proposal of innovations and value creation. Industry-related environmental policy was originally intended to control emissions in various environments. It was widely thought that corrective technical measures at the end of the pipe would sufficiently reduce environmental impact. However, as we have seen through the years, this is insufficient in stopping progressive environmental degradation and also lacks flexibility for an evolving industry. On the one hand, it is necessary to take a quantitative leap in this approach, including the expression of risk; on the other hand, environmental considerations must be included in the entire range of industrial management. That means we must consider environmental impact within all phases of production, marketing, use and end of life once a product's life is over. Figure 1.1 presents an overview of these conceptually related methods in the environmental pillar of sustainability management. They will be explained in this chapter starting with overall strategies in the area of sustainable development, using well-established literature that can be seen as the fundaments of sustainability management.

1.2 OVERALL STRATEGIES

For the world to make substantial progress toward becoming a safer planet, it is necessary to introduce sustainability considerations in all aspects of industrial management practices for all phases of production, marketing, use, and end of life of a product. Based on these reflections, different general objectives have been formulated as programs that intend to encompass the idea of good environmental management sustainability as shown in Figure 1.1.

1.2.1 Sustainable Development

Sustainable development is understood as satisfying the needs of the present generation without compromising the needs of future generations. Sustainability takes into account three aspects:

1. *Economic*: we need economic growth to assure our material welfare.
2. *Environmental*: we need to minimize environmental damage, pollution, and exhaustion of resources.
3. *Social*: the world's resources should be shared more equitably between the rich and the poor.

Agenda 21 is a strategic document that was adopted by the United Nations Conference on Environment and Development (UNCED) in Rio de Janeiro in 1992. At the Rio Summit, or Earth Summit, as it is also known, representatives from 179 nations gathered in what would become the

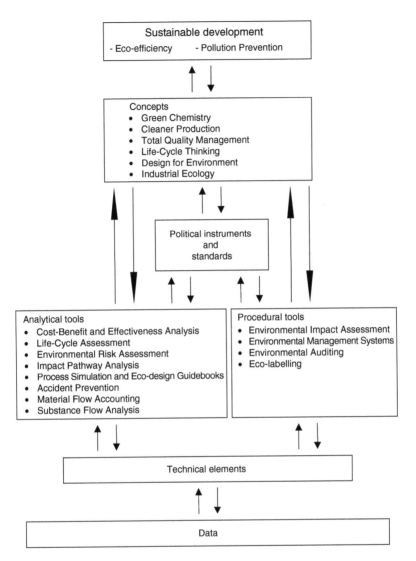

FIGURE 1.1 Conceptually related methods in environmental sustainability management. (Adapted from De Smet, B. et al., *Life-Cycle Assessment and Conceptually Related Programs*, working group draft, SETAC-Europe, Brussels, Belgium, 1996.)

end of a 2-year effort intended to define a model for sustainable development. The Earth Summit was a historical event in which a new global commitment for sustainable development was established. This commitment respected the fact that environmental protection and the development process are indivisible. Based on political commitment and global consensus at its highest level Agenda 21 was an action plan for the 1990s and the early years of the twenty-first century. At the same time, it stood as a global alliance of humankind regarding environment and development, that is, for sustainable development. Agenda 21 is a large document divided into 40 chapters and written to foster an action plan. The goal of this project was to see that development becomes sustainable in social, economic, and environmental terms.

In 2002, the World Summit for Sustainable Development in Johannesburg reviewed the implementation of Agenda 21 over the past 10 years. The world's political situation in 2002 was far different from the one that marked the Rio Earth Summit of 1992. One positive outcome was the new partnership among governments, civil society, industry, and the United Nations (UN) in areas such as corporate responsibility and environmental standards. It is an encouragement to industries to improve their social and environmental performance, taking into account the International Organization for Standardization (ISO) and the Global Reporting Initiative (GRI).

In 2012, the 10-Year Framework of Programs on Sustainable Consumption and Production Patterns (10YFP) was adopted by the Heads of State at the United Nations Conference on Sustainable Development (Rio+20)—as stated in paragraph 226 of the Rio+20 Outcome Document "The Future We Want" (UNCSD, 2012).

The 10YFP is a concrete and operational outcome of Rio+20. It is a global framework of action to enhance international cooperation to accelerate the shift towards sustainable consumption and production (SCP) in both developed and developing countries. The framework supports capacity building, and provides technical and financial assistance to developing countries for this shift. The 10YFP develops, replicates, and scales up SCP and resource efficiency initiatives, at national and regional levels, decoupling environmental degradation and resource use from socio-economic development, and thus increases the net contribution of economic activities to poverty eradication and to social development. It responds to the 2002 Johannesburg Plan of Implementation, and builds on the eight years of work and experience of the Marrakech Process—a bottom-up multi-stakeholder process, launched in 2003 with strong and active involvement from all regions (UNEP, 2012).

Perceiving the dire need for sustainable development, the United Nations and world leaders formulated the Sustainable Development Goals (SDGs) in 2015 as part of the 2030 Agenda for Sustainable Development, a comprehensive framework based on the success of the Millennium Development Goals (MDGs). The goals call for action by all countries, poor, rich, and middle-income, to promote prosperity while protecting Planet Earth and its life support system. Each goal has specific targets to be achieved over the next 15 years. In striving for sustainability, it is important to have inputs from all sectors, societies, and stakeholders. The following 17 SDGs were adopted:

1. No poverty
2. Zero Hunger
3. Good Health and Well-Being
4. Quality Education
5. Gender Equality
6. Clean Water and Sanitation
7. Affordable and Clean Energy
8. Decent Work and Economic Growth
9. Industry, Innovation and Infrastructure
10. Reduced Inequalities
11. Sustainable Cities and Communities
12. Responsible Consumption and Production
13. Climate Action
14. Life Below Water
15. Life on Land
16. Peace, Justice and Strong Institutions
17. Partnerships for the Goals

The philosophy of sustainable development has turned into a valuable guide for many communities that have discovered that traditional methods for planning and development create more social and ethical problems than the ones they solve, and that sustainable development offers them real and long-lasting solutions to consolidate their future. Sustainable development makes possible the efficient use of resources, building of facilities, quality of life protection and enhancement, and the establishment of new businesses to strengthen economies. It may help in building healthy

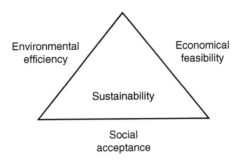

FIGURE 1.2   The three elements of sustainability.

communities capable of sustaining present and future generations. Sustainability can be seen as a triangle with each of its cornerstones representing environmental, economic, and social elements (Figure 1.2). Put simply, sustainability is the balance among these three elements; achieving a steady balance demands equal attention to each element.

1.2.2   ECO-EFFICIENCY

The philosophy of eco-efficiency was first introduced in 1992 in *Changing Course,* a report by the World Business Council for Sustainable Development (WBCSD). In 1993, eco-efficiency was defined in more detail at the First Antwerp Workshop on Eco-Efficiency. The council arrived at the conclusion that eco-efficiency is reached by the delivery of competitively priced goods and services that satisfy human needs and bring quality of life, while progressively reducing ecological impacts and resource intensity throughout the life-cycle to a level that is at least in line with the earth's estimated carrying capacity.

Based on this definition, eco-efficiency may be understood as a philosophy aimed at setting a framework for measuring the degree of sustainable development attained and for which indicators are developed. The WBCSD (1999) has identified seven actions to attain eco-efficiency:

1. Reduce material intensity of goods and services.
2. Reduce energy intensity of goods and services.
3. Reduce toxic dispersion.
4. Enhance materials' ability to be recycled.
5. Maximize sustainable use of renewable resources.
6. Extend product durability.
7. Increase service intensity of goods and services. This philosophy can be applied at a business level; for instance, Dow Chemical has developed a six-point eco-efficiency compass for its eco-innovation efforts that challenges its managers to:
   • Dematerialize to achieve reduction of raw materials, fuels and utilities in the product-service system.
   • Increase energy efficiency to determine where larger quantities of energy are consumed.
   • Eliminate negative environmental impact and reduce and control dispersion of pollutants related to the system's end of life.
   • Redesign products or their use for significant reduction of energy, material consumption, and pollutant emission.
   • Close the loop by means of effective and efficient recycling.
   • Mirror natural cycles by designing the system as a part of a longer natural cycle in which materials taken from nature can be returned to it (UNEP and WBCSD, 2017).

1.2.3 POLLUTION PREVENTION

Pollution prevention (PP2) entails avoiding pollutant production right before pollutants are issued at ends of pipes, through stacks or into waste containers. The principle of prevention says that prevention is better than cure and is related to concepts such as waste reduction, waste minimization, and reduction at source. The PP2 principle may be easier to understand than to implement because establishing the boundaries between wise and unwise prevention procedures is not as easy as it may seem. This principle was created before the emergence of the philosophy of sustainable development. The waste management hierarchy list, established by the 1990 Federal Pollution Prevention Act (U.S.), may serve as reference:

1. Whenever feasible, pollution or waste should be prevented or reduced at the source.
2. If the pollution or waste cannot be prevented, reusing or recycling is the next preferred approach.
3. If the pollution or waste cannot be prevented or recycled, safe treatment must be carried out.
4. Disposal or other release into the environment should be employed as a last resort and accomplished in a safe manner.

Some examples of PP2 measures are (U.S. EPA, 2017a):

- Raw material replacement
- Product replacement
- Process redesign
- Equipment redesign
- Waste recycling
- Preventive maintenance (i.e., pump-end lock leaks)
- Stock minimization to prevent future wastes
- Solvent adsorption or distillation in water and later recycling

1.3 BUSINESS GOALS

In the face of social demands toward sustainable development, businesses may behave mainly in two different ways. Businesses may be content simply with watching regulations and standards or they may undertake a proactive approach by establishing their own sustainability strategies and programs beyond mere implementation of regulations.

1.3.1 IMPLEMENTATION OF REGULATIONS AND STANDARDS

Administrations adopt regulations on the basis of this philosophy: each business must comply with established legislation. ISO and other standards, and voluntary agreements like GRI are not mandatory for businesses, although increasingly more businesses do reporting, according to GRI, and are demanding that their providers use ISO or other certifications.

1.3.2 TOWARD ESTABLISHMENT OF MORE AMBITIOUS STRATEGIES
THAN THOSE GIVEN BY LEGISLATION

Establishing sustainability strategies more ambitious than current legislation results in benefits for businesses as well, although, very often, those benefits will only be noticeable within the mid- or long run. This necessitates investing in the future. In many markets, a greener or more sustainable product sells better than a "regular" product. That is, excellent sustainability management may also contribute to better marketing. Businesses seeking to survive in the global market must not let their competitors have a more advanced sustainability policy; they must always ascertain how competitors stand in comparison to them (benchmarking). One of the criteria currently used by financial assessment organizations is environmental competence. Reviewing development through the past years, we may say that, often, more technologically advanced businesses are also more environmentally competent businesses. Businesses existing at the crest of the wave avoid losing ground before struggling later to regain it. Future benefits include cost reduction, image enhancement, and increased staff motivation. Lubin and Esty (2010) have identified sustainability as a megatrend in their Harvard Business Review paper "The Sustainability Imperative." Specifically, the following advantages can occur:

- Enhancement in optimization and control of energy and raw material consumption
- Optimization in costs derived from waste and emission management and treatment
- Expense reductions in transportation, storage, and packing
- Savings in environmental cleaning and repair tasks derived from accidental leaks
- Reductions in insurance policies for environmental risk
- Better conditions in negotiating bank loans
- Savings in fines from violations of law
- Decreases in accident risk and therefore in derived costs
- Enhanced business image with clients, administration, staff, investors, media, environmental defense organizations, and the general public
- New marketing tool by using an environmental compliance label
- Adoption of an active policy before current legislation and future environmental regulations affect the business
- Increased possibilities to obtain public funding for environmental activities
- Involvement of staff in a system aimed at achieving common goals
- Increased training for staff

1.4 CONCEPTS

The previously mentioned philosophies and businesses goals to establish sustainability strategies and policies are reflected in and can be applied to various concepts.

1.4.1 GREEN CHEMISTRY

Green chemistry is the use of chemistry for PP2. In more detail, it is aimed at designing chemical substances and, at the same time, production processes respectful of the environment. This includes reducing or eliminating use and production of dangerous substances. The concept of green chemistry was coined in 1995 and encompasses all aspects and types of chemical processes such as synthesis, catalysis, analysis, monitoring, reaction separators, and conditions; PP2 minimizes their negative impact on human health and the environment. Therefore, this concept emphasizes the U.S. law concerning chemical substances (U.S. EPA, 2017b).

1.4.2 CLEANER PRODUCTION

The concept of cleaner production (CP) was first introduced by the UNEP (United Nations Environment Programme) in 1989. CP means continuous use of a preventive and integrated environmental strategy. This concept is applied to processes, products, and services to increase eco-efficiency and to reduce risks to the population and the environment. It is intended to preserve energy and raw materials, as well as to eliminate toxic wastes and reduce the amount and toxicity of all emissions and wastes generated in every process. CP demands attitudes different from current ones in order to undertake responsible environmental management in the creation of adequate national policies and assessment of technology options. Currently, administrations tend to demand the use of best available technologies (BAT). This strategy helps reduce costs as much as it does risks, and also identifies new opportunities. The aim of CP is to avoid pollution before it is produced in every process or in the corresponding process chains (UNEP, 2017).

Although eco-efficiency is based on aspects of economic efficiency that have environmental benefits, cleaner production is based on aspects of environmental efficiency with economical benefits. Therefore, the UNEP and the WBCSD are fostering similar concepts and have decided to join efforts. This new initiative combines UNEP interests within the public sector with participation of the industrial sector at the WBCSD. Their first joint action took place at the annual meeting of UNCSD (United Nations Commission on Sustainable Development) in April and May, 1996, in New York, under the title of Eco-Efficiency and Cleaner Production: Charting the Course to Sustainability (UNEP and WBCSD, 2017).

1.4.3 Total Quality Environmental Management

Total quality environmental management (TQEM) derives from quality activities and describes processes following ISO 9000 and other standards. A few of the many businesses that have implemented total quality management (TQM) include Ford Motor Company, Phillips Semiconductor, SGL Carbon, Motorola, and Toyota Motor Company. The following information helps to understand the key elements in the TQM process:

- Total is how each single person and activity within a business is involved.
- Quality is the satisfaction of client demands.
- Management is how quality can and must be managed.
- TQM is a process of quality management that must be implemented in a continuous manner and with the philosophy of permanent enhancement regarding every single activity.

Although ISO 9000 stands as a series of rules for quality assurance, TQM stands as a concept for continuous improvement (Hansen, 2017). The environmental equivalent to the ISO 9000 standards for TQM is the ISO 14000 series of standards applicable to the various phases of environmental management. TQEM means applying TQM to the environmental issues of a business. This is obviously a comprehensive approach because it encompasses the business as a whole plus its management system (Vasanthakumar, 1998).

1.4.4 Life-Cycle Thinking and Management

Life-cycle thinking is a way of addressing sustainability issues and opportunities from a system or holistic perspective and of evaluating a product or service system with the goal of minimizing potential environmental and social impacts over its entire life-cycle (see Figure 1.3) and of maximizing socio-economic value. The concept of life-cycle thinking implies linking individual processes to organized chains starting from a specific function. According to this type of thinking, everyone in the entire chain of a product's life-cycle, from cradle to grave, has a responsibility and a role to play, taking into account all the relevant external effects. From the exploitation of the raw material that will constitute a new product, through all the other processes of extraction, refining, manufacturing, use or consumption, to its reuse, recycling, or disposal, individuals must be aware of the impact of this product on the environment and try to reduce it as much as possible. The impacts of all life-cycle stages need to be considered when making informed decisions on production and consumption patterns, policies, and management strategies (UNEP, 1999).

According to Sonnemann and Margni (2015), Life-Cycle Management (LCM) is a business management concept applied in industrial and service sectors to improve products and services, while

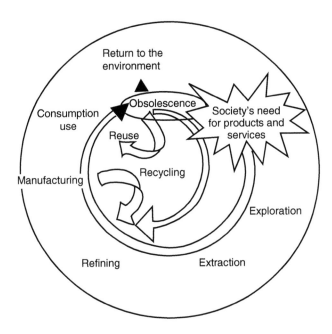

FIGURE 1.3 The life-cycle of a product system. (From UNEP/SETAC, *Background Paper of the UNEP/SETAC Life-Cycle Initiative*, United Nations Environment Program, Paris, France, 2001.)

enhancing the overall sustainability performance of the business and its value chains. It makes life-cycle thinking and product sustainability operational for businesses that are ambitious and are committed to reduce their environmental and socio-economic burden, while maximizing economic and social value. In this regard, LCM is used beyond short-term business success; rather, it aims at taking businesses forward towards long-term achievements and sustainable value creation. So, LCM requires a holistic view and a full understanding of interdependency of businesses in order to support relevant decisions and actions to improve sustainability performance that takes into account both the environmental and social benefits. At the same time, LCM offers a number of value creation opportunities to the business. The term "life cycle management" has been confused with other uses in engineering and manufacturing (Product Lifecycle Management) and in software development (Application Lifecycle Management), in buildings, plants, information management, and so on. Therefore, Sonnemann and Margni (2015) extend the term from focusing on implementation of life-cycle sustainability assessment (LCSA) into business practice to include it as part of sustainable consumption and production (SCP) strategies and policies.

Many people refer to life-cycle management emphasizing "end of life management." A particular subject area for end-of-life management is the study of the limits of recyclability, whose environmental benefit disappears when the energy, materials, and pollution involved in the collection, production, and recycling processes exceed those necessary to produce the product. Extended producer responsibility (EPR) is one aspect of life-cycle thinking. This principle was introduced by Thomas Lindhqvist (2000):

> Extended producer responsibility is an environmental protection strategy to reach an environmental objective of a decreased total environmental impact from a product by making the manufacturer of the product responsible for the entire life-cycle of the product, especially for the take-back, recycling, and final disposal of the product. The extended producer responsibility is implemented through administrative, economic, and informative instruments. The composition of these instruments determines the precise form of the extended producer responsibility.

At least two similar principles exist: responsible care and product stewardship. Since 1988, Responsible Care has helped American Chemistry Council (ACC, 2017) member companies significantly enhance their performance; discover new business opportunities; and improve employee safety, the health of the communities in which they operate, and the environment as a whole, moving us toward a safer, more sustainable future. The product stewardship code is designed to change health, security, and environment protection into an integral part of design, manufacturing, distribution, use, recycling, and disposal. This code reinforces sharing information about adequate product use, storage, and disposal. It has also been designed to include the widest possible extent of the supply chain to help prevent inadequate use likely to harm human health and the environment (CEFIC, 2017).

1.4.5 DESIGN FOR SUSTAINABILITY

Eco-design, design for environment (DfE) or design for sustainability (D4S), allows approaching environmental problems associated with a given product within its phase of design. That is, it implies considering the environmental variable as one of the many product requirements in addition to the conventional design goals: cost, utility, functioning, security, etc. The purpose of DfE is to manufacture products with a lower global environmental load associated with their life-cycle (material/component parts acquisition/purchase, production, distribution, use and end of life). Implementation of the environmental variable in the manufacturing process of the product must be undertaken without sacrifice of the remaining product properties and combining price and quality in a sound manner (Fiksel, 1997).

Eco-design is intended to be the dematerialization and detoxification of the design process. It has to do with taking off the highest possible amount of materials—specifically, dangerous substances—without the product losing its function (Fiksel, 1997; Rodrigo and Castells, 2002; Crul and Diehl, 2009). Because a product life-cycle is decided during the design and production planning phase, influencing it in further phases is much more difficult.

1.4.6 INDUSTRIAL ECOLOGY

Industrial ecology (IE) entails an approximation of industrial systems to natural systems. It deals with the systematic analysis of material and energy flows in industrial systems with the purpose of minimizing waste generation and negative environmental effects (Graedel, 1994).

IE provides a holistic approach of industrial systems, based on its analogy with natural systems: wastes produced by a living being become a source of raw material for another being. Industrial ecology may be defined as a network of industrial systems that cooperate in reusing waste energy and material within the same network; a waste flow from one of the industrial members becomes a source of raw material useful for another member. This industrial approximation is mainly associated with the concepts of eco-park, industrial symbiosis, and industrial clustering. Close cooperation among different industrial systems provides each member with higher efficiency levels in its industrial activity at the same time that it provides a higher level of use of its waste energy and material flows. This cooperation renders significant environmental and economic benefits for each member (Graedel and Allenby, 2002; Clift and Druckmann, 2016).

A specific aspect of IE is the concept of industrial metabolism proposed by Ayres in 1989; this concept consists of the process through which energy and materials flow through industrial systems, starting from the source through various industrial processes to the consumer and to final disposal (Ayres and Ayres, 1996). Although IE is the most comprehensive concept of those presented until now, it is based on the idea that energy and materials flow within local, regional, or global systems; however, systems encompass several processes and are not exclusively intended for a given product or service. According to the International Society for Industrial Ecology (https://is4ie.org/), IE asks how an industrial system works, how it is regulated, and how it interacts with the biosphere.

Then, on the basis of what is known about ecosystems, the goal is to determine how the system could be restructured to make it compatible with the way natural ecosystems function. This way it also includes socioeconomic issues related to sustainable development.

1.5 TOOLS

The preceding concepts, which have been developed to direct sustainability management, are quite abstract. This is why we need tools to transfer them into action and make environmental aspects more concrete, taking into account economical, social, and technological information in the context of sustainable development. The working group on conceptually related methods in the Society of Environmental Toxicology and Chemistry (SETAC) has distinguished the following types of tools (De Smet et al., 1996): political instruments, procedural tools, and analytical tools.

In general, to be applied, the tools need technical elements like dispersion and other pollutant fate models, damage functions, weighting schemes, and available data, e.g., about emissions and resource consumption, as well as technical specifications and geographic location. The application of these tools provides consistent environmental information that facilitates adequate decision-making toward sustainable development. An overview of the conceptually related methods in environmental management is presented in Figure 1.1. Based on the idea of the interaction between these different concepts and tools, a concerted action named CHAINET took place from 1997 to 1999 under the European Union Environment and Climate Program. The mission of this action was to promote the common use of the different tools and to facilitate information exchange among the relevant stakeholders (CHAINET, 1998).

1.5.1 POLITICAL INSTRUMENTS

Generally, political instruments are adopted by political administrations. We will present typical samples of regulations, directives and communications relevant for a given aspect of policies relevant for sustainability management.

1.5.1.1 Regulation on Chemical Substances

Within the framework of green chemistry and chemical substances, we could mention that chemicals legislation in the European Union (EU) aims to protect human health and the environment and to prevent barriers to trade. It consists of rules governing the marketing and use of particular categories of chemical products, a set of harmonized restrictions on the placing on the market and use of specific hazardous substances and preparations, and rules governing major accidents and exports of dangerous substances. The most important achievement at EU level on chemical substances is the regulation on the Registration, Evaluation, Authorization and Restriction of Chemicals (the REACH regulation; EC, 2006) that entered into force in 2007, establishing a new legal framework to regulate the development and testing, production, placing on the market, and use of chemicals, replacing around 40 previous legislative acts. The aim of the REACH regulation is to provide better protection for humans and the environment from possible chemical risks and to promote sustainable development.

1.5.1.2 Roadmap to a Resource Efficient Economy

A range of policies at the European Union (EU) and national level foster resource-efficient and eco-friendly products and raise consumer awareness, such as the EU framework for the eco-design of energy-using products, labelling schemes, and financial incentives granted by Member States to those that buy eco-friendly products. The *Action Plan for Sustainable Consumption and Production and Industry* (EC, 2008) complements and integrates the potential of these different policy instruments, and facilitates new actions where gaps exist. *The Roadmap to a Resource Efficient Europe* (EC, 2011a) outlines how Europe's economy can be transformed into a sustainable one by 2050.

It proposes ways to increase resource productivity and decouple economic growth from resource use and its environmental impact. It illustrates how policies interrelate and build on each other. Areas where policy action can make a real difference are a particular focus, and specific bottlenecks like inconsistencies in policy and market failures are tackled to ensure that policies are all going in the same direction. Crosscutting themes such as addressing prices that do not reflect the real costs of resource use and the need for more long-term innovative thinking are also in the spotlight.

1.5.1.3   Communication on Green Products

In 2013, the European Commission launched a communication on *Building the Single Market for Green Products* (EC, 2013), which contains a package with recommended methods to be used for improving the comparability, reliability, completeness, transparency, and clarity of environmental information, thus, building confidence among consumers, business partners, investors, and other company stakeholders. Among the methods recommended, two methods to measure environmental performance throughout the life cycle (the Product Environmental Footprint [PEF] and the Organization Environmental Footprint [OEF]) are proposed. Their application is voluntary for all potential entities concerned that are among others: Member States' companies, other companies, private organizations, and the financial community.

1.5.1.4   End of Life Policy Oriented to Product Recovery

Adopted July 12, 1996, the European Union's End-of-Life Vehicles (ELV) Directive (2000/53/EC) to promote recycling and provide incentives for environmental-friendly vehicle design represents an early application by the European Union (EU) of extended producer responsibility (EPR) principles for the sustainable life cycle management of products. EU member states must implement this directive with adequate measures to assure manufacturers comply with all or a significant part of the costs resulting from the treatment of end-of-life vehicles, or collect them following the conditions stated in the directive.

1.5.1.5   Policy on Sustainable Materials Management

One of the policies of the OECD that is clearly stating a strong reference to life cycle thinking is Sustainable Materials Management (SMM). It is increasingly recognised as a policy approach that can make a key contribution to green growth and the challenges that are posed by sustained global economic and demographic growth. One of the key challenges of the SMM approach is to effectively address the environmental impacts that can occur along the life cycle of materials, which frequently extends across borders and involves a multitude of different economic actors (OECD, 2012).

1.5.1.6   Special Protection Plans for Specific Areas

The National Coastal Zone Management (CZM) Program is a voluntary partnership (https://oceanservice.noaa.gov/facts/czm.html) between the U.S. government and U.S. coastal states and territories. It is authorized by the Coastal Zone Management Act of 1972 to preserve, protect, develop and, where possible, restore and enhance the resources of the nation's coastal zone.

1.5.2   INTERNATIONAL STANDARDS

The International Organization for Standardization (ISO) was founded in 1946 in Geneva, Switzerland. ISO has established non-mandatory international standards for the manufacturing, communication, trade, and administrative sectors. It has also created ISO 9000 for management and quality assurance; this has been adopted by more than 90 countries and implemented by thousands of industries and service providers. Accordingly, for environmental management, ISO has created the ISO 14000 series, a group of standards to foster national and international trade in compliance with international standards to protect the environment and strive for sustainability. In this way, some common guidelines and similarities between environmental management and business management are established for all businesses regardless of size, activity, or geographical location. Several of the ISO 14000 standards refer to the previously mentioned procedural and analytical tools. ISO 14001, the Environmental Management Systems Standard, is the most popular standard of the ISO 14000 family, which also includes standards such as the following:

- ISO 14004: General guidelines on principles, systems, and support techniques
- ISO 14006: Guidelines for incorporating eco-design
- ISO 14015: Environmental assessment of sites and organizations
- ISO 14020: Environmental labels and declarations
- ISO 14031: Environmental performance evaluation
- ISO 14040: Life cycle assessment
- ISO 14046: Water footprint
- ISO 14050: Vocabulary
- ISO 14063: Environmental communication
- ISO 14064: Greenhouse gases
- ISO 19011: Guidelines for environmental management systems auditing

Some of these standards will be reflected in some of the procedural and analytical tools explained later.

1.5.3 Procedural Tools

Important procedural tools described next are the environmental impact assessment, environmental management system, eco-audit, and eco-labels. Some procedural tools come from current legislation and others from international standards. The latter can be applied at process, product, and service levels.

1.5.3.1 Environmental Impact Assessment

The European Directive 2011/92/EU, amended by Directive 2014/52/EU, on Environmental Impact Assessment (EIA) is a further development of the Directive 85/337/EEC from 1985 and applies to a wide range of defined public and private projects, which are defined in Annexes I and II, a set of research papers and technical systems used to estimate the effects on the environment of implementing a given project, work, or activity. The idea behind EIA is to obtain an objective judgment of the consequences due to the impacts generated by accomplishing a given activity. The main part of such an evaluation is the environmental impact study (Coneza, 1997).

Environmental impact research is a technical interdisciplinary assessment aimed at foreseeing, identifying, determining, and correcting the environmental impact or consequences that certain activities may have on the quality of human life and the environment. It has to do with presenting objective reality to determine the influence on the environment of implementing a given project, work, or activity. In sum, EIA is an analytical tool fundamental for information gathering and necessary for submitting an environmental impact declaration.

The different phases of EIA are summarized in Figure 1.4. The first six phases are related to qualitative assessment. During Phase 7, a quantitative assessment is carried out, which partially continues during Phase 8 and Phase 9; in Phase 10 and Phase 11, more simplified results are produced. The first nine phases are related to the environmental impact study (Coneza, 1997).

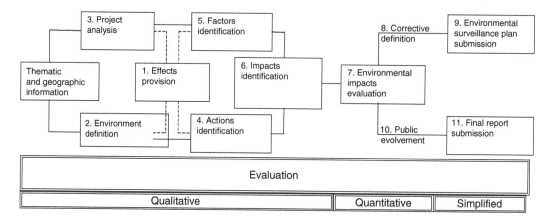

FIGURE 1.4   General structure of an environmental impact assessment.

1.5.3.2   Environmental Management System

The current environmental (or eco-) management and audit scheme (EMAS) is an EU-based system, in line with regulation (EC) No 1221/2009 ("EMAS III") that came into effect in 2010, and is based on Council Regulation (EEC) 1836/1893 from 1993 for the continuous improvement of environmental aspects in businesses. Internationally, it corresponds in many features to ISO 14001, although the latter does not have the same recognition as that of the environmental authorities (Zharen, 1995).

An environmental management system (EMS) is a means of ensuring effective implementation of an environmental management plan or procedures in compliance with environmental policy objectives and targets. A key feature of any effective EMS is the preparation of documented system procedures and instructions to ensure effective communication and continuity of implementation. There are certification systems for EMS, as for the ISO 14001 and EC EMAS schemes (EMAS is now compatible with ISO 14001), which demonstrate that a system is operated to an internationally recognized standard (EC, 2011b). Alternatively, a customized system can be developed addressing the particular needs of the operation. An environmental management system allows businesses to:

- Assure a high level of environmental protection
- Continuously improve their environmental performance
- Obtain competitive advantages out of these improvements
- Communicate their progress with the publication of an environmental declaration showing their efforts

The environmental management department is a recognized instrument in all EU member states, which allows industries non-mandatory adherence to an EU environmental management and audit scheme. EMAS has been developed for organizations involved in industrial activities, energy generation, recycling, and solid and liquid waste treatment. Additionally, it can be applied to other sectors, such as energy, gas and water supply, construction, trade, transportation, financial services, public administrations, entertainment, culture, sports, education, and tourism.

At present, on the international scene, the ISO 14001 standard on environmental management systems is mostly used (Schlemminger and Martens, 2004). This standard is not against that established in regulation (EC) No 1221/2009 and can be seen as a previous step for EMAS adherence. As a standard with international application, the ISO 14001 has a more general nature. In Europe, EMAS enjoys official recognition on the side of political administrations.

1.5.3.3 Eco-Audit

An eco-audit, or environmental audit, is an "independent and methodical test carried out to determine whether the activities and results concerning the environment meet previously established regulations and prove to be adequate for attaining the foreseen goals." As explained in the ISO 19011 standard, environmental management systems auditing is "a process of systematic testing and objective assessment of evidence to determine whether environmental activities, events, conditions and systems, or information about these, conform to audit criteria and communication of results to customers."

Eco-audits are carried out to implement environmental management systems. Audits help acknowledge the current business position in the face of existing legislation. Therefore, the eco-audit may be defined as a management tool to test whether activities and results related to the environment are accomplished; that is, if established goals are attained and standards are met, and if the latter are adequate to attain those goals. It is an important tool to enhance environmental management and has a preventive nature; thus, it is not an inspection and control activity or a witch hunt, but it is intended for problem detection and solution. Environmental audits are generally carried out to accomplish one or several of these goals:

- Determine suitability and effectiveness of an organization's environmental management system to attain environmental goals.
- Provide the audited organization with an opportunity to enhance its environmental management systems and, as a result, contribute to continuous improvement of its environmental performance.
- Check conformity to existing regulations.
- Internally assess the organization's environmental management system within the framework of a given environmental management standard.
- Assess an organization prior to establishing a contractual party relationship with that entity.

1.5.3.4 Eco-Label

The idea of eco-labelling is to guarantee the environmental quality of certain properties or characteristics of the products that obtain the eco-label (Alfonso and Krämer, 1996) in order to provide consumers with better information on green products and promote the design for environment (EC, 1997). An EU eco-label scheme is laid down in Council Regulation 880/92 (EEC) and the Council Regulation 1980/2000 (EEC) on a revised Community eco-label award scheme.

Products and services that meet previously established environmental criteria are allowed to use various official labels for easier recognition. On the one hand, the eco-labelling scheme provides consumers and end-users with enhanced and more reliable information; on the other hand, it fosters design, manufacturing, marketing, use, and consumption of products and services exceeding existing mandatory environmental quality requirements. A product with a lower environmental impact is one that causes lower damage to the environment over its life cycle from extraction via production and use to the end of life. Some examples are paper manufactured without emission of organic chloride compounds, a washing machine with low water and energy consumption, a refrigerator manufactured with recyclable component parts, etc. A service with a lower environmental impact would be a small business (store, repair shop, etc.) whose operation, management or service supply (from the use of environmental quality products to adequate waste management) is respectful of the environment. Examples of other eco-labels are the Blue Angel in Germany and the White Swan in the Nordic countries.

The status of eco-labelling and product information on the international scene is far from coherent, however, particularly among the various stakeholders. At the global level, Ecolabel Index (http://www.ecolabelindex.com) is the largest global directory of ecolabels, currently tracking 465 ecolabels in 199 countries, and 25 industry sectors. These cover product categories ranging from laundry detergents, household cleaners, paints and varnishes, household paper, sanitary items, wood, textiles, white domestic appliances, and garden products, as well as tourism, energy production or efficiency, and services. ISO classified environmental labels and declarations into three categories:

1. Environmental labelling (also called "Label Type I" in ISO 14024); examples: EC eco-label, Nordic Swan and Blue Angel
2. Self-declared environmental claims (also called "Label Type II" in ISO 14021); examples: ozone-friendly label, green dot and animal-cruelty-free
3. Environmental declarations based on life-cycle assessments (also called "Label Type III" in ISO/TR 14025); only example: environmental product declaration (EPD) promoted by the governments of Sweden, Norway, Canada, Korea, and Japan, and companies such as ABB and Daimler.

Related environmental product declarations gain popularity as tools, especially for business-to-business communication, and have great potential to be used widely by institutional buyers in their efforts for green procurement. In general, for the communication to consumers, one overall environmental Type I label based on a single indicator is considered the most effective option to influence consumer choices. However, consumers are likely to be interested in more detailed environmental information for durable goods such as cars or electronics; a Type III label might be provided to influence the purchase for these types of items. ISO/TS 14027 provides principles, requirements, and guidelines for developing, reviewing, registering, and updating Product Category Rules (PCR) within a Type III environmental declaration or footprint communication program based on life-cycle assessment (LCA).

1.5.4 ANALYTICAL TOOLS

The following analytical tools are relevant methods for environmental management:

- Life-cycle assessment (LCA) is a tool standardized according to ISO series 14040 for product-oriented environmental impact assessment and will be further explained in Chapter 2.
- Environmental risk assessment (ERA) and impact pathway analysis (IPA) are the tools generally used for the impact analysis in site-specific environmental impact assessment. These tools will be described in more detail in Chapter 4.
- Cost-benefit analysis (CBA) and cost-effectiveness analysis (CEA) are techno-economic tools to support decision-making towards sustainability. They refer to environmental costs, a topic relevant in some chapters of this book.
- Process simulation (and the related re-engineering) is an important tool for the improvement of industrial processes. It allows foreseeing environmental effects resulting from changes in process design before implementation.
- Accident prevention requires determining the environmental risk that implies installation and operation of an industrial process due to undesirable events. Undesirable events are caused by unforeseen emissions of pollutants for accidental reasons.
- Material Flow Accounting (MFA) and Input–Output Analysis (IOA) have been developed to look at the life cycle of material substances in industrial systems and the environment.

Analytical tools differ depending on the specific aspect of their focus; few of them have been standardized by ISO.

1.5.4.1 Life-Cycle Assessment

In order to consider environmental impacts of a product's life-cycle systematically, the life-cycle assessment (LCA) methodology has been developed. This is the only standardized tool currently used to assess product environmental loads. The steps of LCA are: goal and scope definition, inventory analysis, impact assessment, and interpretation.

The goal, the motivation for research, must be clearly defined from the very beginning because the following phases will be influenced by its early definition. The creation of life-cycle inventories (LCIs) is intended to identify and assess the environmental load—or elementary flows—associated with the full life-cycle of a product, process, or activity. The elementary flows are the exchanges between the life cycle under study and the environment. In the case of a product, the inventory starts at the extraction process of raw materials from the environment, continues in the production, consumption and use of end products, and ends when the product or its derivate turns into waste. Operations such as transportation, recycling, and maintenance must also be considered in the inventory.

Life-cycle impact assessment (LCIA) allows for easier interpretation of environmental information produced during the inventory analysis phase. LCIA includes several phases:

- Classification of elementary flows within the different categories of environmental impact
- Categorization of elementary flows by means of a reference pollutant typical of each environmental impact category
- Normalization of the data obtained from the characterization, dividing it into real or foreseen magnitude for its corresponding impact category within a geographical location and a point in time for reference
- Quantitative or qualitative assessment of the relative significance concerning different categories of impacts

These categories, plus the level of detail and methodology, are chosen depending on the goals and scope of the research.

Following this analysis, more objective and transparent decisions can be made concerning environmental management for the creation of guidelines for new product development and guidebooks to define environmental priorities (SETAC, 1993). See Chapters 2 and 3 for more details.

The term Life-Cycle Sustainability Assessment (LCSA) has emerged to incorporate the economic and social dimensions in addition to the environmental dimension (Heijungs et al., 2010; Traverso et al., 2012; Valdivia et al., 2013). Indeed, according to ISO 14040, "LCA typically does not address the economic or social aspects of a product, but the life cycle approach [...] can be applied to these other aspects." LCSA, therefore, embodies the "triple bottom line" concept of sustainable development (Elkington, 1997) by combining environmental LCA, social LCA, and (often economic) life-cycle costing (LCC) (Kloepffer, 2008; Traverso et al., 2012; Valdivia et al., 2013).

1.5.4.2 Environmental Risk Assessment

The usual point in introducing risk assessment is to emphasize that risk is part of everything we do and that the risk derived from pollutant exposure should be paid the attention it deserves. In the U.S. during the late 1980s, about 460,000 of 2.1 million deaths per year were due to cancer. Without taking the age factor into consideration, the risk of dying from cancer equals 22% ($460,000/2,100,000 = 0.22$). Individuals who smoke one package of cigarettes per day have approximately a 25% risk of dying from heart disease. Meanwhile, the U.S. Environmental Protection Agency (U.S. EPA) intends to control exposure to toxic substances with risk levels ranging from 10^{-7} to 10^{-4} (0.00001%–0.01%) throughout life (Masters, 1991).

Environmental Risk Assessment is a process for determining the probability for negative effects on human health or the environment as a result of exposure to one or more physical, chemical or biological agents. ERA requires knowledge about the negative effects of exposure to chemical substances or materials, as well as knowledge about the intensity and duration necessary for these to cause negative effects on population and the environment. Decision-making within sound risk management entails examining the various choices for risk reduction. The risk assessment scope is generally local; environmental impacts are presented in the form of risk per researched recipient because it is the case with the value for exposure to toxic substances at levels entailing a risk ranging from 10^{-7} to 10^{-4} throughout life.

Environmental risk assessment will be explained in detail in Chapter 4.

1.5.4.3 Impact Pathway Analysis

Contrary to previous approach, the Impact Pathway Analysis (IPA) estimates the overall damage, including even relatively small contributions, in locations more than 1000 km away from the source of emissions. The damage is presented in the form of costs, but because these costs are not included in the price of the product or service responsible for the emissions, they are known as external costs.

An example of an IPA scheme is shown in Figure 1.5. After the location and technology producing the pollutant have been selected, Phase 1 deals with the calculation of emissions and the plant's demand for resources. Phase 2 concerns the distribution of emissions to several recipients. The following phases consider the transition processes and impact assessments, for example, a bad fruit crop or higher incidence of asthma attacks. Once all physical impacts have been calculated, economic assessments are applied to the impacts to estimate damages in currency amounts (e.g., euros, dollars, or yen). Impact pathway analysis is explained as a special form of ERA in Chapter 4.

1.5.4.4 Cost-Effectiveness Analysis and Cost-Benefit Analysis

An efficient emission level is the point at which marginal damages (external costs) are equivalent to marginal abatement costs (internal costs)—that is, both types of costs are neutralized. Again, this point is the efficient level of emissions, a consideration illustrated in Figure 1.6. Because marginal damage costs and marginal abatement costs are equal, both hold the W value at that level of emissions. Marginal damages have their threshold at emission level E, while the uncontrolled emission level is at E_d. It is possible to analyze the results in terms of total values because we know the totals

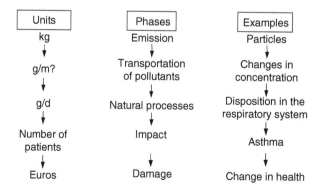

FIGURE 1.5 Impact pathway analysis (IPA) scheme. (Adapted from European Commission—DGX11, *ExternE–Externalities of Energy*, 6 vol., ECSC-EC-EAEC, EUR 16520–162525, DG XII Science, Research and Development, Brussels, Belgium, 1995.)

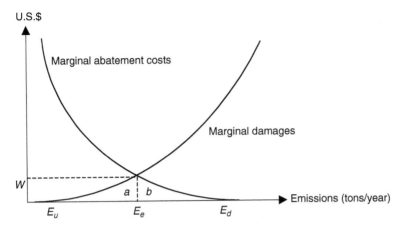

FIGURE 1.6 Efficient emission level.

are the areas under marginal curves. Thus, the triangular area distinguished with an a (marked by points E_u, E_e, and the marginal damage curve) schematizes the existing total damages when emissions are at level E_e, while the b area shows a reduction of the total costs at this level of emissions. The addition of both areas is a measurement of total costs of E_e—tons per year in this case. E_u is the only point at which this addition is minimized, but the a area is not necessarily equal to the b area; this depends on the shape of the marginal curves, which may have wide variations (Field, 1995). The preceding explanations serve as a basis to introduce the cost-effectiveness analysis (CEA) and the cost-benefit analysis (CBA) methods.

CEA considers only the internal costs—that is, the costs resulting from emission reduction technologies. These costs are compared to the reduction of the environmental load due to the economic investment. If we want to invest this money to enhance the environment we will invest it in the most effective choice.

CBA is another economic tool intended to provide decision-making support for long-term investments from a social perspective instead of the business perspective. The field for its application includes the environmental selection of technologies and legislation strategies. CBA is intended to correct dysfunctions caused by market defects. In the sphere of environmental management, the main interest is that external effects be considered external costs. The conversion of damages into costs is based generally on Paretian's theory of well-being, in which individuals facing external effects judge their importance for their quality of life. Preferences may be shown in monetary terms, just as in market choices. In this way, the final assessment of the marginal damage balance (external costs) and marginal abatement costs (internal costs) may be carried out in one monetary unit: currency (dollar, euro, yen), which is a globally accepted reference in decision-making (Dasgupta and Pearce, 1972; Nas, 1996).

1.5.4.5 Process Simulation

Process re-engineering is one of the applications within chemical engineering that shows increases in interest year after year. The vast majority of plants have been built and are currently operating so as to obtain the highest possible return using available equipment and to reduce investment. New process synthesis tools allow analysis, in a relatively simple and fast way, of whether the optimum current process setting is the best and assess possibilities for improvement. To attain this, it is necessary to count on models with an exact and precise representation of equipment operation. In the first place, it is necessary to adjust these models using plant data to ensure successful predictions. After the base case has been obtained, exploring alternative design choices and assessing their applicability through equipment sizing begins. At this point, it is necessary to assess in detail the possibilities for implementing the proposed alternative by means of detailed research of the most promising alternatives (detailed sizing, economic estimate, cost analysis, etc.). Apart from these applications, process simulation has proved to be the fundamental tool for detecting bottlenecks in processes and operation problems, and in obtaining ideal operating conditions for the environment.

1.5.4.6   Accident Prevention

To prevent accidents, it is necessary to determine the environmental risk associated with a facility or process operation resulting from undesirable events or accidents. Undesirable events are those unforeseen events that accidentally cause the emission of pollutants into the environment. Analysis of undesirable events (accidents) must include the following phases:

- Analysis of facilities and information gathering
- Identification of the most representative accident scenarios
- Establishment of accident probability for these scenarios and its evolution
- Establishment of consequences associated with each accident scenario assessed (scope of associated physical effects)
- Establishment of the impact these physical effects have on the environment

Fail trees and event trees are applied when accident prevention assessments are undertaken (AICHE, 1985).

1.5.4.7   Material Flow Accounting and Input–Output Analysis

Material Flow Accounting (MFA) refers to accounting in physical units (usually in tons); the extraction, production, transformation, consumption, recycling, and disposition of materials in a given location (i.e., substances, raw materials, products, wastes, emissions into the air, water, or soil). Within the range of the present work, MFA encompasses methods such as substance flow analysis (SFA) and other types of balance of materials for a given region (Fuster et al., 2002; Brunner and Rechberger, 2016). Examples of flow assessments are:

- Eco-toxic substances such as heavy metals that may cause environmental problems due to their accumulation capacity
- Nutrients such as nitrogen and phosphates due to their critical influence over eutrophication
- Aluminum, the economic use, recycling and reuse of which are to be improved (Bringezu et al., 1997; Fuster et al., 2002)

Material flow analysis (MFA) for countries and regions and derived indicators have been adopted by various statistical offices. They have issued methodological guidance for official reporting and are working to continuously improve the data basis. Eurostat developed the first reporting framework for economy-wide material flow analysis and detailed guidance for recording the basic indicators (Eurostat, 2001). The OECD (2008) within the work program on material flows and resource productivity, edited three reports which provide an overview and detailed insights of economy-wide material flow analysis and related approaches and indicators as well as the state-of-the-art of implementation.

As a part of establishing statistical accounts on a national scale, the input–output analysis (IOA) has been under development since the 1930s. One of the main applications of this analysis is to show the interrelationship of all flows of goods and services within a given economy; it also shows the connection between producers and consumers and interdependence among industries (Miller, 2009). Since 1993, different environmental applications have been designed. Nowadays, this macroeconomic method is frequently applied to environmental analysis (Suh, 2009).

1.5.5 APPLICATION-DEPENDENT SELECTION OF ANALYTICAL TOOLS

According to Wenzel (1998), the governing dimensions for applications of analytical tools such as LCA are site specificity, time scale, and need for certainty, transparency, and documentation. Possible applications can be positioned in relation to these governing dimensions. In the case of LCA, Sonnemann et al. (1999) have examined in which cases LCA is an integrated element of another concept and for which goals or other tools of environmental management should accompany it. They correspond to the following points:

1. *Education and communication*: LCA supplies a potential common ground or basis for discussion and communication. All groups in the society need to understand their individual responsibilities for improvements.
2. *Product development and improvement*: The concept used in the field of environmentally-friendly product (re)design and development is called design for the environment (DfE). LCA provides the information to support it.
3. *Production technology assessment*: LCA helps to ensure that overall reductions are achieved and pollutants are not shifted elsewhere in the life-cycle, although other tools are needed for the assessment of the actual impacts of the technology.
4. *Improving environmental program*: LCA can be particularly effective at identifying sensitive factors such as the number of times a reusable must be returned or possible energy recovery.
5. *Strategic planning for a company's product or service line*: LCA can assist the strategic planning process, especially when coupled with other tools providing economic and risk information.
6. *Public policy planning and legislation*: LCA studies can be used to assure that all relevant environmental information is considered. Because of LCA's restriction to potential impacts, the results should be integrated with data from other tools.
7. *Environment-friendly purchasing support*: LCA is obviously a starting point for eco-labelling, but the current emissions and resource indicators do not give the full picture of sustainable performance. LCA information should be complemented with data from other tools.
8. *Marketing strategies*: By using LCA it is possible to develop an environmental profile of a product or service that can be communicated to the consumers.
9. *Environmental performance and liability evaluation*: The combination of an environmental management system with LCA is an interesting topic for the future. For this reason, it is necessary to use the same pressure and management indicators.

By using the proposed guide in the form of a matrix (Figure 1.7), the sustainability manager should be able to neither overestimate the possibilities of LCA nor be discouraged from using it because of its inherent limitations.

Therefore, we can say that life-cycle thinking is the right concept to evaluate the environmental impacts of a functional unit (product, service, or activity) but that LCA is often not the only tool to consider in a sustainability decision-making context. The practical guide (Figure 1.7) for a sustainability manager helps select the tool that corresponds to a particular application.

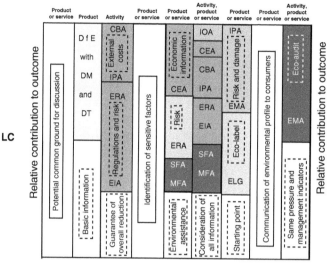

Applications
1. Education and communication.
2. Product development and improvement.
3. Production technology assessment.
4. Improving environmental program.
5. Strategic planning for a company's product or service line.
6. Public policy planning and legislation.
7. Environment-friendly purchasing support.
8. Marketing strategies.
9. Environmental performance and liability evaluation.

Legend
CBA: Cost-Benefit Analysis
EIA: Environmental Impact Assessment
IOA: Input–Output Analysis
CEA: Cost-Effectiveness Analysis
ELG: Eco-labelling
IPA: Impact Pathway Analysis
DfE: Design for Environment
ERA: Environmental Risk Assessment
LCA: Life-Cycle Assessment
DM: Dematerialization
EMA: Environmental Management
MFA: Material Flow Accounting
DT: Detoxification and Audit
SFA: Substance Flow Analysis

FIGURE 1.7 Matrix guide for the inclusion of the life-cycle concept in environmental sustainability management practices. (1) Education and communication; (2) product development and improvement; (3) production technology assessment; (4) improving environmental program; (5) strategic planning for a company's product or service line; (6) public policy planning and legislation; (7) environment-friendly purchasing support; (8) marketing strategies; (9) environmental performance and liability evaluation. **CBA**: cost-benefit analysis; **EIA**: environmental impact assessment; **IOA**: input–output analysis; **CEA**: cost-effectiveness analysis; **ELG**: eco-labelling; **IPA**: impact pathway analysis; **DfE**: design for environment; **ERA**: environmental risk assessment; **LCA**: life-cycle assessment; **DM**: dematerialization; **EMA**: environmental management; **MFA**: material flow accounting; **DT**: detoxification and audit; **SFA**: substance flow analysis.

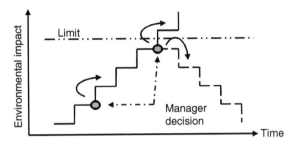

FIGURE 1.8 Decision-making situation in sustainability management.

1.6 EXAMPLE: DECISION-MAKING SITUATION IN SUSTAINABILITY MANAGEMENT

Political administrations or business managers very often make decisions of relevance for sustainability. As can be seen in Figure 1.8, their decisions may enhance or worsen the environmental performance of a chemical substance, process, product, or whole region.

A risk is that, after several decisions, the acceptable level of resource consumption and pollution is exceeded. In the end, in the case of business managers, their decisions may cause problems with the administration, neighbors of their facilities, or the consumers of their product. In the case of administrations, decisions may raise protests among the population and lose elections. In general, managers lack sufficient time to apply environmental sustainability management tools. They entrust corresponding projects or assessments to internal and/or external specialists (consultants), and base the obtained results to support and justify their decision-making under rational arguments. When managers do not master environmental sustainability management tools, they are at risk of choosing a specialist who applies a certain tool that will render a result subject to the methodology on which it is based. Consider, for example, the question of whether to build a new thermal power plant near carbon mines in a very populated area or far from mines in a sparsely populated area.

ERA applied to population exposure would choose the second option, while with an LCA, the first option would seem more appropriate due to reduced transportation. In sum, we must not trust these tools blindly; instead, we must understand their inherent limitations and apply each one, or a combination of them, to the right context.

1.7 CASE STUDY: WASTE INCINERATION AS ENVIRONMENTAL PROBLEM—THE CASE OF TARRAGONA, SPAIN

1.7.1 WASTE INCINERATION AS ENVIRONMENTAL PROBLEM

In recent years, waste incineration has been frequently preferred to other waste treatment or disposal alternatives due to advantages such as volume reduction, chemical toxicity destruction and energy recovery. However, strong public opposition to waste incineration often impedes the implementation of this technology. One of the main reasons for this opposition has been the perception that stack emissions are a real and serious threat to human health (Schuhmacher et al., 1997). In past years, the environmental consequences of incineration processes and their potential impact on public health by emissions of trace quantities of metals and polychlorinated dibenzo-*p*-dioxins (PCDDs) and dibenzofurans (PCDFs), as well as other emission products, have raised much concern. Unfortunately, information presented to the public about health risks of incineration is often incomplete, including only data on PCDD/Fs levels in stack gas samples (Domingo et al., 1999). In order to get overall information on the environmental impact of a municipal solid waste incinerator (MSWI), a wider study must be performed. Next, we present and analyze the case of an MSWI in Tarragona, Spain. This case study is presented with three different alternatives explained in the following section.

1.7.2 MUNICIPAL SOLID WASTE INCINERATOR IN TARRAGONA, SPAIN

Our case study will focus on an MSWI (Servei d'Incineració dels Residus Urbans S.A., or SIRUSA) located in Tarragona (northeastern Spain) that has operated since 1991. In 1997, an advanced acid gas removal system was installed. Thus, two situations (or scenarios) were studied: the operation of the plant during 1996 (later called former situation or scenario 1) and the current operation with the advanced acid gas removal system working (later called current situation or scenario 2).

The incinerator has parallel grate-fired furnaces with primary and secondary chambers. The combustion process is based on Deutsche Babcock Anlagen technology. Each of the furnaces has a capacity of 9.6 tons per hour, which makes approximately 460 tons daily incineration capacity of municipal waste. The temperature in the first combustion chamber varies between 950°C and 1000°C. In the secondary post-combustion chamber, the temperature is 650°C–720°C and the output temperature of the flue gas is 230°C–250°C. The minimum incineration conditions are 2 s of incineration time at 850°C with 6% minimum oxygen excess. The combustion process is controlled by on-line measurements (CO_2, O_2) and visually with the help of TV monitors (Nadal, 1999). The process generates electricity of the steam at a rate of 44.8 tons per hour. About 80% of the total electricity produced is sold and 20% is used for the operation. The scrap metal is collected separately and iron is recycled (STQ, 1998). The incinerated residues are solids. The average composition of the municipal waste is shown in Table 1.1 and a schematic overview of the plant is given in Figure 1.9.

The flue gas cleaning process is a semidry process consisting of an absorber of Danish technology (GSA). The acid compounds of the flue gas, such as HCl, HF, or SO_2, are neutralized with lime, $Ca(OH)_2$. The reaction products are separated in a cyclone and, after that, the gases are treated with injected active carbon to reduce dioxin and furan concentrations. The last cleaning step, a bag filter house, ensures that the total emissions meet the legislative emission limits: Spanish RD 1088/92 Directive and also a regional Catalan Directive 323/1994, which is an improved version of the European 89/369/CEE Directive. The total emissions and other process data are presented in Table 1.2.

TABLE 1.1

Waste Composition of the MSWI Plant in Tarragona, Spain[a]

Component	Percentage
Organics	46
Paper and cardboard	21
Plastics	13
Glass	9
Metals	3
Ceramics	2
Soil	1
Others	5
Total	100

[a] Average, 1999.

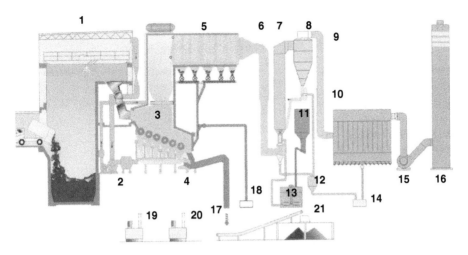

FIGURE 1.9 Scheme of the MSWI plant in Tarragona. (From Nadal, R., Planta incineradora e residuos sólidos urbanos de Tarragona. *Report for Master en Enginyeria i Gestio Ambiental (MEGA)*, Universitat Rovira i Virgili, Tarragona, Spain, 1999.) 1. crane bridge; 2. fans; 3. oven; 4. slag extractor; 5. boiler; 6. combustion gases; 7. reactor; 8. separation cyclone; 9. active coal; 10. recycling; 11. silo; 12. silo; 13. heat pump; 14. fan; 15. fan; 16. chimney; 17. slag treatment; 18. ashes; 19. feeder hydraulic station; 20. extractor hydraulic station; 21. slag.

TABLE 1.2
Overview of Data from the MSWI Plant in Tarragona, Spain

Situation	Without New Filters	With New Filters
Alternative no.	1	2
Production Data		
Produced electricity (MW)	6	6
Electricity sent out (MW)	5.2	4.9
Working hours per year (h)	8,280	8,280
Emission Data		
CO_2[a] (g/Nm^3)	186	186
CO (mg/Nm^3)	40	40
HCl (mg/Nm^3)	516	32.8
HF (mg/Nm^3)	1.75	0.45
NOx (mg/Nm^3)	191	191
Particles (mg/Nm^3)	27.4	4.8
SO_2 (mg/Nm^3)	80.9	30.2
As $(\mu g/Nm^3)$	20	5.6
Cd $(\mu g/Nm^3)$	20	6.6
Heavy metals[b] $(\mu g/Nm^3)$	450	91
Ni $(\mu g/Nm^3)$	30	8.4
PCDD/Fs (ng/Nm^3) as toxicity equivalent (TEQ)	2	0.002
Materials		
In		
CaO (t/yr)	0	921
Cement (t/yr)	88.5	518
Diesel	148.8	148.8
		(Continued)

TABLE 1.2 (*Continued*)
Overview of Data from the MSWI Plant in Tarragona, Spain

Situation	Without New Filters	With New Filters
Out		
Slag (t/yr)	42,208	42,208
Scrap for treatment (t/yr)	2,740	2,740
Ashes for treatment (t/yr)	590	3,450
Ashes for disposal (t/yr)	767	4,485
Plant Data		
Gas volume (Nm³/h)	90,000	90,000
Gas temperature (K)	503	503
Stack height (m)	50	50
Stack diameter[c] (m)	1.98	1.98
Latitude (°)[d]	41.19	41.19
Longitude (°)[d]	1.211	1.211
Terrain elevation (m)	90	90

[a] Corresponds to the measured value, not to the adjusted one used in the LCA study (see Chapter 2).

[b] Heavy metals is a sum parameter in the form of Pb equivalents of the following heavy metals: As, B, Cr, Cu, Hg, Mn, Mo, Ni, Pb, and Sb. Cd is considered apart for its toxic relevance and As and Ni for their carcinogenic relevance.

[c] Although there are two stacks with 1.4 m, due to the limitations of the dispersion models used, one stack with a diameter of 1.98 was considered.

[d] Initially the data were in UTM, the Mercator transversal projection. The conversion was made using the algorithm in http://www.dmap.co.uk/ll2tm.htm.

The emissions are also controlled by the local authorities (Delegació Territorial del Departament de Medi Ambient de la Generalitat de Catalunya.). Thus, the plant is under continuous, real-time control, which guarantees independent information on the emissions.

Material inputs and outputs for sub-processes important for the process chain and taken into account with their site-specific data in the LCA (see Chapter 2) are mentioned in Tables 1.2 and 1.3. An overview of the process chain and the origin of the material is given in Figure 1.10.

On the basis of the model described by Kremer et al. (1998) and a spreadsheet version by Ciroth (1998), a modular steady-state process model with several enhancements has been created by Hagelüken (2001), in cooperation with the Environmental Management and Analysis (AGA) Group of the Universitat Rovira i Virgili in Tarragona, Spain. The MS–Excel-based model takes into account the elementary waste input composition and relevant plant data, such as plant layout and process specific constants.

In the model, the steam generator consists of grate firing and heat recovery systems and a regenerative air preheater. Energy production is calculated using the heating value of the waste input and the state points of the steam utilization process. For the macro-elements (C, H, N, O, S, Cl, and F), the flue gas composition is determined by simple thermodynamic calculation of the combustion, taking excess air into account. The heavy metals, however, are calculated on the basis of transfer coefficients (Kremer et al., 1998). Emissions of CO and TOC depend on the amount of flue gas. For the emissions of NO_x and PCDD/Fs, empirical formulas are used. Because acid-forming substances like S, Cl, and F are partly absorbed by basic ash components, the total amount of SO_2, HCl, and HF in the flue gas is reduced respectively. The flue gas purification consists of an electrostatic precipitator, a two-stage gas scrubber for the removal of acid gases (using NaOH and $CaCO_3$ for neutralization), a denitrogenation unit (DeNOx with selective catalytic reduction using NH_3) and an entrained flow absorber with

TABLE 1.3
Overview of Inputs and Outputs of Scenario 3

Overall Transport (tkm/yr)[a]

Municipal waste	2,762,406
Slag	1,032,278
CaO	224,080
Ammonia	6,859
Scrap treatment	2,236,524
Ash treatment	71,497
Ash disposal	2,439,849
Cement	134,058

Electricity (TJ /yr)

Consumption	18.41
Production	290.11

Materials (t/yr)

In

CaO	2241
Ammonia	686
Cement	1341

Out

Slag	32,259
Scrap for treatment	2,094
Ashes for treatment	8,937
Ashes for disposal	11,618

Emissions

Flue gas (Nm^3/h)	96,000
As ($\mu g/Nm^3$)	0.035
Cd ($\mu g/Nm^3$)	1.50
Ni ($\mu g/Nm^3$)	0.51
NOx (mg/Nm^3)	58
PM10 (mg/Nm^3)	0.45

[a] tkm is equivalent to a mass of 1 t (1000 kg) transported 1 km.

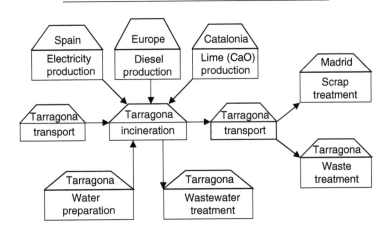

FIGURE 1.10 Overview of the MSWI process chain.

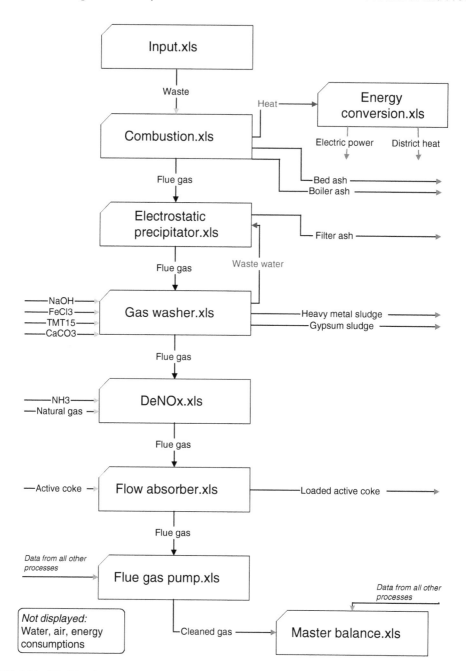

FIGURE 1.11   Workbook structure of the molecular spreadsheet model. (From Hagelüken, M., *Effects of Different Models of Municipal Solid Waste Incinerators on the Results of Life-Cycle Assessments, Student Research Project*, cooperation of TU Berlin and Universitat Rovira i Virgili, Tarragona, Spain, 2001.)

active carbon injection for the removal of dioxins and heavy metals. The plant is a semidry type; all wastewater is evaporated in a spray dryer after the heat exchanger.

The processes and calculations are distributed to several MS–Excel workbooks. The processes represented by the workbook files are linked together by their input–output sheets. The division into workbooks and their major dependencies are shown in Figure 1.11.

Based on the modular model, a future scenario, scenario 3, to be used in Chapter 11, was created for an MSWI similar to the current plant in Tarragona, but with DeNOx as an additional gas cleaning system. An overview of the calculated inputs and outputs for the SIRUSA waste incineration plant is given in Table 1.3. All the calculated emissions are lower than the current situation 2 scenario of the MSNI in Tarragona, Spain. Also, the corresponding transport distances are presented because they are necessary for the estimation of environmental damages within industrial process chains in Chapter 10.

1.8   QUESTIONS AND EXERCISES

Several typical situations for sustainability management decisions are given. Students are asked to think of a tool, or a combination of tools, for environmental sustainability management in order to apply them to a manager's or consultant's situation.

1. The population neighboring an industrial area is becoming increasingly sensitive to pollution issues. Business A wants to protect itself against possible population claims for pollution due to their industrial activity.
2. Business B wants to make sure the implementation of a new gas treatment in an existing plant will be environmentally sustainable.
3. Business C is developing a new electronic product that will be introduced to the market within a year.
4. Business D is developing a new detergent that will be introduced to the market within a few years.
5. Business E is developing a marketing campaign under the slogan of "Join the Struggle against Climate Change" to increase sales of its product manufactured with fully renewable resources.
6. Business F intends to undertake an environmental enhancement for its galvanization process.
7. Business G intends to purchase a civil responsibility insurance policy for its waste outlet.
8. Administration H must decide whether it will grant a license for the construction of a new 50-km highway linking city X with city Y.
9. Administration I is searching for a location at which to build an urban solid waste incineration plant.
10. A committee of experts is asked to provide a recommendation whether a subsidy for electric car is justified from an environmental point of view in country J.

REFERENCES

ACC (American Chemistry Council). (2017), Responsible Care, https://responsiblecare.americanchemistry.com, (accessed on August 15, 2017).

AICHE (American Institute of Chemical Engineers). (1985), *Guidelines for Hazard Evaluation Procedures*, prepared by Batelle Columbus Division for the Center for Chemical Process Safety, New York.

Alfonso, P. and Krämer, L. (1996), *Derecho Medioambiental de la Unión Europea*, McGraw-Hill, Madrid, Spain.

Ayres, R.U. and Ayres, L.W. (1996), Industrial ecology, in Ayres, R.U. and Simonis, U.E. (Eds.), *Industrial Metabolism: Restructuring for Sustainable Development*, United Nations University Press, New York, 1994.

Bringezu, S., Fischer-Kowalski, M., Kleijn, R., and Palm, V. (1997), Regional and national material flow accounting—From paradigm to practice of sustainability, *Proceedings of the ConAccount Workshop*, Leiden, the Netherlands.

Brunner, P.H. and Rechberger, H. (2016), *Handbook of Material Flow Analysis*, CRC Press, Pensacola, FL.

CEFIC (The European Chemical Industry Council). (2017), Product Stewardship, http://www.cefic.org/Industry-support/Responsible-Care-tools-SMEs/Product-stewardship/, (accessed on August 15, 2017).

CHAINET (European Network on Chain Analysis for Environmental Decision Support). (1998), *Definition Document, sCML*, Leiden University, Leiden, the Netherlands.

Ciroth, A. (1998), *Beispielhafte Anwendung der Iterativen Screening –Ökobilang-Diplomarbeit Institut für Technischen Umweltschutz, Fachgebeit Abfallvermeidung*, Technische Universität, Berlin, Germany.

Clift, R. and Druckmann, A. (2016), *Taking Stock of Industrial Ecology*, Springer, Dordrecht, the Netherlands.

Coneza, F. (1997), *Guía Metodológica para la Evaluación del Impacto Ambiental*, Ediciones Mundi-Prensa, Madrid, Spain.

Crul, M.R.M. and Diehl, J.C. (2009), *Design for Sustainability (D4S)*, UNEP and TU Delft, Delft, the Netherlands.

Dasgupta, A.K. and Pearce, D.W. (1972), *Cost-Benefit Analysis—Theory and Practice*, MacMillan Press, Cambridge, UK.

De Smet, B., Hemming, C., Baumann, H., Cowell, S., Pesso, C., Sund, L., Markovic, V., Moilanen, T., and Postlethwaithe, D. (1996), *Life-Cycle Assessment and Conceptually Related Programs, Working Group Draft*, SETAC-Europe, Brussels, Belgium.

Domingo, J.L., Schuhmacher, M., Granero, S., and Llobet, J.M. (1999), PCDDs and PCDFs in food samples from Catalonia, Spain. An assessment of dietary intake, *Chemosphere*, 38, 3517–3528.

EC (European Commission). (1995), *ExternE—Externalities of Energy*, 6 vol., EUR 16520–162525. DG XII Science, Research and Development, Brussels, Belgium.

EC (European Commission). (1997), Guidelines for the application of life-cycle assessment in the EU eco-label award scheme. *A Report Prepared for the EC by the "Groupe des Sages,"* Brussels, Belgium.

EC (European Commission). (2006), *Regulation 1907/2006/EC concerning the Registration, Evaluation, Authorization and Restriction of Chemicals*, Brussels, Belgium.

EC (European Commission). (2008), *Action Plan for Sustainable Consumption and Production and Industry*, Brussels, Belgium.

EC (European Commission). (2011a), *Roadmap to a Resource Efficient Europe*, Brussels, Belgium.

EC (European Commission). (2011b), *EMAS Fact Sheet*, Brussels, Belgium.

EC (European Commission). (2013), *Communication on Single Market for Green Products Initiative*, Brussels, Belgium.

Elkington, J. (1997), *Cannibals with Forks: The Triple Bottom Line of 21st Century Business*, New Society Publishers, Gabriola Island, Canada.

Eurostat. (2001), *Economy-wide Material Flow Accounts and Derived Indicators: A Methodological Guide*, Eurostat, Luxembourg, UK.

Field, B.C. (1995), *Economía Ambiental—Una Introducción*, McGraw-Hill, Santafé de Bogotá, Columbia.

Fiksel, J. (1997), *Ingeniería de Diseño Medioambiental*, McGraw-Hill, Madrid, Spain.

Fuster, G., Schuhmacher, M., and Domingo, J.L. (2002), Human exposure to dioxin and furans. Application of the substance flow analysis to health risk assessment, *Environ. Int. Sci. Pollut. Res.*, 9(4), 241–249.

Graedel, T. (1994), Industrial ecology—Definition and implementation, in Socolow, R. et al. (Eds.), *Industrial Ecology and Global Change,* Cambridge University Press, New York, Chap. 3.

Graedel, T.E. and Allenby, B.R. (2002), *Industrial Ecology*, Prentice Hall, Upper Saddle River, NJ.

Hagelüken, M. (2001), *Effects of Different Models of Municipal Solid Waste Incinerators on the Results of Life-Cycle Assessments, Student Research Project*, cooperation of TU Berlin and Universitat Rovira i Virgili, Tarragona, Spain.

Hansen, D.A. (2017), *Total Quality Management*, TQM, Tutorial Help Page, http://www.flowhelp.com/tqm/ tqm.html, (accessed on August 15, 2017).

Heijungs, R., Huppes, G., and Guinée, J.B. (2010), Life cycle assessment and sustainability analysis of products, materials and technologies. Toward a scientific framework for sustainability life cycle analysis, *Polymer Degrad. Stabil.*, 95, 422–428.

Kloepffer, W. (2008), Life cycle sustainability assessment of products, *Int. J. Life Cycle Assess.*, 13, 89–95.

Kremer, M., Goldhan, G., and Heyde, M. (1998), Waste treatment in product-specific life-cycle inventories, *Int. J. LCA*, 3(1), 47–55.

Lindhqvist, T. (2000), *Extended Producer Responsibility in Cleaner Production: Policy Principle to Promote Environmental Improvements of Product Systems*, IIIEE, Lund University, Lund, Sweden.

Lubin, D.A. and Esty, D.C. (2010), *The Sustainability Imperative*, Harvard Business Review, https://hbr.org/2010/05/the-sustainability-imperative, (accessed on August 15, 2017).

Masters, G. (1991), *Introduction to Environmental Engineering and Science*, Prentice Hall, Englewood Cliffs, NJ.

Miller, R.E. (2009), *Input–Output Analysis: Foundations and Extensions*, Cambridge University Press, Cambridge, UK.

Nadal, R. (1999), Planta incineradora e residuos sólidos urbanos de Tarragona. *Report for Master en Enginyeria i Gestio Ambiental (MEGA)*, Universitat Rovira i Virgili, Tarragona, Spain.

Nas, T.F. (1996), *Cost-Benefit Analysis—Theory and Application*, SAGE Publications, New York.

OECD (Organization for Economic Cooperation and Development). (2008), *Measuring Material Flows and Resource Productivity: The OECD Guide*, Paris, France.

OECD (Organization for Economic Cooperation and Development). (2012), *Sustainable Materials Management: Making Better Use of Resources*, Paris, France.

Rodrigo, J. and Castells, F. (2002), *Electrical and Electronic Practical Ecodesign Guide*, Universitat Rovira i Virgili, Tarragona, Spain.

Schlemminger, H. and Martens, C.-P. (2004), *German Environmental Law for Practitioners*, Kluwer Law International, The Hague, the Netherlands.

Schuhmacher, M., Meneses, M., Granero, S., Llobet, J.M., and Domingo, J.L. (1997), Trace element pollution of soils collected near a municipal solid waste incinerator: Human health risk, *Bull. Environ. Contam. Toxicol.*, 59, 861–867.

SETAC (Society of Environmental Toxicology and Chemistry). (1993), *Guidelines for Life-Cycle Assessment—A Code of Practice*, Sesimbra/Portugal SETAC Workshop report, SETAC Press, Pensacola, FL.

Sonnemann, G. and Margni, M. (2015), *Life Cycle Management, Series–LCA Compendium–The Complete World of Life Cycle Assessment*, Springer, Dordrecht, the Netherlands.

Sonnemann, G.W., Castells, F., and Rodrigo, J. (1999), Guide to the inclusion of LCA in environmental management, *8th Congreso Mediterráneo de Ingeniería Química*, Expoquimia, Barcelona.

STQ (Servei de Tecnologia Química). (1998), *Análisis del Ciclo de Vida de la Electricidad Producida por la Planta de Incineración de Residuos Urbanos de Tarragona*, technical report, Universitat Rovira i Virgili, Tarragona, Spain.

Suh, S. (2009), *Handbook of Input-Output Economics in Industrial Ecology*, Springer, Dordrecht, the Netherlands.

Traverso, M., Asdrubali, F., Francia, A., and Finkbeiner, M. (2012), Towards life cycle sustainability assessment: An implementation to photovoltaic modules. *Int. J. Life Cycle Assess.*, 17, 1068–1079.

U.S. EPA. (2017a), Pollution Prevention P2, https://www.epa.gov/p2, (accessed on August 15, 2017).

U.S. EPA. (2017b), *Green Chemistry*, https://www.epa.gov/greenchemistry, (accessed on August 15, 2017).

UNCSD. (2012), *The Future We Want, Rio+20 Outcome document*, Rio de Janeiro, Brazil.

UNEP and WBCSD. (2017), *Eco-Efficiency and Cleaner Production: Charting the Course to Sustainability*, http://enb.iisd.org/consume/unep.html, (accessed on August 1, 2018).

UNEP. (1999), Towards the global use of life-cycle assessment, Technical report, United Nations Environment Program, Paris, France.

UNEP. (2012), *The 10 Year Framework of Programs on Sustainable Consumption and Production*, United Nations Environment Program, Paris, France.

UNEP. (2017), *Sustainable Consumption and Production Branch*, http://www.unepie.org/scp/, (accessed on August 15, 2017).

UNEP/SETAC. (2001), *Background Paper of the UNEP/SETAC Life-Cycle Initiative*, United Nations Environment Program, Paris, France.

Valdivia, S., Ugaya, C.M.L., Hildenbrand, J., Traverso, M., Mazijn, B., and Sonnemann, G. (2013), A UNEP/SETAC approach towards a life cycle sustainability assessment—Our contribution to Rio+20. *Int. J. Life Cycle Assess.*, 18, 1673–1685.

Vasanthakumar, N.B. (1998), *Total Quality Environmental Management: an ISO 14000 Approach*, Quorum Books, Westport, CT.

WBCSD. (1999), Eco-efficiency indicators and reporting, technical report, *Working Group on Eco-Efficiency Metrics & Reporting*, Geneva, Switzerland.

Wenzel, H. (1998), Application dependency of LCA methodology—key variables and their mode of influencing the method. *Int. J. LCA*, 3(5), 281–288.

Zharen, W.M. (1995), *ISO 14000—Understanding the Environmental Standards*, Government Institutes, Rockville, MD.

2 Life-Cycle Assessment

Dieuwertje Schrijvers, Michael Tsang, Francesc Castells,
and Guido Sonnemann

CONTENTS

2.1   INTRODUCTION

2.1.1   CONCEPT OF LIFE-CYCLE ASSESSMENT

Life-cycle assessment (LCA) of a product comprises the evaluation of the environmental effects produced during its entire life-cycle, from its origin as a raw material until its end, usually as a waste. This concept goes beyond the classical concept of pollution from the manufacturing steps of a product, taking into account the "upstream" and "downstream" steps. These steps can be illustrated by using the life cycle of a chair as an example.

Let us imagine that our chair would be manufactured in polished wood fixed by iron screws, and that the seat would be made of a low-density foam layer covered by polyamide fabric. If we carried out an evaluation of the chair based only on its manufacturing stage, the study would show insignificant environmental impact. This would be justified by the simplicity of the production process, in many cases reduced to electricity consumption, dust generation and wood waste production due to the assembly of all components. However, according to the life-cycle approach, we must consider all the previous operations carried out in order to transform natural resources into the intermediate products that will make up the chair. In this case, the study would start with primary activities like wood planting, iron mining and crude oil extraction and continuing through fabric manufacturing to the final assembly of the chair. Moreover, we must include later stages such as use and final disposal from the point of view of environmental impact. This means that we need to evaluate each aspect related to natural resource consumption or waste releases from the entire life cycle of the chair. The most recognized and well-accepted method of carrying out environmental assessment of products and services along their life cycles is the methodology of LCA. This chapter will present an overview of its conceptual framework, common applications and importance for eco-design and environmental management solutions.

In this framework, LCA is a tool to evaluate the environmental performance of products (SETAC, 1993; UNEP, 1996). LCA focuses on the entire life cycle of a product, from the extraction of resources and processing of raw material through manufacture, distribution, and use to the final processing of the disposed product. Throughout all these stages, extraction and consumption of resources (including energy) and releases into air, water, and soil are identified and quantified. Subsequently, the potential contribution of these flows to several types of environmental impacts is assessed and evaluated (Curran, 1996; EEA, 1998).

2.1.2   HISTORY OF LIFE-CYCLE ASSESSMENT

An overview of the early history of LCA can be found in Assies (1992), Vigon et al. (1993), Pedersen (1993), Boustead (1992), and Castells et al. (1997). It is not easy to determine exactly when studies related to the methodology that would later be known as life-cycle assessment started. In the opinion of Vigon et al. (1993), one of the first studies was H. Smith's, whose calculations of energy requirements for manufacturing final and intermediate chemical products entered the public domain in 1963. Later, other global studies such as those by Meadows et al. (1972) and the Club of Rome (1972) predicted the effects of an increase in population and energy and material resources. These predictions (which foretold fast consumption of fossil fuels and the climate changes resulting from it), together with the oil crisis of the 1970s, encouraged more detailed studies, focused mainly on the optimum management of energy resources. As explained by Boustead and Hancock (1979), the necessity of solving material balance in the process in order to undertake such assessments required the inclusion of raw material consumption as well as waste generation. The so-called "energy assessments" date back to these years, according to Assies (1992), Boustead (1974), and IFIAS (1974). Although these studies focused basically on the optimization of energy consumption, they also included estimations on emissions and releases. More references about these assessments can be found in Boustead (1992).

Vigon et al. (1993) highlighted the 1969 Coca-Cola™ study carried out by the Midwest Research Institute (MRI) aimed at determining the type of container with the lowest environmental effect. However, Assies (1992) considers MRI's assessment conducted by Hunt (1974) for the U.S. Environmental Protection Agency (EPA) in order to compare different drink containers to be the forerunner of LCA studies. This study uses the term "resource and environmental profile analysis" (REPA) and is based on the analysis of a system following the production chain of the researched products from "cradle" to "grave" in order to quantify the use of resources and emissions to the environment. The study was also to develop a procedure enabling comparison of the environmental impacts generated by those products.

In 1979, the SETAC (Society of Environmental Toxicology and Chemistry), a multidisciplinary society of professionals with industrial, public, and scientific representatives, was founded. One of SETAC's goals was, and continues to be, the development of LCA methodology and criteria. In the same year, Boustead and Hancock (1979) published a study describing the methodology of energy assessment with the idea of making energy treatment more systematic and establishing criteria to compare various energy sources.

In 1984, the EMPA (Swiss Federal Laboratories for Materials Testing and Research) conducted research that added the effects on health to emission studies and took into account a limited number of parameters, thus simplifying assessment and decision-making. Products were assessed on the basis of their potential environmental impact expressed as energy consumption, air and water pollution, and solid wastes. It also provided a comprehensive database with access to the public that, according to Assies (1993), catalyzed the implementation of LCA (EMPA, 1984; Druijff, 1984).

Due in part to higher access to public data and an increased environmental awareness of the population, new LCA projects were developed in the industry as well as in the public sector. Such growth resulted in the "launch" of this subject at an international level in 1990. That year, conferences about LCA were held in Washington, D.C. (organized by the World Wildlife Fund and sponsored by the EPA), Vermont (organized by SETAC), and Leuven, Belgium (organized by Procter & Gamble).

Growth in the number of studies, as well as in organizations devoted to this subject matter, allowed the publication of works intended to standardize the criteria to be applied in LCA studies. Among these were Fava et al. (1991), Heijungs et al. (1992), Boustead (1992), Fecker (1992), Vigon et al. (1993), SETAC (1993), and Guinée et al. (1993a, 1993b).

In June 1993, the ISO created the Technical Committee 207 (ISO/TC 207) with the goal of developing international norms and rules for environmental management. The fifth of the six subcommittees created, the LCA SC5, was assigned standardization within the field of LCA. Its aim is to prevent the presentation of partial results or data of questionable reliability from LCA studies for marketing purposes, thus ensuring that each application is carried out in accordance with universally valid structure and features. As a result of this work, we rely on the ISO standards 14040 and 14044, of which the latest versions were published in 2006 (ISO 2006a, 2006b).

In the 1990s, the annual conferences by SETAC and the Working Groups on LCA played a paramount role in developing this methodology to its current status. For an overview on the results of this work, see Udo de Haes et al. (1999, 2002a).

In May 2000, the United Nations Environment Programme (UNEP) and SETAC signed a letter of intent and established the Life-Cycle Initiative (UNEP/SETAC Life-Cycle Initiative, 2002; Udo de Haes et al., 2002b). This initiative builds on the ISO 14040 standards and aims to establish approaches with best practice for a life-cycle economy, corresponding to the call of governments in the Malmö declaration of 2000. This initiative allowed laying the foundations for LCA methodology to be used in a practical manner by all product and service sectors around the globe.

The beginning of the twenty-first century is a period of further elaboration of the LCA methodologies (Guinée et al., 2011). In their overview of the development of LCA in recent years, Guinée et al. show that LCA modelling methods become more technically advanced and cover more societal problems—not only environmental, but also social and economic issues. This results into the shift of environmental LCA to broader Life-Cycle Sustainability Assessments (LCSA). At the same time, LCA is more and more used in policy decisions on national and regional levels, such as of the European Commission. A good definition of the sustainability questions at hand and the identification of suitable LCSA modelling tools to answer these questions are ongoing challenges.

2.1.3 Common Uses of Life-Cycle Assessment

In a first approach, the uses of LCA can be classified as general and particular:

General:
- Compare alternative choices
- Identify points for environmental improvement
- Adopt a more global perspective of environmental issues, to avoid problem shifting
- Contribute to the understanding of the environmental consequences of human activities
- Establish a picture of the interactions between a product or activity and the environment as quickly as possible
- Provide support information so that decision-makers can identify opportunities for environmental improvements

Particular:
- Define the environmental performance of a product during its entire life-cycle
- Identify the most relevant steps in the manufacturing process related to a given environmental impact
- Compare the environmental performance of a product with that of other products giving a similar service

The use of LCA allows defining the environmental profile of a product throughout its life cycle. Thus, the consumption of natural resources or releases into air, water, and soil can be identified, quantified, and expressed in terms of impacts on the environment. LCA does not necessarily need to be applied to the entire life-cycle of a product. In many cases, this kind of evaluation is applied to a single process such as a car assembly or to a service such as raw material transportation. Depending on the context, LCA is useful as a conceptual framework or as a set of practical tools. "Life-cycle thinking" can stimulate creativity and ability to see the extensive dimensions of a problem. In terms of strategic management, a business can find important product improvements, new approaches to process optimization, and, in some cases, radically new ways of meeting the same need (only with a new product or a service) while carrying out an LCA. In this context, LCA can be seen as a support tool in decision-making processes. In addition, life-cycle management (LCM) allows an integrated approach to minimizing environmental loads throughout the life cycle of a product, system, or service.

From a different point of view, LCA can be applied in establishing public policy. Sustainable development has been included as a major item on most governmental agendas already since the Rio Summit in 1992 and gained refreshed momentum after the Paris Agreement in 2015. Also, the UN General Assembly formulated 17 Sustainable Development Goals in 2015 that should inspire future developments (Bey, 2018). It is obvious that the LCA approach must be used to ensure that actions toward a more sustainable future will have the desired effect. In this framework, the main governmental applications regarding LCA are product-oriented policies, deposit-refund programs (including waste management policies), subsidy taxation, and general process-oriented policies. Finally, anything we do to make LCA useful will not really help unless the world believes it is efficient. For this reason, LCA experts admit the necessity of giving more information about LCA issues in order to increase credibility of the tool and gain greater acceptance from the public. A great interest exists about what other people think of the discipline and its implications for the future.

Approaches to consumption have been valuable to the analysis of current conditions and have promoted novel strategies for future development. This has exposed the limitations of isolated production-focused strategies. What is urgently needed is to change the systems of production and consumption in an integrated way, which is also acknowledged in the UN Sustainable Development Goals (Sonnemann et al., 2018). The Life-Cycle Initiative mentioned in Section 2.1.2 is going in that direction, which means a life-cycle approach is needed for changing unsustainable consumption and production patterns.

2.2   LCA FRAMEWORK AND THE ISO 14000 SERIES

The ISO standardized the technical framework for the life-cycle assessment methodology in the 1990s, which has been updated in 2006. On this basis, according to ISO 14040 (2006a), LCA consists of the following steps (Figure 2.1):

- Goal and scope definition
- Inventory analysis
- Impact assessment
- Interpretation

LCA is not necessarily carried out in a single sequence. It is an iterative process in which subsequent rounds can achieve increasing levels of detail (from screening LCA to full LCA) or lead to changes in the first phase prompted by the results of the last phase.

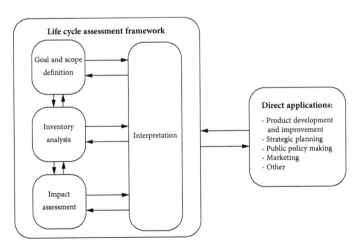

FIGURE 2.1   The phases of LCA adapted from ISO 14040:2006.

Where ISO 14040 (2006a) provides the general framework for LCA, ISO 14044 (2006b) provides more detailed guidance for determining the goal and scope of an LCA study, establishing a life-cycle inventory (LCI), conducting the life-cycle impact assessment phase, and interpreting the results produced by an LCA. Moreover, technical guidelines (e.g., ISO/TR 14049 [2012]) illustrate how to apply the standards. A short introduction of the LCA framework is provided in this section. More details are given in Section 2.3.

2.2.1   Goal and Scope Definition

The goal and scope definition is designed to obtain the required specifications for the LCA study. During this step, the strategic aspects concerning questions to be answered and identifying the intended audience are defined. To carry out the goal and scope of an LCA study, the practitioner must follow some procedures:

1. Define the purpose of the LCA study, ending with the definition of the functional unit, which is the quantitative reference for the study.
2. Define the scope of the study, which embraces two main tasks:
   • Establish the spatial limits between the product system under study and its neighborhood, which will be generally called "environment."
   • Detail the system through drawing up its unit processes flowchart, taking into account a first estimation of inputs from and outputs to the environment (the elementary flows or burdens to the environment).
3. Define the data required, which includes a specification of the data necessary for the inventory analysis and for the subsequent impact assessment phase.

2.2.2   Inventory Analysis

The Life-Cycle Inventory (LCI) analysis collects all the data of the unit processes within a product system and relates them to the functional unit of the study. In this case, the following steps must be considered:

1. Data collection, which includes the specification of all input and output flows of the processes within the product system (i.e., product flows, flows to other unit processes, and elementary flows from and to the environment)
2. Normalization to the functional unit, which means that all data collected are quantitatively related to one quantitative output of the product system under study; usually, 1 kg of material is chosen, but often other units such as a car or 1 km of mobility are preferable
3. Allocation, which means the distribution of emissions and resource extractions within a given process throughout its different products (e.g., petroleum refining providing naphtha, gasolines, heavy oils, etc)
4. Data evaluation, which involves a quality assessment of the data (e.g., by eventually performing a sensitivity analysis)

The result of the inventory analysis, consisting of the elementary flows related to the functional unit, is often called the life-cycle inventory table.

2.2.3   Impact Assessment

The Life-Cycle Impact Assessment (LCIA) phase aims at making the results from the inventory analysis more understandable and more manageable in relation to human health, the availability of resources, and the natural environment. To accomplish this, the elementary inputs and outputs of the product system that were identified in the LCI will be converted into a smaller number of indicators. The mandatory steps to be taken in this regard are:

1. Select and define impact categories, which are classes of a selected number of environmental impacts (global warming, acidification, etc).
2. Classify by assigning the results from the LCI to the relevant impact categories.
3. Characterize by aggregating the inventory results in terms of adequate factors (so-called characterization factors) of different types of substances within the impact categories; therefore, a common unit is defined for each category. The results of the characterization step are known as the environmental profile of the product system.

More details will be given in Chapter 3.

2.2.4   INTERPRETATION

The interpretation phase aims to evaluate the results from the inventory analysis or impact assessment and compare them with the goal of the study defined in the first phase. The following steps can be distinguished within this phase:

1. Identification of the most important results of the LCI and impact assessment
2. Evaluation of the study's outcomes, consisting of a number of the following routines: completeness check, sensitivity analysis, uncertainty analysis, and consistency check
3. Conclusions, recommendations, and reports, including a definition of the final outcome, a comparison with the original goal of the study, drawing up recommendations, procedures for a critical review, and the final reporting of the results

The results of the interpretation may lead to a new iteration round of the study, including a possible adjustment of the original goal.

2.3   GOAL AND SCOPE DEFINITION

Section 2.2 briefly described the different steps of an LCA according to ISO 14040 and ISO 14044. This section describes how to run an LCA goal and scope definition in practice.

2.3.1   PURPOSE OF AN LIFE-CYCLE ASSESSMENT STUDY

The main purpose of the study must be clearly defined at the very beginning because it has a strong influence on further steps. If the study is designed to compare a product with another product that has already been submitted to an LCA, the structure, scope, and complexity of the first product's LCA must be similar to those of the other product so that a reliable comparison can be made. If the aim is to analyze the environmental performance of a product to determine its present status and to enable future improvements, the LCA study must be organized by carefully dividing the manufacturing process into well-defined sections or phases in order to identify afterwards which parts of the process are responsible for each environmental effect. The purpose of the study should be described with regard to different levels: the reason to conduct the LCA, the intended application of the results of the LCA, and the perspective from which the LCA is conducted.

2.3.1.1   The Reason to Conduct the Study

Azapagic and Clift (1999) distinguish two types of goals that can motivate an LCA or (part of the) life cycle of a product:

- To evaluate the environmental performance of the production process of a product from "cradle-to-factory gate" as a guide for environmental management (i.e., *process-oriented LCA*)
- To guide environmental management of the entire product life cycle by providing LCA data for downstream users of the product (i.e., *product-oriented LCA*)

Process-oriented LCAs could give relevant information if the aim of the LCA is to evaluate the environmental performance of production sites. This would be interesting for companies that would like to improve the environmental performance of a production or waste treatment process, or that would like to do an organizational LCA of which the production process takes part. Process-oriented LCAs could also provide information for environmental taxation that is targeted at production sites. An example of a situation that could ask for a process-oriented LCA is the comparison of two different waste treatment options for a product, such as landfilling and incineration. The subject of a process-oriented LCA could be formulated as "the production of 1 kg of product" or "the treatment of 1 kg of waste."

Product-oriented LCAs can inform downstream users of the products (e.g., for larger LCA studies where the product is used as an intermediate material or for marketing purposes). The subject of a product-oriented LCA could be formulated as "the demand for 1 kg of product."

2.3.1.2   The Intended Audience

The LCA might be conducted for—among others—internal use within a company, for marketing purposes or for communication to governmental institutes. The intended audience influences the level of detail that is necessary for the LCA as well as the presentation of the results.

2.3.1.3   The Intended Application

The results of LCAs can be used for multiple purposes. The application areas that are often referred to are:

- Quantifying the impacts of a product or process for use in other LCA studies or for environmental labeling. Environmental labeling could show customers or collaborators what the environmental impact of the product is that they consume.
- Identifying opportunities to improve the environmental performance of products or processes at various points in their life cycle. A hotspot analysis can identify which products or processes are the largest contributors to environmental impacts. Improvement options could focus on these contributors.
- Making a decision between multiple alternative options. The main goal of an LCA is often to compare two alternative products or services that provide the same function.

Often, the results of the LCA are used in multiple applications. For example, if a hotspot analysis is done first, and improvement options are identified, a comparative LCA is still necessary to confirm that the improvement options result in a lower environmental impact related to the product.

2.3.1.4   The Perspective of the Life-Cycle Assessment

Developments in the LCA domain have indicated the existence of various types of LCA goals, which would require different modeling procedures. Tillman (2000) proposed a distinction between retrospective, accounting, or cause-oriented LCAs—which are now known as attributional LCAs—and prospective, effect-oriented LCAs studying the effects of changes (i.e., consequential LCAs).

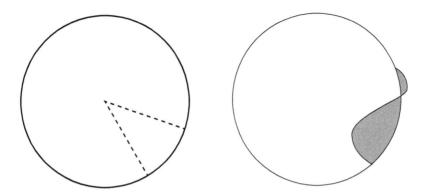

FIGURE 2.2 Conceptual difference between attributional (left) and consequential (right) LCA approaches, where the circles represent the total environmental burdens. (From Weidema, B.P., *Market Information in Life Cycle Assessment*, Copenhagen, Denmark, 2003; Sonnemann, G.W. and Vigon, B.W., *Global Guidance Principles for Life Cycle Assessment Databases—A Basis for Greener Processes and Products*, Paris, France, 2011.)

Although LCAs that have been conducted in the nineties are often a mix of attributional and consequential LCA studies, attributional LCA (a-LCA) is the approach that is often considered as the "traditional" approach for LCA. The purpose of an attributional approach is to identify which environmental impacts can be directly associated with the functional unit of a product system (Heijungs, 1997; Ekvall et al., 2005). An attributional model therefore aims to first identify which products and processes have been or will be directly involved with, for example, the production of the required materials and the treatment of wastes. These products and processes are represented by data on actual suppliers of the products, or, if products are bought on a regional or global market, average data of the suppliers to the market are used (Sonnemann and Vigon, 2011). Subsequently, it is identified which environmental impacts that are caused by these products and processes should be attributed to the functional unit, which is determined by the allocation procedure that is applied (see Section 2.4.3). The ideal application of an attributional LCA would result in a distribution of all global environmental burdens over all final products, as illustrated by the left picture of Figure 2.2 (Sonnemann and Vigon, 2011).

The consequential approach is applied to obtain information on the direct and indirect changes in environmental impacts due to a decision that typically results in a change in the production quantity of, or the demand for, a product. Only the processes affected by the change are considered; these are the marginal processes (Tillman, 2000; Weidema, 2003; European Commission, 2010; Sonnemann and Vigon, 2011). De Camillis et al. (2013) state that in c-LCA, the "[b]aseline scenario is the World as it is, now or in the future, without any action. The question that the approach aims is to answer is 'what are the net impacts associated to a change (in a product system) relative to the baseline scenario, where that change does not take place?'" This is depicted in the right picture of Figure 2.2. The processes that are affected by the change in demand for a product are not necessarily directly involved with the production of this product, which would be modelled in an attributional LCA. This can be illustrated by the use or production of by-products. By-products are co-products of a production process that provide a relatively small economic revenue. This revenue by itself is generally not sufficient to make the production process operate. An example is the production of agricultural waste due to the cultivation of a crop. This biomass could be used to provide energy in another product system. However, an additional demand for this energy will not result in an additional cultivation of crops. Therefore, the process that is affected by the demand for this biomass is not the additional production of biomass, but rather the avoided incineration of this biomass in the field. A consequential LCA can therefore also model *avoided* environmental impacts.

The processes that are affected by a changing demand for a product can be influenced by actors in the supply chain of this product that choose specific technologies or suppliers. However, this influence is limited. If these technologies and suppliers are constrained in their production capacity, market mechanisms can take place that lead to indirect environmental effects in other supply chains. To come back to the example of agricultural waste, it is possible that this waste is already fully used in other supply chains. An additional demand for this waste could lead to a decreased availability for other product systems in which other sources for energy now must be used. As a result, the consequential LCA will model environmental impacts that are not directly related to the product that is used. Besides, the user of the waste has no influence on the alternative energy source that is used by competing users of the waste, and cannot directly be held accountable for these impacts. For this reason, an attributional approach is often preferred over a consequential approach by companies, as attributional LCAs only include environmental impacts that are directly related to the product under study and for which the company could be held accountable. The application of LCA as described in this book is only based on an attributional approach.

2.3.2 The Functional Unit

The functional unit is the central concept in LCA; it is the measure of the performance delivered by the system under study. This unit is used as a basis for calculation and usually also as a basis for comparison between different systems fulfilling the same function. Table 2.1 presents examples of functional units related to the function performed by different systems.

An important point regarding the functional unit concerns the function carried out by the system. When different alternatives for manufacturing products or providing services are possible, the functional unit must be clear and constantly enable a sound comparison of the options considered. For example, let us evaluate the environmental impact produced by the transportation service of a person from Barcelona to Paris, cities separated by 1000 km. The system's function is clear: transfer a passenger. Nevertheless, the transfer can be done by different modes, except by ship.

Orange juice provides another good example. When the function of the system under study is orange juice consumption, the production of orange juice, its transport, processing, packaging, distribution, storage, sewage treatment, and final disposal, are considered. If the aim is to compare two different processes of juice production, 1 or 1000 L of orange juice will serve as the functional unit, taking into account that only the manufacturing system presents different alternatives. However, if the aim was to compare the use of different types of packaging systems, the functional unit should be the packaging of 1 L of orange juice.

Flowers are a classic example because people usually want "a bunch of flowers," rather than "750 g of flowers" or "flowers for 1 week." Thus, the functional unit should be defined as accurately as possible, considering that it should comprise the selected products and their end use, and that it is compatible with the nature of the application.

TABLE 2.1
Examples of Functional Units

Class of Products, Process, and Services	System Function	Functional Unit
Goods use	Light generation	lumen-hour
	Laundry washing	5 kg washed clothes
Process	Gasoline production	m^3 produced/h
	Liquid effluents treatment	t of removed COD/day
Transportation	Goods transport	tkm[a]
	Passengers transferring	Person km

[a] 1000 kg transported 1 km.

In practice, the functional unit must be measurable and, when two products with different life spans are compared (e.g., a match and a lighter), it is important that the period of use be considered for its establishment.

Finally, from the functional unit, the reference flow can be derived. This flow is often expressed in a quantity of a product that is required from a product system to fulfill a certain function. For example, the reference flow of the functional unit "packaging of 1 L of orange juice" could be "1 glass bottle with a useful volume of 1 L."

2.3.3 The System Boundaries

Within the goal and scope definition of an LCA, it is crucial to define the system boundaries. They define the range of the system under study and determine the processes and operations it comprises, such as primary material extraction, manufacturing, and waste disposal. In this sub-step, the inputs and outputs to be taken into account during the LCA study must be established. Ideally, processes are included until flows in and out of the product system only comprise elementary flows. Typical elements of a product system that fall within the system boundaries are the acquisition of raw materials, distribution, energy consumption, use and maintenance of products, waste disposal, recycling processes, and the manufacturing, use, and disposal of ancillary materials and capital equipment that are required to produce the functional unit (ISO, 2006a); in practice, capital equipment is not always taken into account (ICCA, 2016).

The definition of system boundaries can be carried out according to the following criteria: life-cycle boundaries, geographical boundaries, and environmental load boundaries.

2.3.3.1 Life-Cycle Boundaries

Let us assume that the life span of a product is composed of the life-cycle stages shown in Figure 2.3. The product life cycle could result in different system boundaries: if the entire life from the primary material extraction until the final disposal is considered, the limits will be defined as "cradle to grave." When the destination of a product is not known, the analysis can be stopped after manufacture and the limits will be cradle to gate.

In a situation of mature LCA practices, each life-cycle step will carry out its own gate-to-gate analysis and the entire cradle-to-grave process will be the result of the composition of a set of gate-to-gate systems.

2.3.3.2 Geographic Boundaries

These boundaries consider geographic limits to establish the limits of the product system. They can be considered life-cycle boundaries when the different life-cycle steps are confined in some region. These criteria are well recommended in cases of site-specific studies of LCA, as will be discussed later.

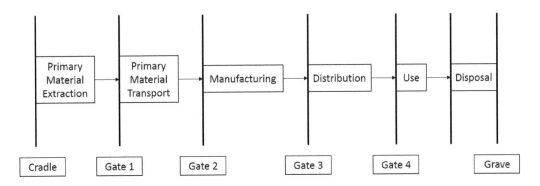

FIGURE 2.3 Product life-cycle stages.

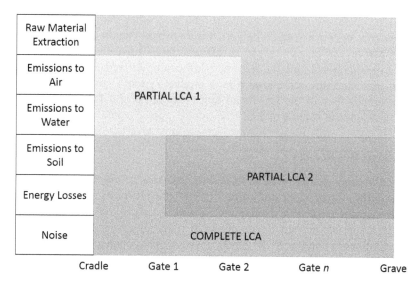

FIGURE 2.4   Boundaries in LCA.

2.3.3.3   Environmental Load Boundaries

Examples of environmental loads (EL)—or elementary flows in later ISO terminology—are renewable and nonrenewable raw materials, emissions to air, water, and soil, energy losses, radiation, and noise. An LCA can be carried out considering the entire list of inputs and outputs (complete LCA) or taking into account only emissions to air and water (partial LCA). In Figure 2.4, partial LCA 1 considers only emissions to air and water and is carried out from the beginning (cradle) until gate 2. Partial LCA 2 takes into account only emissions to soil and energy losses and goes from gate 1 to the end of life (grave).

2.3.3.4   Foreground and Background Subsystems

In LCA, often the distinction is made between the foreground and the background subsystems. The foreground refers to all processes that are directly influenced by the functional unit of the LCA study (Azapagic and Clift, 1999). Data collection of the foreground system is relatively straightforward, as the LCA practitioner often is familiar with or has direct access to these processes. Flows to and from the foreground system are elementary or intermediate flows (Figure 2.5). Intermediate flows are products that are used as input materials which are often supplied via a material market

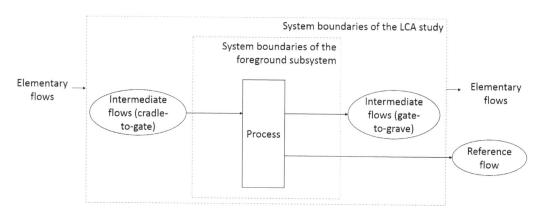

FIGURE 2.5   Indication of the foreground subsystem in a cradle-to-gate LCA study.

(Azapagic and Clift, 1999). Therefore, the production process of these products are part of the background system, and these processes are not always known—let alone influenced—by the user. Waste flows that need further waste treatment are outgoing intermediate flows. The differentiation between the foreground and the background subsystems is important, as the difference between attributional and consequential modeling lies mainly in the background subsystem.

2.3.4 DATA REQUIREMENTS

The quality of data used in the life-cycle inventory is naturally reflected in the quality of the final result of LCA. In this frame, it is important that the data quality be described and assessed in a systematic way that allows other practitioners to understand and reproduce the actual LCA. Initial data quality requirements must be done with the assistance of the following parameters:

- Time-related coverage concerns the age of the data that is acceptable for the LCA study. For example, if the study aims to evaluate the environmental impacts of a novel technology, old data is not likely to give valuable results.
- Geographical coverage refers to the geographic area from which data for a process unit should be collected (local, regional, national, continental, or global) in order to satisfy the purpose of the study.
- Technology coverage reflects the nature of the technology used in process units. Depending on the product system and the goal of the study, a technological mix could be used, or specific technologies could be more relevant. The technology coverage can consider, for example, a weighted average of the actual process mix, the best available technology, or the worst operating unit.

The goal and scope considers also the method of data collection—for example, collected from specific sites or from published sources—and whether the data are to be measured, calculated (e.g., by material or energy balances) or estimated by using other process units with similar operational conditions to that presented by the system under study as a proxy.

Data quality indicators such as precision, completeness, representativeness, consistency, and reproducibility should be taken into consideration in a level of detail depending on the premises of the goal and scope definition step. Chapter 9 presents and discusses some techniques of data analysis applied in order to assess the uncertainty of the data to be used in the life-cycle inventory.

2.4 LIFE-CYCLE INVENTORY

2.4.1 INTRODUCTION

Within LCA, life-cycle inventory (LCI) is considered the step in which all the environmental loads or elementary flows generated by a product or activity during its life-cycle are identified and evaluated. Environmental loads are defined here as the flows emitted to or removed from the surroundings that cause potential or actual harmful effects. Within this definition can be found raw materials consumption, emissions to air, water, and soil, radiation, noise, vibration, odors, etc.—factors commonly known as environmental pollution. Environmental loads must be quantifiable and relevant. During the inventory analysis, it can happen that the goal and scope definition needs refinement due to the increased knowledge about the product system.

To prepare an LCI, each environmental load (EL) generated by the process must be added to the ELs due to material and energy inputs and waste outputs. The result is assigned to the product. Thus, the inventory basically consists of an environmental load balance in which the ELs assigned to a product are the sum of ELs assigned to inputs and the ELs generated by the process. To illustrate this procedure, the CO_2 assigned to the product of Figure 2.6 is the sum of the CO_2 emitted by

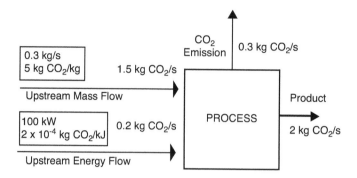

FIGURE 2.6 CO_2 assigned to a process in the life cycle of a product.

the process plus the CO_2 produced during the production of the raw materials input and the generation of the energy input.

The development of an LCI can be divided into the development of a process flow diagram and data collection, the application of allocation criteria, and the calculation of environmental loads. Each of these sub-steps will be briefly discussed next.

2.4.2   PROCESS FLOW DIAGRAM AND DATA COLLECTION

Data must be collected based on a process flow diagram of the system under study according to the defined life-cycle boundaries. A typical example of a flow diagram applied to chair manufacturing is shown in Figure 2.7. Data collection is the most time-consuming task in an LCA study; establishing qualitative and quantitative information concerning the process and its elementary flows requires a lot of work. The data collection can build on data from different kinds of data sources, which can be divided into four main categories, as presented in Table 2.2.

When using data from an electronic database or literature, it is important to ensure that they concern the relevant processes and come from secure sources, and are updated and in accordance with the goal and the system range previously established. For example, a database that includes current processes is of limited use in a study that aims to identify the potential impacts of a product in a future situation, as processes in the database might be outdated in 20 years.

Experience with data collection shows large differences in the availability of input and output data. Input data are the most readily available because energy and raw material consumption are

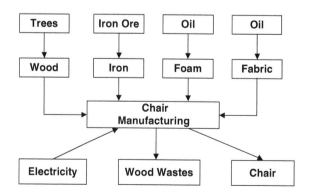

FIGURE 2.7   Simplified process flow diagram for chair manufacturing.

TABLE 2.2
Various Types of Data Sources

Data Sources	
Electronic databases	Several databases—such as Ecoinvent and GaBi—provided by commercial and public software and Internet sources on LCA
Literature data	Scientific papers, public reports, and existing LCA studies
Unreported data	Provided by companies, laboratories, authorities, and correlated sources
Measurements and/or computations	Calculated or estimated where data are nonexistent or should be improved

Source: Hauschild, M. and Wenzel, H., *Environmental Assessment of Products—Scientific Background* (Vol. 2), Chapman & Hall, London, UK, 1998.

registered by companies. Also, for cases of companies with uniform production profiles, energy consumption per product unit can be calculated on the basis of the company's total energy consumption. For non-uniform profiles, energy consumption must be estimated for each individual process.

Output data, with the exception of the main product and sometimes some by-products, are difficult to find. This difficulty is, in many cases, due to the absence of control registers of all releases, and the impossibility of allocating the existing data to the individual product. This feature is typically dependent on the size of the company in the study. Nevertheless, as recommended by Hauschild and Wenzel (1998), the problem of the availability of output data can be avoided, in some cases, by carrying out mass and energy balances from some inputs in order to calculate output values. This can be entirely adequate in many cases, and sometimes, even better than using data from direct measurements of the releases or emissions.

Finally, all data must be well specified concerning type and amount. Because the data collected are often originally intended for other purposes, they must be processed. Units must be converted to a standard set, preferably to SI units, and the data normalized (i.e., expressed in relation to a given output from a stage or operation comprised by the product system). The previous procedure of data normalization for each step of the system is helpful and will make the following step of the calculation of environmental loads easy.

2.4.3 Application of Allocation Criteria

In LCA, the term "allocation" means distribution of environmental loads. If we consider a manufacturing process for only one product, there is no allocation problem, because all the environmental loads must be assigned to that product. In a very common case in process industries, the same process delivers several products, so allocation criteria need to be created in order to distribute the environmental load.

Guinée et al. (2004) state that the functional flows of the system must be identified. A functional flow is "any of the flows of a unit process that constitute its goal, viz. the product outflows (including services) of a production process and the waste inflows of a waste treatment process." If a process yields not only the functional unit, but also other functional flows, the process is multifunctional (Figure 2.8). Guinée et al. (2004) identify three typical multifunctional situations:

1. *Co-production*: a multifunctional process having more than one functional outflow and no functional inflow

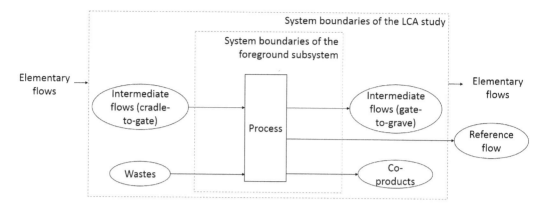

FIGURE 2.8   Typical multifunctional process that produces the reference flow and co-products, and provides a waste treatment service.

2. *Combined waste processing*: a multifunctional process having no functional outflow and more than one functional inflow
3. *Recycling*: a multifunctional process having one or more functional outflows and one or more functional inflows (including cases of combined waste processing and co-production simultaneously)

When the multifunctional process—or a product produced by this process—is compared to a process that only provides one functional reference flow, an allocation procedure must be applied in order to identify the environmental load of the product or service of interest (Figure 2.8).

ISO 14044 provides a stepwise procedure for applying allocation:

1. Step 1: Wherever possible, allocation should be avoided by:
   a. dividing the unit process to be allocated into two or more sub-processes and collecting the input and output data related to these sub-processes, or
   b. expanding the product system to include the additional functions related to the co-products […]
2. Step 2: Where allocation cannot be avoided, the inputs and outputs of the system should be partitioned between its different products or functions in a way that reflects the underlying physical relationships between them—that is, they should reflect the way in which the inputs and outputs are changed by quantitative changes in the products or functions delivered by the system.
3. Step 3: Where physical relationship alone cannot be established or used as the basis for allocation, the inputs should be allocated between the products and functions in a way that reflects other relationships between them. For example, input and output data might be allocated between co-products in proportion to the economic value of the products.

Subdivision of the unit process, which is the first preference of the ISO hierarchy, is, in practice, rarely possible due to a lack of detailed data or does not entirely eliminate the multi-functionality issue. Therefore, most LCA practitioners avoid allocation by system expansion. Let us look at a hypothetical LCA study of two alternative waste treatment options for plastic: landfilling and incineration with the recovery of energy. These two treatment processes do not provide entirely the same functions. Besides the treatment of waste, the incineration process produces electricity and heat as co-products. This process, therefore, falls in the category of "recycling" as classified previously.

System expansion can be applied by expanding the initial functional unit of "treating 1 kg of plastic waste" to "treating 1 kg of plastic waste and producing 4 MJ of heat and 3 MJ of electricity." The landfilling scenario can now be made comparable to the incineration scenario by adding alternative sources of electricity and heat to the product system. Instead of adding the inventory of the alternative energy sources to the landfilling scenario, many practitioners subtract this inventory from the product system of the incineration process, as this would provide the same net results in a comparison of the two waste treatment options. This is, however, debatable in an attributional LCA, as this implies that the production from the alternative energy source is *avoided*, which would only be modeled in a consequential LCA (Schrijvers et al., 2016). ISO 14044:2006 does not give specific recommendations with regard to attributional and consequential LCA, and, therefore, implicitly allows this operation.

Large functional units are not always desirable—for example, when the purpose of the LCA is to provide background data for another study in a product-oriented LCA. In that case, step 2 of the ISO allocation hierarchy could be considered: partitioning according to underlying physical relationships. However, relevant underlying physical relationships can only be identified when the production ratio of the co-products can be adjusted (Frischknecht, 2000; Weidema, 2001). An often used example is the co-production of sodium hydroxide and chlorine from the separation of sodium chloride. The production quantity of sodium hydroxide cannot be increased without also increasing the production quantity of chlorine. Therefore, no relevant underlying physical relationships could be identified that could be used as a basis for allocation. In this case, other relationships should be identified (step 3). The relative economic revenue that each co-product generates is most often used as a basis for allocation in attributional LCA (Pelletier et al., 2014). Also, in consequential LCA, the market mechanisms that are assumed to take place after the production of co-products could be considered as economic relationships (Geyer et al., 2015).

Another allocation method that is sometimes applied to recycled products or low-value by-products is the cut-off approach. In this method, all environmental load is attributed to the primary product or the main co-product, and the recycled product or the by-product is considered burden-free. This approach is not explicitly mentioned in the ISO hierarchy, but it could be considered as a simplification of an economic allocation when the recycled product or the by-product generates a very low economic revenue compared to the other product(s) (Schrijvers et al., 2016).

2.4.4 ENVIRONMENTAL LOADS CALCULATION

After data collection and the selection of the allocation criteria, a model of environmental loads calculation is set up for the product system. One of the most efficient alternatives using an eco-vector will be presented next.

According to Castells et al. (1995), in the LCI analysis, the assignment of the environmental loads to the different flows of a process and the realization of the corresponding balance are carried out by a methodology based on an eco-vector. The eco-vector (v) is a multidimensional mathematical operator in which each dimension or element corresponds to a specific EL. In an LCA study, each elementary flow is associated with an eco-vector with information about natural resource depletion and/or emissions generated along the product system under study.

In this framework, each ingoing mass flow of the system (kg/s) has an associated eco-vector (v) whose elements are expressed in specific units. The most common units are kilograms of pollutant per kilogram of product, in cases of mass units, and kilojoule per kilogram of product regarding energy units. ELs that cannot be expressed in terms of mass or energy (e.g., radiation or acoustic intensity) are transduced, in terms of eco-vector, as ELs per product mass unit (EL/kg product).

An important aspect of the use of eco-vectors to calculate environmental load is that each eco-vector must be expressed in units that can be accumulated in order to carry out material and energetic balances.

$$v_m = \begin{bmatrix} (kg/kg) \, or \, (EL/kg) \\ \text{Renewable Raw Material} \\ \text{Non-renewable Raw Material} \\ \text{Emissions to Air} \\ \text{Emissions to Water} \\ \text{Emissions to Soil} \\ \text{Energy Losses} \\ \text{Radiation} \\ \text{Noise} \\ \text{Other Environmental Loads} \end{bmatrix} \qquad (2.1)$$

Equation 2.1 shows a mass eco-vector (v_m) in which different kinds of environmental loads are grouped. Examples of database lists with environmental loads that constitute elements of the eco-vectors of Equation 2.1 can be found in, for example, Boustead, (2005a, 2005b).

The product of any process mass flow M [kg/s] and its corresponding eco-vector ($v_{m,M}$) gives the rate of pollutants (P)—expressed in kg/s or EL/s—generated by this mass flow until the life-cycle phase of the system. It is shown in Equation 2.2:

$$P = M * v_{m,M} \qquad (2.2)$$

In parallel, an eco-vector (v_e) is defined for the energy flows. The elements of (v_e), in turn, are expressed in specific energy bases (e.g., kilograms of pollutant per kilojoule). The rows of (v_e) have analogous elements compared to those of the mass eco-vector, as presented by Equation 2.3:

$$v_e = \begin{bmatrix} (kg/kJ) \, or \, (EL/kJ) \\ \text{Renewable Raw Material} \\ \text{Non-renewable Raw Material} \\ \text{Emissions to Air} \\ \text{Emissions to Water} \\ \text{Emissions to Soil} \\ \text{Energy Losses} \\ \text{Radiation} \\ \text{Noise} \\ \text{Other Environmental Loads} \end{bmatrix} \qquad (2.3)$$

The product of an energy flow E [kW] and its corresponding vector ($v_{e,E}$) gives the pollutant flow (P_e) regarding an energetic flow of the system in study. The equation to calculate P_e is shown in the following equation:

$$P_e = E * v_{e,E} \qquad (2.4)$$

The use of Equations 2.2 and 2.4 makes it possible for the environmental loads of the mass and energy flow, both measured in the same units, to be handled together, once the pollutant flows obtained by these treatments are expressed, respectively, in terms of natural resource consumption rate and emission rate.

In this framework, each of the system's inputs has an associated eco-vector and its content must be distributed to the output of the system. The balance of each of the elements of the eco-vector must be closed. This means that total amount of output from a process is equal to the pollutant quantity that entered with the inlets plus the amount of pollutant generated during its operation.

To enable this balance, the output is divided into products and wastes. In order to differentiate both classes, a convention establishes that the waste flows have eco-vectors with negative elements corresponding to the pollutants they contain. The environmental load of the input and waste flows must be distributed among the products of the process.

In this way, the LCI or the balance of environmental loads of the product system under study is carried out similar to a material balance. Thus, in the case of the whole and complex plant, this can be divided into its units or subsystems, and the system of equations obtained for each of them is solved in order to calculate the eco-vectors for every intermediate or final product. The solution of the equation's systems allows detailed knowledge of the origin of the pollution, which can be assigned to each product of a plant.

The balances are carried out in a similar way for discontinuous processes, only changing the basis of the computation. For example, instead of considering a pollutant rate, the calculations are carried out in mass of pollutant per mass of obtained product. An illustration of a generic discontinuous system taken with n inputs of raw materials and energy and n outputs of products and waste releases is presented in Figure 2.9.

The algorithm resulting from the global balance of EL is given as follows:

$$\sum_{i=1}^{n}(P_i \cdot v_{m,pi}) = \sum_{i=1}^{n}(IP_i \cdot v_{m,IP_i}) + \sum_{i=1}^{n}(IE_i \cdot v_{e,IE_i}) - \sum_{i=1}^{n}(W_i \cdot v_{m,W_i}) \qquad (2.5)$$

where:
IP_i is the mass inputs
IE_i is the energy inputs
P_i is the outputs (products and by-products)
W_i is the wastes
$v_{m,e}$ is the mass and energy eco-vectors of the flows

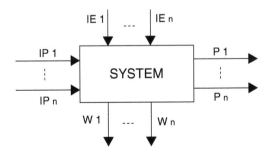

FIGURE 2.9   Representation of a generic product system.

The only unknowns in Equation 2.5 are the eco-vectors associated with the products. If only one product is assumed, the correspondent eco-vector will be calculated by:

$$v_{m,p} = \frac{\sum IP_i \cdot v_{m,IP_i} + \sum IE_i \cdot v_{i,IE_i} - \sum W_i \cdot v_{m,W_i}}{P} \tag{2.6}$$

In the next section, we present a simple example to learn how to calculate an LCI of a product. Nevertheless, many LCA tools—e.g., Simapro, OpenLCA, Umberto, and GaBi—in the market can help the user carry out an LCI of a big variety of systems and products.

2.5   EXAMPLE: SCOPE DEFINITION AND INVENTORY CALCULATION

Following the example of life-cycle assessment of the chair presented in Section 2.1.1, we will illustrate the LCA phases according to ISO 14040 and 14044. Considering the steps of Section 2.3, we conduct a product-oriented LCA in order to be able to inform potential users of the chair about its environmental performance. Clients buying the chair should be able to use the LCA results to make a choice between this chair and an alternative. The results will also be used to inform the chair manufacturer about potential future improvement options. We are interested in the environmental impacts for which the (user of the) chair could be held accountable. Therefore, an attributional LCA is conducted. As a functional unit, we will consider "comfortable sitting over a period of 5 years." The reference flow for this functional unit is one fictional chair: a Tarraco 53 model, with a total weight of 5.3×10^3 g composed of the following materials:

- *Wood*: 4852 g (frame and seat)
- *Iron*: 10 g (screws)
- *Foam*: 124 g (seat)
- *Fabric*: 117 g (seat)

2.5.1   SCOPE OF THE STUDY

In this case, we will consider a simplified cradle-to-gate life cycle taking the raw materials from its source as wood from pine trees, iron from ore, foam from polyurethane, and fabric from polyamide. The processes that are considered within the system boundaries are the processes and flows that are presented in the simplified process flow diagram of Figure 2.7.

2.5.2   INVENTORY CALCULATION

In this simplified LCA, only data relative to the material and energy inputs, and wastes, according to Figure 2.9, will be considered.

In the LCI, data were collected based on existing datasets obtained from Tool for Environmental Analysis and Management (TEAM®). All the input and output flows were referred to the chosen functional unit, that is, one chair. No allocation is necessary, as we have considered that all the manufacturing facilities only produce this chair.

Let us consider an example of the calculation of environmental load assigned to this chair. The environmental load balance will be based on the scheme presented in Figure 2.10. Material and energy inputs and outputs associated with the manufacturing of the chair (1 functional unit, FU) are

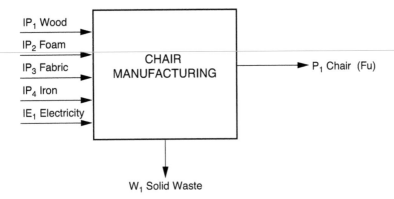

FIGURE 2.10   Environmental load balance in the manufacturing of a chair.

TABLE 2.3
Input and Output Energy and Material Flow
Assigned to a Chair

Flow	Variable	Units	Value
Wood	IP1	kg	4.852
Foam (polyurethane)	IP2	kg	0.124
Fabric (polyamide)	IP3	kg	0.117
Iron	IP4	kg	0.010
Electricity	IE1	kWh	4.0
Solid waste	W1	kg	0.485
Chair	P1	FU	1

---

characterized in Table 2.3. To simplify the application example, a condensed eco-vector containing the following environmental loads is selected:

- Raw materials:
  - Coal (in ground)
  - Natural gas (in ground)
  - Oil (in ground)
- Emissions to air:
  - Carbon dioxide (CO_2, fossil)
  - Nitrogen oxides (NO_x as NO_2)
  - Sulfur oxides (SO_x as SO_2)
- Emissions to water:
  - BOD5 (biochemical oxygen demand)
  - COD (chemical oxygen demand)
  - Nitrates (NO_3^-)
- Emissions to soil (not further specified)

Based on data from software TEAM® and Frischknecht et al. (1996), we obtain the eco-vectors shown in Table 2.4.

TABLE 2.4
Eco-Vectors of Input and Output Streams of Chair Manufacturing System

Stream		IP1	IP2	IP3	IP4	IE1	W1
Vector		$v_{m, IP1}$	$v_{m, IP2}$	$v_{m, IP3}$	$v_{m, IP4}$	$v_{e, IE1}$	$v_{m, W1}$
EL		Wood	Foam	Polyamide	Iron	Electricity	Solid waste
Load	Units	EL/kg	EL/kg	EL/kg	EL/kg	EL/kW.h	EL/kg
Coal	kg	7.93×10^{-2}	3.89×10^{-1}	7.46×10^{-1}	6.87×10^{-1}	1.39×10^{-1}	0
Natural gas	kg	3.27×10^{-2}	8.32×10^{-1}	1.41	9.10×10^{-2}	4.07×10^{-2}	0
Crude oil	kg	5.10×10^{-2}	7.38×10^{-1}	8.12×10^{-1}	5.50×10^{-2}	2.88×10^{-2}	0
CO_2	kg	1.05	4.06	7.00	2.05	5.27×10^{-1}	0
NO_x	kg	1.80×10^{-3}	1.78×10^{-2}	2.60×10^{-2}	3.60×10^{-3}	9.50×10^{-4}	0
SO_x	kg	2.01×10^{-3}	1.91×10^{-2}	2.90×10^{-2}	4.50×10^{-3}	2.57×10^{-3}	0
BOD5	kg	9.26×10^{-7}	4.64×10^{-4}	3.60×10^{-3}	1.70×10^{-4}	1.34×10^{-7}	0
COD	kg	1.45×10^{-5}	2.78×10^{-3}	1.50×10^{-2}	4.63×10^{-4}	1.99×10^{-6}	0
NO_3^-	kg	5.00×10^{-6}	6.30×10^{-3}	3.00×10^{-2}	6.90×10^{-6}	7.70×10^{-7}	0
Emissions to soil	kg	1.50×10^{-1}	5.95×10^{-1}	3.05×10^{-1}	7.60×10^{-1}	8.04×10^{-2}	-1

TABLE 2.5
Calculation of EL Associated with the Chair as a Function of EL of Input and Output Streams

EL	Value Units	IP1.$v_{m,IP1}$ (kg)	IP2.$v_{m,IP2}$ (kg)	IP3.$v_{m,IP3}$ (kg)	IP4.$v_{m,IP4}$ (kg)	IE1.v_{elE1} (kg)	W1.v_{mw1} (kg)	v_{mP} (kg)
Coal	kg	3.85×10^{-1}	4.82×10^{-2}	8.72×10^{-2}	6.87×10^{-3}	5.58×10^{-1}	0	1.08
Natural gas	kg	1.59×10^{-1}	1.03×10^{-1}	1.65×10^{-1}	9.10×10^{-4}	1.63×10^{-1}	0	5.90×10^{-1}
Crude oil	kg	2.47×10^{-1}	9.15×10^{-2}	9.50×10^{-2}	5.50×10^{-4}	1.15×10^{-1}	0	5.50×10^{-1}
CO_2	kg	5.09	5.04×10^{-1}	8.19×10^{-1}	2.05×10^{-2}	2.11	0	8.55
NO_x	kg	8.71×10^{-3}	2.20×10^{-3}	3.04×10^{-3}	3.60×10^{-5}	3.80×10^{-3}	0	1.78×10^{-2}
SO_x	kg	9.75×10^{-3}	2.37×10^{-3}	3.39×10^{-3}	4.50×10^{-5}	1.03×10^{-2}	0	2.58×10^{-2}
BOD5	kg	4.49×10^{-6}	5.75×10^{-5}	4.21×10^{-4}	1.70×10^{-6}	5.35×10^{-7}	0	4.85×10^{-4}
COD	kg	7.04×10^{-5}	3.44×10^{-4}	1.76×10^{-3}	4.63×10^{-6}	7.96×10^{-6}	0	2.18×10^{-3}
NO_3^-	kg	2.43×10^{-5}	7.81×10^{-4}	3.51×10^{-3}	6.90×10^{-8}	3.08×10^{-6}	0	4.32×10^{-3}
Emissions to soil	kg	7.28×10^{-1}	7.38×10^{-2}	3.57×10^{-2}	7.60×10^{-3}	3.22×10^{-1}	-0.49	1.65

Each element of the eco-vector is presented in units of environmental load, EL, per unit of mass or energy depending on whether we are dealing with a mass or energy stream. W1 is a waste stream, representing the wood wastes generated by the chair manufacture. Provided this stream is an elementary flow, its values are –1 or 0, depending on whether the corresponding environmental load of the eco-vector is present. To close the environmental balance assigning the values of the waste eco-vectors, $v_{m,W1}$, to the product P of the system, the non-zero elements of $v_{m,W1}$ must be negatives (-1).

By the application of Equation 2.6, the value of the eco-vector, $v_{m,P}$, corresponding to the environmental load associated with the chair can be obtained. Table 2.5 shows the calculation of the environmental loads as a function of inputs and outputs.

These results show the procedure to calculate the environmental load assigned to a product as a function of the environmental data of the different process inputs and the loads generated by the process. From Table 2.5, it is possible to determine the relative contribution of each input to the total value of the corresponding environmental loads, as presented in Table 2.6.

TABLE 2.6
Relative Percentage Contribution of Input and Output Streams in Total EL Assigned to the Chair

EL	Stream	Wood	Foam	Fabric	Iron	Electricity	Manufacture	Chair
Coal	kg	35.5	4.4	8.0	0.6	51.5	0.0	100.0
Natural gas	kg	26.9	17.5	27.8	0.2	27.6	0.0	100.0
Crude oil	kg	45.0	16.6	17.3	0.1	21.0	0.0	100.0
CO_2	kg	59.6	5.9	9.6	0.2	24.7	0.0	100.0
NO_x	kg	48.9	12.4	17.1	0.2	21.4	0.0	100.0
SO_x	kg	37.8	9.2	13.1	0.2	39.7	0.0	100.0
BOD5	kg	0.9	11.9	86.7	0.4	0.1	0.0	100.0
COD	kg	3.2	15.8	80.4	0.2	0.4	0.0	100.0
NO_3^-	kg	0.6	18.1	81.2	0.0	0.1	0.0	100.0
Emissions to soil	kg	44.0	4.5	2.2	0.5	19.5	29.3	100.0

2.6 CASE STUDY: LIFE-CYCLE ASSESSMENT STUDY OF A MUNICIPAL SOLID WASTE INCINERATION IN TARRAGONA, SPAIN

For the MSWI plant introduced in Chapter 1, an LCA study (STQ 1998) was developed as an Excel spreadsheet model on the basis of the Code of Practice (SETAC 1993) and according to the steps of an early version of ISO 14040 (1997). The inventory was based on providers' information, literature data on raw materials, and a detailed analysis of the incineration process.

2.6.1 GOAL AND SCOPE DEFINITION OF THE MUNICIPAL SOLID WASTE INCINERATION LIFE-CYCLE ASSESSMENT STUDY

The objective of the study was to identify, evaluate, and compare the environmental loads derived from the electricity production by the municipal waste incinerator of Tarragona in Scenarios 1 (the old situation) and 2 (the current situation, as described in Chapter 1) in order to analyze the environmental efficiency of the investment in an advanced acid gas removal system. Next, the project is described according to the points indicated in ISO 14040 (1997).

The main function of the incineration process is to reduce the volume and toxicity of the municipal waste treated. The production of electric energy must be seen as an added value to the incineration process. For simplicity, and because the objective of the study is the analysis of the electric energy generated in the incineration process, the functional unit selected is "1 TJ of produced electricity."

The study comprises all the processes from the municipal waste disposal in containers to the landfill of the final waste, as shown in Figure 2.11. Consequently, the following processes are considered: transport of the municipal waste to the incinerator, combustion, gas treatment and ashes removal, as well as slag disposal (including transport to the final localization). The final step, with its emissions associated to the landfill, is not analyzed. The incineration process has been divided into subsystems:

1. Waste incineration plant with combustion process and including gas treatment
2. Water treatment (treatment process applied in the ash bath, demineralization process applied in the kettles by osmosis and refrigeration process by means of a tower)
3. Ash treatment (ionic ashes and waste from the gas treatment filters)
4. Scrap treatment (iron waste recycling process)

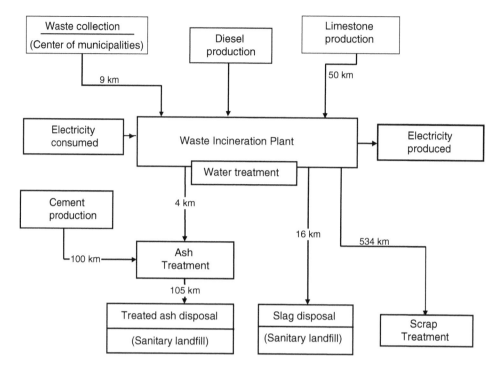

FIGURE 2.11 Processes considered in the LCA study within the boundaries of the system (including transport distances).

The latter were analyzed as processes that contribute to the functional unit of the incineration process. Literature values have been used for raw material production and the transport process (16-t truck).

Due to the study's objective, the electricity produced is considered to be the only useful measurable product to which all environmental loads are assigned. In other words, the function of "treatment of waste" that is also provided by the product system is not considered here. The fact that this additional service is not considered results in an overestimation of the environmental loads that are attributed to the recovered electricity.

2.6.2 DATA USED IN LIFE-CYCLE INVENTORY

In order to carry out the study, three types of data were used:

- Literature data of environmental loads for the raw material used in the analyzed processes. The considered source was the report referred to as ETH 1996 (Frischknecht et al., 1996). The exactness of the report and the agreement of these data with the particular situation in Spain determine the data quality.
- Real data of consumption and emissions associated with the incineration process, average values from 1996 (Scenario 1) and average values for 2 months with the advanced acid gas removal system in operation (Scenario 2). The data quality can be considered reliable because they have been obtained directly from the process.
- Real data of consumption and emissions associated with the waste treatment processes, obtained by visits and questionnaires answered by the treatment companies. The reliability of the data delivered by the companies depends on the available information and the degree of collaboration.

2.6.3   Assumptions and Limitations

The principal assumptions are:

- The incineration plant operates 345 days a year and 24 h each day.
- The journey to the destination and the return of the truck are considered for the analysis of the environmental loads associated with the transport, assuming the same EL for the empty truck as for the full one.
- The internal consumption of electricity is covered by properly produced electric energy. The importation of electricity is necessary only in cases of an operation stop. By this, the environmental loads associated with the electric energy consumed during the process is the result of the total inventory analysis (need of iterative computation).
- In the case of scrap-metal treatment, the generated useful product has been classified as an avoided production of iron within the raw materials.[1]

The main limitations of the study are:

- There is neither an analysis nor a total characterization of the municipal waste entering the system.
- The environmental loads associated with the emissions of the final waste disposal in a landfill has not been considered.
- It was not possible to simulate the following products used in the incineration process because of an information lack in the databases consulted: ferric chloride, active carbon and additives used in the osmosis process.

2.6.4   Results of the Inventory Analysis

The results for the life-cycle inventory analysis are shown in Table 2.7, including the emissions to air and water.

2.6.4.1   Raw Material Consumption

The current situation (Situation 2) is unfavorable for all of the 14 analyzed parameters that are considered, according to Frischknecht et al. (1996) due to the higher consumption of raw materials (especially cement, for the higher waste quantity per produced TJ, and CaO, for the advanced gas treatment) and more transport activity because of the higher raw material consumption and waste quantity.

2.6.4.2   Energy Consumption

The current situation is unfavorable due to the higher energetic consumption per produced TJ because of the additional energy consumption of the advanced gas treatment system.

2.6.4.3   Air Emissions

From the 37 analyzed parameters, the current situation is unfavorable for all except nine: arsenic, cadmium, dioxins and furans, dust, sulfur dioxide, hydrochloric acid, hydrofluoric acid, nickel, and other heavy metals. These are basically the parameters reduced by the operation of the advanced gas treatment system.

---

[1] Although the modeling of avoided processes is debatable in an attributional LCA, this is often done in attributional LCA studies and was common practice before the strict separation between "attributional" and "consequential" LCAs was recognized within the LCA domain.

TABLE 2.7

Results of LCI Analysis from MSWI in Tarragona, Spain

Emissions to Air	Unit	Electricity Sit. 1[a] (TJ)	Electricity Sit. 2[b] (TJ)	Diff.[c]
Aldehydes	kg	4.54×10^{-5}	5.92×10^{-5}	−30%
Ammonia	kg	2.82×10^{-2}	3.70×10^{-2}	−31%
As	kg	1.17×10^{-1}	3.96×10^{-2}	66%
Benzene	kg	1.32×10^{-1}	1.71×10^{-1}	−30%
Benzo(a)pyrene	kg	4.94×10^{-5}	5.97×10^{-5}	−21%
Cd	kg	1.13×10^{-1}	3.98×10^{-2}	65%
CO	kg	2.75×10^{-5}	3.01×10^{2}	−9%
CO_2	kg	2.33×10^{5}	2.57×10^{5}	−10%
C_xH_y aromatic	kg	9.22×10^{-4}	1.28×10^{-3}	−39%
Dichloromethane	kg	5.00×10^{-5}	5.40×10^{-5}	−8%
Dust	kg	1.71×10^{2}	1.61×10^{2}	−8%
Ethanol	kg	2.60×10^{-3}	3.56×10^{-3}	6%
Ethene	kg	2.10	2.55	−21%
Ethylbenzene	kg	1.47×10^{-2}	1.78×10^{-2}	−21%
Formaldehyde	kg	9.10×10^{-3}	1.61×10^{-2}	−77%
H_2S	kg	1.85×10^{-2}	2.91×10^{-2}	−57%
Halon-1301	kg	2.25×10^{-3}	2.66×10^{-3}	−18%
HCl	kg	2.89×10^{3}	1.95×10^{2}	93%
Heavy metals	kg	2.65	7.17×10^{-1}	73%
HF	kg	9.93	2.80	72%
Methane	kg	3.32×10^{1}	4.92×10^{1}	−48%
N_2O	kg	1.71	2.06	−20%
Ni	kg	1.80×10^{-1}	6.51×10^{-2}	64%

Emissions to Water	Unit	Electricity Sit. 1[a] (TJ)	Electricity Sit. 2[b] (TJ)	Diff.[c]
AOX	kg	1.15×10^{-3}	1.36×10^{-3}	−18%
Aromatics	kg	1.73×10^{-1}	2.06×10^{-1}	−19%
As	kg	5.13×10^{-3}	1.04×10^{-2}	−103%
B	kg	1.17×10^{-2}	1.51×10^{-2}	−29%
Ba	kg	9.27×10^{-1}	1.27	−37%
BOD	kg	5.07	8.93×10^{-2}	98%
Cd	kg	9.30×10^{-3}	8.17×10^{-2}	12%
COD	kg	1.51×10^{1}	1.37	91%
Cr	kg	3.03×10^{-2}	5.48×10^{-2}	−81%
Cu	kg	1.39×10^{-2}	2.73×10^{-2}	−96%
Dissolved subst.[d]	kg	1.01	2.14	−112%
Hg	kg	7.14×10^{-5}	2.73×10^{-5}	62%
Mn	kg	7.00×10^{-2}	1.28×10^{-1}	−83%
Mo	kg	8.40×10^{-3}	1.64×10^{-2}	−95%
NH_3	kg	5.23×10^{-1}	6.13×10^{-1}	−17%
Ni	kg	1.34×10^{-2}	2.69×10^{-2}	−101%
Nitrates	kg	6.75×10^{-1}	8.06×10^{-1}	−19%
Pb	kg	2.56×10^{-2}	3.55×10^{-2}	−39%
Phosphate	kg	1.51×10^{-1}	3.10×10^{-1}	−105%
Sb	kg	4.43×10^{-5}	7.10×10^{-5}	−60%
SO_4^-	kg	2.31×10^{1}	3.74×10^{1}	−62%
Undissolved subst.[d]	kg	2.68×10^{1}	2.94×10^{1}	−10%
TOC	kg	3.22×10^{1}	3.82×10^{1}	−19%

(Continued)

TABLE 2.7 (Continued)
Results of LCI Analysis from MSWI in Tarragona, Spain

Emissions to Air	Unit	Electricity Sit. 1a (TJ)	Electricity Sit. 2b (TJ)	Diff.c
Non methane VOC	kg	6.98×10^1	8.31×10^1	-19%
NO$_x$	kg	1.23×10^3	1.33×10^3	-8%
PAH	kg	7.37×10^{-4}	1.02×10^{-3}	-38%
Phenol	kg	7.39×10^{-6}	9.93×10^{-6}	-34%
Phosphate	kg	1.94×10^{-3}	3.88×10^{-3}	-100%
Pentane	kg	5.51×10^{-1}	6.54×10^{-1}	-19%
Propane	kg	4.35×10^{-1}	5.57×10^{-1}	-28%
Propene	kg	3.07×10^{-2}	4.49×10^{-2}	-46%
SO$_2$	kg	5.08×10^2	2.56×10^2	50%
TCDD-Equiv.	ng	1.12×10^7	1.43×10^4	100%
Tetrachloromethane	kg	2.55×10^{-5}	3.63×10^{-5}	-42%
Toluene	kg	6.68×10^{-2}	8.39×10^{-2}	-26%
Trichloromethane	kg	1.93×10^{-6}	3.07×10^{-6}	-59%
Vinylchloride	kg	1.19×10^{-5}	1.89×10^{-5}	-59%

Emissions to Water	Unit	Electricity Sit. 1a	Electricity Sit. 2b	Diff.c
Bauxite (ore)	kg	3.59×10^1	4.27×10^1	-19%
Clay	kg	2.12×10^2	1.01×10^3	-376%
Coal	kg	1.50×10^3	3.15×10^3	-110%
Copper (ore)	kg	2.71	3.36	-24%
Iron (ore)	kg	-1.32×10^4	-1.38×10^4	-5%
Lignite	kg	6.56×10^2	8.86×10^2	-35%
Limestone (ore)	kg	1.03×10^3	1.62×10^4	-1473%
Natural gas	Nm^3	1.06×10^2	5.48×10^2	-417%
Nickel (ore)	kg	1.00	1.19	-19%
Oil	kg	5.79×10^3	6.86×10^3	-18%
Silica (ore)	kg	8.74×10^3	1.04×10^4	19%
Uranium (ore)	kg	5.69×10^{-2}	7.43×10^{-4}	-31%
Water	kg	3.57×10^5	6.41×10^5	-80%
Wood	kg	3.35×10^1	4.99×10^1	-49%

Note: All environmental loads referred to the functional unit, 1 TJ of electricity produced.

[a] Sit. 1 = Scenario 1.
[b] Sit. 2 = Scenario 2.
[c] Diff. = (Scenario 1 − Scenario 2)/Scenario 1.
[d] Subst. = substance.

2.6.4.4   Water Emissions

From the 23 analyzed parameters, the current situation is unfavorable for all except 4: BOD, COD, Cd, and Hg. These are basically the parameters reduced by the current operation that works without water emissions into the sewage.

2.6.5   COMPARISON WITH SPANISH ELECTRICITY MIX

The electricity produced by MSWI, Servei d'Incineració dels Residus Urbans S.A. (SIRUSA) is fed into the Spanish electricity net. One ton of waste produces about 1 MJ electric energy in the current situation with advanced flue gas treatment. This can be seen as an energy benefit because SIRUSA replaces the electricity production of a conventional electrical power plant. On one hand, this saves resources; on the other hand, some of the emissions are reduced in comparison to the Spanish electricity mix.

Therefore, the results of the MSWI LCA were compared with data on the Spanish electricity mix. This comparison depends strongly on CO_2 emissions. If the neutrality of CO_2 from renewable resources is considered, the life-cycle inventory results of the Spanish electricity mix will not change remarkably, but in a life-cycle study of municipal waste incineration, the high content of renewable materials that are burned will provoke different results. Thus, adapting the LCA methodology (Sonnemann et al., 1999), the total amount of CO_2 in the incineration process needed to be distributed in two parts between the carbon-containing wastes: one for waste originating in renewable resources and the other for waste with its origin in fossil fuels. Therefore, it is considered that the waste contains 13% plastics, as the only component with its origin in fossil fuels, and that the carbon content of plastics is 56.43%, according to U.S. EPA (1996).

The quantitative comparison was made with the software TEAM. The database integrated in TEAM was used for the emissions of Spanish electricity production. The calculation was normalized to the functional unit. Table 2.8 compares the results of Scenario 2 with the data of the Spanish Electricity mix given by Frischknecht et al. (1996) for selected priority air pollutants. Positive values represent higher emissions of the average electricity production in Spain than in the SIRUSA plant. This means that emissions of CO, heavy metals, Ni, particles, and SO_2 are much higher in the conventional electrical power plants. The negative values present pollutants that are lower for the average Spanish electricity mix than in the MSWI incineration process chain of Tarragona.

TABLE 2.8
Comparison of the Differences between the LCI Data of 1 TJ Electricity Produced by the MSWI and the Spanish Mix, Indicating the Relevance of the Consideration of Electricity Benefit in Such an LCA Study

Pollutant	Unit	MSWI Situation 2 "With Filters" TJ	Electricity Spain TJ	Difference TJ	%
(a) Arsenic (As)	kg	3.96×10^{-2}	2.99×10^{-2}	9.75×10^{-3}	−24.6
(a) Cadmium (Cd)	kg	3.98×10^{-2}	1.30×10^{-2}	2.68×10^{-2}	−67.4
(a) Carbon dioxide (CO_2, fossil)	kg	2.57×10^{5}	1.63×10^{5}	9.43×10^{4}	−36.7
(a) Carbon monoxide (CO)	kg	3.01×10^{2}	1.33×10^{3}	-1.03×10^{3}	341.2
(a) Heavy metals (sum)	kg	7.17×10^{-1}	3.68	−2.97	413.7
(a) Hydrogen chloride (HCl)	kg	1.95×10^{2}	5.06×10^{1}	1.44×10^{2}	−74.1
(a) Nickel (Ni)	kg	6.51×10^{-2}	2.54×10^{-1}	-1.89×10^{-1}	290.3
(a) Nitrogen oxides (NO_x as NO_2)	kg	1.33×10^{3}	3.32×10^{2}	9.98×10^{2}	−75.0
(a) Particles (unspecified)	kg	1.61×10^{2}	7.72×10^{2}	-6.11×10^{2}	379.8
(a) Sulfur oxides (SO_x as SO_2)	kg	2.56×10^{2}	9.29×10^{2}	-6.73×10^{2}	262.9

2.7 CASE STUDY: ENERGY PRODUCTION USING THIRD GENERATION ORGANIC PHOTOVOLTAIC SOLAR PANELS

As mentioned earlier in this chapter, LCI can be built from pre-existing databases (e.g., Ecoinvent), primary data (i.e., measurements), secondary data (e.g., industry reports), or modelled (i.e., estimated). Often, a mixture of these approaches will be used in any single project. When discussing the topic of new and emerging technologies, a problem of readily available LCI data can become an obvious barrier to carrying out one's LCA. Here, a case-study involving the evaluation of organic photovoltaics (OPV) is presented. OPVs are a so-called 3rd-generation photovoltaic (PV) technology that are mostly confined to the lab-scale and proof-of-concept stages of technological maturity. Additionally, OPVs employ engineered nanomaterials (ENM) (e.g., fullerenes), which are themselves another emerging technology. Thus, the following case-study demonstrates how an LCI can quickly become a large data gathering process, whereby a mixture of foreground and background data are required to build novel inventories for emerging technologies.

2.7.1 GOAL AND SCOPE DEFINITION OF THE OPV STUDY

The LCA was conducted according to International Organization for Standardization (ISO) 14040:2006 and 14044:2006 guidelines and the International Energy Agency's (IEA) recommendations for implementing LCAs for PV technology (Fthenakis et al., 2011). A cradle-to-grave assessment was completed, comparing the life-cycle impacts for polymer-based OPV technologies with conventional silicon technology.

The life-cycle stages and impacts from raw materials extraction, materials processing, and manufacturing of the PV panels were considered, while the use and end-of-life considerations were excluded from this LCI and LCA. The functional unit was defined as 1 watt-peak[2] (Wp), or in other words, the amount of OPV panel required to produce 1 Wp of power. The OPV panel was assumed to require 200 cm² of surface area per Wp power produced at a module efficiency of 5% (Roes et al., 2009).

All foreground inventory data are explained in detail in the following paragraphs, while relevant background data was taken from the attributional Ecoinvent v2.2 (The Ecoinvent Association, Zurich) life-cycle inventory database. Capital equipment (e.g., buildings for solar panel production) was excluded from the OPV inventory as such environmental burdens are often negligible when considering the entire life-cycle and lifetime of a product (US Environmental Protection Agency, 2006). Where applicable, co-products resulting from the end-of-life treatment option of the solar panels (e.g., electricity from incineration) were handled as avoided products using system expansion. Energy production from incineration was assumed to replace an average European medium-voltage electricity production mix (RER) defined by Ecoinvent v2.2. Only electrical energy, as opposed to thermal energy was considered. Using an energy conversion efficiency half that of thermal energy, all potential thermal energy was converted to electrical energy.

2.7.2 LIFE-CYCLE INVENTORY DATA AND CALCULATIONS (INCLUDING ASSUMPTIONS)

The OPV technology was based on a typical polymer-based bulk heterojunction (BHJ[3]) that employs the fullerene derivative phenyl-C_{61}-butyric acid methyl ester (PCBM) and the polymer poly(3-hexylthiphene) (P3HT) in its active layer (Thompson and Frechet, 2007). The general geometry of

---

[2] Also known as the nominal power of a solar cell, 1 Wp is the maximum amount of power the cell can produce under standardized conditions.

[3] A BHJ device is one in which the donor (e.g., PCBM) and acceptor (e.g., P3HT) materials exist in an interwoven matrix to maximize their surface area interface.

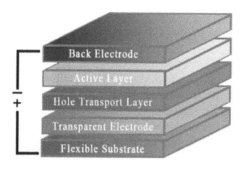

FIGURE 2.12   General layout and geometry of an organic photovoltaic solar panel.

TABLE 2.9

Generalized Account of *Direct* Material Components and Energy Requirements for Producing 1 m² of PCBM:P3HT-based Polymer OPV Solar Panel

Materials	Amount	Notes	References
Aluminum	0.995 g	Back electrode	(Roes et al. 2009; Anctil et al. 2013)
PET	74.0 g	Substrate	(Roes et al. 2009)
FTO	1.80 g	Applied as an FTO solution (see Tsang et al. 2015)	(Aukkaravittayapun et al. 2006; Roes et al. 2009; García-Valverde et al. 2010)
PET	133 g	Lamination	(Roes et al. 2009)
Lithium Fluoride	6.00E-5 g	Back electrode	(Roes et al. 2009)
Molybdenum oxide	0.240 g	Hole transport layer	(Roes et al. 2009; García-Valverde et al. 2010; Espinosa et al. 2011; Anctil et al. 2013)
Chlorobenzene	7.66 g	Solvent for active layer application	(Espinosa et al. 2011)
P3HT	0.235 g	Active layer	(Roes et al. 2009; García-Valverde et al. 2010; Espinosa et al. 2011; Anctil et al. 2013)
PCBM	0.205 g	Active layer	(Roes et al. 2009; García-Valverde et al. 2010; Espinosa et al. 2011; Anctil et al. 2013)
Energy	**Amount**	**Notes**	**References**
Electricity	5.13 MJ	Annealing	(García-Valverde et al. 2010; Espinosa et al. 2012)
Electricity	2.56 MJ	For printing of panel components	(Roes et al. 2009)
Electricity	0.0850 MJ	For lamination of the panel	(Roes et al. 2009)

Note: Indirect, upstream, or auxiliary (e.g., additional solvents) material and energy requirements, as well as emissions, are not reported here.

the solar cell is depicted in Figure 2.12 and the main material and energy requirements are listed in Table 2.9. The OPV panel is roll-to-roll (R2R) printed under normal atmosphere conditions on a flexible substrate made from polyethylene terephthalate (PET), using a transparent fluorine doped tin oxide (FTO) electrode on the light-collecting face, an active layer composed of PCBM:P3HT, a hole transport layer made of molybdenum oxide (MoO_3), a back electrode made from aluminum with a thin layer of lithium fluoride, encapsulation and lamination layers composed of PET, silica and various epoxy resins, and, lastly, silver metallization paste for the interconnections.

2.7.2.1   Substrate

The substrate—the foundation to which all other components are placed—of an organic solar device can be either a rigid glass or a flexible substrate. Production using a flexible substrate was

TABLE 2.10

Inventory for the Production of Extruded Plastic Film Adapted from the Ecoinvent Process for Extrusion of Polyethylene Terephthalate (Frischknecht et al. 2007)

Plastic Film, Pet, at Plant	Reference Flow Amount: 1 kg			
Inventory Item	Amount	Units	Other	Inventory Flow Source
Extrusion, plastic film, RER	1.0	kg		Ecoinvent
Polyethylene terephthalate, granulate, amorphous, at plant, RER	1.03	kg		Ecoinvent
Transport, freight, rail, RER	0.6	t*km	Based on a distance traveled of 600 km by train.	Ecoinvent
Transport, lorry 16-32t, EURO3, RER	0.1	t*km	Based on a distance traveled of 100 km by truck.	Ecoinvent

Source: Frischknecht, R. et al., *Overview and Methodology; Final Report Ecoinvent Data 2.0*, Swiss Centre for Life Cycle Inventories, Dubendorf, Switzerland, 2007.

chosen due to its roll-to-roll compatibility, using polyethylene terephthalate (PET) which is the most successful substrate used to produce flexible organic solar cells (Krebs, 2009a). For 1 m² of such a substrate, 0.074 kg of PET was required (Roes et al., 2009). PET substrate production was modeled using the Ecoinvent process for "extrusion, plastic film" using the input "polyethylene terephthalate, granulate, amorphous, at plant" as the substrate source material (Table 2.10).

2.7.2.2 Transparent (Front) Electrode

The transparent electrode—needed on the light collecting face of the PV module in order to allow electromagnetic radiation to reach the active layer of the module—commonly used in OPV devices is indium tin oxide (ITO) given its high level of optical transparency and electrical conductivity (Krebs, 2009b). However, large-scale use of indium is less attractive due to cost, scarcity and to a lesser extent toxicity (Lizin et al., 2013; Kumar & Zhou, 2010; Samad et al., 2011). Fluorine doped tin oxide (FTO) was chosen as a suitable replacement due to favorable qualities such as high availability, cheapness, and stability (Krebs, 2009b; Samad et al., 2011; Kim et al., 2008). An FTO solution was assumed to be produced from tin tetrachloride pentahydrate, ammonium fluoride, ethanol, and water and heated to 60°C (see Table 2.11) (Samad et al., 2011; Aukkaravittayapun et al., 2006). Only the energy needed to raise the temperature of water and ethanol were considered. The materials and energy to produce tin tetrachloride pentahydrate was based on stoichiometric reactions (Graf, 2007). Wastes were not estimated from these processes. The amount of FTO solution applied was 0.042 liters per m² and was calculated as an average of the values reported for ITO in Garcia-Valverde et al. (2010) and Roes et al. (2009). The FTO wasM applied via sputtering in a roll-to-roll process (Garcia-Valverde et al., 2010; Roes et al., 2009).

2.7.2.3 Active Layer

In a traditional silicon solar cell, the active layer—where the production of charges occurs—consists of p- and n-doped silicon layers (Zeman et al., 2011; Yang & Uddin, 2014). Organic photovoltaic devices do not use silicon in their active layer, but rather a combination of organic donor and acceptor materials. The bulk heterojunction (BHJ) is an interlacing morphology of donor:acceptor which maximizes the opportunity for charge separation. The archetype donor:acceptor BHJ combination has been P3HT as the donor and PCBM as the acceptor, (Krebs, 2009a; Kroon et al., 2008; Li et al., 2012).

TABLE 2.11

Inventory for the Production of Fluorine-Doped Tin Oxide Substrate

FTO Substrate, Sputtered, at Plant	Reference Flow Amount: 1 m²			
Inventory Item	**Amount**	**Units**	**Other**	**Inventory Flow Source**
Electricity, medium voltage, production RER, at grid	24	MJ		Ecoinvent
FTO solution, at plant	0.034	L	The amount of FTO solution used is calculated using the Sn and F amounts that correspond to the weights of indium and tin in ITO conducting films.	This study
Plastic film, pet, at plant	0.074	kg		This study
Transport, freight, rail, RER	0.018	t*km	Based on a distance traveled of 600 km by train.	Ecoinvent
Transport, lorry 16-32t, EURO3, RER	0.0037	t*km	Based on a distance traveled of 100 km by truck.	Ecoinvent

FTO Solution, at Plant	Reference Flow Amount: 1 kg		**Adapted from Aukkaravittayapun et al. (2006) for the Production of an FTO Solution Using the Precursors $SnCl_4$ and NH_4F Mixed with 80:20 $EtOH:H_2O$**	
Ammonium fluoride, at plant	0.014	kg		This study
Ethanol from ethylene, at plant, RER	0.63	kg		Ecoinvent
Heat, heavy fuel oil, at industrial furnace 1 MW, RER	298.19	kg		Ecoinvent
Tin tetrachloride pentahydrate, at plant	0.078	kg		This study
Transport, freight, rail, RER	0.43	t*km	Based on a distance traveled of 600 km by train.	Ecoinvent
Transport, lorry 16-32t, EURO3, RER	0.072	t*km	Based on a distance traveled of 100 km by truck.	Ecoinvent
Water, deionized, at plant, CH	0.20	kg		Ecoinvent

Ammonium Fluoride, at Plant	Reference Flow Amount: 1 kg		**Estimated from Ullmans Encyclopedia of Industrial Chemistry 2007 Using the Feedstocks Ammonia and Anhydrous Fluoride (Zapp et al. 2007)**	
Ammonia, liquid, at regional storehouse, RER	0.46	kg		Ecoinvent
Heat, heavy fuel oil, at industrial furnace 1 MW, RER	0.57	MJ		Ecoinvent
Hydrogen fluoride, at plant, GLO	0.54	kg		Ecoinvent
Transportation, freight, rail, RER	0.60	t*km	Based on a distance traveled of 600 km by train.	Ecoinvent
Transport, lorry 16-32t, EURO3, RER	0.10	t*km	Based on a distance traveled of 100 km by truck.	Ecoinvent

Tin Tetrachloride Pentahydrate, at Plant	Reference Flow Amount: 1 kg		**Estimated Using Stoichiometric Calculations Following $SnCl_4 + 5H_2O = SnCl_4.5H_2O$**	
Tin tetrachloride, at plant	0.74	kg		This study
Water, ultrapure, at plant, GLO	0.26	kg		Ecoinvent

(Continued)

TABLE 2.11 (*Continued*)
Inventory for the Production of Fluorine-Doped Tin Oxide Substrate

Tin Tetrachloride, at Plant	Reference Flow Amount: 1 kg		Adapted from Ullman's Encyclopedia of Industrial Chemistry Following the Reaction of Tin and Chlorine (Graf 2007)	
Inventory Item	Amount	Units	Other	Inventory Flow Source
Chlorine, liquid, production mix, at plant, RER	0.54	kg		Ecoinvent
Heat, heavy fuel oil, at industrial furnace 1 MW, RER	0.13	MJ		Ecoinvent
Tin, at regional storage, RER	0.46	kg		Ecoinvent
Transport, freight, rail, RER	0.6	t*km	Based on a distance traveled of 600 km by train.	Ecoinvent
Transport, lorry 16-32t, EURO3, RER	0.1	t*km	Based on a distance traveled of 100 km by truck.	Ecoinvent
Plastic Film, Pet, At Plant	Reference Flow Amount: 1 kg		The Production of Extruded Plastic Film Adapted from the Ecoinvent Process for Extrusion of Polyethylene Terephthalate (Frischknecht et al. 2007)	
Extrusion, plastic film, RER	1.0	kg		Ecoinvent
Polyethylene terephthalate, granulate, amorphous, at plant, RER	1.03	kg		Ecoinvent
Transport, freight, rail, RER	0.6	t*km	Based on a distance traveled of 600 km by train.	Ecoinvent
Transport, lorry 16-32t, EURO3, RER	0.1	t*km	Based on a distance traveled of 100 km by truck.	Ecoinvent

PCBM can be produced in different ways including plasma and pyrolysis techniques. Although plasma techniques can produce a more discrete distribution of fullerenes, this technique also produces smaller quantities of fullerenes, while pyrolysis can produce large quantities of various-sized fullerenes (Anctil & Fthenakis, 2012). Our inventory includes the production of PCBM via the pyrolysis technique using toluene[4] as a feedstock reported by Anctil et al. (2012) (see Table 2.12). P3HT is a region-regular polymer derived from 3-bromothiophene and bromohexane (Garcia-Valverde et al., 2010). Data for bromothiophene production was taken from the inventory presented in Garcia-Valverde et al. (2010), while thiophene production was calculated based on stoichiometric calculations (see Table 2.13) (Swantson, 2007). The energy input for annealing the active layer was estimated from Espinosa et al. as 5.07 MJ per m[2] and includes the energy for drying the electron transport and hole transport layers (Espinosa et al., 2012). Amounts of 0.21 g PCBM and 0.235 g P3HT per m[2] solar cell were estimated as an average of previously reported values, (Espinosa et al., 2011; Garcia-Valverde et al., 2010; Anctil et al., 2010; Roes et al., 2009). Deposition of the P3HT:PCBM layer was modeled via gravure printing in a roll-to-roll process (Yang et al., 2013; Kopola et al., 2011) using chlorobenzene as the solvent (Garcia-Valverde et al., 2010).

---

[4] Although Anctil et al. report that production of PCBM using tetralin as a feedstock has lower impacts than toluene, toluene was chosen as a conservative choice given this is a prospective study.

TABLE 2.12

Inventory for the Production of Fullerene Nanomaterials

PCBM, Purified 99%, at Plant	Reference Flow Amount: 1 kg		Functionalization of C60 into PCBM Adapted from the Description in Anctil et al. (2013)	
Inventory Item	Amount	Units	Other	Inventory Flow Source
Disposal, hazardous waste, 25% water, to waste incineration, CH	0.05	kg		Ecoinvent
Methanol, at regional storage, CH	11.87	kg		Ecoinvent
PCBM, Unpurified, at plant	1.05	kg		This study
Solvent regeneration	237.5	kg		This study
Transport, freight, rail, RER	7.12	t*km	Based on a distance traveled of 600 km by train.	Ecoinvent
Transport, lorry 16-32t, EURO3, RER	1.19	t*km	Based on a distance traveled of 100 km by truck.	Ecoinvent
Methanol, at regional storage, CH	8.9	kg		Ecoinvent

PCBM, Unpurified, at Plant	Reference Flow Amount: 1 kg		Functionalization of C60 into PCBM Adapted from the Description in Anctil et al. (2013). The Authors Deviate from Anctil et al. by Accounting for C60 Regeneration. Often Large Amounts of C60 Are Not Converted to PCBM and This Would Not Be Cost Effective Unless That Amount Was Regenerated	
C60 regeneration	1.21	kg		This study
C60, fullerenes, at plant	2.12	kg		This study
Disposal, hazardous waste, 25% water, to waste incineration, CH	54.17	kg	Estimated as: 90% of the methyl 4 benzoybutyrate p-tosylhydrazone; 100% of the pyridine-compounds; 10% of the oDCB; 100% of the sodium hydroxide as sodium chloride; 10% of the toluene; 10% of the monochlorobenzene.	Ecoinvent
Electricity, medium voltage, production RER, at grid	159.23	MJ		Ecoinvent
Methyl 4-Benzobutyrate p-tosylhydrazone, at plant	2.01	kg		This study
Monochlorobenzene, at plant, RER	0.087	kg		Ecoinvent
Nitrogen, liquid, at plant, RER	18.90	kg		Ecoinvent
o-Dichlorobenzene, at plant RER	219.63	kg		Ecoinvent
pyridine-compounds, at regional storage, RER	39.50	kg		Ecoinvent
Sodium methoxide, at plant, GLO	0.30	kg		Ecoinvent
Solvent regeneration	347.00	kg		This study
Steam, for chemical processes, at plant, RER	495.34	kg		Ecoinvent
Tap water, at user, RER	26,418.13	kg		Ecoinvent
Toluene, liquid, at plant, RER	0.11	kg		Ecoinvent
Transport, freight, rail, RER	111.84	t*km	Based on a distance traveled of 600 km by train.	Ecoinvent
Transport, lorry 16-32t, EURO3, RER	18.64	t*km	Based on a distance traveled of 100 km by truck.	Ecoinvent

(*Continued*)

TABLE 2.12 (*Continued*)
Inventory for the Production of Fullerene Nanomaterials

C60 Regeneration	Reference Flow Amount: 1 kg		Adapted from the Purification Step of C60 in Anctil et al. (2013) Using 1 L Xylene per 15 g of C60. Assumes an 85% Recovery Rate. This Process Is Used in the PCBM Production Step	
Inventory Item	**Amount**	**Units**	**Other**	**Inventory Flow Source**
Electricity, medium voltage, production RER, at grid	52.02	MJ		Ecoinvent
Solvent regeneration	172.8	kg		Ecoinvent
Transport, freight, rail, RER	38.36	t*km	Based on a distance traveled of 600 km by train.	Ecoinvent
Transport, lorry 16-32t, EURO3, RER	6.40	t*km	Based on a distance traveled of 100 km by truck.	Ecoinvent
Xylene, at plant, RER	63.94	kg	It is assumed that 95% of the xylene is recovered from the solvent regeneration process.	Ecoinvent

C60 Fullerenes, at Plant	Reference Flow Amount: 1 kg		Production C60 Fullerenes Adapted from Anctil et al. (2013). It Is Based on Their Pyrolysis Production Route Using Toluene as a Feedstock. C60 Is Coproduced with C70 and Allocation Is Applied according to the Masses of Each Produced	
Electricity, medium voltage, production RER, at grid	802.8	MJ		Ecoinvent
Oxygen, liquid, at plant, RER	109.7	kg		Ecoinvent
Solvent Regeneration	456.4	kg		This study
Toluene, liquid, at plant, RER	137.1	kg		Ecoinvent
Transport, freight, rail, RER	228.3	t*km	Based on a distance traveled of 600 km by train.	Ecoinvent
Transport, lorry 16-32t, EURO3, RER	38.13	t*km	Based on a distance traveled of 100 km by truck.	Ecoinvent
Xylene, at plant, RER	24	kg	The authors assume this value reported by Anctil et al for xylene is a reduced value which has taken into account the regenerated xylene.	Ecoinvent
Carbon dioxide (Emissions to air)	128.11	kg	Estimated waste production.	Ecoinvent
Disposal, hazardous waste, 25% water, to hazardous waste incineration, CH	8.09	kg	Estimated waste production.	Ecoinvent

Solvent Regeneration	Reference Flow Amount: 1 kg		Adapted from Geisler et al. (2004) Using Steam, Nitrogen, and Cooling Water. Assumption That This Represents the Inventory for a Process Recovering 95% of Solvent	
Electricity, medium voltage, production RER, at grid	0.20	MJ		Ecoinvent
Nitrogen, liquid, at plant, RER	0.01	kg		Ecoinvent
Steam, for chemical processes, at plant, RER	1.50	kg		Ecoinvent
Tap water, at user, RER	80.00	kg		Ecoinvent

(*Continued*)

TABLE 2.12 (*Continued*)

Inventory for the Production of Fullerene Nanomaterials

Methyl 4-Benzobutyrate p-tosylhydrazone, at Plant	Reference Flow Amount: 1 kg		Production of Methyl 4 Benzoybutyrate p-Tosylhydrazone as Outlined in Anctil et al. (2013) Using the Precursors Methyl 4-Benzobutyrate and p-Toluenesulfonyl Hydrazine. The Authors Deviate from Anctil et al. by Accounting for Methanol Regeneration	
Inventory Item	Amount	Units	Other	Inventory Flow Source
Disposal, hazardous waste, 25% water, to waste incineration, CH	0.15	kg		Ecoinvent
Electricity, medium voltage, production RER, at grid	3.38	MJ		Ecoinvent
Heat, heavy fuel oil, at industrial furnace 1 MW, RER	1.06	MJ	Energy needed for distillation of methanol for regeneration. No other inventory data were expected of this regeneration process.	Ecoinvent
Methanol, at regional storage, CH	1.39	kg		Ecoinvent
Methyl 4-Benzobutyrate, at plant	0.57	kg		This study
p-toluenesulfonyl hydrazine, at plant	0.62	kg		This study
Transport, freight, rail, RER	1.64	t*km	Based on a distance traveled of 600 km by train.	Ecoinvent
Transport, lorry 16-32t, EURO3, RER	0.28	t*km	Based on a distance traveled of 100 km by truck.	Ecoinvent
Methanol, at regional storage, CH (avoided product)	7.22	kg	The amount of methanol regenerated.	Ecoinvent

Methyl 4-Benzobutyrate, at Plant	Reference Flow Amount: 1 kg		Production of Methyl 4 Benzoybutyrate as Outlined in Anctil et al. (2013) Using the Precursors 4-Benzobutyric Acid and Methanol. The Authors Deviate from Anctil et al. by Accounting for Methanol Regeneration	
4-Benzobutyric acid, at plant	1.01	kg		This study
Electricity, medium voltage, production RER, at grid	0.71	MJ		Ecoinvent
Heat, heavy fuel oil, at industrial furnace 1 MW, RER	5.54	MJ	Energy needed for distillation of methanol for regeneration. No other inventory data were expected of this regeneration process.	Ecoinvent
Methanol, at regional storage, CH	8.03	kg		Ecoinvent
Steam, for chemical processes, at plant, RER	1.20	kg		Ecoinvent
Transport, freight, rail, RER	0.54	t*km	Based on a distance traveled of 600 km by train.	Ecoinvent
Transport, lorry 16-32t, EURO3, RER	0.09	t*km	Based on a distance traveled of 100 km by truck.	Ecoinvent
Methanol, at regional storage, CH (avoided product)	7.22	kg	The amount of methanol regenerated.	Ecoinvent

(Continued)

TABLE 2.12 (*Continued*)
Inventory for the Production of Fullerene Nanomaterials

4-Benzobutyric Acid, at Plant	Reference Flow Amount: 1 kg		Adapted from Anctil et al. (2013) Using the Precursors Glutaric Acid, Benzene, Steam, and Aluminum Chloride	
Inventory Item	Amount	Units	Other	Inventory Flow Source
Aluminum chloride, at plant	0.83	kg	Original estimate seemed low and was increased to require 6.25 moles.	This study
Benzene, at pant, RER	3.15	kg		Ecoinvent
Electricity, medium voltage, production RER, at grid	0.70	MJ		Ecoinvent
Heat, heavy fuel oil, at industrial furnace 1 MW, RER	35.03	MJ	Energy for maleic acid recovery and distillation recycling.	Ecoinvent
Maleic anhydride, at plant	0.70	kg	Substitute for glutaric acid.	Ecoinvent
Steam, for chemical process, at plant, RER	1.18	kg		Ecoinvent
Transportation, freight, rail, RER	0.69	t*km	Based on a distance traveled of 600 km by train.	Ecoinvent
Transport, lorry 16-32t, EURO3, RER	0.11	t*km	Based on a distance traveled of 100 km by truck.	Ecoinvent
Disposal, hazardous waste, 25% water, to hazardous waste incineration, CH	0.49	kg	Aluminum hydroxide waste reported by Anctil et al.	Ecoinvent

Aluminum Chloride, at Plant	Reference Flow Amount: 1 kg		Estimated From Ullmans Encyclopedia of Industrial Chemistry 2007 for the Chlorination of Aluminum Oxide (Hudson et al. 2007)	
Aluminum, production mix, at plant, RER	0.20	kg		Ecoinvent
Chlorine, liquid, production mix, at plant, RER	0.40	kg		Ecoinvent
Transportation, freight, rail, RER	0.36	t*km	Based on a distance traveled of 600 km by train.	Ecoinvent
Transport, lorry 16-32t, EURO3, RER	0.06	t*km	Based on a distance traveled of 100 km by truck.	Ecoinvent
Heat, waste	2.25	MJ	This chemical reaction is highly exothermic, producing around 300 kJ/moll of aluminum chloride produced.	Ecoinvent

2.7.2.4 Hole Transporter

The hole transport layer is responsible for carrying the "electron hole"—created when the mobile electron travels through the PCBM layer to the back electrode—through the P3HT to the transparent electrode (Koster, 2007). Poly(3,4-ethylenedioxythiophene):poly(styrenesulfonate) (PEDOT:PSS) is a common hole transport layer used in OPV cells, but this material has been shown to have low stability in ambient conditions (Girotto et al., 2011; Jiang et al., 2010). Instead, molybdenum trioxide[5] (MoO_3) was chosen as the hole transport layer (Lizin et al., 2013; Jiang et al., 2010; Voroshazi

[5] MoO_3 does not interfere with the light absorption profile (350–700 nm) of the active layer.

TABLE 2.13

Inventory for the Production of Poly(3-hexylthiophene-2,5-diyl) or P3HT

Poly(3-hexylthiophene-2,5-diyl), at Plant	Reference Flow Amount: 1 kg		Reaction of Thiophene, Bromine, and Hexane. Amounts of Inputs and Energy Adapted from García-Valverde et al. (2010)	
Inventory Item	Amount	Units	Other	Inventory Flow Source
Bromine, at plant	8.97	kg		This study
Electricity, medium voltage, production RER, at grid	131.07	MJ		Ecoinvent
Heat, heavy fuel oil, at industrial furnace 1 MW, RER	1.15	GJ		Ecoinvent
Hexane, at plant, RER	2.58	kg		Ecoinvent
Thiophene, at plant	7.81	kg		This study
Transport, freight, rail, RER	11.61	t*km	Based on a distance traveled of 600 km by train.	Ecoinvent
Transport, lorry 16-32t, EURO3, RER	1.94	t*km	Based on a distance traveled of 100 km by truck.	Ecoinvent
Water, ultrapure, at plant, GLO	11,136.15	kg		Ecoinvent

Bromine, at plant	Reference Flow Amount: 1 kg		Adapted from Ecoinvent 3 Description but Using Ecoinvent 2 Data (Frischknecht et al. 2007)	
Bromine	1.00	kg		Ecoinvent
Chemical plant, organics, RER	4.0E-10	number of items		Ecoinvent
Chlorine, liquid, production mix, at plant, RER	0.60	kg		Ecoinvent
Steam, for chemical processes, at plant, RER	40.07	kg		Ecoinvent
Sulfuric acid, liquid, at plant, RER	0.057	kg		Ecoinvent
Transport, freight, rail, RER	0.99	t*km	Based on a distance traveled of 600 km by train.	Ecoinvent
Transport, lorry 16-32t, EURO3, RER	0.17	t*km	Based on a distance traveled of 100 km by truck.	Ecoinvent
Water	0.44	m^3		Ecoinvent
Sulfate (Emissions to water)	0.055	kg		Ecoinvent
Water vapor (Emissions to air)	0.44	kg		Ecoinvent

Thiophene, at Plant	Reference Flow Amount: 1 kg		Production of Thiophene (C_4H_4S) from Butane (C_4H_{10}) and Sulfur (CS_2). The Inputs and Outputs of Production Process Were Adapted from the Patent US 3939179 A and from the Ullmann's Encyclopedia of Industrial Chemistry (Swantson 2007)	
Aluminium oxide, at plant, RER	0.40	kg		Ecoinvent
Butanes from butenes, at plant, RER	0.69	kg		Ecoinvent
Chromium oxide, flakes, at plant, RER	0.80	kg		Ecoinvent
Heat, heavy fuel oil, at industrial furnace 1 MW, RER	1.81	MJ		Ecoinvent
Hydrogen sulfide, H_2S, at plant, RER	1.22	kg		Ecoinvent
Secondary Sulphur, at refinery, RER	1.52	kg		Ecoinvent
Transport, freight, rail, RER	1.58	t*km	Based on a distance traveled of 600 km by train.	Ecoinvent
Transport, lorry 16-32t, EURO3, RER	0.26	t*km	Based on a distance traveled of 100 km by truck.	Ecoinvent

TABLE 2.14
Inventory for the Production of Molybdenum Oxide

Molybdenum(VI) Oxide, at Plant	Reference Flow Amount: 1 kg		Estimation of the Production of Molybdenum VI Oxide Based on Stoichiometric Calculations Involving Molybdenite and Oxygen Heated Together at 700°C	
Inventory Item	Amount	Units	Other	Inventory Flow Source
Heat, heavy fuel oil, at industrial furnace 1 MW, RER	0.76	MJ		Ecoinvent
Molybdenite, at plant, GLO	1.11	kg		Ecoinvent
Oxygen, liquid, at plant, RER	0.39	kg		Ecoinvent
Transport, freight, rail, RER	0.9	t*km	Based on a distance traveled of 600 km by train.	Ecoinvent
Transport, lorry 16-32t, EURO3, RER	0.15	t*km	Based on a distance traveled of 100 km by truck.	Ecoinvent
Sulfur dioxide (emissions), to air	0.89	kg		Ecoinvent

et al., 2011). Oxides of molybdenum do not occur in large amounts in nature, but rather as sulfides (e.g., molybdenite). Production of MoO_3 was modeled by roasting molybdenite at high temperatures in the presence of oxygen (see Table 2.14) (Kim et al., 2009). The amount of MoO_3 was estimated to be 0.24 g per m^2 (Espinosa et al., 2011; Garcia-Valverde et al., 2010; Anctil et al., 2010; Roes et al., 2009). Deposition of MoO_3 was modeled via gravure printing in a roll-to-roll process as has been demonstrated for PEDOT:PSS (Krebs, 2009a; Yang et al., 2013; Kopola et al., 2011).

2.7.2.5 Back Electrode

In forward structure OPVs, the back electrode is the cathode to which the negative charges flow. Aluminum was used as the back electrode (Krebs, 2009a). A layer of lithium fluoride was deposited along with aluminum to provide efficient operation (Table 2.15) (Brabec et al., 2002) estimated from Roes et al. (2009).

2.7.2.6 Encapsulation and Lamination

Framing of the polymer solar cell is not considered in this inventory. Roes et al. (2009) describe an OPV cell with additional barrier layers and hard coats (Table 2.16). They propose the consumption of 133 g PET and 0.44 g silica (SiO_2) for the barriers, while 0.99 g of an epoxy-silica nanocomposite is used as the hard coat (Roes et al., 2009), and 9.25 g of an epoxy resin as an adhesive (Roes et al., 2009) (Table 2.17).

TABLE 2.15
Inventory of Production of Lithium Fluoride

Lithium Fluoride, Layer, Application	Reference Flow Amount: 1 item (i.e., per m²)		Adapted from Roes et al. (2009)	
Inventory Item	Amount	Units	Other	Inventory Flow Source
Electricity, medium voltage, production RER, at grid	0.46	MJ		Ecoinvent
Heat, heavy fuel oil, at industrial furnace 1 MW, RER	1.08	MJ		Ecoinvent
Lithium fluoride, at plant, CN	6.0E-8	kg		Ecoinvent

TABLE 2.16

Inventory for Production of Lamination and Encapsulation of OPV Panel

Lamination Flexible Solar Module, at Plant	Reference Flow Amount: 1 m²		Adapted from Roes et al. (2009)	
Inventory Item	Amount	Units	Other	Inventory Flow Source
Electricity, medium voltage, production RER, at grid	0.047	MJ		Ecoinvent
Epoxy resin, liquid, at plant, RER	0.0093	kg		Ecoinvent
Epoxy silica nanocomposite, at plant	0.00099	kg		This study
Heat, natural gas, at industrial furnace > 100 kW, RER	0.039	MJ		Ecoinvent
Polyethylene terephthalate, granulate, amorphous, at plant, RER	0.13	kg		Ecoinvent
Silicon product, at plant, RER	0.00044	kg		Ecoinvent
Transport, freight, rail, RER	0.12	t*km	Based on a distance traveled of 600 km by train.	Ecoinvent
Transport, lorry 16-32t, EURO3, RER	0.019	t*km	Based on a distance traveled of 100 km by truck.	Ecoinvent
Epoxy Silica Nanocomposite, at Plant	**Reference Flow Amount: 1 kg**		**Adapted from the Production of Silica Sol Used in Roes et al. (2009)**	
Heat, heavy fuel oil, at industrial furnace 1 MW, RER	23	MJ		Ecoinvent
Sodium silicate, furnace process, pieces, at plant, RER	3.9	kg		Ecoinvent
Sulphuric acid, liquid, at plant, RER	0.66	kg		Ecoinvent
Water, ultrapure, at plant, GLO	40	kg		Ecoinvent

TABLE 2.17

Inventory for Printing of All Layers of the OPV Panel

Gravure Printing (Energy)	Reference Flow Amount: 1 Items (i.e., per m²)		Adapted from Roes et al. (2009)	
Inventory Item	Amount	Units	Other	Inventory Flow Source
Electricity, medium voltage, production RER, at grid	0.46	MJ		Ecoinvent
Heat, heavy fuel oil, at industrial furnace 1 MW, RER	1.08	MJ		Ecoinvent
Toluene (emissions), to air	0.022	kg	Emissions from toluene, which is used as a solvent in some of the layer applications.	Ecoinvent

2.7.2.7 Deposition

Framing of the polymer solar cell is not considered in this inventory. Roes et al. (2009) describe an OPV cell with additional barrier layers and hard coats (Table 2.16). They propose the consumption of 133 g PET and 0.44 g silica (SiO_2) for the barriers, while 0.99 g of an epoxy-silica nano-composite is used as the hard coat (Roes et al., 2009), and 9.25 g of an epoxy resin as an adhesive (Roes et al., 2009).

2.8 QUESTIONS AND EXERCISES

1. What are the main advantages of LCA?
2. What are the main steps of LCA and in which ISO standard are they considered?
3. What is the difference between an attributional and a consequential LCA?
4. In which steps of LCA are each of the following different functions carried out: (a) allocation; (b) selection and definition of impact categories; (c) identification of the most important results of the LCIA?
5. What information should be given at the end of the interpretation phase of an LCA?
6. Design the LCA framework scheme for the case of an old fabric factory building with a main structure made of steel and reinforced concrete, which has been in use for 50 years, and has subsequently been transformed into a department store. Consider that the department store has been operating for 20 years and that the building is finally knocked down. The owner of the department store invested an additional amount, taking some marketing initiatives adapting solar PV panels, a modern air conditioning system, and an extra investment in aesthetic comfort and design features.
7. Explain the differences between reusing, recycling, and disposal of a product. How would the corresponding life-cycle inventory change?
8. List all possible applications of LCA to the chemical industry.
9. Compare the use of virgin paper and recycled paper. Compare, from an environmental point of view, the use of recycled paper and the use of paper obtained from paper pulp. The recycled paper has been obtained from paper wastes transported in 16-t trucks from a distance of 1000 km. The total emissions of carbon dioxide, CO_2, and the generated solids in kg of them are used as comparison parameters. It is desirable to compare the derived effects of purchasing a 400-g pack of new writing paper at the corner stationery shop with the effects derived from driving to buy a similar pack of recycled paper at an establishment 1 km away. Only the ELs due to raw materials and transportation are considered. Use the following environmental loads:

EL	EL/kg of Recycled Paper	EL/kg of Pulp Paper	EL/km (Car)	EL/tkm (16-t Truck)
kg CO_2	1.03	1.61	2.16×10^{-1}	3.46×10^{-1}
kg Particles (unspecified)	7.06×10^{-2}	1.73×10^{-1}	7.50×10^{-6}	7.19×10^{-2}

tkm is equivalent to a mass of 1 t (1000 kg) transported 1 km.

10. Compare the use of paper towels for drying hands with the use of a hand-dryer supplied with electricity from a gas thermal power plant or from a wind turbine. The considered comparison parameters are the emissions of carbon dioxide, CO_2, and sulfur dioxide, SO_2. Suppose that the weight of a paper towel is 7 g and that the electric hand-dryer is 2000 W and works for 30 s. With these data, determine the best way, from an environmental point of view, to dry the hands. The ELs related to the paper and the use of electricity are assumed to be the following:

EL	Units	Paper EL/kg	Electricity Gas Thermal Power Plant EL/kWh	Electricity Wind Turbine EL/kWh
Lignite	kg	1.10×10^{-1}	1.21×10^{-3}	1.68×10^{-3}
Coal	kg	1.51×10^{-1}	7.06×10^{-2}	1.10×10^{-2}
Natural gas	Nm^3	7.97×10^{-2}	2.09×10^{-1}	2.25×10^{-3}
Crude oil	t	1.84×10^{-4}	2.35×10^{-6}	4.82×10^{-6}
Water	kg	2.60×10^3	3.40	5.15×10^1
CO_2 (air)	kg	1.46	1.42	3.3×10^{-2}
SO_x (air)	kg	1.04×10^{-2}	2.07×10^{-4}	1.37×10^{-4}
NO_x (air)	kg	3.03×10^{-3}	2.16×10^{-3}	7.1×10^{-5}
CH_4 (air)	kg	2.04×10^{-3}	1.99×10^{-3}	1.1×10^{-4}
HCl (air)	kg	1.30×10^{-4}	1.00×10^{-6}	5.8×10^{-6}

Electricity consumption: Ec (kWh); Ec = (2000 (J/s)/1000); (30/3600) = 1.67×10^{-2} kWh.

REFERENCES

Anctil, A. & Fthenakis, V. (2012). Life cycle assessment of organic photovoltaics. In: Fthenakis, V. (Ed.), *Third Generation Photovoltaics*. InTech, Rejeka, Croatia.

Anctil, A., Babbit, C., Landi, B., & Raffaelle, R.P. (2010). Life cycle assessment of organic solar cell technologies. *Photovoltaics Specialists Conference PVSC 2010 35th IEEE*. Honolulu, Hawaii.

Anctil, A., Babbitt, C.W., Raffaelle, R.P., & Landi, B.J. (2012). Material and energy intensity of fullerene production. *Environmental Science and Technology*, 45(6): 2353–2359.

Anctil, A., Babbitt, C.W., Raffaelle, R.P., & Landi, B.J. (2013). Cumulative energy demand for small molecule and polymer photovoltaics. *Progress in Photovoltaics Research and Applications*, 21: 1541–1554.

Assies, J.A. (1992). State of art. Life-cycle assessment, pp. 1–20, Workshop Report, December 1991, SETAC, Leiden, the Netherlands.

Aukkaravittayapun, S., Wongtida, N.M., Kasecwatin, T., Charojrochkul, S., Unnanon, K., & Chindaudom, P. (2006). Large scale F-doped SnO_2 coating on glass by spray pyrolysis. *Thin Solid Films*, 496(1), 117–120.

Azapagic, A. & Clift, R. (1999). Allocation of environmental burdens in co-product systems: Product-related burdens (Part 1). *Journal of Cleaner Production,* 7: 101–119. doi:10.1016/S0959-6526(98)00046-8.

Bey, N. (2018). Life cycle management. In: Hauschild, M.Z., Rosenbaum, R.K., & Olsen, S.I. (Eds.), *Life Cycle Assessment—Theory Practice*, Springer, Dordrecht, the Netherlands, p. 1216.

Boustead, I. (1974). Resource implications with particular reference to energy requirements for glass and plastic milk bottles. *Journal of the Society of Dairy Technology*, 27(3): 159–165.

Boustead, I. & Hancock, G. (1979). *Handbook of Industrial Energy Analysis*. John Wiley and Ellis Horwood, New York, Chap. 3.

Boustead, I. (1992). Eco-balance methodology for commodity thermoplastics. A report for the European Center for Plastics in the Environment, Brussels.

Boustead, I. (2005a). *Eco-profiles of the European Plastics Industry: Low Density Polyethylene (LDPE)*. Plastics Europe, Brussels, Belgium.

Boustead, I. (2005b). *Eco-profiles of the European Plastics Industry: Polyethylene Terephthalate (PET) (Amorphous grade)*. PlasticsEurope, Brussels, Belgium.

Brabec, C., Shaheen, S., Winder, C., & Sariciftci, S. (2002). Effect of LiF/ Metal electrodes on the performance of plastic solar cells. *Applied Physics Letters*, 80(7): 1288–1290.

Castells, F., Aelion, V., Abeliotis, K.G., & Petrides, D.P. (1995). An algorithm for life-cycle inventory. AICHE *Symposium Series on Pollution Prevention via Process and Product* Modifications, 90(303): 151–160.

Castells, F., Alonso, J.C., & Garreta, J. (1997). Antecedentes y estado actual del Análisis de Ciclo de Vida. *Ingeniería Química*, 29(339): 163–171.

Club of Rome. (1972). A blueprint for survival. *Ecologist*, 2(1): 1–44.

Curran, M.A. (1996). *Environmental Life-Cycle Assessment*. McGraw-Hill, New York.

De Camillis, C., Brandão, M., Zamagni, A., & Pennington, D. (2013). Sustainability assessment of future-oriented scenarios: A review of data modelling approaches in Life Cycle Assessment. doi:10.2788/95227.

Druijff, E.A. (1984). *Milieurelevante Produktinformatie (environmental information on products)*. CML mededelingen 15, Leiden, The Netherlands.

EEA (European Environmental Agency). (1998). Life-cycle assessment (LCA)—A guide to approaches, experiences and information sources, *Environmental Issues Series* No. 6., European Environment Agency, Copenhagen, Denmark.

Ekvall, T., Tillman, A.M., & Molander, S. (2005). Normative ethics and methodology for life cycle assessment. *Journal of Cleaner Production*, 13: 1225–1234. doi:10.1016/j.jclepro.2005.05.010.

EMPA. (1984). Eidgenössische Materialprüfungs- und Forschungsanstalt. Ökobilanzen von Packstoffen. Schriftenreihe Umweltschutz 24. Herausgegeben vom Bundesamt für Umweltschutz (BUS), Bern, Switzerland.

Espinosa, N., Garcia-Valverde, R., Urbina, A., & Krebs, F. (2011). A life cycle analysis of polymer solar cell modules prepared using roll-to-roll methods under ambient conditions. *Solar Energy Materials & Solar Cells*, 95(5): 1293–1302.

Espinosa, N., Hosel, M., Angmo, D., & Krebs, F.C. (2012). Solar cells with one-day energy payback for the factories of the future. *Energy & Environmental Science*, 5(1): 5117–5133.

European Commission. (2010). *ILCD Handbook—General Guide for Life Cycle Assessment—Provisions and Action Steps. Joint Research Centre, Ispra, Italy*.

Fava, J.A., Denison, R., Jones, B., Curran, M.A., Vigon, B., Selke, S., & Barnum, J. (1991). *SETAC Workshop Report: A Technical Framework for Life-cycle Assessments*, August 18–23, 1990. Smugglers Notch, VT, Washington, DC, January 1991.

Fecker, I. (1992). How to calculate an ecological balance? Report No. 222, EMPA, St. Gallen, Dübendorf, Switzerland.

Frischknecht, R., Bollens, U., Bosshart, S., Ciot, M., Ciseri, L., Doka, G., Hischier, R., Martin, A., Dones, R., and Gantner, U. (1996). *Ökoinventare von Energiesystemen: Grundlagen für den ökologischen Vergleich von Energiesystemen und den Einbezug von Energiesystemen in Ökobilanzen für die Schweiz*, 3rd ed., ETH Zürich: Gruppe Energie-Stoffe-Umwelt, PSI Villigen: Sektion Ganzheitliche Systemanalysen.

Frischknecht, R. (2000). Allocation in life cycle inventory analysis for joint production. *MIIM LCA PhD Club*, 179:85–95. doi:10.1065/lca2000.02.013

Frischknecht, R., Jungbluth, N., Althaus, H. et al. (2007). *Overview and Methodology; Final Report Ecoinvent Data 2.0*. Swiss Centre for Life Cycle Inventories, Dubendorf, Switzerland.

Fthenakis, V., Frischnecht, R., Rugei, M., Kim, H., Alsema, E., Held, M., & de Wild-Scholten, M. (2011). *Methodology Guidelines on Life Cycle Assessment of Photovoltaic Electricity* (2nd ed.). International Energy Agency PVPS, Paris, France.

Garcia-Valverde, R., Cherni, J., & Urbina, A. (2010). Life cycle analysis of organic photovoltaic technologies. *Progress in Photovoltaics: Research and Applications*, 18(7): 535–558.

Geisler, G., Hofstetter, T.B., & Hungerbuhler, K. (2004). Production of fine and speciality chemicals: Procedure for the estimation of LCIs. *International Journal of Life-Cycle Assessment*, 2: 101–113.

Geyer, R., Kuczenski, B., Zink, T., & Henderson, A. (2015). Common misconceptions about recycling. *Journal of Industrial Ecology* 20: 1010–1017. doi:10.1111/jiec.12355.

Girotto, C., Voroshazi, E., Cheyns, D., Heremans, P., & Rand, B. (2011). Solution processed MoO_3 thin films as a hole injection layer for organic solar cells. *Applied Materials and Interfaces*, 3: 3244–3247.

Graf, G.G. (2007). Tin, tin alloys, and tin compounds. In: Elvers, B. (Ed.), *Ulmann's Encyclopedia of Industrial Chemistry* (7th ed., pp. 1–35). Wiley-VCH, Weinheim, Germany.

Guinée, J.B., Udo de Haes, H.A., & Huppes, G. (1993a). Quantitative life-cycle assessment of products, 1: Goal definition and inventory. *The Journal of Cleaner Production*, 1(1): 3–13.

Guinée, J.B., Heijungs, R., Udo de Haes, H.A., & Huppes, G. (1993b). Quantitative life-cycle analysis of products, 2: Classification, valuation and improvement analysis. *The Journal of Cleaner Production*, 1(2): 81–91.

Guinée, J.B., Heijungs, R., & Huppes, G. (2004). Economic allocation: Examples and derived decision tree. *The International Journal of Life Cycle Assessment*, 9: 23–33. doi:10.1007/BF02978533.

Guinée, J.B., Heijungs, R., Huppes, G. et al. (2011). Life cycle assessment: Past, present, and future. *Environmental Science & Technology*, 45: 90–96. doi:10.1021/es101316v.

Hauschild, M. & Wenzel, H. (1998). *Environmental Assessment of Products—Scientific Background* (Vol. 2), Chapman & Hall, London, UK.

Heijungs, R. (1997). Economic drama and the environmental stage—Formal derivation of algorithmic tools from a unified epistemological principle. Dissertation, Rijksuniversiteit Leiden, Leiden, the Netherlands.

Heijungs, R., Guinée, J.B., Huppes, G., Lankreijer, R.M., Udo de Haes, H.A., & Wegener-Sleeswijk, A. (1992). Environmental life-cycle assessment of products—Guide and backgrounds, technical report, CML, University of Leiden, Leiden, the Netherlands.

Hudson, K.L., Misra, C., Perrotta, A.J., Wefers, K., & Williams, F. (2007). Aluminum oxide. In: Elvers, B. (Ed.), *Ulmann's Encyclopedia of Industrial Chemistry* (7th ed., pp. 1–40). Wiley-VCH, Weinheim, Germany.

Hunt, R.G. (1974). *Resource and Environmental Profile Analysis of Nine Beverage Container Alternatives*. Midwest Research Institute for U.S. EPA (Contract 68-01-1848), Washington, DC.

ICCA. (2016). *How to Know If and When It's Time to Commission a Life Cycle Assessment*. International Council of Chemical Associations, Brussels, Belgium.

IFIAS (International Federation of Institutes for Advanced Study). (1974). *Energy Analysis Workshop on Methodology and Conventions*. Guldsmedshyttan, Örebro, Sweden.

International Organization for Standardization. (2006). 14044:2006 *Environmental Management—Life Cycle Assessment—Requirements and Guidelines*. ISO, Geneva, Switzerland.

International Organization for Standardization. (2006). *ISO 14040:2006 Environmental Management—Life Cycle Assessment—Principles and Framework*. ISO, Geneva, Switzerland.

ISO 14040. (1997). *Environmental Management—Life-Cycle Assessment—Principles and Framework, Technical Standard*, ISO, Geneva, Switzerland.

ISO 14040. (2006a). *Environmental Management—Life Cycle Assessment—Principles and Framework*. The International Organization for Standardization (ISO), Geneva, Switzerland.

ISO 14044. (2006b). *Environmental Management—Life Cycle Assessment—Requirements and Guidelines*. The International Organization for Standardization (ISO), Geneva, Switzerland.

ISO/TR 14049. (2012). *Environmental Management—Life Cycle Assessment—Illustrative Examples on How to Apply ISO 14044 to Goal and Scope Definition and Inventory Analysis*. The International Organization for Standardization (ISO), Geneva, Switzerland.

Jiang, C., Sun, X., Zhao, D., Kyaw, A., & Li, Y. (2010). Low work function metal modified ITO as cathode for inverted polymer solar cells. *Solar Energy Materials and Solar Cells*, 94(10): 1618–1621.

Kim, B.-S., Lee, H.-I., Choi, Y.-Y., & Kim, S. (2009). Kinetics of the oxidative roasting of low grade mongolian molybdenite concentrate. *Materials Transactions*, 50(11): 2669–2674.

Kim, H., Kushto, G., Auyeung, R., & Piqué, A. (2008). Optimization of F-Doped SnO_2 electrodes for OPV devices. *Applied Physics*, 93: 521–526.

Kopola, P., Aernouts, T., Sliz, R., Guillerez, S., Ylikunnari, M., Cheyns, D., & Maaninen, A. (2011). Gravure printed flexible organic photovoltaic modules. *Solar Energy Materials and Solar Cells*, 95: 1344–1347.

Koster, L.J.A. (2007). The optimal band gap for plastic photovoltaics. *SPIE Newsroom*. doi:10.1117/2.1200701.0528.

Krebs, F. (2009a). Fabrication and processing of polymer solar cells—A review of printing and coating techniques. *Solar Energy Materials and Solar Cells*, 93(4): 394–412.

Krebs, F. (2009b). Roll-to-roll fabrication of monolithic large area polymer solar cells free from indium-tin-oxide. *Solar Energy Materials and Solar Cells*, 93(9): 1636–1641.

Kroon, R., Lenes, M., Hummelen, J., Blom, P.W., & De Boer, B. (2008). Small bandgap polymers for organic solar cells. *Polymer Reviews*, 48: 531–582.

Kumar, A. & Zhou, C. (2010). The race to replace tin-doped indium oxide: Which material will win. *ACS Nano*, 4(1): 11–14.

Li, C.-Z., Yip, H.-L., & Jen, A. (2012). Functional fullerenes for OPVs. *Journal of Materials Chemistry*, 22: 4161–4177.

Lizin, S., van Passel, S., De Schepper, E., Maes, W., Lusten, L., Manca, J., & Vanderzande, D. (2013). Life cycle analyses of organic photovoltaics: A review. *Energy and Environmental Science*, 6: 3136–3149.

Meadows, D.H., Meadows, D.L., Randers, J., & Behrens, W. (1972). *The Limits to Growth: A Report for the Club of Rome's Project on the Predicament of Mankind*, Universe Books, New York.

Pedersen, B. (1993). *Environmental Assessment of Products*, UETP-EEE.

Pelletier, N., Ardente, F., Brandão, M. et al. (2014). Rationales for and limitations of preferred solutions for multi-functionality problems in LCA: Is increased consistency possible? *The International Journal of Life Cycle Assessment*. doi:10.1007/s11367-014-0812-4.

Roes, A., Alsema, E., Blok, K., & Patel, M. (2009). Ex-ante environmental and economic evaluation of polymer PV. *Progress in Photovoltaics*, 17(6): 372–393.

Samad, W., Salleh, M.M., Shafiee, A., & Yarmo, M. (2011). Structural, optical and electrical properties of FTO thin films deposited using inket printing technique. *Sains Malaysiana*, 40(3): 251–257.

Schrijvers, D.L., Loubet, P., & Sonnemann, G. (2016). Developing a systematic framework for consistent allocation in LCA. *The International Journal of Life Cycle Assessment*, 21: 976–993.

SETAC (Society of Environmental Toxicology and Chemistry). (1993). *Guidelines for Life-Cycle Assessment—A Code of Practice*, Sesimbra/Portugal SETAC workshop report, SETAC Press, Pensacola, FL.

Sonnemann, G.W., Castells, F., & Rodrigo J. (1999). Guide to the inclusion of LCA in environmental management, *8th Congrés Mediterrani d'Enginyeria Química, Expoquimia, Barcelona, Spain*, pp. 10–12, November 1999.

Sonnemann, G., Gemechu, E.D., Sala, S. et al. (2018). Life cycle thinking and the use of LCA in policies around the world. In: Hauschild, M.Z., Rosenbaum, R.K., & Olsen, S.I. (Eds.), *Life Cycle Assessment*. Springer, Cham, Switzerland, pp. 429–463.

Sonnemann, G.W. & Vigon, B.W. (2011). *Global Guidance Principles for Life Cycle Assessment Databases—A Basis for Greener Processes and Products*. Paris, France.

STQ (Servei de Tecnologia Química). (1998). Análisis del ciclo de vida de la electricidad producida por la planta de incineración de residuos urbanos de Tarragona, technical report, Universitat Rovira i Virgili, Tarragona, Spain.

Swantson, J. (2007). Thiophene. In: Elvers, B. (Ed.), *Ulmann's Encyclopedia of Industrial Chemistry* (7th ed., pp. 25889–25901). Wiley-VCH, Weinheim, Germany.

Thompson, B.C. & Frechet, J.M. (2007). Polymer–fullerene composite solar cells. *Angewandte Chemie International Edition*, 47(1): 58–77.

Tillman, A.-M. (2000). Significance of decision-making for LCA methodology. *Environmental Impact Assessment Review*, 20: 113–123. doi:10.1016/S0195-9255(99)00035-9.

Tsang, M.P., Sonnemann, G.W., & Bassani, D.M. (2015). A comparative human health, ecotoxiciy, and product environmental assessment on the production of organic and silicon solar cells. *Progress in Photovoltaics Research and Applications*, (Early Online View). doi:10.1002/pip.2704.

Udo de Haes, H.A., Finnveden, G., Goedkoop, M. et al. (2002a). *Towards Best Available Practice in Life-Cycle Impact Assessment*. SETAC Press, Pensacola, FL.

Udo de Haes, H.A., Finnveden, G., Goedkoop, M. et al. (2002b). UNEP/SETAC life-cycle initiative: Background, aims and scope. *The International Journal of Life Cycle Assessment*, 7(4): 192–195.

Udo de Haes, H.A., Jolliet, O., Finnveden, G., Hauschild, M., Krewitt, W., & Mueller-Wenk, R. (1999). Best available practice regarding categories and category indicators in life-cycle impact assessment— Background document for the Second Working Group on LCIA of SETAC–Europe, *The International Journal of Life Cycle Assessment*, 4: 66–74; 167–174.

UNEP (Industry and Environment). (1996). LCA—What is it and how to use it? Technical report, United Nations Environment Program, Paris, France.

UNEP/SETAC Life-Cycle Initiative. (2002). *Background Paper of the UNEP/SETAC Life-Cycle Initiative*, UNEP, DTIE, Paris, France.

U.S. EPA. (1996). *MSW Factbook 3.0. EPA*. Office of Solid Waste, Washington, DC.

U.S. EPA. (2006). *Life Cycle Assessment: Principles and Practice*. Cincinnati, OH.

Vigon, B.W., Tolle, D.A., Corneby, B.W., Lotham, H.C., Harrison, C.L., Boguski, T.L., Hunt, R.G., & Sellers, J.D. (1993). *Life-cycle Assessment: Inventory Guidelines and Principles, Conducted by Battelle and Franklin Associates for the EPA, Ltd.* Office of Research and Development, Washington, DC, EPA/600/R-92/245.

Voroshazi, E., Verreet, B., Aernouts, T., & Heremans, P. (2011). Long term operational lifetime and degradation analysis of P3HT PCBM photovoltaic cells. *Solar Energy Materials and Solar Cells*, 95(5): 1303–1307

Weidema, B.P. (2001). Avoiding co-product allocation in life-cycle assessment. *The Journal of Industrial Ecology*, 4: 11–33.

Weidema, B.P. (2003). *Market Information in Life Cycle Assessment*. Copenhagen, Denmark.

Yang, J., Vak, D., Clark, N. et al. (2013). Organic photovoltaic modules fabricated by an industrial gravure printing proofer. *Solar Energy Materials and Solar Cells*, 109: 47–55.

Yang, X., & Uddin, A. (2014). Effect of thermal annealing on P3HT:PCBM bulk heterojunction organic solar cells: A critical review. *Renewable and Sustainable Energy Reviews*, 30: 324–336.

Zapp, K.-H., Wostrbrock, K.-H., Schafer, M. et al. (2007). Ammonium compounds. In: Elvers, B. (Ed.), *Ulmann's Encyclopedia of Industrial Chemistry*, (7th ed., pp. 1–26). Wiley-VCH, Weinheim, Germany.

Zeman, M., Isabella, O., Solntsev, S. & Jäger, K. (2011). Modelling of thin-film silicon solar cells. Solar Energy Materials & Solar Cells, 520(3): 1096–1101.

3 Life-Cycle Impact Assessment

Guido Sonnemann, Michael Tsang, and Francesc Castells

CONTENTS

3.1   INTRODUCTION

The life-cycle inventory offers product-related environmental sustainability information consisting basically of a quantified list of environmental loads (raw material consumption, air and water emissions, wastes, etc.) that give the amount of pollutants to be assigned to the product. However, the environmental damage associated with them is not yet known.

Let us consider, for example, well-known air pollutants such as sulfur dioxide, SO_2, nitrogen dioxide, NO_2, and hydrogen chloride, HCl, that generate an environmental impact known as acid rain. The capacity of these pollutants to acidify the atmosphere can be measured by the potential to generate H^+ protons, so the acid concentration could be multiplied by a corresponding factor to obtain a global value of H^+ protons equivalent. In this way, an environmental impact category has been measured based on inventory data. The same occurs with air emissions: carbon dioxide, methane, nitrogen oxides, halocarbons, etc. contribute to Earth's global warming and cause the well-known greenhouse effect, measured in CO_2 equivalents. Thus, a new type of impact category, global warming potential, is introduced from inventory data.

Thus, the life-cycle impact assessment (LCIA) is introduced as the third step of life-cycle assessment (LCA), described in ISO 14044 (2006). The purpose of LCIA is to assess a product system's life-cycle inventory (LCI) to understand its environmental significance better. Thus, LCIA provides information for interpretation—the final step of the LCA methodology.

Jointly, with other LCA steps, the LCIA step provides a wide perspective of environmental and resource issues for product systems by assigning life-cycle inventory results to impact categories. For each impact category, impact potentials are selected and category indicator results are calculated. The collection of these results defines the LCIA profile of the product system, which provides information on the environmental relevance of resource use and emissions associated with it. In the same way as LCA as a whole, LCIA builds up a relative approach based on the functional unit.

On the other hand, to compare the potentials for different impacts, it is necessary to evaluate the seriousness of the environmental impact categories relative to one another. This can be expressed by a set of weighting factors—one factor per impact category within each of the main category groups. The weighted impact potential, WP(j), can be calculated by multiplying the normalized impact potential or resource consumption, NP(j), by the weighting factor, WF(j), associated with the impact category.

3.2   PHASES OF LIFE-CYCLE IMPACT ASSESSMENT

The general framework of the LCIA phase is composed of several mandatory elements that convert life-cycle inventory results into indicator results. In addition, there are optional elements for normalization, grouping and weighting of the indicator results and data quality analysis techniques. The LCIA phase is only one part of a total LCA study and should be coordinated with other phases of LCA. An overview of the mandatory and optional elements in LCIA is given in Figure 3.1.

The mandatory LCIA elements are (ISO 14044, 2006):

• Selection of impact categories indicators and models
• Classification of environmental loads within the different categories of environmental impact
• Characterization of environmental loads by means of a reference pollutant typical of each environmental impact category

These categories, plus the level of detail and methodology, are chosen depending on the goals and scope of the research.

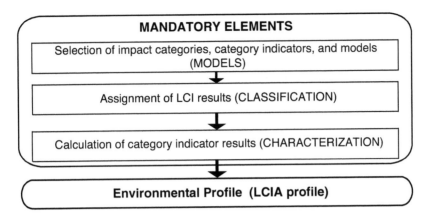

FIGURE 3.1 Mandatory elements of LCIA. (Based on ISO 14044, *Environmental Management—Life Cycle Assessment—Requirements and Guidelines*, The International Organization for Standardization (ISO), Geneva, Switzerland, 2006.) (available at www.afnor.fr).

Optional elements and information can be used depending on the goal and scope of the LCA study (ISO 14044, 2006):

- Calculating the magnitude of category indicator results relative to reference values (normalization) means that all impact scores (contribution of a product system to one impact category) are related to a reference situation.
- The indicators can be grouped (sorted and possibly ranked).
- Weighting (across impact categories) is a quantitative comparison of the seriousness of the different resource consumption or impact potentials of the product, aimed at covering and possibly aggregating indicator results across impact categories.
- Data quality analysis serves to better understand the reliability of the LCIA results.

The use of models is necessary to derive the characterization factors. The applicability of these factors depends on the accuracy, validity and characteristics of the models used. For most LCA studies, no models are needed because existing impact categories, indicators and characterization factors can be selected.

Models reflect the cause–effect chain, also called environmental mechanism or impact pathway, by describing the relationship among the life-cycle inventory results, indicators and, if possible, category endpoints or damage indicators. For each impact category, the following procedure is proposed by ISO:

1. Identification of the category endpoints
2. Definition of the indicator for given category endpoints
3. Identification of appropriate LCI results that can be assigned to the impact category, taking into account the chosen indicator and identified category endpoints
4. Identification of the model and the characterization factors

Figure 3.2 illustrates the relationship among the results of the life-cycle inventory analysis, indicators and category endpoints for one impact category for the example of acidification. It clearly shows where a model is needed. These items are explained in detail in this chapter.

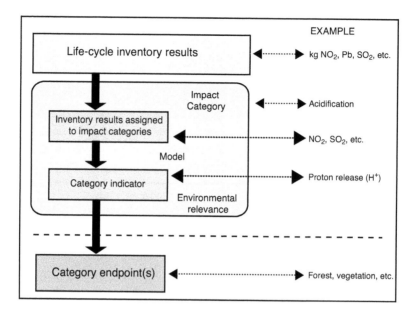

FIGURE 3.2 The concept of indicators. (Based on ISO 14044, *Environmental Management—Life Cycle Assessment—Requirements and Guidelines*, The International Organization for Standardization (ISO), Geneva, Switzerland, 2006.) (available at www.afnor.fr).

3.3 IMPACT CATEGORIES

An impact category is defined as a class representing environmental issues of concern into which life-cycle inventory results may be assigned. As has been previously mentioned, Udo de Haes et al. (1999) have proposed classifying impacts in input- and output-related categories. Input refers to environmental impacts associated with material or energy inputs to the system and output corresponds to damages due to emissions or pollutants, vibrations, or radiation. Table 3.1 gives an overview of input and output impacts currently used in LCIA with a proposal of possible indicators.

Some of the impact categories mentioned in Table 3.1, such as climate change and stratospheric ozone depletion, have a global effect; others, such as photo-oxidant formation or acidification, have

TABLE 3.1
Impact Categories and Possible Indicators

Impact Categories	Possible Indicator
Input-related categories	
Extraction of abiotic resources	Resource depletion rate
Extraction of biotic resources	Replenishment rate
Output-related categories	
Climate change	kg CO_2 as equivalence unit for GWP
Stratospheric ozone depletion	kg CFC-11 as equivalence unit for ODP
Human toxicity	HTP
Eco-toxicity	Aquatic eco-toxicity potential (AETP)
Photo-oxidant formation	kg ethene as equivalence unit for photochemical ozone creation potential (POCP)
Acidification	Release of H^+ as equivalence unit for AP
Nutrification	Stoichiometric sum of macronutrients as equivalence unit for the nutrification potential (NP)

Source: Udo de Haes, H.A. et al., *Int. J. LCA*, 4, 66–74, 167–174, 1999. With permission.

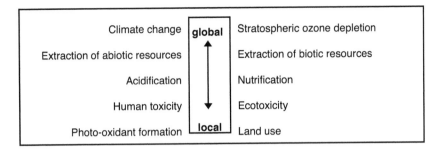

FIGURE 3.3 The need for spatial differentiation in different impact categories. (From UNEP DTIE, *Evaluation of Environmental Impact in Life-Cycle Assessment*, UNEP publication, Paris, France, 2003.)

a local effect. This highlights the need for spatial differentiation in the fate and exposure analysis in different impact categories. Figure 3.3 shows the global–local impacts for the different impact categories, each of which is described next.

3.3.1 EXTRACTION OF BIOTIC AND ABIOTIC RESOURCES

This impact category includes the extraction of different types of nonliving material from the natural environment. It is possible to distinguish three different subcategories: extraction of (1) deposits (e.g., fossil fuels and mineral ores), (2) funds (e.g., groundwater, sand and clay), and (3) flow resources (e.g., solar energy, wind and surface water). Examples of used indicator categories are: rareness of resources, energy content of resources, mineral concentrations, degree of use of flow resources in relationship to the size of the flow, total material requirement, and indicators related to other categories, such as energy requirement or land use. Extraction of biotic resources is mainly related to the extraction of specific types of biomass from the natural environment. The rareness and regeneration rate of the resources is generally used as indicator (Udo de Haes et al, 1999).

3.3.2 CLIMATE CHANGE: GLOBAL WARMING POTENTIAL

Most of the radiant energy received by the Earth as short-wave radiation is reflected directly, re-emitted from the atmosphere, or absorbed by the Earth's surface as longer infrared wave radiation (IR). This natural greenhouse effect is increased by manmade emissions of substances or particles that can influence the Earth's radiation balance, thus raising the planet's temperature.

Many of the substances emitted to the atmosphere as a result of human activities contribute to this manmade greenhouse effect and must be classified in this impact category. Listed in order of importance, they are (Hauschild and Wenzel, 1998):

- CO_2 (carbon dioxide)
- CH_4 (methane)
- N_2O (nitrous oxide or "laughing gas")
- Halocarbons (hydrocarbons containing chlorine, fluorine or bromine)

The potential contribution to global warming is computed with the aid of a procedure that expresses the characteristics of a substance relative to those of the other gases. The Intergovernmental Panel of Climate Change (IPCC) has developed a characterization factor system that can weight the various substances according to their efficiencies as greenhouse gases (Huijbregts et al., 2016). This system can be used in political efforts to optimize initiatives to counter manmade global warming.

TABLE 3.2
GWP for Some Substances Depending on the Time Horizon

Substance	Formula	Lifetime Years	GWP (kg CO_2 eq./kg substance)	
			20 Years	100 Years
Carbon dioxide	CO_2	No single lifetime can be given	1	1
Methane (biogenic)	CH_4	12.4	84	34
Methane (fossil)	CH_4	12.4	85	36
Nitrous oxide	N_2O	121	264	298

Source: Huijbregts, M.A.J. et al., ReCiPe2016: A harmonized life cycle impact assessment method at mid-point and endpoint level. RIVM Report 2016-0104. Bilthoven, the Netherlands, 2016; IPCC., Climate change 2013: The physical science basis, in Stocker, T.F. et al. (Eds.), *Contribution of Working Group I to the Fifth Assessment Report of the Intergovernmental Panel on Climate Change.* Cambridge University Press, Cambridge, UK, p. 1535, 2013.

The system classifies these substances according to their global warming potential (GWP), which is calculated as the anticipated contribution to warming over a chosen time period from a given emission of the substance, divided by the contribution to warming from emission of a corresponding quantity of carbon dioxide (CO_2). Multiplying a known emission of greenhouse gas by the relevant GWP yields the magnitude of the CO_2 emission that, under the chosen conditions, will result in the same contribution to global warming: the emission of the greenhouse gas expressed as CO_2 equivalents.

CO_2 was chosen as a reference substance by the IPCC because it makes the most significant contribution to the manmade greenhouse effect. The expected contribution in terms of warming from a greenhouse gas is calculated based on knowledge of its specific IR absorption capacity and its expected lifetime in the atmosphere. The GWP is internationally accepted and well documented, and provides characterization factors for the substances encountered in an LCA. Table 3.2 presents examples of GWP values for direct contribution of three substances mentioned previously: CO_2, CH_4, and N_2O.

3.3.3 STRATOSPHERIC OZONE DEPLETION

Human activities have caused an increase of substances as different chloride and bromide-containing halocarbons, especially CFCs, tetrachloromethane, 1,1,1-trichloroethane, HCFCs, halons and methyl bromide, involved in the breakdown of ozone in the stratosphere. A common characteristic of these compounds is that they are chemically stable because they can survive long enough to reach the stratosphere, where they can release their content of chlorine and bromide under the influence of UV radiation (Solomon and Albritton, 1992).

The Earth's atmosphere receives ultraviolet radiation from the sun, but not at full intensity. Ozone molecules in the atmosphere absorb large quantities of UV radiation. The reduction of the ozone layer supposes that more UV radiation reaches the surface of the earth and causes damage, especially to plants, animals and humans.

Table 3.3 presents a list of factors to calculate the stratospheric ozone depletion potential of different chemical substances expressed in kilograms of CFC-11 (Freon 11) equivalent as a reference.

TABLE 3.3
Characterization Values for Ozone Layer Depletion

Substance	ODP100 (kg CFC11-eq/kg Substance)
CFC-11	1
CFC-12	0.587
CFC-113	0.664
CFC-114	0.27
CFC-115	0.061
Halon-1301	14.066
Halon-1211	8.777
Halon-2402	14.383
CCl4	0.895
H_3CCl_3	0.178
HCFC-22	0.045
HCFC-123	0.011
HCFC-124	0.022
HCFC-141b	0.134
HCFC-142b	0.067
HCFC-225ca	0.022
HCFC-225cb	0.033
CH_3Br	0.734
Halon-1202	1.892
CH_3Cl	0.022
N_2O	0.011

Source: Huijbregts, M.A.J. et al., ReCiPe2016: A harmonized life cycle impact assessment method at midpoint and endpoint level. RIVM Report 2016-0104. Bilthoven, the Netherlands, 2016; WMO, Scientific assessment of ozone depletion: 2010, Global Ozone Research and Monitoring Project-report no. 52. World Meteorological Organization, Geneva, Switzerland, 2011.

3.3.4 HUMAN TOXICITY

Chemical emissions such as heavy metals, persistent organic substances (POPs), volatile organic compounds (VOCs) and others may lead to direct human exposure (inhalation or drinking water) or to indirect exposure (food consumption). Apart from their toxicity, these substances are all persistent as common characteristics (low degradability in the environment and ability to bioaccumulate). In contrast to other impact categories, e.g., global warning and ozone depletion, no common internationally accepted equivalency factors for toxic compounds express the substances' "impact potentials." For the calculation of equivalency factors, considerations about fate and transport, exposure assessment and human toxicity have been considered.

A frequently used indicator for evaluating human health effects of a functional unit is human toxicity potential (HTP) (Guinée et al., 1996; Hertwich et al., 2001). HTP is a site-generic impact potential that is easy to apply; however, it has a limited environmental relevance because it is based on a multimedia environmental fate model that assumes uniformly mixed environmental compartments. In other words, it represents the behavior of chemicals in a uniform world model environment.

Two HTP methods developed by the Center of Environmental Sciences at Leiden University (CML) (Heijungs et al., 1992) and within the Danish Environmental Design of Industrial Products (EDIP) (Hauschild and Wenzel, 1998) will be considered in the case study (MSWI). The HTP of the EDIP method has the unit m^3 and expresses the volume to which the substance emitted must be diluted in order to avoid toxic effects as a consequence of the emission in question in the relevant compartment. The HTP of the CML method is dimensionless. The HTP for every pollutant "p" (HTP_p) is calculated using the human toxicity factor (HTF_p) for every pollutant and the mass of every pollutant (M_p); it is shown in the following equation:

$$HTP_p = HTF_p \cdot M_p \tag{3.1}$$

The HTF_p is expressed in units of m^3/kg in the EDIP method (Hauschild and Wenzel, 1998) and in $-/kg$ for the CML method (Heijungs et al., 1992). The overall HTP for the functional unit is then the sum of all HTP_p as seen in the next equation (expression of overall HTP for the functional unit):

$$HTP = \sum HTP_p \tag{3.2}$$

Table 3.4 shows the HTP for the pollutants considered in the case study (MSWI). It should be mentioned that ozone, nitrate and sulfate are not considered in these HTFs due to the unavailability of the mass of these substances in the life-cycle inventory because they are not directly emitted but formed during dispersion into the atmosphere. Particulate matter with apparent diameter lower than 10 μm (PM_{10}) is also not included because no HTF is available.

3.3.5   ECO-TOXICITY

Eco-toxic substances are those toxic to organisms in a manner that affects the functioning and structure of the ecosystem in which the organism lives and, as result, affects the health of the ecosystems.

TABLE 3.4

Human Toxicity Potential from the CML and EDIP Methods for Different Substances

Pollutant	CML (–/kg)	EDIP (m^3/kg)
As	4700	9.5×10^9
Benzo(a)pyrene	17	5.0×10^{10}
Cd	580	1.1×10^{11}
Ni	0.014	6.7×10^7
NO_x	0.78	2.0×10^6
SO_2	1.2	1.3×10^6

Sources: CML—Heijungs, R. et al., Environmental life-cycle assessment of products—guide and backgrounds, technical report, CML, University of Leiden, the Netherlands, 1992; EDIP—Hauschild, M. and Wenzel, H., *Environmental Assessment of Products—Scientific Background*, Vol. 2, Chapman & Hall, London, UK, 1998.

They are characterized by their persistence (low degradability) in the environment and their ability to bioaccumulate in organisms. Substances such as toxic heavy metals (Cd, Pb, Hg), persistent organic compounds (dioxins and furans, PCDD/Fs, polycyclic aromatic hydrocarbons, PCBs, etc.), and organic substances (PVC, etc.) that are emitted into the environment can accumulate in organisms and cause different types of damage. The target system is not one organism, as in human toxicity, but a variety of organisms (fauna and flora, entire ecosystems). This makes the assessment even more complex.

In contrast to other environmental impact categories in the LCIA, the impacts of these types of substances are not based in one individual mechanism but in a large number (genotoxicity, inhibition of specific enzymes, etc).

3.3.6 PHOTO-OXIDANT FORMATION

Human activities can increase air concentrations of photo-oxidant substances that can affect the health of living organisms and human beings. These substances can arise via photochemical oxidation of volatile organic compounds (VOCs) and carbon monoxide (CO) emitted by human activities in the troposphere.

The photo-oxidants include a large number of unstable substances formed when VOCs react with different oxygen compounds and oxides of nitrogen (NO_x). The most important oxygen compounds are hydroxyl radicals, OH·. Among the most important photo-oxidants are ozone and peroxyacetyl nitrate (PAN). The transformation of VOCs and CO to ozone requires, apart from the reactive forms of oxygen, sunlight and NO_x, which have a catalytic effect. The potential contribution to photochemical ozone formation is described by its maximum incremental reactivity (MIR) in the American literature and by its photochemical ozone creation potential (PCP) in Europe.

3.3.7 ACIDIFICATION

Combustion processes contribute greatly to the air emission of contaminants as NO_x and SO_2. In contact with water, these oxides are converted to acids (nitric acid, HNO_3) and sulfuric acid (H_2SO_4). Once deposited (by dry and wet deposition), these chemicals may lead to exceeding the acid buffer capacity of the soil and water, generating degradation of terrestrial and aquatic ecosystems. The presence of NH_3 (emitted primarily from agricultural soil) increases the potential uptake of SO_2 in drops of water in clouds and rains by the formation of $(NH_4)_2SO_4$, and thus affects the forest in which SO_2 is deposited.

The principal effect of acidification of the environment is the loss of health, especially among conifers in many forests. The acidification of lakes can lead to dead fish. On the other hand, metals, surface coatings and mineral building materials exposed to air conditions are attacked by the air and acid rain, leading to patrimonial and economic loss of historic monuments.

3.3.8 NUTRIFICATION

Emission of salt nutrients by human activities involves a big impact in the environment. The eutrophication process in lakes, watercourses, and open coastal waters is due to excessive quantities of nutrient salts emitted by man, and, consequently, results in increased production of planktonic algae and aquatic plants, which leads to a reduction in the quality of water. The process of decomposition of dead algae consumes important oxygen and causes a loss of water quality. Agriculture has been identified as the most significant source of nitrogen loading. Wastewater treatment plants and fish farming are the most predominant causes of phosphorus emissions.

3.4   AREAS OF PROTECTION

The set of category indicators resulting from the life-cycle inventory configures and defines the environmental diagnosis associated with product manufacture or any other activity. The impact indicators are associated with environmental damages corresponding to areas of protection (AoP) or sectors of the environment to be protected.

In the first report of the Second SETAC Working Group on Life-Cycle Impact Assessment (Udo de Haes et al., 1999), an AoP was defined as a class of category endpoints. In ISO 14042 three of these classes are mentioned: human health, natural environment and natural resources. Another term used is the expressive "safeguard subject" introduced by Steen and Ryding (1992). It is important to note that these two terms convey the same message: they relate to the category endpoints as physical elements, not as societal values. Thus, following this terminology, the human right to life or economic welfare cannot be an AoP or a safeguard subject; neither can respect for nature or cultural values.

However, the concept of AoPs enables a clear link with the societal values that are the basis for the protection of the endpoints concerned. Table 3.5 gives an overview of the AoPs with underlying societal values, as presented by Udo de Haes and Lindeijer (2001). Because the AoPs are the basis for the determination of relevant endpoints, their definition implies value choices. Thus, there is no one scientifically correct way to define a set of AoPs (Udo de Haes et al., 2002a).

Udo de Haes and Lindeijer (2001) propose to differentiate among the sub-AoPs' life support functions, natural resources and biodiversity, and natural landscapes within the AoP natural environment. Life support functions concern the major regulating functions of the natural environment, which enable life on Earth (human and nonhuman). These particularly include the regulation of the Earth's climate, hydrological cycles, soil fertility and the bio–geo–chemical cycles. Like manmade environments (materials, buildings, crops, livestock) and natural resources, the life support functions are of functional value for society. From a value perspective, these are fundamentally of another nature than those of AoPs with intrinsic value to society, particularly those connected with human health, biodiversity and natural landscapes, works of art, monuments and manmade landscapes. An overview of the classification of AoPs according to societal values is presented in Figure 3.4.

TABLE 3.5
Assignment of Societal Values to AoP

Societal Values	Human/Manmade	Natural
Intrinsic values	Human health Manmade environment (landscapes, monuments, works of art)	Natural environment (biodiversity and natural landscapes)
Functional values	Manmade environment (materials, buildings, crops, livestock)	Natural environment (natural resources) Natural environment (life support functions)

Source: Reprinted with permission from Udo de Haes, H.A. and Lindeijer, E., in *Towards Best Available Practice in Life-Cycle Impact Assessment*, Udo de Haes et al., ©2001 SETAC Press, Pensacola, FL.

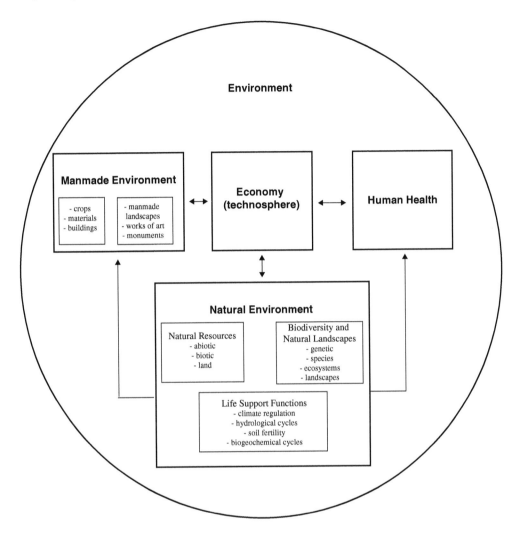

FIGURE 3.4 Classification of AoPs according to societal values. Arrows pointing both ways express interactions between economy and AoPs. Other arrows indicate main relationships between AoPs. (Reprinted with permission from Udo de Haes, H.A. and Lindeijer, E., in *Towards Best Available Practice in Life-Cycle Impact Assessment*, Udo de Haes et al., Eds., ©2001 SETAC Press, Pensacola, FL.)

3.5 MIDPOINT AND ENDPOINT INDICATORS

The terms midpoint and endpoint refer to the level within the environmental mechanism at which the respective effects are characterized. In general, it is assumed that an indicator defined closer to the environmental intervention will result in more certain modeling and that an indicator further away from the environmental intervention will provide environmentally more relevant information (i.e., more directly linked to society's concerns and the areas of protection). Although midpoints and endpoints can be overlapped in some cases, midpoint indicators are used to measure a substance's potency of effect, which, in most cases, is characterized by using a threshold, and does not take into account the severity of the expected impact. Figure 3.5 shows a schematic illustration of the definition of midpoint and endpoint levels (Olsen et al., 2001).

FIGURE 3.5 Schematic illustration of the definition of midpoint and endpoint levels. (Reprinted from *Environ. Impact Assess. Rev.*, 21, Olsen, S.I. et al., Life-cycle impact assessment and risk assessment of chemicals—A methodological comparison, 385–404, ©2001, with permission from Elsevier.)

According to Udo de Haes and Lindeijer (2001), historically, the midpoint approaches have set the scene in LCIA; some prominent examples include the thematic approach (Heijungs et al., 1992), the Sandestin workshop on LCIA (Fava et al., 1993), the Nordic LCA guide (Lindfors et al., 1995), the eco-indicator 95 method (Goedkoop, 1995) and the EDIP model (Wenzel et al., 1997). They also have mostly structured the way of thinking and examples chosen in ISO 14040 (2006).

Since the middle of the 1990s, the endpoint approach has been set on the agenda (Udo de Haes and Lindeijer, 2001). Particularly in LCA studies that require the analysis of tradeoffs between and/or aggregation across impact categories, endpoint-based approaches are gaining popularity. Such methodologies include assessing human health and ecosystem impacts at the endpoint that may occur as a result of climate change or ozone depletion, as well as other categories traditionally addressed using midpoint category indicators. The endpoint approach already has a longer history, particularly in the EPS (environmental priority strategy) approach from Steen and Ryding (Steen and Ryding, 1992; Steen, 1999); however, it has received a strong impetus from the eco-indicator 99 approach (Goedkoop and Spriensma, 1999). In Japan, impact assessment models are currently developed according to this approach (Itsubo and Inaba, 2000), which starts from the main values in society, connected with areas of protection. From these values and connected endpoints the modeling goes back to the emissions and resource consumption (Udo de Haes and Lindeijer, 2001).

Figure 3.6 shows the steps that can be involved if a practitioner wishes to take an LCA study from the inventory stage to valued scores via midpoints and endpoints in the impact assessment. Not all possible environmental loads can be considered in the inventory because data are not available for all of them. Based on the inventory table, two different routes to arrive at valued scores, representing the routes taken when using midpoint and endpoint approaches, are presented

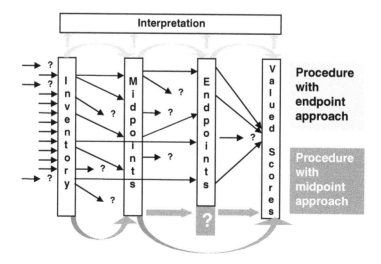

FIGURE 3.6 Some basic differences between the midpoint (lower row of swinging arrows) and the endpoint approach (upper row of swinging arrows). (From Bare, J.C. et al., *Int. J. LCA*, 5, 319–326, 2000. With permission.)

(Bare et al., 2000). On the one hand, the impact categories that can be expressed in the form of midpoints are directly presented as valued scores; on the other hand, as far as possible according to current knowledge, the impacts are expressed in the form of endpoints by relating the midpoints to endpoints or by modeling effects directly from the inventory to the endpoints. Then several endpoints can be aggregated to a valued score if the selected weighting scheme allows it.

At the moment, the availability of reliable data and sufficiently robust models to support endpoint modeling remains quite limited. Uncertainties may be very high beyond well-characterized midpoints. As a result, a misleading sense of accuracy and improvement over the midpoint indicators can be obtained. One of the biggest differences between midpoint and endpoint approaches is the way in which the environmental relevance of category indicators is taken into account. For midpoint approaches, the environmental relevance is presented as a qualitative relationship, while endpoint modeling can facilitate more informed and structured weighting (Bare et al., 2000; UNEP DTIE, 2003).

3.6 WEIGHTING: SINGLE INDEX APPROACHES

3.6.1 INTRODUCTION

Weighting (in ISO terminology) or valuation (in SETAC workgroup terminology) is the phase of LCIA that involves formalized ranking, weighting and, possibly, aggregation of the indicator results into a final score across impact categories. Weighting or valuation inherently uses values and subjectivity to derive, respectively, a rank order and then weighting factors with values supporting the aggregation into a final score. Three types of weighting along similar lines are used:

- Monetary methods, such as mediation costs, willingness to pay, etc.
- Sustainability and target methods, such as in the distance-to-target procedure
- Social and expert methods

The results of an LCIA in the impact categories explained earlier can be difficult to interpret in certain cases because they may be contradictory. In these cases, it would be helpful to have one single score.

The prioritization of impact categories often depends on political targets or business strategies. Weighting is necessary to obtain a single index of environmental performance of a functional unit. However, the weighting across impact categories is the most critical and controversial step in LCIA—that is, a quantitative comparison of the seriousness of the different resource consumption or impact potentials of the product, aimed at covering, and possibly aggregating, indicator results across impact categories.

The weighting methods in LCIA to obtain a single index can be distinguished and classified according to five types of concepts (Udo de Haes, 1996). Table 3.6 presents a description of these concepts, indicating their advantages and disadvantages. In this frame, no simple truth can decide what works best.

Examples for the proxy approach are the sustainable process index (SPI; Sage, 1993) and the material-intensity per-service unit (MIPS) (Schmidt-Bleek, 1994). MIPS is a measure of the environmental impact intensities of infrastructures, goods and services. Materials and fuels are aggregated by mass and energy content. Important cases for the distance-to-target methods are eco-scarcity (Braunschweig et al., 1994), eco-indicator 95 (Goedkoop, 1995) and EDIP (Hauschild and Wenzel, 1998). Eco-scarcity is a Swiss method that has also been adapted by Chalmers University of Technology to suit Swedish conditions. Its units are ECO points per gram of emission or per MJ of energy. Panel approaches have been used, for instance, by the German EPA (Schmitz et al., 1994) and in the eco-indicator 99 weighting step (Goedkoop and Speedesma, 1999). A similar approach, the multi-criteria evaluation (MCE), has been proposed for LCA by Powell and Pidgeon (1994). The abatement technology concept has been used in the method developed by the Tellus Institute (1992). It consists of an evaluation of internal environmental costs by means of the most adequate technology to fulfill the legal requirements. Monetization has been

TABLE 3.6

Comparison of Concepts for Weighting across Impact Categories

Type of Concept	Description	Costs	Advantages	Disadvantages
I Proxy	Selection of one parameter for the representation of the total impact	No	Simple application	Parameter is only a bad approximation of total impact
II Distance to target	Standard or environmental objectives established by the authorities as reference	No	The reference value is accepted if it exists	No accepted reference value for comparison of different impact categories
III Panels	Consideration of the different opinions of experts and/or the general society	No	Achievement of a value that is accepted by a group	Result depends on composition of the panel and/or selected individuals
IV Abatement technology	Efforts to reduce pollution by technological means as reference	Internal	The efforts can be expressed by costs that are known	Internalized costs do not correspond to external costs
V Monetization	Expression of environmental damages in monetary values	External	Attempt to estimate the actual damage costs	External costs can only be estimated

used as a weighting scheme in some of the damage-oriented methods like environmental priority strategies (EPS) (Steen, 1999) and in the uniform world model (UWM; Rabl et al., 1998).

In addition to the weighting scheme used, single index approaches can be differentiated according to whether impact potentials are the basis for the weighting. For instance, this is the case for eco-indicator 95 and EDIP, but not for Tellus and eco-scarcity, in which directly weighting factors are applied.

In the ongoing methodology development of LCIA, panel methods are increasingly important; a tendency also exists to reflect the emission–effect relation more accurately. In turn, proxy indicators "energy" and "mass displacement" (as a measure of energy and resource intensity) and monetization methods based on damage or abatement cost are also acquiring relatively increasing importance.

Next, the eco-indicator 95 will be further explained as a single index method.

3.6.2 Eco-Indicator 95 as Example of a Single Index Approach

Eco-indicators are numbers that express the total environmental load of a product or process. The eco-indicator 95 (Goedkoop, 1995) is one of the weighting methods based on the "distance to target" in the same way as the similarly structured EDIP method (Hauschild and Wenzel, 1998). The steps to achieve a weighting are:

1. Determine the relevant effects caused by a process or product.
2. Determine the extent of the effect; this is the normalization value. Divide the effect by the normalization value. This step determines the contribution of the product to the total effect. This is done because it is not the effect that is relevant, but rather the degree to which the effect contributes to the total problem. An important advantage of the normalization stage is that all the contributions are dimensionless.
3. Multiply the result by the ratio between the current effect and the target value for that effect. The ratio, also termed the reduction factor, may be seen as a measure of the seriousness of the effect.
4. Multiply the effect by a so-called subjective weighting factor to link fatalities, health and ecosystem impairment.

An overview of the principle of eco-indicator 95 is given in Figure 3.7. The problem, of course, lies in determining the weighting factors—the subjective damage assessment phase. The eco-indicator

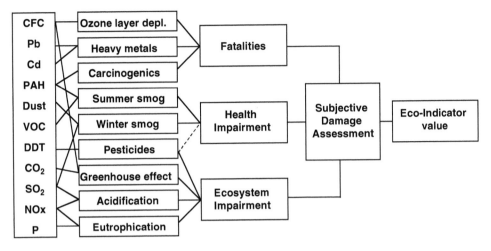

FIGURE 3.7 Overview of the structure of eco-indicator 95. (From Goedkoop, M.J., Eco-indicator 95—final report, NOH report 9523, Pré Consultants, Amersfoort, the Netherlands, 1995. With permission.)

95 uses the so-called distance-to-target principle to determine weighting factors. The underlying premise is that a correlation exists between the seriousness of an effect and the distance between the current and target levels. Thus, if acidification must be reduced by a factor of 10 in order to achieve a sustainable society, and smog by a factor of 5, acidification is regarded as twice as serious. The reduction factor is the weighting factor.

To establish a correlation between these damage levels and the effects, a detailed study of the actual state of the environment in Europe was carried out within the eco-indicator 95 project. The resulting data were used to determine the level of an environmental problem and by which factor the problem must be reduced to reach an acceptable level. Table 3.7 lists the weighting factors and the criteria applied.

TABLE 3.7
Weighting Factors in Eco-Indicator 95

Environmental Effect	Weighting Factor	Criterion
Greenhouse effect	2.5	0.1°C rise every 10 years, 5% ecosystem degradation
Ozone layer depletion	100	Probability of 1 fatality per year per million inhabitants
Acidification	10	5% ecosystem degradation
Eutrophication	5	Rivers and lakes, degradation of an unknown number of aquatic ecosystems (5% degradation)
Summer smog	2.5	Occurrence of smog periods, health complaints, particularly among asthma patients and the elderly, prevention of agricultural damage
Winter smog	5	Occurrence of smog periods, health complaints, particularly among asthma patients and the elderly
Pesticides	25	5% ecosystem degradation
Airborne heavy metals	5	Lead content in children's blood, reduced life expectancy and learning performance in an unknown number of people
Waterborne heavy metals	5	Cadmium content in rivers, ultimately impacts people (see airborne)
Carcinogenic substances	10	Probability of 1 fatality per year per million people

Source: Goedkoop, M.J., Eco-indicator 95—final report, NOH report 9523, Pré Consultants, Amersfoort, the Netherlands, 1995. With permission.

3.7   DAMAGE-ORIENTED METHODS

3.7.1   INTRODUCTION

All damage-oriented methods try to assess the environmental impacts, not in the form of impact potentials, but at the damage level, or, "further down" in the cause–effect chain. In the case of human health effects, for example, this means not as HTP, but as cancer cases. In order to illustrate the theory behind these damage-oriented methods, the eco-indicator 99 methodology (Goedkoop and Spriensma, 1999) and the uniform world model (Rabl et al., 1998) are introduced. Another method based on the same principles has been developed by Steen (1999).

The described approaches use particular weighting methods, especially to evaluate the damage to human health. Eco-indicator 99 applies the cultural theory and disability-adjusted life years (DALY) concept, using estimates of the number of years lived disabled (YLD) and years of life lost (YOLL), while the Uniform World Model (UWM) is based on monetization of environmental damages.

3.7.2   CULTURAL THEORY

Hofstetter (1998) proposes using the sociocultural viability theory (Thompson et al., 1990), called cultural theory, to deal with the problem of modeling subjectivity. Based on this theory, Goedkoop and Spriensma (1999) distinguish five extreme value systems, which are illustrated in Figure 3.8. The most important characteristics of the five extreme archetypes can be summarized in the following way:

1. *Individualists* are free from strong links to group and grid. In this environment, all limits are provisional and subject to negotiation. Although they are relatively free of control by others, they are often engaged in controlling others.
2. *Egalitarians* have a strong link to the group, but a weak link to their grid. No internal role differentiation exists in this environment and relations between group members are often ambiguous; conflicts can occur easily.

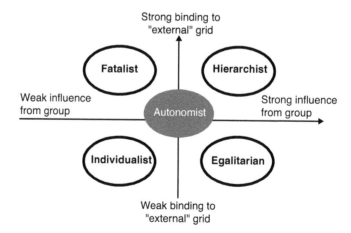

FIGURE 3.8   The grid-group dependency of the five extreme archetypes distinguished in cultural theory. The autonomist has no fixed position in this figure because it does not have social relations and should be seen as floating over the archetypes. (From Goedkoop, M. and Spriensma, R., The eco-indicator-99. A damage-oriented method for life-cycle impact assessment, Pré Consultants, Amersfoort, the Netherlands, 1999. With permission.)

3. *Hierarchists* have a strong link to group and grid. In this environment, people control others and are subject to control by others. This hierarchy creates a high degree of stability in the group.

4. *Fatalists* have a strong link to grid, but not to group. Although these people act as individuals they are usually controlled by others who influence their conception of destiny.

5. *Autonomists* are assumed to be the relatively small group that escapes the manipulative forces of groups and grids.

There is sufficient evidence to assume that the representatives of the first three extreme archetypes have distinctly different preferences as to modeling choices that must be made. Therefore, they are relevant for decision-making (Table 3.8). The last two archetypes cannot be used. The fatalist tends to have no opinion on such preferences because he is guided by what others say, and the autonomist cannot be captured in any way because he thinks independently.

Only the hierarchist, egalitarian and individualist perspectives are relevant for decision-making and can be defined as the default scenarios, which are proposed as extreme cases if no other scenarios based on more specific information are available (Weidema et al., 2002).

The real value of sociocultural viability theory is that a wide range of basic attitudes and assumptions can be predicted for the three remaining extreme archetypes: hierarchist, individualist and egalitarian. (Figure 3.8 specifies some of the many different characteristics per archetype.) Therefore, the eco-indicator 99 methodology uses these three perspectives to facilitate analysis of the relative contribution of the different damage category indicators to one endpoint.

3.7.3 THE DISABILITY-ADJUSTED LIFE YEARS CONCEPT FOR HUMAN HEALTH IMPACT

The disability-adjusted life years (DALY) concept, developed by Murray and Lopez (1996) for the World Health Organization (WHO) and World Bank, aggregates health effects leading to death or illness. Health effects leading to death are described using the years of life lost (YOLL) indicator, which includes all fatal health effects such as cancer or death due to respiratory health effects. Respiratory health effects are further divided into acute and chronic death. Acute death means the immediate occurrence of death due to an overdose of a certain pollutant; chronic death accounts for health effects that lead to a shorter life expectancy. In order to derive the number of life years lost due to a fatal disease, statistics are used, especially from the WHO. These statistics can show at what age and with which probability death occurs due to a certain cancer type or respiratory health effect. Combining these statistics and the dose–response and exposure–response functions (see Chapter 4), it can be calculated how many years of life are lost due to the concentration increase of a certain pollutant.

The DALY concept includes not only the mortality effects, but also morbidity. Morbidity describes those health effects that do not lead to immediate death or to a shorter period of life, but which account for decreased quality of life and for pain and suffering. Cough, asthma or hospitalizations

---

TABLE 3.8
Attitudes Corresponding to the Three Cultural Perspectives Used in the Eco-Indicator 99

Archetype	Time Perspective	Manageability	Required Level of Evidence
Hierarchist	Balance between short and long term	Proper policy can avoid many problems	Inclusion based on consensus
Individualist	Short term	Technology can avoid many problems	Only proven effects
Egalitarian	Very long term	Problems can lead to catastrophe	All possible effects

Source: Goedkoop, M. and Spriensma, R., The eco-indicator-99. *A Damage-oriented Method for Life-cycle Impact Assessment, Pré Consultants*, Amersfoort, the Netherlands, 1999. With permission.

due to different pollutants refer to this indicator. The morbidity health effects are expressed in years lived disabled (YLD). Value choices must be made to weight the pain or suffering during a certain period of time against premature death. Depending on the severity of the illness, suffering and pain, the weighting factor for morbidity is between 0 and 1. A weighting factor of 0.5 means that 1 year of suffering is supposed to be as severe as half a year of premature death. The DALY indicator is, then, the result of the addition of both indicators, with DW the relative disability weight and L representing the duration of the disability:

$$DALY = YOLL + YLD = YOLL + DW \times L \tag{3.3}$$

Often, a pollutant contributes to more than one health effect and a certain health effect can lead to morbidity and mortality. Cancer, for instance, often leads to a period of suffering and pain before death occurs and therefore contributes to YLD and YOLL.

The value of YOLL or YLD does not depend only on the pollutant and the type of disease. Because value choices are necessary for weighting, YLD and YOLL strongly depend on the attitude of the person carrying out the weighting step. Moreover, a year of life lost at the age of 20 and a year of life lost at the age of 60 are not equally appreciated in every socioeconomic perspective, according to the cultural theory. For instance, one cultural theory discounts years of life lost in the same way that discounting is done in finances. Therefore, a year of life lost in the future is worth less than a year of life lost today. Another cultural perspective judges every YOLL to be equally important independently of the age when it occurs (Hofstetter, 1998). If one looks at the units YOLLs, YLDs and DALYs, it can be said that the overall damage for cancer is determined by mortality effects (YOLL), while morbidity effects (YLD) can be neglected. For respiratory health effects, however, morbidity plays an important role.

3.7.4 MONETIZATION OF ENVIRONMENTAL DAMAGES

Lately, acceptance of the approach of valuing health and environmental impacts in monetary units for policy-oriented decision support, which is based on the theory of neoclassical welfare economics, has been growing. In the U.S., cost-benefit analysis (CBA) is mandatory for evaluation of various environmental policy measures. In Europe, use of CBA to justify new equipment regulation has also been increasing. The consideration of health and environmental impacts within a CBA requires quantification of health and environmental impacts as far as possible on the endpoint level to facilitate a subsequent valuation.

If a company or public administration must choose between one technological solution and another, money is a very important parameter. The cost-benefit analysis has been developed to support long-term decisions from a societal point of view, in contrast to a company perspective. In particular, the field of application includes the evaluation of regulatory measures with a huge influence on the environment and the selection of general public environmental strategies. The CBA intends to convert the cost and benefits of regulatory measures, public environmental strategies, etc. to monetary units (Nas, 1996). The basic principle behind this purpose from economic science is to arrange the disequilibria caused by imperfections of the market in the economic optimum between public and private interests. Therefore, it is necessary to quantify the effects of the analyzed plans on society economically. Because these effects can be environmental damages, they refer to effects on the environment. Methods for their monetization allow for estimating external environmental costs or externalities. They are called external because they are not considered in conventional accounting methods (Dasgupta and Pearce, 1972).

The CBA facilitates efficient management of resources for the whole society. When the results indicate that, as a consequence of the project, negative effects to third parts such as atmospheric pollution or generation of dangerous wastes dominate, the public administration intervenes. Some of the interventions the government can undertake to neutralize the negative effects are to establish

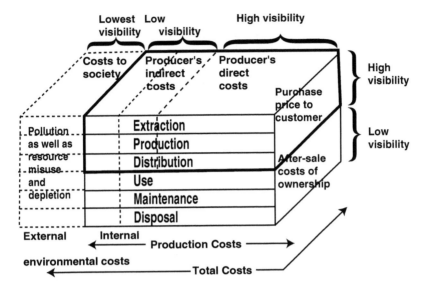

FIGURE 3.9   Types of production and environmental costs and their visibility.

emission thresholds or taxes related to activities that provoke the damages. The CBA methodology consists of four phases (Nas, 1996):

- Identification of the relevant costs and benefits
- Assignment of monetary values to the costs and benefits
- Comparison of the costs and benefits in the form of monetary units generated along the lifetime of the project
- Final decision about the viability of the project and, if appropriate, adoption of necessary interventions by the public administration

Figure 3.9 gives an overview of all the costs generated in the life cycle of a product and its visibility. The total costs are divided into two main types: production and environmental. The costs with a lot of visibility are the direct ones of the producer included in the selling price to the client and generated in the phases from extraction to distribution. These are the conventional costs for raw materials, energy and salaries. The costs in the second half of the life cycle until the disposal are less visible; these are the costs related to ownership after buying the product.

The indirect costs of the producer do not have much visibility; they consist of pollution abatement costs, actions to reduce the accident risk at the working place and other measures not directly necessary to manufacturing the product. The first part of the environmental costs, the internal environmental costs or abatements costs, belongs to the producer's indirect costs in the first half of the life-cycle. They are internal costs from an environmental point of view because the polluter pays them. Moreover, in each phase of the product life cycle, costs to society are generated in the form of pollution as well as misuse and depletion of resources. These costs are the second part of environmental costs—the external environmental costs or externalities—and have very little visibility.

Thus, two types of environmental costs can be distinguished:

- Internal environmental costs or abatement costs are those a company pays to reduce its environmental loads to, at least, under the legal threshold, e.g., the installation and maintenance of gas filters.
- External environmental costs or externalities are emissions and other environmental loads caused to society, e.g., increase of asthma cases; to obtain them the monetization of environmental damage estimates is necessary.

The conversion of environmental damages in external costs is called monetization. With the external environmental costs or externalities at one's disposal, it is possible to internalize these costs and calculate the total cost of a product. Theoretically, this is the price a product needs to be consistent with the market. In a figurative sense, it could be considered the amount that must be paid to maintain the planet in equilibrium, apart from the amount paid to the producer. More practically, it means that, with the monetization, environmental damages can be introduced into the equations of economic balances and that monetization gives support to solving the allocation problems of public funds for the protection of life.

However, lacking a common reference for comparison of different impact endpoints inevitably involves a value judgment. Monetization is just one option; therefore, the following critical points must be mentioned against the monetization of environmental damages:

- On a more fundamental level, there are doubts whether the monetary evaluation of human health and the environment is ethically defendable.
- The assignment of economic values to human health and the environment is not necessarily a guarantee for a sustainable development; they are considered insufficient for the prescription of environmental policies.

It is out of the scope of this book to take the part of a particular point of view. The decision-maker must make the choice. Monetization methods will be briefly presented according to the state of the art; other weighting schemes exist and some of them are presented in this chapter.

In principle, two fundamental concepts exist in the science of environmental economics for the monetization of environmental damages:

- In the direct measurement of damages, the costs are directly quantified in the market, for example, costs of illness (COI).
- In the case of environmental impacts that individuals consider damages but which cannot be measured directly in the market, another perspective is taken. It is considered that the function of the willingness to pay (WTP) for the reduction of the emission is equal to the marginal damage function for the increment of emissions.

With regard to environmental damages, the most important concept is the value of statistical life (VSL). The loss of a statistical life is defined as the increment of the number of deaths expressed as 1/certain number of inhabitants. This corresponds to the probability to die by a factor of $1/n$, where n corresponds to a certain number of inhabitants as a reference group. The focus of scientists evaluating the statistical life can be distinguished between the WTP approach and the human capital concept in which a salary not received essentially assesses the statistical life. When using VSL to evaluate the death of a person due to environmental damages, the age of the person is not taken into account. Therefore, the YOLL principle has been established. It is possible to estimate YOLLs based on VSL if data on the age of the reference group affected by environmental damages are available.

Another important point for the monetization of environmental damages is the discount rate. Discounting is the practice of giving a lower numerical value to benefits in the future than to those in the present. This fact has many consequences when applied to the monetization of environmental damages because these often occur in the near or even far future.

The damage evaluated today (X_0) that will occur in t years is quantified by Equation 3.4, which means, for example, that at a 10% discount rate (r), $100 today is comparable to $121 in 2 years' time. Different stakeholders have broadly discussed the question of the most adequate discount rate.

$$Xt = X_0 \cdot (1 + r)^t \qquad (3.4)$$

The accurate economic evaluation of environmental damages depends not only on the monetization method chosen and the discount rate used, but also on the question of how far economic damage values can be transferred from one place to another once they have been determined. For example, it is difficult to decide which modifications should be done in order to use results of U.S. studies in the EU.

In the case study in this book, monetization is done by the following methods for economic valuation of damages used by the European Commission (1995):

- The direct estimation of damage costs is the most evident evaluation method. Here, the external damage costs that are measurable in the market are taken into consideration, which facilitates the valuation of an important part of the impacts. It allows obtaining an under-borderline of the total environmental cost, although other types of costs exist.
- The WTP method tries to answer the question of how much one is prepared to pay to reduce emissions. This method is considered an adequate measure of preference. For example, with certain decisions, such as buying a car with or without an airbag, individuals give a price to their lives.
- Discounting corresponds to weighting on the level of intergenerational equity, which means that the interests of future generations must be taken into account. However, because in practice it is not possible to measure the values of future generations, the discount rates applied can be understood as the true social discount rate minus the rate of appreciation of the value. This consideration justifies the use of a discount rate below rates observed in capital markets.

With these methods, the types of monetary values obtained are, for instance:

- Mediation costs, i.e., costs of illness
- Productivity loss/company's accounting data, i.e., wage loss
- Economic valuation of a VSL on the basis of:

$$VSL = \frac{\Sigma \ WTP}{\Delta p} \qquad (3.5)$$

where:
   VSL is the value of statistical life
   WTP is the willingness to pay
   Δp is the change in probability of death

The conventional approach for valuing mortality is based on the estimation of the WTP for a change in the risk of death (Δp), allowing calculation of VSL by dividing the WTP by the change in risk. A meta-analysis of valuation studies from Europe and North America undertaken in ExternE suggests a mean VSL of 3.1 Mio Euro at a 3% discount rate, derived from accident studies according to the ExternE project (Mayerhofer et al., 1997).

Most of the valuation studies are based on a context in which the individuals involved are exposed to an accidental risk, leading to a loss of life expectancy of about 30–40 years; thus, the transfer of results to the air pollution context is problematic. Increased mortality from air pollution is mainly expected to affect old people in poor health, leading to a loss of life expectancy between some few days (harvesting effect due to a high pollution episode—acute mortality) and some few years (resulting from long-term exposure to increased levels of air pollution—chronic mortality). An alternative valuation approach that seems to better reflect the context of mortality related to air pollution is to value a change in risk in terms of the willingness to pay for life years and to derive a value of a life year lost (VLYL). Because little empirical evidence on the WTP for LYLs exists, the ExternE study has developed a theoretical framework to calculate the VLYL from the VSL. Assuming, for simplicity, that the value of a life year is independent of age, a relationship between the VSL and the VLYL is established (Krewitt et al., 1999).

Rabl et al. (1998) indicate that based on this assumption, a VLYL corresponds to approximately $100,000. In principle, a discount rate of 3% is applied throughout the case study of this book. Based on the uncertainty analysis in Chapter 5, this book will try to compare the uncertainties due to this valuation step with other sources of uncertainties in environmental impact analysis.

3.7.5 Eco-Indicator 99 as Approach Using Cultural Theory and Disability-Adjusted Life Years

Eco-indicator 95 was based on the distance-to-target approach; however, this method has been criticized because it offers no clear-cut objective way to define sustainable target levels. Thus, the subjectivity of the weighting factors used contributed to the development of a new damage-oriented approach: Eco-Indicator 99 (Goedkoop and Spriensma, 1999).

To calculate the eco-indicator score, three steps are necessary:

1. Inventory of all relevant emissions, resources extraction and land-use in all processes that form the life-cycle of a product, which is the standard procedure in life-cycle assessment as described in Chapter 2
2. Calculation of the damages these flows cause to human health, ecosystem quality and resources
3. Weighting of these three damage categories

To simplify the weighting procedure, damage categories were identified, and as a result, new damage models were developed that link inventory results into three damage categories: damage to (1) human health, (2) ecosystem quality, and (3) resources. A brief description of these three damages follows. Figure 3.10 gives an overview of the eco-indicator 99 method.

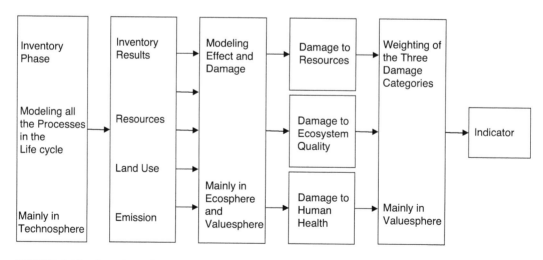

FIGURE 3.10 Overview of the eco-indicator 99 method. The term "sphere" is used to indicate that the method integrates different fields of science and technology. (From Goedkoop, M. and Spriensma, R., The eco-indicator-99. A damage-oriented method for life-cycle impact assessment, Pré Consultants, Amersfoort, the Netherlands, 1999. With permission.)

3.7.5.1 Damage to Human Health

Damage models were developed for respiratory and carcinogenic effects, the effects of climatic change, ozone layer depletion and ionizing radiation. In these models for human health, four steps are used:

- Fate analysis links an emission to a temporary change in concentration.
- Exposure analysis links concentration changes to a dose.
- Effect analysis links the dose to a number of health effects, such as occurrences and types of cancers or respiratory effects.
- Damage analysis links health effects to DALYs for humans, using estimates of the number of YLD and YOLL.

3.7.5.2 Damage to Ecosystem Quality

The entire damage category consists of ecotoxicity and acidification/eutrophication.

Ecotoxicity is expressed as the percentage of all species present in the environment living under toxic stress. The potentially affected fraction (PAF) is used (Van de Meent and Klepper, 1997) as an indicator and corresponds to the fraction of a species exposed to a concentration equal to or higher than the no-observed-effect concentration (NOEC). It is a measure for toxic stress and, in fact, is not a real damage.

Acidification and *eutrophication* are treated as one category. To evaluate the damage to target species in natural areas, the probability of occurrence (POO; Wiertz et al., 1992) is used. The eco-indicator 99 translates this concept to potentially disappeared fraction (PDF) = 1 − POO. Local damage on occupied or transformed areas and regional damage on ecosystems are taken into account. For land use, the PDF is used as an indicator and all species are considered target species. Damages to ecosystem quality are expressed as percentage of species disappeared in a certain area due to environmental load (PDF). The PDF is then multiplied by the area size and the time period to obtain damage. For one specific emission, this procedure is repeated for the concentrations in all relevant environmental receiving compartments separately (water, agricultural soil, industrial soil, natural soil). Finally, the damages in potentially affected fraction (PAF) expressed in m^2yr of the different compartments can be added up, resulting in the total damage (Hamers et al., 1996). Table 3.9 shows

TABLE 3.9

PDF Calculation for Emissions to Air and Resulting Damage in Natural Soil for 1 kg Pollutant Emissions in Europe

Calculation Step	Calculation Procedure	Result
Emission to air in Europe	10,000 kg/d standard flow	1.0×10^{-6} kg/m^2/yr
Concentration increase (ΔC) in natural soil	EUSES	6.96×10^{-7} mg/L
No effect concentration (NOEC terrestrial)	Geometric mean NOECs	1.04 mg/L
Hazard unit (HU) increase	$\Delta HU = \Delta C/NOEC$	6.69×10^{-7}
PAF/HU at Combi-PAF = 24% (European average)	Slope factor = 0.593 · (PAF/ΔH)	
PAF increase in natural soil for 10,000 kg/d in Europe	$\Delta PAF = \Delta HU \cdot 0.593$	4.13×10^{-7}
PAF increase in natural soil for 1 kg/yr in Europe	$\Delta PAF/(10,000 \cdot 365)$	1.130×10^{-13}
PAFm2 yr in natural soil (2.16×10^6 km^2)	$(1.13 \times 10^{-13}) \cdot$ surface area natural soil	0.244 PAFm^2 · yr

Source: Goedkoop, M. and Spriensma, R., The eco-indicator 99. *A Damage-oriented Method for Life-cycle Impact Assessment, Pré Consultants*, Amersfoort, the Netherlands, 1999. With permission.

an example of a calculation procedure given for an emission to air and the resulting damage in natural soil. The damages in PAFm²yr of the different compartments can be added up, resulting in the total damage in Europe.

3.7.5.3   Damage to Resources

With respect to damage category resources, the eco-indicator methodology only models mineral resources and fossil fuels. Chapman and Roberts (1983) developed an assessment procedure for the seriousness of resource depletion based on the energy needed to extract a mineral in relation to the concentration. Until now, no accepted unit to express damages to resources has been found.

For minerals, geostatistical models are used to analyze the relation between availability and quality of minerals and fossil fuels. This step could be described as resource analysis in analogy with the fate analysis. In this case, the "decrease" of a concentration as a result of an extraction is modeled.

For fossil fuels, surplus energy is based on future use of nonconventional resources, especially oil shale and tar sands. In this case, the model for the surplus energy is constructed by means of descriptions of the typical characteristics of the fossil resources and with data on the increased extraction energy for nonconventional resources.

3.7.5.4   Weighting in the Eco-Indicator 99 Method

With respect to weighting step, different schemes for the evaluation of environmental damages have been developed. The most fundamental problem in damage estimations is that the final outcome often refers to value choices and, thus, the weighting scheme of the decision-maker. A single truth simply does not exist as long as value choices are necessary. For example, in the case of the YLD and YOLL (previously seen), because value choices are necessary for weighting, they strongly depend on the attitude of the person carrying out the weighting step. Moreover, a YOLL at the age of 20 and a YOLL at the age of 60 are not equally appreciated in every socioeconomic perspective. Therefore, the DALY concept is linked to the Cultural Theory earlier described. Moreover, a panel approach is used as another weighting scheme.

3.7.5.5   The Panel Approach and Graphical Representation

Because weighting should represent the views of society or groups of stakeholders, the panel approach and the revealed preference were used by eco-indicator 99. The procedure was developed by Mettier (1998), based on previous experiences with panel and Delphi methods in LCA (Udo de Haes, 1996), and consists of a five-part questionnaire to be answered by the panel:

1. An introduction containing a brief description of the purpose, the outline and intended application of the eco-indicator 99 methodology, and a description of the damage categories (human health, ecosystems health and resources)
2. Ranking of the three damage categories (in order of decreasing importance)
3. Assigning weights
4. Linkages to cultural perspectives
5. Background questions (age, sex, etc.)

The results given by the panel can be represented on a triangle graphic (Figure 3.11).

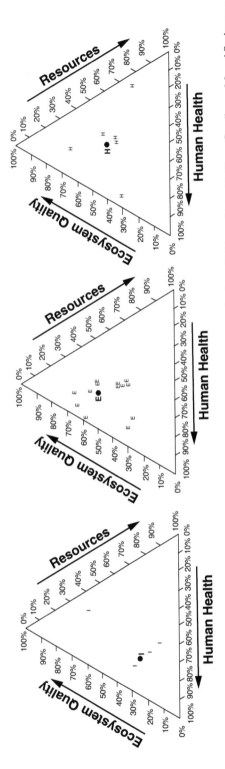

FIGURE 3.11 Triangle graphic for weighting per the three cultural perspectives: individualist = I; egalitarian = E; hierarchist = H. (From Goedkoop, M. and Spriensma, R., The Eco-Indicator 99. A damage-oriented method of life-cycle impact analysis, Pré Consultants, Amersfoort, the Netherlands, 1999. With permission.)

3.7.6 Uniform World Model

Based on the Impact Pathway Analysis (IPA) studies on a European level in the ExternE Project (see in Chapter 4 for more details), Rabl et al. (1998) compared the results of detailed site-specific calculations for more than 50 electric power stations and MSWIs all over Europe and introduced the Uniform World Model (UWM) with the Equations (3.6) and (3.7) for:

1. Primary pollutants

$$D = D_{uni} = \frac{f_{CR} \cdot \rho_{uni}}{\kappa_{uni}} \cdot Q \qquad (3.6)$$

where:

$D = D_{uni}$ = (uniform) damage [cases/a]
f_{CR} = slope of concentration–response function
    (dose–response or exposure–response) [cases/persons*a*μg/m³]
ρ_{uni} = uniform receptor density [1.05E-04persons/m²]
κ_{uni} = uniform removal velocity [m/s]
Q = emission [μg/s]

2. Secondary pollutants

$$D_{2uni} = D_{uni} = \frac{f_{CR2} \cdot \rho_{uni}}{\kappa_{2uni,eff}} \cdot Q_1 \qquad (3.7)$$

where:

D_{2uni} = uniform damage due to secondary pollutant [cases/a]
f_{CR2} = slope of concentration–response function for secondary pollutant
    (dose–response or exposure–response) [cases/persons*a*μg/m³]
$\kappa_{2uni,eff}$ = effective uniform removal velocity for secondary pollutant[m/s]
Q_1 = emission of primary pollutant [μg/s]

The slope of functions states the incremental number of cases (e.g., hospitalizations per concentration increment). Table 3.10 shows typical removal velocity values as obtained in different IPA studies.

Even though the assumption that the removal velocity κ_{uni} is universal may not appear very realistic, especially for near-point sources, Rabl et al. (1998) found that the deviation is surprisingly small.

TABLE 3.10
Typical Removal Velocity Values for Different Pollutants

Primary Pollutants	$\kappa_{2uni,\,eff}$ (m/s)
NO$_2$ → nitrates	0.008
SO$_2$ → sulphates	0.019
Secondary Pollutants	κ_{uni} (m/s)
PM10	0.01
SO$_2$	0.01
CO	0.001
Heavy metals	0.01
PCDD/Fs	0.01

Source: Rabl, A. et al., *Waste Manage. Res.*, 16, 368–388, 1998.

TABLE 3.11

European Health Damage Costs Calculated with the Uniform World Model

Pollutant	Cost (mU.S.$/kg Emitted Pollutant)	Multiplier for Site (Rural ↔ Urban)	Multiplier for Stack Emissions (Height 250 ↔ 0 m, t, v[a])
CO	2.07	?	?
NO_x via Nit	$1.69 \leftrightarrow 10^4$	$\approx 0.7 \leftrightarrow 1.5$	≈ 1.0
SO_2 tot	$1.22 \leftrightarrow 10^4$	$\approx 0.7 \leftrightarrow 1.5$	≈ 1.0
PM 10	$1.36 \leftrightarrow 10^4$	$\approx 0.3 \leftrightarrow 3$	$\approx 0.6 \leftrightarrow 2.0$
As	$1.50 \leftrightarrow 10^5$	$\approx 0.3 \leftrightarrow 3$	$\approx 0.6 \leftrightarrow 2.0$
Cd	$1.83 \leftrightarrow 10^5$	$\approx 0.3 \leftrightarrow 3$	$\approx 0.6 \leftrightarrow 2.0$
Cr	$1.23 \leftrightarrow 10^5$	$\approx 0.3 \leftrightarrow 3$	$\approx 0.6 \leftrightarrow 2.0$
Ni	$2.53 \leftrightarrow 10^3$	$\approx 0.3 \leftrightarrow 3$	$\approx 0.6 \leftrightarrow 2.0$
PCDD/Fs	$1.63 \leftrightarrow 10^{10}$	$\approx 0.3 \leftrightarrow 3$	$\approx 0.6 \leftrightarrow 2.0$

Source: Rabl, A. et al., *Waste Manage. Res.*, 16, 368–388, 1998.

[a] t is the temperature.

m is the 10^{-3}.

v is the velocity.

The reason is that the total damage is dominated by regional damages, which occur sufficiently far away from the source, where the pollutant is well diluted and the difference of the model from real conditions is negligible.

Thus, it is plausible that these results are fairly representative and that the UWM can be a useful tool for a first estimate within an order of magnitude of damage estimates expressed as external costs, monetized according to the guidelines of the European Commission (1995). Table 3.11 presents the results computer with the UWM. The multipliers indicate how much the costs can change with site (rural and urban) and stack conditions (height, temperature and exhaust velocity).

3.8 SOPHISTICATION AND CHOICES IN LIFE-CYCLE IMPACT ASSESSMENT

Sophistication in LCIA has been an important topic for scientific discussion (Bare et al., 1999; UNEP DTIE, 2003). Sophistication is considered to be the ability to provide very accurate and comprehensive reports that reflect the potential impact of the stressors to help decision-making in each particular case. In language more consistent with recent ISO publications, the practitioners of LCA are faced with the task of trying to determine the appropriate level of sophistication in order to provide a sufficiently comprehensive and detailed approach to assist in environmental decision-making. Sophistication has many dimensions and, depending upon the impact category in LCIA, may simulate the fate and exposure, effect, and temporal and spatial dimensions of the impact. It has the ability to reflect the environmental mechanism with scientific validity (Fava et al., 1993, Udo de Haes, 1996; Owens et al., 1997; Udo De Haes et al., 1999).

Traditionally, LCIA uses linear modeling, takes the effects of the substances into account (but not their fate and background concentrations) and aggregates the environmental consequences.

All of this allows for calculating potential impact scores, but not actual damages. Therefore, the appropriate level of sophistication of LCIA involves quite a number of issues. A major point concerns the extension of the characterization modeling to include not only the effects of the substances, but also their fates. Another issue concerns a possible differentiation in space and time. Studies can include impact models that use data at world level and do not specify time periods; in contrast, more recent options involve spatial differentiation of impacts and distinguish between different time periods. A further point concerns the type of modeling. More sophisticated possibilities arise that take background levels of substances into account and make use of nonlinear dose–response functions. An important question here is whether these are real science-based thresholds, or whether these thresholds are always of political origin. A further question relates to the role and practicality of including uncertainty analysis. Although sensitivity analysis is increasingly included in LCA studies, this is not yet the case for uncertainty analysis. Finally, there are the questions of how to apply these different options for sophistication of LCIA, which applications can afford to keep it simple, and for which applications a more detailed analysis is needed (Bare et al., 1999; UNEP DITE, 2003).

The important issue of deciding the appropriate level of sophistication is typically not addressed in LCA. Often, determination of the level of sophistication is based on considerations that may be appropriate for a scientific point of view, but which include practical reasons for limiting sophistication (e.g., the level of funding). A discussion of the most appropriate ways of determining sophistication will include (Bare et al., 1999; UNEP DTIE, 2003):

- Study objective
- Inventory data and availability of accompanying parameters
- Depth of knowledge and comprehension in each impact category
- Quality and availability of modeling data
- Uncertainty and/or sensitivity analysis
- Level of financial resources

A profound comparison of existing LCIA methods was performed by Hauschild et al. (2013) for the establishment of recommended LCIA models for the European context. Taking this work as a starting point, Hauschild et al. (2018) provide a complete and updated qualitative comparison of widely used LCIA methods available in current t LCA software, from which the practitioner can choose.

The ISO 14040/14044 standards by principle do not provide any recommendations about which LCIA method should be used, but some organizations do recommend the use of a specific LCIA method or parts of it. The European Commission has established specific recommendations for midpoint and endpoint impact categories by systematically comparing and evaluating all relevant existing approaches per category, leading to the recommendation of the best available approach (EC-JRC, 2011). This effort resulted in a set of characterization factors, which is directly available in all major LCA software as the ILCD method. Some methods with a stronger national focus are recommended by national governmental bodies for use in their respective country, such as LIME in Japan, or TRACI in the US (Hauschild et al., 2018).

3.9   INTERPRETATION

To conclude the LCIA step, the practitioner must carry out analysis and interpretation of its results in order to evaluate the environmental performance of the product or activity under investigation. The actual assessment of the environmental profile of the product takes place during the evaluation. The nature of the assessment is determined by the goal step of the study. Usually, this will be a comparative assessment. Other examples include providing information about the environmental performance of the product regarding some function, product regulation by government agencies, benchmarking and comparing a product with one or more possible alternatives of its redesign.

The interpretation is an independent step when the goal of the LCA is to find options to improve a product. During the improvement analysis, environmental LCA-based product information is used to make recommendations about the optimization of its manufacturing (including actions of processes or product design) or changes concerning its use by the consumer, e.g., washing at low temperature.

In any case, some priorities need to be established in order to guide the work of the practitioner. In this frame, questions like "What is more important at this moment?" or "What comes first: dealing with the greenhouse effect or with photo-oxidation formation?" or "In terms of LCI, should the first action be to reduce the CO_2 emissions or the COD (chemical oxygen demand) generation?" define the type of evaluation to be carried out during the interpretation step.

The LCIA generates an environmental profile of the product consisting of a certain number of impact potentials that help to compare product alternatives. It depends on the specific case if it will then be possible to draw a conclusion without further weighting. In principle, this is only possible when all of the impact potentials of a product alternative are better than those of the other product (Heijungs et al., 1992).

However, in many cases, one product alternative will present a better environmental performance for some impact potentials but present worse on others. In cases like this, the impact potentials will have to be rated in order to make an assessment. Usually, two methods can be used for this: qualitative multi-criteria analysis and quantitative multi-criteria analysis. As presented by Heijungs et al. (1992), both methods include methodological as well as procedural aspects. The procedural aspects are largely concerned with issues such as who will undertake the evaluation and what information is provided to those concerned.

In the qualitative method, a panel rates the better and poorer impact potentials (see eco-indicator 99 example in Section 3.7.5). The advantage of this method is that all involved parties can express their points of view, furnishing a multidisciplinary perspective to data interpretation. A clear disadvantage is the loss of uniformity inherent to the method: when two different persons assess a set of two environmental profiles, their results can be highly different.

A quantitative multi-criteria assessment is based on weighting factors established by the explicit weighting of the impact potentials.

In this event, an important point about LCA application must be remarked no matter which interpretation method is selected. This methodology is a powerful instrument of support regarding the evaluation of environmental and human health impacts; however, in many cases, its results can be useful without an appreciation of the reliability and validity of the information. In this framework, a quantitative sensitivity analysis must be performed to assess the effect of the key assumptions on the final results, to check the data whose quality is suspected or unknown, to show if the study results are highly dependent on particular sets of inputs, and to evaluate life-cycle effects of changes being considered (Consoli et al., 1993).

Conclusions should only be drawn on study results with consideration of the data variability and resulting variability of the findings. Chapter 5 will discuss this subject more thoroughly and present alternative methods to carry out a qualitative sensitivity analysis.

3.10 EXAMPLE: COMPARISON OF PET AND GLASS FOR MINERAL WATER BOTTLES

This example addresses the question of what is better from an environmental point of view: consuming mineral water in nonreturnable small plastic bottles made of PET or in returnable glass bottles. For the sake of simplicity, as parameter for the comparison, only the greenhouse effect (GE) is to be considered. The basis of the calculation is 1 L of mineral water consumed in small bottles. Calculations must be made taking into account the life-cycle of the two types of bottles and the environmental load of water and bottle transportation. The following assumptions are considered:

- In the case of the plastic bottles, the impact related to bottle manufacturing is not taken into account. The same holds for transportation because the bottles are manufactured in the bottling plant. Empty glass bottles are delivered from the glass factory in 16-t trucks.
- The bottled water is delivered from the bottling plant to the wholesaler in 16-t trucks and from the wholesaler to the retail trader by van. The glass bottles are returned by van to the wholesaler and from the wholesaler to the bottle manufacturer in 16-t trucks.
- The impact of cleaning the bottles is not considered.

For the calculations the following data must be used:

	PET	Glass
Bottle weight (g)	20	237
Bottle capacity (L)	0.33	0.25
Number of uses	1	20
Distance from bottling plant to wholesaler (km)	50	50
Distance from wholesaler to retail trader (km)	20	20
Distance from bottle manufacturer to bottling plant (km)	—	100

Regarding GWP, the following EL must be considered:

		Production		Van <3.5 t	Transport
EL	Units	PET (1 kg)	Glass (1 kg)	(1 tkm)	16-t Truck (tkm)
CO_2	kg	3.45	9.68×10^{-1}	1.54	3.46×10^{-1}
CH_4	kg	1.17×10^{-2}	2.32×10^{-3}	2.61×10^{-3}	5.34×10^{-4}

tkm equivalent to a mass of 1 t (1000 kg) transported 1 km.

For the calculation of the greenhouse effect, consider the following impact factors:

Compound	Factor
CO_2	1
CH_4	62

Solution:

Weight of bottling material associated with 1 L of water consumption:

PET: 20/0.33/1000 = 0.0606 kg

Glass: 237/0.25/20/1000 = 0.0474 kg

Transport TP-PET (tkm) assigned to 1 L of water consumption in PET bottles
1/transport in 16-t truck
Only one trip from bottling plant to the wholesaler
(TP-PET)$_{16t}$ = ((1000 + 60.61)/10^6)0.50 = **0.053 tkm**
2/transport in **Van <3.5 t**
One trip from wholesaler to the retail trader
(TP-PET)$_{van}$ = ((1000 + 60.61)/10^6)0.20 = **0.0212 tkm**
Transport TP-glass (tkm) assigned to 1 L of water consumption in glass bottles
1/transport in **16-t truck**
1a/one trip of empty bottles from glass manufacturer to bottling plant
(TP-glass)$_{16t,a}$ = (47.4/10^6)0.100 = 0.00474 tkm
1b/one trip of full bottles from bottling plant to the wholesaler
(TP-glass)$_{16t,b}$ = ((1000 + (237/0.25))/10^6)0.50 = 0.0974 tkm
1c/one return trip of empty bottles from wholesaler to the bottling plant
(TP-glass)$_{16t,c}$ = ((237/0.25)/10^6)0.50 = 0.0474 tkm
Total (TP-glass)$_{16t}$ = 0.00474 + 0.0974 + 0.0474 = **0.150 tkm**
2/transport in **Van < 3.5 t**

2a/one trip from wholesaler to the retail trader
(TP-glass)$_{Van,a}$ = ((1000 + (237/0.25))/10^6)*20 = 0.0390 tkm
2b/one return trip of empty bottles from retail trader to the wholesaler
(TP-glass)$_{Van,b}$ = ((237/0.25)/10^6)*20 = 0.0190 tkm
Total (TP-glass)$_{van}$ = 0.0390 + 0.0190 = **0.058 tkm**

The respective amounts (kg) of CO_2 and CH_4 assigned to the consumption of 1 L of mineral water bottled in PET or glass are calculated by multiplying the corresponding mass of bottling material (kg) and transport intensity (tkm) by the EL per unit of material and transport previously given as data:

EL	PET	PET Bottle Transport	Total	Glass	Glass Bottle Transport	Total
CO_2 (kg)	2.09×10^{-1}	5.11×10^{-2}	$\mathbf{2.60 \times 10^{-1}}$	4.59×10^{-2}	1.41×10^{-1}	$\mathbf{1.87 \times 10^{-1}}$
CH_4 (kg)	7.09×10^{-4}	8.37×10^{-5}	$\mathbf{7.93 \times 10^{-4}}$	1.10×10^{-4}	2.31×10^{-4}	$\mathbf{3.41 \times 10^{-4}}$

The contribution in percentage is presented as follows:

EL	PET	PET Bottle Transport	Total	Glass	Glass Bottle Transport	Total
CO_2 (kg)	80.4	19.6	100.0	47.6	52.4	100.0
CH_4 (kg)	89.4	10.6	100.0	57.1	42.9	100.0

The amount of equivalent CO_2 is calculated by multiplying the mass of methane (CH_4) by the corresponding factor and adding it to the mass of CO_2.

CO_2 Equivalent	PET	Glass
CO_2	0.260	0.187
CH_4	0.049	0.021
CO_2 TOTAL kg	0.309	0.208

Conclusion: Regarding the greenhouse effect, it is preferable to consume mineral water in glass bottles, in consideration of the assumptions made and the data provided.

3.11 CASE STUDY: APPLICATION OF LCIA METHODS IN THE MSWI PROCESS CHAIN LCA

The results of the application of the Eco-Indicator 95 method to the MSWI process chain LCA (see Chapter 2) are shown in Table 3.12. Scenario 2, the current operation of the incineration plant after the installation of an advanced gas treatment system, is associated with a higher global warming potential (GWP) and nutrification potential (NP) than Scenario 1, i.e., the former operation, because

TABLE 3.12
Differences in Impact Categories according to Eco-Indicator 95

	GWP	ODP	POCP	NP	AP	Pb equiv.	PAH equiv.	SO$_2$ equiv.	Eco-Ind. 95
Difference^a (%)	−10.3	−18.5	−19.5	−7.5	65.4	64.9	63.0	38.6	59.9

^{a} (Scenario 1 − Scenario 2)/Scenario 1.

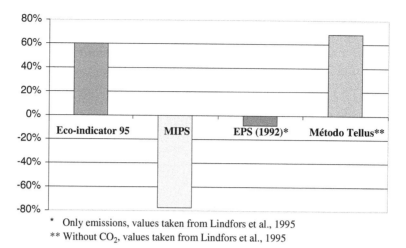

* Only emissions, values taken from Lindfors et al., 1995
** Without CO_2, values taken from Lindfors et al., 1995

FIGURE 3.12   Comparison of different single-index LCIA methods.

it has a higher CO_2 and NO_x emission per produced TJ due to additional energy consumption for the advanced gas treatment system. Scenario 2 also has a higher ozone depletion potential (ODP) and photochemical ozone creation potential (POCP) caused by the higher contribution of the transport.

Scenario 2 is more favorable than Situation 1 in cases of acidification potential (AP) and winter smog (SO_2 equivalent) due to the reduction of HCl, SO_2, and dust. This scenario is also favorable for heavy metals (Pb equivalent) and carcinogenic substances (PAH equivalent) because they are removed by the advanced gas treatment system.

The global environmental evaluation according the Eco-Indicator 95 is positive for the installation of the advanced gas treatment system. The method assigns especially high weightings to impacts reduced by the advanced gas treatment system (mainly acidification and heavy metals). Therefore, it can be concluded that the installation of an advanced gas treatment decreases stack emissions and the related inventory and impact assessment data, but increases the majority of the other environmental loads considered because of higher raw material and energy consumption per produced TJ as well as more transport activity. Nevertheless, the overall environmental efficiency measured according Eco-Indicator 95 clearly improves.

Furthermore, the results obtained with Eco-Indicator 95 methods are comparable with results obtained from other LCIA methods: MIPS (Schmidt-Bleek, 1994), EPS (Steen and Ryding, 1992) and the method of the Tellus Institute (1992). Figure 3.12 shows overall results for the difference between Scenario 1 without and Scenario 2 with an advanced gas system.

It can be observed that selected methods corresponding to different weighting approaches do not deliver results with the same tendency. Two methods show an improvement and two others show a worsening of the environmental performance of the process chain under study. This observation questions the validity of an approach based only on a value to measure the environmental performance. However, above all, this kind of comparison is difficult because the number and types of environmental loads considered in each method vary significantly.

According to EU environmental policy, the installation of advanced gas treatment systems is obligatory in order to reduce the emission of gases such as SO_2 and HCl, as well as particulate matter, PCDD/Fs and heavy metals. The results of the Eco-Indicator 95 and the Tellus methods are in agreement with this policy, whereas the MIPS and EPS indicate the contrary. Using MIPS, such a result is found because more raw materials are necessary for the emission reduction technologies. In the EPS results obtained, more than 96% of the total is caused by CO_2. In the same way as explained for the Eco-Indicator 95 results, the contribution of the heavy metals decreases, while the values for NO_x and CO_2 increase because they are not eliminated and the overall energy efficiency declines.

3.12   CASE STUDY: APPLICATION OF LCIA METHODS TO THE LCA ON ENERGY PRODUCTION USING OPV TECHNOLOGY COMPARED WITH CONVENTIONAL TECHNOLOGY

An LCIA was applied to the OPV case-study introduced in Chapter 2. Impacts (Table 3.13) were calculated using ReCiPe v1.0.5 midpoint (H) impact categories (Goedkoop et al., 2009). Cumulative energy demand (CED) was calculated according to the approach outlined in the Ecoinvent methods (Hichier et al., 2010). The LCIA was conducted in OpenLCA v1.4, an open source life-cycle software.

The results of the LCIA for each impact category are compiled in Table 3.14. Compared with the conventional multi-crystalline silicon PV panels (m-Si), the environmental and human health impacts resulting from the cradle-to-gate production of 1 Wp were, on average, 92% lower for the OPV panels.

Figure 3.13 shows the relative contribution to the OPV impacts from each component or process associated with the functional unit (1 Wp). One benefit of conducting an LCA on an emerging technology is that it is possible to quickly identify which of these components or processes contributes the greatest share of the environmental impact. This low-hanging fruit can then be adjusted or removed from the technology altogether. For instance, results demonstrate that the FTO substrate constitute a significant overall burden across all impact categories. Particularly, this has to do with the energy consumed during the sputtering process. It is therefore possible to identify an alternative

---

TABLE 3.13

Life-Cycle Impact Assessment Categories Considered in This Study. ReCiPe 2008 Midpoint (H) Impact Categories Were Explicitly Used below Except for the Toxicity Impacts Which Were Removed and Replaced with USEtox Impact Indicators. Additionally, the Cumulative Energy Demand Indicator Was Used.

ReCiPe Impact Category	Abbreviation	Unit
Agricultural land occupation	(ALO)	$m^2 \cdot$ yr (agricultural land)
Climate change	(CC)	kg (CO_2 to air)
Cumulative Energy Demand	(CED)	MJ-eq
Fossil fuel depletion	(FD)	kg (oil)
Freshwater ecotoxicity	(FET)	CTUe
Freshwater eutrophication	(FE)	kg (P to freshwater)
Human toxicity	(HT)	CTUh
Ionising radiation	(IR)	kg (U_{235} to air)
Marine eutrophication	(ME)	kg (N to freshwater)
Mineral resource depletion	(MRD)	kg (Fe)
Natural land transformation	(NLT)	m^2 (natural land)
Ozone depletion	(OD)	kg (CFC-115 to air)
Particulate matter formation	(PMF)	kg (PM_{10} to air)
Photochemical oxidant formation	(POF)	kg (NMVOC6 to air)
Terrestrial acidification	(TA)	kg (SO_2 to air)
Urban land occupation	(ULO)	$m^2 \cdot$ yr (urban land)
Water depletion	(WD)	m^3 (water)

TABLE 3.14

Absolute Life-Cycle Impacts for Each Organic Photovoltaic Cell Considered in This Study. Values Are Reported with Their Respective Reference Unit.

Impact Category	Reference Unit	OPV	m-Si
Agricultural land occupation	$m^2 \cdot$ yr (agricultural land)	1.97E−03	5.38E−02
Climate change	kg (CO_2 to air)	1.15E−01	1.09E+00
Cumulative Energy Demand	MJ-eq	2.60E+00	2.18E+01
Fossil fuel depletion	kg (oil)	3.79E−02	3.33E−01
Freshwater ecotoxicity	CTUe	6.02E−01	6.81E+00
Freshwater eutrophication	kg (P to freshwater)	8.29E−05	5.34E−04
Human toxicity	CTUh	3.21E−08	0.388E−07
Ionising radiation	kg (U_{235} to air)	7.01E−02	2.96E−01
Marine eutrophication	kg (N to freshwater)	2.95E−05	4.92E−04
Mineral resource depletion	kg (Fe)	2.94E−02	8.99E−02
Natural land transformation	m^2 (natural land)	1.70E−05	2.27E−04
Ozone depletion	kg (CFC-115 to air)	5.71E−09	8.96E−08
Particulate matter formation	kg (PM_{10} to air)	1.48E−04	1.38E−03
Photochemical oxidant formation	kg (NMVOC6 to air)	2.56E−04	4.35E−03
Terrestrial acidification	kg (SO_2 to air)	4.43E−04	3.90E−03
Urban land occupation	$m^2 \cdot$ yr (urban land)	4.12E−04	5.99E−03
Water depletion	m^3 (water)	7.18E−01	2.93E+01

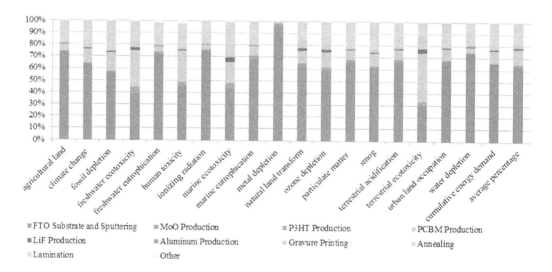

FIGURE 3.13 Contribution from each life-cycle stage and cradle-to-gate process as a percent contribution to the overall environmental and human health impacts.

deposition process that might be more preferable from an environmental point of view. This was done by substituting the energy requirements for sputtering with that of an inkjet deposition process. Samad et al. (2011) demonstrate the application of fluorine-doped tin oxide solutions onto glass substrates using inkjet printing. As a prospective analysis, the approach presented by Samad et al. was assumed to be compatible with printing onto a flexible PET substrate (Huang et al., 2010) and R2R compatible. Electricity usage for inkjet printing is estimated as 21.3 kJ for a 1 m² substrate (Roes et al., 2009).

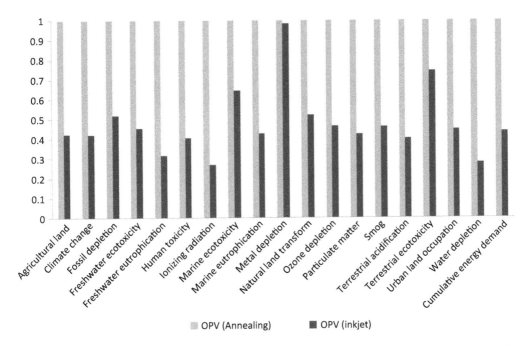

FIGURE 3.14 Comparison of the environmental and human health impacts for an OPV panel that employs sputtering to deposit the transparent electrode (default) and one that employs inkjet printing (FTOinkjet).

The change in environmental and human health impacts from this modification is shown in Figure 3.14. One salient change is in the cumulative energy demand (CED) which decreased 56% in the inkjet substitution production route. This was anticipated given that the energy consumption for inkjet printing was nearly 3-fold lower than for sputtering. Apart from the CED, there were impact reductions ranging between a minimal 2% for metal depletion to a maximum 72% for ionizing radiation. Such results reinforce the conclusions of Lizin et al. who expressed the need for LCAs to not depend on evaluations of the CED alone (Lizin et al., 2013), since other environmental and human health impacts will be affected differently.

As was the case while building the LCI in Chapter 2, the application of the LCIA on an emerging technology can present a barrier to carrying out one's LCA. Such barriers present themselves in two ways during the LCIA: (1) a lack of readily available data (e.g., toxicological data), and (2) a lack of readily available models that appropriately convert inventory flows for emerging materials (e.g., ENM) into their corresponding environmental and/or human health impacts. In the current case study there was no consideration for (1) the emissions of ENM (e.g., fullerenes or TiO_2) during the manufacturing process nor did (2) the USEtox models include any toxicological characterization factors for those potential emissions. Correspondingly, it is not known how these aspects contribute to the final evaluation of the OPV. These data and methodological gaps are explored further in Chapters 6 and 7.

3.13 QUESTIONS AND EXERCISES

1. Build a cause-effect diagram associating inventory results with impact categories, category indicators and category endpoints for each case, starting from a life-cycle inventory for: CO_2, SO_2, CH_4, N_2O, NO_2 and Pb.
2. Associate possible midpoint and/or endpoint indicators with the following impact categories: stratospheric ozone depletion; acidification; eco-toxicity; and human toxicity.

3. Explain the main differences between midpoint and endpoint indicators in LCIA. Try to explain these differences by using some examples.

4. Summarize the main advantages and disadvantages of using endpoint approaches in LCA.

5. Estimate the GWP of the air emissions, including direct impacts to the environment (from 1,1,1-TCA) and indirect impacts from fossil fuel-based energy use (CO_2, N_2O), and the percentage distribution for each chemical. Use the following data for air emission (20,000 kg 1,1,1-TCA/h).

Chemical	m_i (kg/h)	GWP
TCA	13	100
CO_2	10	1
N_2O	0.18	310

Discuss the result comparing the effect resulting from the direct impacts and those from indirect impacts due to the fossil fuel–based energy used. Considering the possibility of a renewable energy use (biomass-based fuels), what could be concluded regarding the effects to the global warming of the actual process?

6. Describe the scheme of the eco-indicator 95 structure of LCA for eutrophication and carcinogenics.

7. The application of eco-indicator 95 has been sometimes criticized because of its failure to be an objective assessment tool at the weighting phase level. Explain briefly the main faults of eco-indicator 95 and the advantages introduced by eco-indicator 99.

8. Summarize the weighting methods proposed for use in LCIA and briefly describe their advantages and disadvantages.

9. How many times should the glass bottle be recycled to generate the same amount of equivalent CO_2 as the plastic bottle? Answer: 7 times.

10. Compare plastic sheets to paper from cellulose pulp.
    According to NASA, the ozone's layer hole above Antarctica has been increasing. If, on September 19, 1998, it was measured 27.2×10^6 km² and, 2 years later, on September 3, this surface had increased to 28.3×10^6 km², determine what would have a worse effect on the ozone's layer: using a 5-g paper pulp sheet or a 5-g low density polyethylene (LDPE) sheet. The comparison must be based on the measurement of the ozone depletion potential (ODP) expressed as kilograms of CFC-11 equivalent. The needed environmental loads (ELs), in this case, emissions of chlorfluorinated compounds (CFCs) to the air for paper and polyethylene production and the corresponding impact factors, are given in the following table:

EL	Unit	Paper (EL/kg)	LDPE (EL/kg)
Halon 1301	kg	7.15×10^{-8}	6.85×10^{-7}
CFC-11	kg	2.37×10^{-9}	4.06×10^{-9}
CFC-114	kg	6.24×10^{-8}	1.07×10^{-7}
CFC-12	kg	5.08×10^{-10}	8.73×10^{-10}
CFC-13	kg	3.19×10^{-10}	5.48×10^{-10}
HCFC-22	kg	5.66×10^{-10}	9.60×10^{-10}

Impact factors:

Compound	Unit	Factor
Halon-1301	kg	1.6×10^1
CFC-11	kg	1
CFC-114	kg	8.0×10^{-1}
CFC-12	kg	1
CFC-13	kg	1
HCFC-22	kg	5.5×10^{-2}

11. Compare the use of two different plastics: PP and PVC. A company endeavors to compare the environmental impact due to the industrial use of two polymers: polypropylene (PP) and polyvinyl chloride (PVC) as raw material for manufacturing piping. It considers it necessary to take into account two variables: the raw material and its transport. Two alternatives are supposed:
    a. Buying PP piping from a distributor who receives the product by train from a factory 1000 km away and uses a van to deliver the product to a consumer who is 200 km from the distributor
    b. Buying PVC tubes from a distributor who acquires the product by a 16-t truck from a factory 700 km away and uses a van to deliver the product to a consumer who is 50 km from the distributor

To compare the environmental behavior of each product, the company makes use of three parameters: kilograms of consumed oil, GWP measured as kilograms of equivalent CO_2, and acidification (acid potential, AP) measured as kilograms of equivalent SO_2.

Determine for each alternative the value of these parameters related to environmental impact caused by the use of 1 kg of polymer and determine which of these kinds of plastics would be better used from an environmental point of view. The necessary ELs and impact factor data are given in the following table.

EL	Unit	PP (1 kg)	PVC (1 kg)	Van (tkm)	16-t Truck (tkm)	Train (tkm)
Oil	kg	1.82	1.07	4.71×10^{-1}	$1,06 \times 10^{-1}$	8.73×10^{-3}
CO_2	kg	3.11	3.22	1.54	3.46×10^{-1}	5.41×10^{-2}
SO_2	kg	2.22×10^{-2}	1.92×10^{-2}	3.39×10^{-3}	7.3×10^{-4}	2.15×10^{-4}
NO_2	kg	4.44×10^{-2}	6.70×10^{-3}	8.72×10^{-2}	3.52×10^{-3}	2.89×10^{-4}
CH_4	kg	1.07×10^{-2}	8.05×10^{-3}	2.61×10^{-3}	5.34×10^{-4}	1.06×10^{-4}
HCl	kg	1.65×10^{-4}	3.26×10^{-4}	3.48×10^{-5}	6.05×10^{-6}	5.43×10^{-3}

tkm equivalent to a mass of 1 t (1000 kg) transported 1 km.

Environmental impact factors:

GWP		AP	
Component	Factor	Component	Factor
CO_2	1	SO_x	1
CH_4	62	NO_x	0.7
		HCl	0.88

REFERENCES

Bare, J.C., Hofstetter, P., Pennington, D.W., and Udo de Haes, H.A. (2000), Midpoints versus endpoints—The sacrifices and benefits, *Int. J. LCA*, 5, 319–326.

Bare, J.C., Pennington, D.W., and Udo de Haes, H.A. (1999), Life-cycle impact assessment sophistication— International workshop, *Int. J. LCA*, 4(5), 299–306.

Braunschweig, A., Foerster, R., Hofstetter, P., and Müller-Wenk, R. (1994), *Evaluation und Weiterentwicklung von Bewertungsmethoden fuer Oekobilanzen—Erste Ergebnisse*. IÖW-Diskussionsbeitraege Nr. 19, IÖW-HSG, St. Gallen, Switzerland.

Chapman, P.F. and Roberts, F. (1983), *Metal Resources and Energy, Butterworth's Monographs in Materials*. Butterworth-Heinemann Ltd, Oxford, UK.

Consoli, F., Allen, D., Boustead, I. et al., Eds. (1993), *Guidelines for life-cycle assessment: A 'code of practice'*. From the workshop held at Sesimbra, Portugal, 31 March–3 April 1993, SETAC, Brussels, Belgium/ Pensacola, FL.

Dasgupta, A.K. and Pearce, D.W. (1972), *Cost-Benefit Analysis—Theory and Practice*, Macmillan Press, Cambridge, UK.

EC (European Commission). (1995), *ExternE—Externalities of Energy*, 6 vols., EUR 16520-162525. DG XII Science, Research and Development, Brussels, Belgium.

EC-JRC—European Commission-Joint Research Centre—Institute for Environment and Sustainability. (2011), *International Reference Life Cycle Data System (ILCD): Handbook Recommendations for Life Cycle Impact Assessment in the European Context—Based on Existing Environmental Impact Assessment Models and Factors*, 1st ed., EUR 24571 EN. Publication Office of the European Union, Luxemburg, UK.

Fava, J., Consoli, F., Denison, R., Dickson, K., Mohin, T., and Vigon, B., (1993), *A Conceptual Framework for Life-Cycle Impact Assessment*, SETAC Press, Pensacola, FL.

Frischknecht, R., Bollens, U., Bosshart, S., Ciot, M., Ciseri, L., Doka, G., Hischier, R., Martin A. (ETH Zürich), Dones, R., and Gantner, U. (PSI Villigen). (1996), *Ökoinventare von Energiesystemen—Grundlagen für den ökologischen Vergleich von Energiesystemen und den Einbezug von Energiesystemen in Ökobilanzen für die Schweiz*. 3rd ed., ETH Zürich: Gruppe Energie-Stoffe-Umwelt, PSI Villigen: Sektion Ganzheitliche Systemanalysen, 1996.

Goedkoop, M. and Spriensma, R. (1999), The eco-indicator 99. *A Damage-oriented Method for Life-cycle Impact Assessment, Pré Consultants*, Amersfoort, the Netherlands.

Goedkoop, M., Heijungs, R., Huijbregts, M., Schryver, D., Struijs, J., and Van Zelm, R. (2009), A life cycle impact assessment method which comprises harmonised category indicators at the midpoint and the endpoint level; First edition Report I: Characterisation. Report, Amersfoort, the Netherlands.

Goedkoop, M.J. (1995), Eco-indicator 95—final report, NOH report 9523, Pré Consultants, Amersfoort, the Netherlands.

Guinée, J., Heijungs, R., van Oers, L., Van de Meent, D., Vermeire, T., and Rikken, M. (1996), LCA impact assessment of toxic releases—Generic modeling of fate, exposure, and effect for ecosystems and human beings with data for about 100 chemicals, RIVM report number 1996/21. Bilthoven, the Netherlands.

Hamers, T., Aldenbergc T., and van de Meent, D., (1996), Definition report indicator effects toxic substances (I_{tox}), RIVM report number: 60712001.

Hauschild, M., Rosenbaum, R., and Olsen, S-I. (2018), *Life Cycle Assessment—Theory and Practice*, Springer, Cham, Switzerland.

Hauschild, M. and Wenzel, H. (1998), *Environmental Assessment of Products—Scientific Background*, Vol. 2, Chapman & Hall, London, UK.

Hauschild, M., Goedkoop, M., Guinée, J.B. et al. (2013), Identifying best existing practice for characterization modeling in life cycle impact assessment. *Int. J. Life Cycle Assess.*, 18, 683–697.

Heijungs, R., Guinée, J.B., Huppes, G., Lankreijer, R.M., Udo de Haes, H.A., and Wegener-Sleeswijk, A. (1992), Environmental life-cycle assessment of products—Guide and backgrounds, technical report, CML, University of Leiden, Leiden, the Netherlands.

Hertwich, E.G., Mateles, S.F., Pease, W.S., and McKone, T.E. (2001), Human toxicity potentials for life-cycle assessment and toxic release inventory risk screening, *Environ. Toxicol. Chem.*, 20, 928–939.

Hischier, R., Weidema, B., ALthaus, H.-J., Bauer, C., Doka, G., Dones, R., and Nemecek, T. (2010), *Implementation of Life Cycle Impact Assessment Methods*. Ecoinvent Centre, Dubendorf, Switzerland.

Hofstetter, P. (1998), *Perspectives in Life-Cycle Impact Assessment—a Structured Approach to Combine Models of the Technosphere, Ecosphere and Valuesphere*, Kluwer Academic Publishers, London, UK.

Huang, X., Yu, Z., Huang, S., Li, D., Luo, Y., and Meng, Q. (2010). Preparation of FTO film on PET substrate. *Mater. Lett.*, 64(15), 1701–1703.

Huijbregts, M.A.J., Steinmann, Z.J.N., Elshout, P.M.F., Stam, G., Verones, F., Vieira, M.D.M., Hollander, A., and Van Zelm, R. (2016), ReCiPe2016: A harmonized life cycle impact assessment method at midpoint and endpoint level. RIVM Report 2016-0104. Bilthoven, the Netherlands.

IPCC. (2013), Climate change 2013: The physical science basis. In: Stocker, T.F., Qin, D., Plattner, G.K., Tignor, M., Allen, S.K., Boschung, J., Nauels, A., Xia, Y., Bex, V., and Midgley, P.M. (Eds.), *Contribution of Working Group I to the Fifth Assessment Report of the Intergovernmental Panel on Climate Change.* Cambridge University Press, Cambridge, UK, p. 1535. doi:10.1017/CBO9781107415324.

ISO 14040. (2006), *Environmental Management—Life Cycle Assessment—Principles and Framework.* The International Organization for Standardization (ISO), Geneva, Switzerland.

ISO 14044. (2006), *Environmental Management—Life Cycle Assessment—Requirements and Guidelines.* The International Organization for Standardization (ISO), Geneva, Switzerland.

Itsubo, N. and Inaba, A. (2000), Definition of safeguard subjects for damage oriented methodology in Japan, *Presented at the 4th Eco-Balance Conference*, Tsukuba, Japan.

Krewitt, W., Holland, M., Truckenmüller, A., Heck, T., and Friedrich, R. (1999), Comparing costs and environmental benefits of strategies to combat acidification and ozone in Europe, *Environ. Econ. Policy Stud.*, 2, 249–266.

Lindfors, L.-G., Christiansen, K., Virtanen, Y., Juntilla, V., Janssen, O.J., Rønning, A., Ekvall, T., and Finnveden, G. (1995), Nordic guidelines on life-cycle assessment, *Nord* 1995: 20, Nordic Council of Ministers, Copenhagen, Denmark.

Lizin, S., van Passel, S., De Schepper, E., Maes, W., Lusten, L., Manca, J., and Vanderzande, D. (2013). Life cycle analyses of organic photovoltaics: A review. *Energy Environ. Sci.*, 6, 3136–3149.

Mayerhofer, P., Krewitt, W., and Friedrich, R. (1997), Extension of the accounting framework, Final report. ExternE Core Project. IER, Universität Stuttgart, Stuttgart, Germany.

Mettier, T. (1998), Der Vergleich von Schutzguetern—Ausgewaehlte Resultate einer Panel-Befragung. In: Hofstetter, P., Mettier, T., and Tietje, O. (Eds.), *Ansaetze zum Vergleich von Umweltschaeden, Nachbearbeitung des 9.* Diskussionsforums Oekobilanzen vom 4. ETZ Zuerich, Zuerich, Switzerland.

Murray, C.J.L. and Lopez, A.D. (1996), *The Global Burden of Disease*, vol. 1 of *Global Burden of Disease and Injury Series*, WHO, Harvard School of Public Health, World Bank. Harvard University Press, Boston, MA.

Nas, T.F. (1996), *Cost-Benefit Analysis—Theory and Application*, SAGE Publications, New York.

Olsen, S.I., Christensen, F.M., Hauschild, M., Pedersen, F., Larsen, H.F., and Toerslov, J. (2001), Life-cycle impact assessment and risk assessment of chemicals—A methodological comparison, *Environ. Impact Assess. Rev.*, 21, 385–404.

Owens, J.W., Amey, E., Barnthouse, E. et al. (1997), *Life-Cycle Impact Assessment—State of the Art*, SETAC Press, Pensacola, FL.

Powell, J.C. and Pidgeon, S. (1994), Valuation within LCA: A multicriteria approach. In: SETAC-Europe (Eds.), *Integrating Impact Assessment into LCA*, SETAC publication, Brussels, Belgium.

Rabl, A., Spadaro, J.V., and McGavran, P.D. (1998), Health risks of air pollution from incinerators—A perspective, *Waste Manage. Res.*, 16(4), 368–388.

Roes, A., Alsema, E., Blok, K., and Patel, M. (2009), Ex-ante environmental and economic evaluation of polymer PV. *Prog Photovoltaics*, 17(6), 372–393.

Sage, S. (1993), Industrielle Abfallvermeidung und deren Bewertung am Beispiel der Leiterplattenherstellung. Ph. D. thesis. Institute for Process Engineering, TU Graz, Austria.

Samad, W., Salleh, M.M., Shafiee, A., and Yarmo, M. (2011), Structural, optical and electrical properties of FTO thin films deposited using inkjet printing technique. *Sains Malays.*, 40(3), 251–257.

Schmidt-Bleek, F.B. (1994), *Wieviel Umwelt Braucht der Mensch? MIPS das Mass für ökologisches Wirtschaften*, Birkhauser Verlag, Berlin, Germany.

Schmitz, S., Oels, H.J., and Tiedemann, A. (1994), *Ökobilanz für Getränkeverpackungen*, UBA-Texte 52/95, Umweltbundesamt, Berlin, Germany.

Solomon, S. and Albritton, D.L. (1992), Time-dependent ozone depletion potentials for short-and long-term forecasts, *Nature*, 357, 33–37.

Steen, B. (1999), Systematic approach to environmental priority strategies in product development (EPS), version 2000—Models and data of the default method, Chalmers University of Technology, Göteborg, Sweden.

Steen, B. and Ryding, S.O. (1992), The EPS enviro-accounting method, report for Federation of Swedish Industries, Swedish Environmental Research Institute, Göteborg, Sweden.

Tellus Institute. (1992), The tellus packaging study—Inventory of material and energy use and air and water emissions from the production of packing materials, technical report, Tellus Institute, Boston, MA.

Thompson, M., Ellis, R., and Wildavsky, A. (1990), *Cultural Theory*, Westview Print, Boulder, CO.

Udo de Haes, H., Jolliet, O., Finnveden, G. et al. (2002a), *Towards Best Available Practice in Life-Cycle Impact Assessment*, SETAC-Europe publication, Brussels, Belgium.

Udo de Haes, H.A. and Lindeijer, E. (2001), The conceptual structure of life-cycle impact assessment, final draft for the second working group on impact assessment of SETAC-Europe (WIA-2). In: Udo de Haes et al. (2002b), *Towards Best Available Practice in Life-Cycle Impact Assessment*, SETAC-Europe publication, Brussels, Belgium.

Udo de Haes, H.A. (1996), *Towards a Methodology for Life-Cycle Impact Assessment*, SETAC-Europe publication, Brussels, Belgium.

Udo de Haes, H.A., Joillet, O., Finnveden, G., Hauschild, M., Krewitt, W., and Mueller-Wenk, R. (1999), Best available practice regarding categories and category indicators in life-cycle impact assessment, background document for the Second Working Group on LCIA of SETAC-Europe, *Int. J. LCA*, 4, 66–74, 167–174.

UNEP DTIE. (2003), *Evaluation of Environmental Impact in Life-Cycle Assessment*, UNEP publication, Paris, France.

Van de Meent, D. and Klepper, O. (1997), Mapping the potential affected fraction (PAF) of species as an indicator of generic toxic stress, RIVM report number: 607504001.

Weidema, B.P., Ekvall, T., Pesonen, H.L., Rebitzer, G., and Sonnemann, G.W. (2002), *Scenario Development in LCA. Report of the SETAC-Europe LCA Working Group Scenario Development in LCA*, SETAC Press, Pensacola, FL.

Wenzel, H., Hauschild, M., and Alting, L. (1997), *Environmental Assessment of Products*. Vol. 1, *Methodology, Tools and Case Studies in Product Development*, Chapman & Hall, London, UK.

Wiertz, J., van Dijk, B., and Latour, J.B. (1992), MOVE. Vegetatie-module; de kans op voorkomen van 700 plantensoorten als functie van voch, pH, nutrienten en zout. RIVM report number: 711901006, Bilthoven, the Netherlands.

WMO. (2011), Scientific assessment of ozone depletion: 2010, Global Ozone Research and Monitoring Project-report no.52. World Meteorological Organization, Geneva, Switzerland.

4 Environmental Risk Assessment

Marta Schuhmacher, Montserrat Mari,
Michael Tsang, and Guido Sonnemann

CONTENTS

4.1   INTRODUCTION

This chapter deals with the concept of risk assessment as a fundamental basis of environmental management. The chapter begins with a general introduction of the risk concept and is followed by a detailed explanation of the four steps of the risk assessment process: hazard identification, exposure assessment, dose–response assessment and, finally, risk characterization for human health and ecological exposure. The sources of data for the exposure assessment such as the environmental monitoring, as well as the most updated fate and exposure models are discussed. The dose–response and exposure–response functions for carcinogenic and noncarcinogenic effects are explained giving special attention to the emengent *in silico* and *in vitro* methods, which have recently become more important due to mandates such as the European Union's REACH Directive. In addition, current hot risk assessment issues such as: risk due to mixtures, risk due to all exposure through lifespan also known as "exposome," risk to indoor pollution or to engineered nanomaterials (ENMs) are also addressed. Some examples are introduced and different practical problems related to them are given at the end of the chapter.

Another approach is also explained: the impact pathway analysis (IPA), which is shown as an alternative way of analysis in cases for which risk assessment (RA) results have to be converted into damage estimations. IPA is used to assess the impacts produced by some of the processes of the lifecycle of a process or service, with a higher level of detail than that obtained from a conventional LCA. Because the impact caused by a pollutant emitted at a specific site depends on site-specific parameters such as population density, meteorological data, etc., the tools used by risk assessment, such as fate, transport and exposure assessment, are included in the analysis. In this manner, risk assessment and LCA can be linked.

IPA is presented in this chapter as an application of environmental risk assessment (ERA) but with a huge potential for use within the life-cycle impact assessment (LCIA). The IPA will be thus taken up again in later chapters for different applications, development of methodology and examples.

4.2   RISK ASSESSMENT

The term "risk" has different meanings depending on different contexts. For a layperson, it embodies the concepts of severity and probability of outcome. For example, people do not consider death by asteroid impact very risky, primarily because the likelihood of such an occurrence is perceived to be very small. Similarly, death from an accident or a fall at home is not appreciated as a significant risk because these do not normally connote a lethal injury, and their severity seems to be within an individual's control. Death and injury from attack by strangers is widely feared as a high risk because of the apparent frequency of such occurrences as reported by the news media. Risk implies not only some adverse result, but also uncertainty. Risk changes as information becomes more specific—a golfer has greater risk of death by lightning than the population as a whole, whether this is perceived as likely or not. The risk from an injury at home or being struck by lightning can be calculated because these events actually happen. In contrast, assessment of risk attributable to low levels of environmental contaminants is an uncertainty exercise.

People use the term risk in everyday language to mean "chance of disaster." When used in the process of risk assessment, it has specific definitions; the most commonly accepted is "the combination of the probability, or frequency, of occurrence of a defined hazard and the magnitude of the consequence of the occurrence" (Royal Society, 1992). On the other hand, hazard can be defined as "the potential to cause harm" and also as "a property or situation that in particular circumstances could lead to harm" (Royal Society, 1992).

The risk assessment is applied in a wide range of professions and academic subjects. Engineers "risk assess" bridges to determine the probability and effect of failure of components; social welfare workers "risk assess" their clients to evaluate the likelihood of the recurrence of antisocial behavior. Risk assessment has become a commonly used approach in examining environmental problems. It is used to examine risks of very different natures.

Environmental contamination problems are complex issues with worldwide implications. Risks to human and ecological health as a result of toxic materials or their introduction into the environment are a matter of great interest to modern society. The effective management of environmental contamination problems has therefore become an important environmental aim that will remain a growing social issue for the next years.

The foundations for risk assessment methodologies have traditionally been based on the examination of effects to human health, but much more emphasis is now placed on all types of environmental damage. In comparison to human health risk assessment, which is a relatively new field, risk assessment for ecological effects is very much in its infancy and the field is constantly developing.

Environmental risk assessment (ERA) consists of evaluating the probability that adverse effects on the environment or human health occur or may occur as a consequence of exposure to physical, chemical or biological agents. Evaluation of environmental risk requires knowledge of adverse effects that might be caused by exposure to chemical substances or materials, as well as of the intensity and duration necessary to produce adverse effects on the environment, including the population.

Risk assessment is a tool used to organize, structure and compile scientific information in order to help identify existing hazardous situations, anticipate potential problems, establish priorities and provide a basis for regulatory controls and/or corrective actions. It can also be used to determine and measure the effectiveness of corrective measures or remedial actions. A key underlying principle of risk assessment is that some risks are tolerable—a reasonable and even sensible view, considering the fact that nothing is wholly safe per se. In fact, whereas large amounts of toxic substances may be of major concern, simply detecting a hazardous chemical in the environment should not necessarily be a cause for alarm. The intrinsic knowledge of the physical–chemical properties of pollutants, biodegradability, and potential of bioaccumulation or potential effects of the chemical substances is necessary for the evaluation of environmental risk. Moreover, it is necessary to carry out a detailed evaluation of the emission sources, as well as the fate, transport and distribution in the different media. Due to all this, the analysis of environmental samples in the laboratory and the application of mathematical models are vital (EC 2003).

4.3   FRAMEWORK OF ENVIRONMENTAL RISK ASSESSMENT

ERA is a systematic procedure used to evaluate potential hazards introduced by pollutant emissions in human health and the environment. This risk assessment process entails a sequence of actions outlined as follows:

1. *Hazard identification*: identification of the adverse effect that a substance has an inherent capacity to cause
2. *Exposure assessment*: estimation of the concentrations/doses to which human populations (i.e., workers, consumers and individuals exposed indirectly via the environment) or environmental compartments (aquatic environment, terrestrial environment and air) are or may be exposed
3. *Dose–response assessment*: estimation of the relationship between dose, or level of exposure to a substance, and the incidence and severity of an effect
4. *Risk characterization*: estimation of the incidence and severity of the adverse effects likely to occur in a human population or environmental compartment due to actual or predicted exposure to a substance, i.e., the quantification of that likelihood

Figure 4.1 shows a framework for human and ecological risk assessment. The EU has provided a Technical Guidance Document (TGD) that supports legislation on assessment of risks of chemical substances to human health and the environment (EC 2003). It is based on the Technical Guidance Document in support of the Commission Directive 93/67/EEC on risk assessment for new notified substances and the Commission Regulation (EC) No. 1488/94 on risk assessment for existing substances, published in 1996. This guidance was refined by taking into account the experience gained when using it for risk assessments of about 100 existing substances and hundreds of

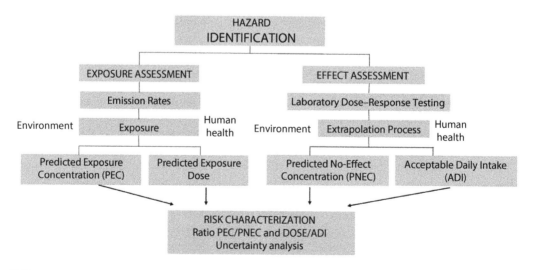

FIGURE 4.1   Framework of environmental and human health risk assessment.

new substances. Furthermore, it has been extended to address some of the needs of the Biocidal Products Directive (Directive 98/8/EC of the European Parliament and of the Council). The U.S. Environmental Protection Agency has produced different risk assessment guidelines: Generic Ecological Assessment Endpoints (GEAE) for Ecological Risk Assessment (EPA/630/P-02/004F), Guidelines for Ecological Risk Assessment (EPA/630/R-95/002F. FR 63(93) 26846–26924), the Guideline for Exposure Assessment (EPA/600Z-92/001. FR 57: 22888–22938) and the Guidelines for Carcinogen Risk Assessment (EPA/630/P-03/001F), among others.

4.4 HAZARD IDENTIFICATION

The first step in a human health and environment risk assessment is to determine whether exposure of humans and ecosystems to chemicals is likely to have any adverse effect.

4.4.1 HUMAN HEALTH

The human health hazard identification involves an evaluation of whether a pollutant can cause an adverse health effect in humans. The process is a qualitative risk assessment that examines the potential for exposure and the nature of the adverse effect expected. The information used in hazard identification includes human, animal and mechanistic evidence; therefore, the risk assessor must evaluate the quality of the evidence, the severity of the effects, and whether the mechanisms of toxicity in animals are relevant to humans. The result is a scientific judgment of whether a particular adverse health effect in humans is caused by a chemical at certain concentrations. This is the work of toxicologists and epidemiologists, who study the nature of the adverse effects caused by toxic agent and the probability of their occurrence.

4.4.2 ECOSYSTEMS

The design of an ecological risk assessment program for an environmental contamination problem typically involves a process to define the common elements of populations and ecosystems that may be affected; this then forms a basis for the development of a logical framework that can be used for risk characterization. First, the development of an ecological risk assessment includes the identification of one or several ecological assessment endpoints—a very important point because the different types of ecosystems have unique combinations of physical, chemical, and biological characteristics and thus may respond to contamination in unique ways. The physical and chemical structure of an ecosystem determines how contaminants affect its resident species and biological interactions may determine where and how the contaminants are distributed in the environment and which species are exposed to particular concentrations. Some examples of populations and ecosystems evaluated in an ERA are:

- Terrestrial ecosystems are classified depending on the vegetation types that dominate the plant community and terrestrial animals.
- Wetlands are areas in which topography and hydrography create a zone of transition between terrestrial and aquatic environments.
- In freshwater ecosystems, the dynamics of water temperature and movement of water can affect the availability and toxicity of contaminants.
- Marine ecosystems are of primary importance because of their vast size and critical ecological functions.
- Estuaries support a multitude of diverse communities and are important breeding grounds for numerous fish, shellfish and bird species.

Assessment endpoints, mentioned in Chapter 3 with regard to life-cycle impact assessment, are explicit expressions of the actual environmental value to be protected. The main criteria used in the selection of assessment endpoints include their ecological relevance, their susceptibility to the stressor, and whether they represent management goals (to include a representation of societal values). Ecological resources are considered susceptible when they are sensitive to human-induced stressors to which they are exposed. Delayed effects and multiple stressor exposures add complexity to evaluations of susceptibility. Conceptual models need to reflect these factors. If a species is unlikely to be exposed to the stressor of concern, it is inappropriate as an assessment endpoint.

To evaluate every species that may be affected by an environmental contamination problem is not feasible. Therefore, the selected target of indicator species will normally be chosen in an ERA study. Then, by using reasonably conservative assumptions in the overall assessment, it is rationalized that adequate protection of selected indicator species will enable protection for all other environmental species as well.

A guiding criterion for the selection of ERA target species considers if they are:

- Threatened, endangered, rare or of special concern
- Valuable for several purposes of interest to human populations (i.e., of economic and societal value)
- Critical to the structure and function of the particular ecosystem that they inhabit
- Indicators of important changes in the ecosystem
- Of relevance for species at the site and its vicinity

4.5   EXPOSURE ASSESSMENT

Exposure assessment is the determination of the concentration/doses to which human populations or environmental compartments are or may be exposed. An exposure assessment is designed to estimate the magnitude of actual and potential receptor exposures to environmental contaminants, as well as the frequency and duration of these exposures, the nature and size of the populations potentially at risk (i.e., the risk group), and the pathways by which the risk group are or may be exposed.

The following steps must be taken in a typical exposure analysis for an environmental contamination problem:

- Determination of the concentrations of the chemicals of concern in each medium to which potential receptors are or may be exposed
- Estimation of the intakes of the chemicals of concern, using the appropriate case-specific exposure parameter values

The exposure assumptions election can be very difficult and is one of the critical elements of an ERA. Efforts have been made to standardize the process of exposure assessment, but the best approach remains to tailor the exposure assessment to the particular characteristics of the study. For instance, risk experts should visit the study area, if possible, and contact relevant agencies and individuals to assemble information regarding the habits and activities of local populations.

In 2005, the exposure assessment approach went a step beyond when the *exposome* concept was defined by Wild (2005) aiming to capture all exposures that potentially affect health and disease. *Exposome* is defined as all environmental exposures from conception throughout the lifespan. Therefore, *exposome* includes the cumulative measure of exposures to both chemical and non-chemical agents such as diet, stress and sociobehavioral factors. The *exposome* includes not only the more traditional measures of exposure (e.g., traditional biomonitoring, environmental monitoring) but also measurements, which rely on different technologies (e.g., "omics") resulting in an expensive set of protocols. In addition, assessing the association between many exposures and health raises statistical challenges since it involves the efficient analysis of a huge amount of data. Therefore, on-going *exposome* projects are trying to overcome those technical and statistical challenges. In any case, these studies will be key from a public health perspective for a better understanding of the environmental risk factors and for improving prevention strategies.

4.5.1 EXPOSURE ASSESSMENT DATA

One of the most important steps in the RA process is the determination of potential exposure. Exposure estimation involves combining environmental measured or predicted concentrations (by means of environmental models) for target chemicals with certain assumptions about the activity patterns of the receptors. However, the availability of representative and reliable monitoring data and/or the amount and detail of the information necessary to derive realistic exposure levels will vary in each case, so assumptions are usually required.

Whenever possible, high-quality and relevant measured exposure data should be used in risk characterization. Measured exposure data may be obtained from industry monitoring programs, particularly for occupational exposure, or other biological or environmental monitoring studies. Recently, different types of sensors have been used to monitor air pollutants, especially in industry, where concentrations are high, with the challenge now being to make them more sensitive to low concentration levels (Kumar et al. 2016). However, when not available, models can be used to estimate chemical concentrations in different compartments of the environment or the human body.

4.5.1.1 Environmental and Biological Monitoring

As a first step, the available data must be assessed with regard to their reliability. The confidence in measured exposure concentrations is determined by the adequacy of techniques, strategies and quality standards applied for sampling analysis and protocol. Second, whether the data are representative must be established. The type, location, duration and frequency of sampling should be evaluated. The selected representative measured data need to be allocated to specific exposure scenarios to allow meaningful exposure assessment.

The types of monitoring can be classified as follows:

- Biological monitoring allows actual measurement of exposure and accurate assessment of likely health outcomes. It involves analyzing human biological samples (i.e., blood, urine, hair, nails, or breast milk) for the presence of target chemicals.
- Environmental monitoring allows actual measurement of exposure and accurate assessment of likely ecological outcomes. It involves analyzing environmental samples (i.e., air, grass, soils, fish or shellfish) for the presence of target chemicals.

Biomonitoring is useful in assessing occupational exposures to airborne chemicals because workplaces typically involve exposure to a single or only a few chemicals at relatively high concentrations (in contrast to typical environmental concentrations) and exposure activity is well known. Although biomonitoring is a useful method, some disadvantages can be found. The main advantages and disadvantages of biomonitoring monitoring are:

Advantages:
1. Defines environmental exposure accurately and precisely
2. Identifies associated health effects in a good way
3. Improves the determination of susceptibility to target pollutants

Disadvantages:
1. Biomarkers integrate all routes and sources of exposure; thus, it is impossible to distinguish whether the exposure is due to the chemicals in air, water, or food.
2. A distinction between variations in the exposed populations, such as health status and individual lifestyle, cannot be made.
3. The timing of sample collection in relation to exposure is critical for the successful measurement of a biomarker.

The best biomarker would be one that is chemical-specific, measured well in trace quantities, measurable in easily sampled biological media or by noninvasive techniques (i.e., urine, hair or nails), and well correlated with a previous exposure. For instance, a good biomarker to assess the municipal solid waste incinerator (MSWI) emissions of our case study would ideally be associated with a chemical unique to the emissions, easily monitored in the stack, and associated only with inhalation exposure. Inorganic tracer chemicals for MSWI emissions include antimony, arsenic, beryllium, cadmium, chromium, lead, mercury, nickel and tin. Organic tracer chemicals include benzo(a)pyrene, polychlorinated biphenyls and dioxins. None of these chemicals is good for biomonitoring because each one exists naturally in the environment, so exposure may occur naturally via air, water, soil, and food. Although the organic compounds are not naturally occurring, they are inadvertently produced as an impurity in the manufacture of many chemicals or as a by-product of many combustion processes (including not only the industrial ones, but also traffic), and are thus considered to be ubiquitous in the environment.

4.5.1.2 Fate and Exposure Models

When monitoring data are not available, fate and transport models can be used to estimate the distribution of chemicals released in the environment. A chemical's final distribution in the media and its respective concentrations are the result of numerous highly complex and interacting processes that are not easy to estimate. Once a chemical has been emitted to a medium (air, water, or soil), it is distributed in the environment. The distribution is not normally restricted to one environmental compartment, but is partitioned among different compartments. Therefore, a pollutant emission may cause impacts in one or more environmental compartments and humans may be exposed through all those impacted environmental compartments. Also, pollutants can enter the food chain and become a risk to human beings through food ingestion. This distribution depends on the specific physical-chemical properties of the pollutant and of the properties and characteristics of the medium to which the emission is released. On the other hand, chemicals can suffer different processes (i.e., degradation, biodegradation, metabolization, transformation, dissociation, hydrolytic process, etc.) in the environment. In this way, fate and transport models can help resolve how a chemical will be distributed in the different media and which transformation it may suffer.

Two different modeling approaches exist: (1) multimedia fate and exposure modeling and (2) specific single-medium models. Integrated multimedia fate and exposure models represent the distribution of a chemical among different compartments and the transfer of chemicals through various exposure routes to a species of interest. For human toxicity, the models calculate a potential dose, which is indicative of the level of impact expected. For ecological toxicity, the models calculate environmental compartment concentrations or potential doses for animals at different levels of the food chain. The examination of the total exposure is a comprehensive evaluation most accurately carried out for micropollutants (organics and heavy metals) by the use of multimedia modeling. For macropollutants (SO_2, NO_x and particles) single-medium models are generally applied.

Table 4.1 consists of selected models that may be applied to some aspects of risk assessment and environmental management problems. The choice of one particular model over another will generally be specific to the problem. Environmental fate models determine the concentration in different compartments (air, surface water, sediments) through the solution of mass balance equations describing the release, transformation, and intercompartmental distribution of a pollutant. Exposure pathway models calculate the exposure of an organism via a stated pathway resulting from a given environmental concentration. These factors take into account transfer factors, uptake rates, such as the rate of inhalation, partitioning and bioconcentration factors, and environmental concentrations.

Numerous model classification systems with different complexities exist in practice; these are broadly categorized as analytical or numerical models, depending on the degree of mathematical sophistication involved in their formulation. Analytical models are models with simplifying underlying assumptions, often sufficient and appropriate for well-defined systems for which extensive data are available and/or for which the limiting assumptions are valid. Whereas analytical models

TABLE 4.1
Environmental Models Applicable to Risk Assessment and Environmental Management Problems

Model	Description/Developer	General Features
	Fate and Transport Environmental Models	
Qwasi Released in 1983. Updated in 2014. It is freely available from the Trent University CEMC web site (www. trentu.ca/cemc).	Quantitative Water Air Sediment Interaction Trent University CEMC	• Fate in water systems model (water, sediment, biota) • Ecotoxicological effects • Organic and inorganic chemicals • Local Scale • Steady and unsteady • No uncertainty estimation
ChemCAN Released in 1996, Version 4.95. Last Version 6.00 released in September 2003 is freely available http:// www.trentu.ca/academic/ aminss/envmodel/models/ CC600.html	Developed by the Canadian Environmental Modelling Centre Trent University	• Fate multimedia model (air, surface water, soil, bottom sediment, groundwater, coastal water and terrestrial plants) • Organic compounds and non-volatile compounds • Regional • Steady state • No uncertainty estimation
CSOIL2000	Developed by Dutch National Institute for Public Health and the Environment (RIVM) to derive intervention values for contaminated soil	• Fate multimedia model (soil, air, crops, drinking water) • Metals, organic and inorganic contaminants. • Local scale • No uncertainty estimation
Simple Box (latest version: Version 4.0, 2012) It is feely available: download the SimpleBox4.0 tool	Developed by Dutch National Institute for Public Health and the Environment (RIVM)	• Environmental fate model (air, water, soil, sediments and vegetation) • Organic chemicals • Ecotoxicological effects • 5 scales: regional, continental, global consisting of: arctic, moderate and topic zones) • Steady state and unsteady • No uncertainty estimation
B4N SimpleBox4nano Released: 2016. It is feely available: download the SimpleBox4.0-nano tool	Developed by Dutch National Institute for Public Health and the Environment (RIVM)	• Environmental fate model (air, soil, water, and sediment) • Nanocolloids • 5 scales: regional, continental, global consisting of arctic, moderate and topic zones) • First order kinetics • No uncertainty estimation
MendNano Released: 2014. http://nanoinfo.org/	Developed by the University of California	• Environmental fate model (air, soil, water, sediment, vegetation canopy and biota) • ENM • Minimum region size 1 km^2 • Dynamic • Emission rates based on LCA approaches (LearNano)

(Continued)

TABLE 4.1 (*Continued*)

Environmental Models Applicable to Risk Assessment and Environmental Management Problems

Model	Description/Developer	General Features
	Fate and Transport and Human Health Risk	
CalTOX Released in 1994. Last update 1997. Download: https://www.dtsc.ca.gov/AssessingRisk/ctox_dwn.cfm	Developed by the California Department of Toxic Substances Control	• Multimedia fate model • Exposure evaluation (inhalation, ingestion and dermal contact) • Human and ecological toxicity • Different human populations • Organic and inorganic compounds • Site-Specific • Dynamic • Monte Carlo uncertainty estimation
EUSES (latest version: Version 2.1.2, 2012) EUSES (2.1) and a manual to the program can freely be downloaded from the internet: (https://ec.europa.eu/jrc/en/scientific-tool/european-union-system-evaluation-substances).	European Union System for the Evaluation of Substances by the European Chemical Bureau (UCB): Commissioned by the EU to the Dutch National Institute for Public Health and the Environment (RIVM)	• Fate multimedia (air, soil, surface water, sediment, biota) and transformation model • The environmental fate calculations are based on the fate models SimpleBox • Human and ecological toxicity • Organic and inorganic substances • Local, regional, continental scale • Uncertainty estimation
XtraFOOD model Released: 2006	Developed by the Flemish Institute for Technological Research (VITO)	• Fate multimedia model focused on the transfer of pollutants to the primary terrestrial food chain • Human exposure and risk to contaminated food products • Different human populations • Organic compounds and heavy metals • Steady state • Uncertainty estimation
ConsExpo (Latest version 4.1. At the end of 2016 a renewed program will be launched with new applications such as exposure assessments for nomaterials.) Download: http://www.rivm.nl/en/Documents_and_publications/Scientific/Models/Download_page_for_ConsExpo_software	Developed by Dutch National Institute for Public Health and the Environment (RIVM) ConsExpo is a set of coherent, general models that enables the estimation and assessment of exposure to substances from consumer products that are used indoors and their uptake by humans	• Exposure to indoor consumer products • Broad set of consumer products • Uncertainty estimation
4 FUN MERLIN-Expo tool (Last version 2016) Download: http://merlin-expo.eu/	Developed within the 4-FUN project (EU 7th Framework Programme)	• Fate multimedia integrated to PBPK model • Human toxicity

(Continued)

TABLE 4.1 (*Continued*)
Environmental Models Applicable to Risk Assessment and Environmental Management Problems

Model	Description/Developer	General Features
		• Different human populations • Organic and inorganic chemicals • Steady state and time-varying simulations uncertainty estimation
	Fate and Transport, Human Health Risk and LCI Models	
USEtox (latest version: Version 2.01, February 2016) Download: http://www.usetox.org/model/download	Endorsed by the UNEP/SETAC Life Cycle Initiative	• Multimedia fate and LCI model • Human and ecotoxicological impacts of chemicals • Characterization factors for human toxicity and freshwater ecotoxicity • Metals and organics • Continental and global scales
Globox Latest version 1.0; September 14, 2010 Download: http://www.universiteitleiden.nl/en/research/research-output/science/cml-globox	Universiteit Leiden	• Multimedia fate and exposure model and spatially differentiated LCA characterization factors • Human and environmental risk assessment • LCA characterization factors on a global scale • Global scale (different countries/territories and seas/oceans are considered with its own set of homogeneous compartments • Multimedia transport and degradation calculations are largely based on the European Union model EUSES version 2.0

may be enough for some situations, numerical models (with more stringent underlying assumptions) may be required for more complex configurations and systems.

The development of the ERA methodology was an important step forward with the inclusion of such models for pollutant fate calculations. Figure 4.2 shows a general overview of the exposure assessment by multimedia modeling for human health. This figure shows the influence of human activities on the environment and illustrates the connections of the different compartments as well as the impact pathways within the total exposure. This figure describes the main goal of multimedia modeling: to evaluate the fate of a pollutant in the environment and to calculate the total exposure due to the different sources.

In our case study, starting with air emissions from the MSWI plant, the transportation and distribution of the pollutants may go over the air, deposit into soil or surface water from where they can leach into ground water or reach the other bordering compartments. Thus, humans may be directly exposed to pollutants through all those environmental compartments. In addition, pollutants can enter the food chain of human beings and animals. With the consumption of agricultural products, i.e., plants or animals, humans may be also exposed indirectly to pollutants.

The procedure of deriving an exposure level by applying model calculations must be transparent and the input data or default values used for the calculations should be documented. Nowadays,

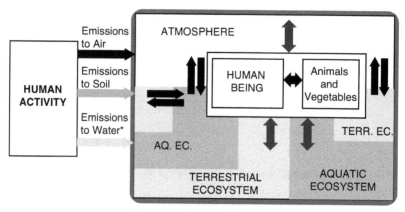

*Emissions to superficial and underground water

FIGURE 4.2 Cause–effect chain for ecosystem and human health as basis for exposure assessment by multimedia modeling.

a large number of different models are available to describe an exposure situation, so the choice of the most appropriate model for the specific substances and scenario should be made and explained. The choice of which model will be used for specific applications depends on numerous factors; choosing a more complex model over a simple one will not necessarily ensure a better solution in all situations. In fact, because a model is a mathematical representation of a complex system, usually some degree of mathematical simplification must be made about the system being modeled. Due to the complexity of natural systems, it is usually not possible to obtain all the input parameters, so data limitations must be weighted appropriately when choosing a model. Here, ERA is confronted with the same problem as that of the LCA described previously, because both are methods based on system analysis. Ultimately, the type of model selected will be dependent on the overall goal for the assessment, complexity of the problem, type of contaminants of concern and nature of the impacted and threatened media, as well as the type of corrective actions considered in the investigation.

4.5.2 Human Health Exposure Assessment

After estimation of the increasing concentrations of pollutants in air, water, soil, and food in the affected regions (by means of monitoring and/or multicompartmental or single modeling), an exposure assessment must be carried out. The exposure assessment phase of human health risk assessment is the estimation of the rates at which chemicals are absorbed by potential receptors. Because most potential receptors can be exposed to chemicals from a variety of sources and/or in different environmental media, an evaluation of the relative contributions of each medium and/or source to total pollutant intake could be critical in a multi-pathway exposure analysis. In fact, the accuracy of exposure characterization could be a main determinant for the validity of a risk assessment.

Humans may be exposed to substances in their workplaces (occupational exposure), due to use of consumer products (consumer exposure) and indirectly via the environment and even during fetal or early life, as it is nowadays considered by the *exposome* approach. Different types of individuals and ages may be required to characterize the population at greatest risk. Since the developing fetus is particularly vulnerable to potential environmental hazards and since exposures during the critical in utero period may have a lifetime impact, the pregnancy period is an important starting point in characterizing the life course *exposome*. However, since *exposome* approach is still in development, frequent exposure evaluations only consider adults and children separately because adulthood provides the longest period of exposure and childhood accentuates some exposure routes (such as the incidental ingestion of soil) and potential sensitivities due to higher ratios of intake to body weight.

Another paramount issue in terms of exposure to pollutants is the indoor exposure (see Section 4.9.1 Indoor Air Quality). Commonly, human health risk has been evaluated taking into account only outdoor pollutant concentrations. However, indoor exposure has been gaining importance during the last years since; in industrialized societies, people spend around 80%–90% of their time indoors. Diffusion of outdoor air into buildings contributes to a mixture of indoor and outdoor pollutants and resulting indoor exposure levels, depending on ventilation, air conditioning and on the indoor-outdoor temperature gradient. Indoor environments also have a wide and varied range of primary sources of potentially harmful substances (e.g., environmental tobacco smoke, cooking and heating with natural gas or solid fuels) which are independent of the outdoor environment, but can modify a resident's exposure substantially since they are often within immediate personal space.

In a first step of the exposure assessment, the probability of an exposure of the population to the substances under consideration must be evaluated. Exposure levels and concentrations for each exposed population need to be evaluated based on available measured data and/or modeling. A contaminant can enter the body using any of three pathways: ingestion, inhalation or by contact with the skin (dermal or other exterior surfaces such as eyes). Once in the body, it can be absorbed and distributed to various organs and systems. The toxic may then be stored, for example in fat, as in the case of PCDD/Fs, or it may be eliminated from the body by transformation into other substances. The biotransformation process usually provides metabolites more readily eliminated from the body than the original chemicals; however, metabolism can also convert chemicals into more toxic forms.

On the other hand, environmental exposure to chemicals can be direct (as a result of exposure to the media where the emission directly takes place) or indirect (as a result of exposure to a media in which the pollutants arrive by transport for another media where the emission takes place). Thus, all derived exposure levels should be representative of the exposure situation they describe. The duration and frequency of exposure, routes of exposure, human habits and practices need to be considered. Furthermore, the spatial scale of exposure (e.g., personal, local, regional levels) must be taken into account.

The quantitative process of estimating exposure is straightforward.

With the exception of the inhalation pathway, exposure is normally estimated as the rate of pollutant contact per unit of body weight:

$$\text{Dose} \equiv \frac{\text{Concentration} \cdot \text{Contact Rate} \cdot \text{Frequency}}{\text{Body weight}} \tag{4.1}$$

where *Dose* is the rate of exposure, *Concentration* is the level of pollutant in a particular environmental media, *Contact rate* is the amount (per time) of the media contacted, *Frequency* is a measure of how often (and over what period) exposure occurs, and *Body weight* is the weight of the individual.

In the case of inhalation however, the amount of chemical that reaches a target site is not a simple function of inhalation rate and body weight and, therefore, those parameters are not used to estimate inhalation dose. In fact, the interaction of the inhaled contaminant with respiratory tract is affected by other factors such as species-specific relationships of exposure concentrations (ECs) to deposited/delivered doses and physicochemical characteristics of the inhaled contaminant. Accordingly, the EPA's superfund program updated their approach for determining risk from inhaled chemicals to be consistent with the inhalation dosimetry methodology (EPA 2009).

When detailed information on the activity patterns of a receptor at a site is available, risk assessors can use these data to estimate the EC for either noncarcinogenic or carcinogenic effects. The activity pattern data describes how much time a receptor spends, on average, in different microenvironments (MEs), each of which may have a different contaminant concentration level. By combining data on the contaminant concentration level in each ME and the activity pattern data, the risk assessor can calculate a time-weighted average EC for a receptor. This approach may also be used to address exposures to contaminants in outdoor and indoor environments at sites where both indoor and outdoor samples have been collected (EPA 2009).

$$EC_j = (CA_i \times ET_i \times EF_i) \times ED_j / AT_j \qquad (4.2)$$

where:

EC_j ($\mu g/m^3$) = average exposure concentration for exposure period j
CA_i ($\mu g/m^3$) = contaminant concentration in air in ME i
ET_i (hours/day) = exposure time spent in ME i
EF_i (days/year) = exposure frequency for ME i
ED_j (years) = exposure duration for exposure period j
AT_j (hours) = averaging time = $ED_j \times 24$ hours/day $\times$ 365 days/year

For some exposure routes, the individual term of doses may include multiple parameters. For example, in estimating dermal pollutant intake during swimming, the contact rate is calculated as the product of (1) the surface area of the skin, (2) a chemically specific permeability, and (3) the density of water. Exposure parameters are generally selected as a mix of typical and high-end values to afford an overall conservative bias. Although situation-specific values are always preferable, they are seldom available and often impractical to develop. Default values have been established for many parameters and some conventions have been yielded. For example, an average adult body weight of 70 kg is routinely used in dose calculations. Moreover, exposure profiles are subject to considerable discretion; the difficulty of exposure assessment is to choose a combination of assumptions that satisfies the aim of the assessment and is appropriate for the populations of interest. Implications of parameter variability and uncertainty are difficult to test with deterministic methods; probabilistic techniques such as those described in Chapters 5 and 9 can directly incorporate these aspects.

Risk assessments contain numerous uncertainties that are typically compensated by conservative assumptions designed to bias risk estimates high. Recently, the philosophy has shifted toward the use of less conservatism. Most risk assessments conducted in the late 1980s were centered on extreme situations such as a maximally exposed individual (MEI). An MEI was built to receive (in theory) a level of exposure not likely to be exceeded by any person, a level that would be extremely improbable. More recent guidance, however, has recommended the use of reasonable maximum exposure (RME) scenarios that attempt to work out plausible, high-end exposure estimates. In reality, the difference between MEI and RME scenarios may be one of semantics, because concepts such as plausible, maximum and high-end are too often subjective. Psychologically, however, the shift from MEI to RME implies a movement from the unlikely to the plausible and assigns a greater sense of realism to the risk estimates.

4.5.3 Ecological Exposure Assessment

The objective of the ecological exposure assessment is to estimate the concentration to which an environmental compartment is or may be exposed. For each environmental compartment potentially exposed, the exposure concentrations should be evaluated.

A chemical may be released into the environment and is then subject to physical dispersal into the air, water, soil, or sediment. The chemical may then be transported spatially and into the biota and, perhaps, be chemically or otherwise modified or transformed and degraded by abiotic processes (such as photolysis, hydrolysis, etc.) and/or by microorganisms present in the environment. The resulting transformation may have different environmental behavior patterns and toxicological properties from those of the chemical. Nonetheless, it is the nature of exposure scenarios to determine the potential for any adverse impacts. The amount of a target species' exposure to environmental contamination is based on the maximum plausible exposure concentrations of the chemicals in the affected environmental matrices. The total daily exposure (in mg/kg-day) of target species can be calculated by summing the amounts of constituents ingested and absorbed from all sources (e.g., soil, vegetation, surface water, fish tissue and other target species) as well as those absorbed through inhalation and dermal contacts.

The process for the environmental risk assessment of a substance is based on the comparison of the concentration in the environmental compartment (predicted environment concentration or PEC) with a concentration below which unacceptable effects on organisms will most likely not occur (predicted no effect concentration or PNEC) as shown in Section 4.6.2. Therefore, the aim of exposure assessment for the environment is the evaluation of PEC. It can be derived from available monitoring data and/or model calculations.

Analytical processes used to estimate receptor exposure to chemicals in various contaminated media (such as a wildlife or a game species' daily chemical exposure and the resulting body burden) are similar to those discussed under human health risk assessment.

4.6 DOSE–RESPONSE AND EXPOSURE–RESPONSE FUNCTIONS

Some experts argue that it is unnecessary to determine first if a chemical is hazardous. Their philosophy stems from the first definition of toxicology, given by Paracelsus (1493–1541) over 450 years ago: "All substances are poisons; there is none which is not a poison. The right dose differences a poison and a remedy." In other words, all chemicals have the potential to be hazardous, depending on the dose; therefore, exposure assessment could potentially take the place of hazard identification.

Dose–response and exposure–response evaluation, the third component of the risk assessment process, involves the characterization of the relationship between the dose administered or received and the incidence or severity of an adverse effect in the exposed population or ecosystem. Characterizing the dose–response relationship includes understanding the importance of the intensity of exposure, the concentration vs. time relationship, whether a chemical has a threshold level, and the shape of the dose–response curve. In the determination of a dose response and exposure response, the following aspects need to be considered: the metabolism of a chemical at different doses, its persistence over time, and an estimate of the similarities in disposition of a chemical between humans and animals.

The dose–response and exposure–response functions are based on toxicological dose-oriented and epidemiological exposure-oriented studies. Figure 4.3 proposes to illustrate the difference between the toxicological and the epidemiological approaches. The toxicological approach is based on bioassays (*in vitro* tests) or animal tests that allow determining dose–response functions. In turn, during the last years and in the framework of the Registration, Evaluation, Authorization, and Restriction of Chemicals (REACH) Directive (in line with EC 1907/2006), in the European Union, computational *in silico* models have gained importance. *In silico* models are predictive tools needed to use limited *in vivo* studies and enable *in vitro* data on toxicological effects to be interpreted for comprehensive risk assessment of chemicals. In this way, exposure–response functions, which allow an estimation of human effects depending on the exposure concentration, are calculated. On the other hand, the epidemiological approach uses empirical studies in which correlations are established between exposure situations and observed human effects.

Epidemiological studies focus more on macropollutants responsible for respiratory effects such as SO_2, NO_x and particles, because they usually act together, and it is quite difficult to conduct laboratory assays. On the other hand, bioassays are the foundations to obtain toxicological information for micropollutants, i.e., heavy metals and the huge number of organic compounds like PCDD/Fs or PAHs (polyaromatic hydrocarbons). The dose–response functions permit the determination of risk due to the accumulation of pollutants in the human organism. The risk is a way to foresee the probability of physical impacts. Potential cancer risk factors are determined by both toxicological and epidemiologic approaches and both types of damage functions are important sources of uncertainty, especially due to the extrapolation from high to lower doses.

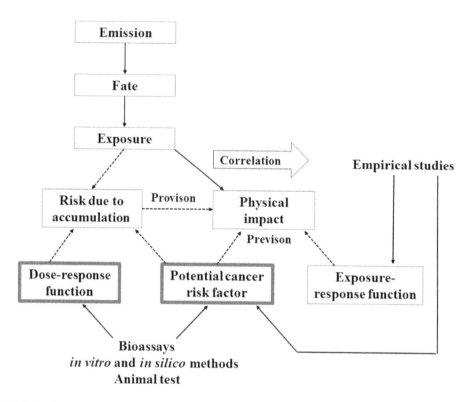

FIGURE 4.3 Comparison of the toxicological approach by bioassays, *in vitro* and *in silico* methods and animal tests and the epidemiological approach using empirical studies to determine damage factors for human health impacts.

4.6.1   HUMAN HEALTH DOSE–RESPONSE AND EXPOSURE RESPONSE

Government agencies are charged with the protection of the public health and environment. To do that, the U.S. Environmental Protection Agency (USEPA), the Agency for Toxic Substances and Disease Registry (ATSDR), and the U.S. Food and Drug Administration (FDA) must have a reference, or comparison value, upon which to base an evaluation of potential health threats posed by any substances or chemicals. The bases, or starting points, for such estimations may have different names or acronyms but they represent more or less the same thing. The risk of carcinogenic pollutants is estimated with a factor of potential cancer or inhalation unit risk; however, the risk of noncarcinogenic pollutants is characterized by a reference dose or reference concentration (inhalation). The values for noncarcinogenic endpoints are called reference doses (RfDs) and inhalation reference concentrations (RfCs) by the USEPA, and oral and inhalation minimal risk levels (MRLs) by ATSDR (Table 4.2).

RfDs, RfCs and MRLs are, very often, construed as rigid threshold limits, above which toxicity is likely to occur. The truth, however, is that these values actually represent levels of a potential toxicant that are highly unlikely to represent any threat to human health over a particular or specified duration of daily exposures. The more frequently these levels are exceeded and the greater the excess, the more likely that some toxic manifestation will occur. Most definitely, these guidance or reference values are not threshold values for the onset of toxicology in any exposed population. Health guidance values must be considered in the context of their intended role as mere screening or trigger values, for which they serve as tools for assisting in the determination of whether further evaluation of a given potential exposure scenario is warranted.

TABLE 4.2

Oral RfDs/MRLs for Noncarcinogenic Effects of Some Different Chemicals

Regulated Substance	RfDs (mg/kg-day)	Oral MRLs (mg/kg-day)	RfC (mg/m³)	Inhalation MRLs
Acetone	0.9	Intermediate 2	3.1E+01	Acute 26 ppm Intermediate 13 ppm Chronic 13 ppm
Cadmium	0.001	Intermediate 0.0005 Chronic 0.0001	1.0E−05	Acute 0.00003 mg/m³ Chronic 0.00001 mg/m³
Chloroform	0.01	Acute 0.3 Intermediate 0.1 Chronic 0.01	9.8E−02	Acute 0.1 ppm Intermediate 0.05 ppm Chronic 0.02 ppm
Tetrachoroethylene	0.006	Acute 0.008 Intermediate 0.008 Chronic 0.008	4.0E−02	Acute 0.006 ppm Intermediate 0.006 ppm Chronic 0.006 ppm
Toluene	0.08	Acute 0.8 Intermediate 0.2	5.0E+00	Acute 2 ppm Chronic 1 ppm
1,1,1-Trichloroethane	2	Intermediate 20	5.0E+00	Acute 2 ppm Intermediate 0.7 ppm
Xylene	0.2	Acute 1 Intermediate 0.4 Chronic 0.2	1.0E−01	Acute 2 ppm Intermediate 0.6 ppm Chronic 0.05 ppm

Source: MRLs from ATSDR (March 2016); RfDs and RfCs from US EPA PRG Tables (May 2016).

Note: Oral RfDs and RfCs and ADIs refer to chronic effects. MRL values refer to acute (1–14 days), intense (15–364 days) or chronic (365 days and longer) effects.

For a general environmental health risk assessment of an unspecified emitted pollutant, it is necessary to divide the population into different groups depending on their sensibility, for instance, babies, children, adults, adults above 65 years of age or people with asthma. A division is especially important for the assessment of noncarcinogenic pollutants with the effect of chronic illness. In the case of carcinogenic pollutants, it is generally sufficient to divide the population into two groups of adults and children (until a certain age) in order to consider the different physical conditions (e.g., ingestion rate or surface of the body) and the different life-styles (e.g., children playing outside). For substances that induce a carcinogenic response, it is always conservatively assumed that exposure to any amount of the carcinogen will create some likelihood of cancer; that is, a plot of response vs. dose is required to go through the origin. Therefore, for noncarcinogenic responses, it is usually assumed that a threshold dose exists, below which no response will occur. As a result of these two assumptions, the dose–response curves and the methods used to apply them are quite different for carcinogenic and noncarcinogenic effects, as suggested in Figure 4.4. Realize that the same chemical may be capable of causing both kinds of responses.

4.6.1.1 Toxicological Information: Carcinogenic Effect

A controversial point in the discussion of carcinogenic effects is the estimation of the dose–response functions for different pollutants. Animal tests administer various doses to observe toxic effect; however, because the doses in these tests are higher than those existing in the environment, the discussion deals with the possibility of an extrapolation from the higher concentration in the animal test to the lower concentration in the environment and the possible effects. The default approach for determining predictive cancer risk recommended by EPA's Guidelines for Carcinogen Risk

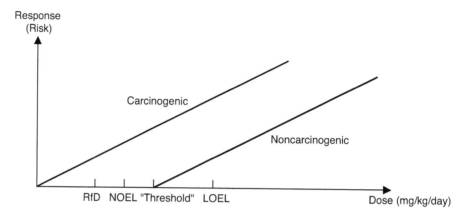

FIGURE 4.4   Dose–response functions for carcinogenic and noncarcinogenic pollutants. This is a schematic presentation; generally these curves are not linear.

Assessment (EPA 2005) is a linear extrapolation from exposures observed in animal or occupational studies. Dose–response assessment for each tumor type is performed in two steps: assessment of observed data to derive a point of departure (POD), followed by extrapolation to lower exposures to the extent that is necessary. The default linear extrapolation approach is generally considered to be conservatively protective of public health, including sensitive subpopulations (EPA 2005). The slope of this line is commonly called slope factor, which is defined by the EPA in the integrated risk information system (IRIS) glossary as an upper bound, approximating a 95% confidence limit, on the increased cancer risk from a lifetime exposure to an agent. This estimate, usually expressed in units of proportion (of a population) affected per mg/kg-day, is generally reserved for use in the low-dose region of the dose–response relationship—that is, for exposures corresponding to risks less than 1 in 100. When the units are risk µg/m^3, it is also called Inhalation Unit Risk (IUR). EPA defines IUR in the IRIS glossary as the upper-bound excess lifetime cancer risk estimated to result from continuous exposure to an agent at a concentration of 1 µg/m^3 in air. The interpretation of inhalation unit risk would be as follows: if unit risk = 2×10^{-6} per µg/m^3, 2 excess cancer cases (upper bound estimate) are expected to develop per 1,000,000 people if exposed daily for a lifetime to 1 µg of the chemical per m^3 of air.

The cancer guidelines emphasize the importance of weighing all of the evidence in reaching conclusions about the human carcinogenic potential of agents. This is accomplished in a single integrative step after assessing all of the individual lines of evidence, which is in contrast to the step-wise approach in the 1986 cancer guidelines. Evidence considered includes tumor findings, or lack thereof, in humans and laboratory animals; an agent's chemical and physical properties; its structure-activity relationships (SARs) as compared with other carcinogenic agents; and studies addressing potential carcinogenic processes and mode(s) of action, either *in vivo* or *in vitro*. Data from epidemiologic studies are generally preferred for characterizing human cancer hazard and risk. However, all of the information discussed previously could provide valuable insights into the possible mode(s) of action and likelihood of human cancer hazard and risk (EPA 2005).

4.6.1.2   Toxicological Information: Noncarcinogenic Effect

A threshold exists for noncarcinogenic substances; that is, any exposure below the threshold would be expected to show no increase in adverse effects above natural background rates. As for carcinogenic substances, in order to get knowledge of this threshold, it is necessary to conduct animal tests by changing the dose. The minimal dose at which a special effect appears is called LOEL (lowest observed effect level). The maximum dose without any special effect to the tested animal represents the NOEL (no observable effect level) (Olsen et al. 2000).

The RfD is defined as an estimation (with uncertainty spanning perhaps an order of magnitude) of the daily exposure of a human population (including sensitive subgroups) that is likely to be without appreciable risk of deleterious effects during a lifetime (EPA 2002). However, it should not be concluded that all doses below a reference dose, or concentration, are acceptable. Despite this caution, most non-cancer health studies adopt the reference dose as an adequate standard to be met. Residual risks associated with doses at or below such standards are not commonly estimated.

The RfD is derived by dividing the NOEL (or LOEL) by uncertainty factors and a modifying factor. Separate adjustment factors are specified for each of several extrapolations, e.g., from average to sensitive individuals, from animal studies to humans, from subchronic to chronic exposure durations, and to account for the quality and breadth of the database. A factor of 10 is usually the default value for the uncertainty factors. Values below 10 are sometimes used when sufficient data and justifications are available. LOEL is only used in the absence of NOEL, and an additional adjustment factor is required in this case to compensate for the lack of a NOEL estimate. For instance, the use of a large number of animals in a study may enhance NOEL certainty. An important point is that the factors may include an inconsistent margin of safety across different chemicals, contributing to varying RfDs, in terms of their conservatism.

Altogether, the values of the RfD represent a dose of approximately 1000 times the value below which it is not possible to observe an adverse effect (NOEL) in animals. The RfD has the unit milligram of a daily intake of the toxic element, which is absorbed in the body, divided by body weight (Table 4.3). In turn, for inhalation exposures, the reference concentration is calculated under the *Inhalation Dosimetry Methodology* (EPA 2009). In this case, the experimental exposures are extrapolated to Human Equivalent Concentration (HEC), and reference concentration (RfC) is calculated by dividing the HEC by uncertainty factors (UF). Different DAF (Dosimetric Adjustment Factor) for the specific site of effects (e.g., respiratory tract region or extra-respiratory) are also applied to calculate the RfC depending on the type of chemical. The DAF is typically based on ratios of animal and human physiologic parameters. The specific DAF used depends on the nature of the contaminant (e.g., particle or gas) and the target site where the toxic effect occurs (e.g., respiratory tract or a location in the body remote from the portal of entry) (EPA 2009).

4.6.1.3 Epidemiological Information

Epidemiology is the study of the incidence rate of diseases in real populations. Epidemiological studies are statistical methods in which causal coherence between environmental pollutant concentration and the occurrence of cases of illness have been thoroughly investigated. Some data

TABLE 4.3

Example of RfD, RfC Values, Oral Potential Factors, and IUR

Regulated Substance	RfD (mg/kg-day)	Oral Cancer Factor (kg/day/mg)	RfC (mg/m³)	IUR (m³/µg)
As	3.0E–04	1.5E–00	1.5E–05	4.3E–03
Cd (Diet)	1.0E–03	–	1.0E–05	1.8E–03
Cr VI	3.0E–03	–	1.0E–04	8.4E–02
Ni (Refinery dust)	1.1E–02	–	1.4E–05	2.4E–04
Hg (Elemental)	–	–	3.0E–04	–
Zn	3.0E–01	–	3.0E–01	–
TCDD	7.0E–10	1.3E+05	4.0E–08	3.8E+01

Source: IRIS, CALEPA.

can be obtained from clinical studies of humans who have inadvertently been exposed to a suspected toxicant; thus, a source of information relating exposure to risk is obtained from epidemiological studies. When attempting to find correlations between elevated rates of incidence of a particular disease in certain groups of people and some measure of their exposure to various environmental factors, an epidemiologist tries to show in a quantitative way the relationship between exposure and risk. Such data should be used to complement animal data, clinical data and scientific analyses of the characteristics of the substances in question.

From studies of former smog pollution episodes (e.g., London smog in the 1950s), it is known that very high ambient pollution concentration is associated with adverse health effects on the same day or on subsequent days. In the last 20 years, numerous well-conducted epidemiological, as well as experimental studies, have confirmed this correlation between exposure to pollutants and the occurrence of health and environmental damages. Hence, they allow establishing a direct link between ambient concentration of certain pollutants and effects on different receptor endpoints, as indicated in Pilkington et al. (1997). Under the assumption of linearity in incremental damage with incremental exposure, slope factors (SF) could be defined. Figure 4.5 shows some possible forms of exposure–response (E–R) functions as they have been found. E–R functions exist for human health effects, damages to material and crops, and harm to ecosystems. Unfortunately, at the moment, sufficient epidemiological data are not available to address human health effects caused by the majority of chemicals. Therefore, the epidemiological approach must be combined with bioassays. Moreover, it should be taken into account that epidemiological data are criticized for providing insights that may be limited to the identification of correlations.

The E–R functions of macropollutants for human health can be subdivided into the seriousness of their effects:

Morbidity: This effect concerns the following primary and secondary pollutants: NO_x, SO_2, NH_3, CO, nitrate aerosol, sulphate aerosol and PM_{10}. Possible impacts are, among others, hospital visits, bronchodilator use and chronic cough.

Mortality: The effect of mortality can be expressed by fatal cases or years of life lost (YOLL), as previously explained in Chapter 3. However, since in recent years researchers have moved from studies based on fatal cases to YOLL (EC 1995, 2000), YOLL is used as endpoint in this study. Table 4.4 illustrates the highly important E–R functions that use YOLL as endpoint.

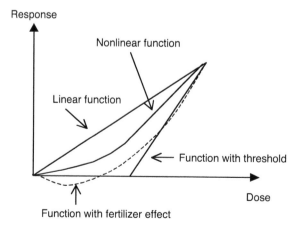

FIGURE 4.5   Possible forms of exposure–response functions.

TABLE 4.4
Mortality Functions Expressed in YOLL due to Concentration Increments (µg/m³)

Receptor	Category	Pollutant	Formula
	Adults: Percentage of Population with Age above 18		
	"chronic" YOLL	PM_{10}/ Nitrates	0.00072*Dc*Pop.*adults
		Sulphates	0.0012*Dc*Pop.*adults
	"acute YOLL"	SO_2	0.0719*Dc*Pop.*b_m/100*adults
		PM_{10}/ Nitrates	0.04*Dc*Pop.*b_m/100*adults
		Sulphates	0.0677*Dc*Pop.*b_m/100*adults
		NO_x	0.0648*Dc*Pop.*b_m/100*adults

Source: IER (Institut für Energiewirtschaft und Rationale Energieanwendung), EcoSense 2.1 Software, University of Stuttgart, Germany, 1998.
Note: b_m: baseline mortality

For example the additional YOLL by sulfates are calculated in the following way:

$$0.0012 \text{ YOLL}/(mg/m^3) * 0.00082 \ (mg/m^3)/$$

$$\text{functional unit} * 423{,}000 \text{ pers.} * 0.75 \text{ adults/pers.} =$$

$$0.312 \text{ YOLL/ TWh}$$

where:
Dc = 0,00082 (µg/m³) concentration increase/functional unit
Pop. = 423,000 pers. (population expressed as persons in region considered)
adults = 0.75 adults/pers. (percentage of population which age above 18 years)

Caution must be exercised in interpreting every epidemiological study because any number of confusing variables may lead to invalid conclusions. For example, a study may be biased because workers are compared with non-workers (workers are usually healthier) or because relative rates of smoking have not been accounted for, or other variables that may be the actual causal agents may not even be hypothesized in the study. As an example of the latter, consider an attempt to compare lung cancer rates in a city with high ambient air pollution levels with rates in a city with less pollution. Suppose the rates are higher in the more polluted city, even after accounting for smoking history, age distribution, and working background. To conclude that ambient air pollution is causing those differences may be totally invalid. Instead, different levels of radon may be in homes, for example, or differences that are causing the cancer variations may exist in other indoor air pollutants associated with the type of fuel used for cooking and heating.

4.6.2 ALTERNATIVE TESTS FOR TOXICOLOGICAL ASSESSMENT

The need to decrease costs and reduce animal suffering for chemical risk assessment has encouraged the use of alternative methods instead of animal tests to predict toxicity. The term *alternative* includes "all procedures which can completely replace the need for animal experiments, reduce the number of animals required, or diminish the amount of distress or pain suffered by animals in meeting the essential needs of man and other animals." These alternative methods can be generally divided into two subgroups: study of toxicity in laboratory tubes on small organisms (*in vitro*) and computational techniques (*in silico*). These techniques have recently become more important due to mandates such as the categorization of the European Union's REACH (Bakhtyari et al. 2013).

4.6.2.1   *In Vitro* Tests

In vitro tests involve experiments that can be done in a tube including biological components, such as organ cultures, tissue slices, primary cells, or cell lines, in such a way that those biological models are extrapolated, taking into account the similarity with the *in vivo* systems (e.g., cel lines from tumor cells to evaluate toxicity inhalation studies). The objective of these studies is to measure the effect, or named endpoint, after the exposure to chemicals. Examples of endpoints can be the release of an enzyme, the activation of a pathway, changes in the cell cycle or even death. The measurement of the endpoint allows defining the concentration–response relationship in the used system. The specificity of the endpoint is a fundamental key point to make a biological system more useful in obtaining relevant information on mechanistic basis of the chemical-cell interactions. *In vitro* tools can be used alone or in batteries.

4.6.2.2   *In Silico* Models

Computational programs can be used to predict the toxicity of chemical compounds. The main driving force behind this trend has been the emergence of new chemical descriptors, algorithms and statistical perspectives. These techniques may be coarsely classified into techniques that mimic human reasoning about toxicological phenomena (Expert Systems) and methods that derive predictions from a training set of experimentally determined data (Data Driven Systems). Expert systems approach is intuitively appealing to most users, because it promises easy access to toxicological knowledge, and some of the most used predictive toxicology software tools are, in fact, Expert Systems (e.g., Derek Nexus and Toxtree). Data Driven Systems are formalized methods for the extraction of prediction models directly from experimental data. These systems vary in sophistication, from the relatively simplistic approach of forming chemical groupings (read-across and grouping) to the more complex development of SARs (qualitative identification of chemical substructures with the potential of being reactive or toxic) and QSARs (quantitative prediction of relative reactivity or toxicity). The SAR model investigates the existence of a relationship between a certain chemical property, such as a fragment, and the effect, such as carcinogenic effect, without assigning a numerical continuous value to the toxicity.

The non-testing method called read-across implies the use of endpoint information for one chemical, called a "source chemical," to make a prediction of the same endpoint for another chemical, called a "target chemical." The source and target chemicals are considered to be similar in some way, usually on the basis of structural similarity. It is assumed that, in general, similar compounds will exhibit similar biological activity. In principle, read-across can be applied to characterize not only human health effects and ecotoxicity, but also physicochemical and fate properties, and it may be performed in a qualitative or quantitative manner, depending on whether the data being used is categorical or numerical in nature. To estimate the properties of a given substance, read-across can be performed in a one-to-one manner (one analogue used to make the estimate) or in a many-to-one manner (two or more analogues used).

The reliability of read-across depends on the selection of appropriate analogues associated with the availability of reliable experimental data. In some cases, it is only possible to identify a limited number of suitable analogues, whereas in other cases, it is possible to build up a larger and more robust chemical group, called a chemical category. A chemical category is a group of chemicals whose physicochemical and human health and/or environmental toxicological and/or environmental fate properties are likely to be similar or follow a regular pattern as a result of structural similarity (or other similarity characteristic). The presence of common behavior or coherent trends in the chemical category is generally associated with a common underlying mechanism of action. In general, the application of read-across between analogues in a chemical category is considered to be more reliable than the application of read-across in a smaller group of analogues (in which trends are not apparent).

4.6.2.3 PBPK/PD

Physiological-based pharmacokinetic/pharmacodynamic (PBPK/PD) models are helpful to incorporate known mechanisms and biological processes into a software for studying different exposure scenarios and their possible risk, which can be finally used for regulatory purposes. These models are increasingly used in the field of human health risk assessment, particularly in quantifying the relationship between measures of external exposure and internal dose. PBPK/PD models are mathematical representations of the human body, where organs are considered as compartments. These models allow the prediction of the theoretical concentration of chemicals along the time in different human tissues, taking into account species-specific physiological, chemical and biochemical parameters. PBPK/PD models are based on a system of differential equations which estimate the concentration or amount of individual substances of sets of chemicals in each body compartment (Nadal et al. 2013). The development of a PBPK/PD model involves three main steps: (1) definition of the model, identifying the organs or tissues involved in the ADME process, (2) data collection regarding initial concentration, tissue volumes, solubility, metabolic constants, etc. and (3) mathematical simulation; generally, mass balance ordinary differential equations are used to explain the change of a chemical in a tissue over time, describing the transfer between compartments, as well as metabolism and excretion process. Subsequently, the model is compared with experimental data available for calibration and validation purposes, and the model is finally modified according to new suppositions and presumptions (Nadal et al. 2013). Due to their potential power, PBPK models have been largely used in pharmacological development and health risk assessment.

4.6.3 Physico-Chemical and Toxicological Data Sources

4.6.3.1 Databases

Currently, there are many databases where you can obtain the physicochemical and toxicological parameters required to perform a risk assessment. It must be borne in mind that the existence of quality control parameters included in the databases is of great importance. This quality control can be accomplished through periodic updating of the database, the inclusion of bibliographical references of the origin of each parameter, peer review bibliography, etc (Rovira et al. 2013).

The following are a selection of physicochemical and toxicological databases, which are freely available online:

- TOXNET (TOXicology Data NETwork) is a cluster of databases covering toxicology, hazardous chemicals, environmental health and related areas of the United States Library of Medicine. The TOXNET includes, among others, toxicological databases such as:
- Chemical Carcinogenesis Research Information System (CCRIS)
- Hazardous Substances Data Bank (HSDB)
- Integrated Risk Information System (IRIS)
- International Toxicity Estimates for Risk (ITER)

Also TOXNET includes ChemIDplus, a dictionary containing synonyms, structures and physico-chemical properties.

- eChemPortal, a global portal to information on chemical substances. eChemPortal allows simultaneous searching of reports and datasets by chemical name and number and by chemical property. Direct links to collections of chemical hazard and risk information prepared for government chemical review programs at national, regional and international levels are obtained. Classification results according to national/regional hazard classification schemes or to the Globally Harmonized System of Classification and Labelling of Chemicals (GHS) are provided when available. The eChemPortal is an effort of the Organisation for Economic

Co-operation and Development (OECD) in collaboration with the European Commission (EC), the European Chemicals Agency (ECHA), the United States, Canada, Japan, the International Council of Chemical Associations (ICCA), the Business and Industry Advisory Committee (BIAC), the World Health Organization's (WHO) International Program on Chemical Safety (IPCS), the United Nations Environment Programme (UNEP) and environmental non-governmental organisations.

- The Agency for Toxic Substances and Disease Registry (ATDSR) produces the "toxic substances profile" for hazardous substances found at National Priorities List (NPL) sites. These hazardous substances are ranked based on frequency of occurrence at NPL sites, toxicity, and potential for human exposure.
- The Carcinogenic Potency Database (CPDB) contains the results of 6,540 chronic, long-term animal cancer tests on 1,547 chemicals. The CPDB provides easy access to the bioassay literature, with qualitative and quantitative analyses of both positive and negative experiments that have been published over the past 50 years in the general literature through 2001 and by the National Cancer Institute/National Toxicology Program through 2004.
- IUCLID (International Uniform ChemicaL Information Database) is a software application to capture, store, maintain and exchange data on intrinsic and hazard properties of chemical substances.
- The Risk Assessment Information System (RAIS) Guidance documents, tutorials, databases, historical information and risk models were all housed and integrated on the RAIS.
- ECOTOX (ECOTOXicology) is a database of single chemical toxicity data for aquatic life, terrestrial plants and wildlife. ECOTOX was created and is maintained by the U.S. EPA, Office of Research and Development (ORD), and the National Health and Environmental Effects Research Laboratory's (NHEERL's) Mid-Continent Ecology Division (MED).

4.6.3.2   Estimation Data Tools

Despite the existence of several databases for certain substances, especially for new chemicals, it is not possible to find physicochemical and/or toxicological parameters to assess the risk. The lack of data is one of the main problems in risk assessment. In the case of emerging pollutants, these lack of knowledge in physicochemical properties and/or toxicological properties are a more evident problem. One solution to solve this problem is the use of quantitative structure–activity relationship (QSAR) or estimation tools. As previously discussed, QSAR models correlate the structure of the substance with their activities (physicochemical properties, environmental fate activity and/or toxicological properties).

The following are a selection of physicochemical and toxicological estimation tools, which are freely available online or to download:

- EPI Suite v4.0. The EPI (Estimation Programs Interface) Suite™ is a suite of physicochemical property and environmental fate estimation programs developed by the EPA's Office of Pollution Prevention Toxics and Syracuse Research Corporation (SRC). Among others, EPI Suite estimates Kow, Koc, Koa, Henry's Law constant, melting and boiling points, aerobic and anaerobic biodegradability of organic chemicals, biodegradation half-life for hydrocarbons and bioconcentration factors.
- Danish (Q)SAR Database. This Danish (Q)SAR database is a repository of estimates from over 70 (Q)SAR models for 166,072 chemicals. The (Q)SAR models encompass endpoints for physicochemical properties, fate, eco-toxicity, absorption, metabolism and toxicity.
- The Toxicity Estimation Software Tool (TEST) has been developed by U.S. EPA program to allow users to easily estimate toxicity using a variety of QSAR methodologies. TEST allows to estimate the value for several toxicity endpoints: 96 hour fathead minnow LC50, 48 hour Daphnia magna LC50, 48 hour Tetrahymena pyriformis IGC50, Oral rat LD50,

Bioaccumulation factor, Developmental toxicity, and Ames mutagenicity. TEST also estimate several physical properties.

- Computer-Assisted Evaluation of industrial chemical Substances According to Regulations (CAESAR) was an EC-funded project, which was specifically dedicated to develop QSAR models for the REACH legislation. Five endpoints with high relevance for REACH have been addressed within CAESAR: bioconcentration factor, skin sensitization, carcinogenicity, mutagenicity and developmental toxicity.
- The Ecological Structure Activity Relationships (ECOSAR) program estimates a chemical's acute (short-term) toxicity and chronic (long-term or delayed) toxicity to aquatic organisms such as fish, aquatic invertebrates, and aquatic plants by using computerized Structure Activity Relationships (SARs).
- OECD QSAR Toolbox was developed by OECD to make (Q)SAR technology readily accessible, transparent and less demanding in terms of infrastructure costs. The Toolbox has multiple functionalities allowing the user to perform a number of operations. The Toolbox potentially covers all relevant regulatory endpoints. Nevertheless, the ability to fill data gaps via the analogue approach or by building chemical categories is limited by the availability of experimental results that can be used to perform read-across or trend analysis.

4.6.4   Ecosystems (Environment)

The procedure for the environmental risk assessment of substances consists of comparing the concentration in the environmental compartments (PEC) with the concentration below which unacceptable effects on organisms will most likely not occur (PNEC).

The PNEC values are usually determined on the basis of results from monospecies laboratory tests or of established concentrations from model ecosystem tests, taking into account adequate safety factors. A PNEC is regarded as a concentration below which an unacceptable effect will most likely not occur. In principle, the PNEC is calculated by dividing the lowest value of short-term L(E)C50 (Median Lethal Concentration) or long-term NOEC (No Observed Effect Concentration) value by an appropriate assessment factor. The assessment factors reflect the degree of uncertainty in extrapolation from laboratory toxicity test data for a limited number of species to the "real" environment. Assessment factors applied for long-term tests are smaller because the uncertainty of the extrapolation from laboratory data to the natural environment is reduced. For this reason, long-term data are preferred over short-term data.

Because aquatic organisms are exposed for a short period to compounds with intermittent release patterns, short-term L(E)C50 values are used to derive a PNECwater for these compounds. For most compounds, data will probably not be present for sediment-dwelling organisms. Appropriate test systems are under development, but standardized guidelines are not yet available. A method to compensate for this lack of toxicity data, known as the equilibrium partitioning method, is used to derive a PNECsed. Toxicity data are also scarce for the soil compartment. When such data are present, they will normally include only short-term studies. In cases in which data are missing, the equilibrium partitioning method can be used to derive a PNECsoil. For the atmosphere, biotic and abiotic effects like acidification are addressed. Due to the lack of suitable data and unavailability of adequate methods to assess both types of effects, a provisional strategy is used.

The main function of risk assessment is the overall protection of the environment. However, certain assumptions are made concerning the aquatic environment that allow an uncertainty extrapolation to be made from single-species, short-term toxicity data to ecosystem effects. It is assumed that ecosystem sensitivity depends on the most sensitive species and that protecting ecosystem structure protects community function. These two assumptions have some important consequences. When the most sensitive species to the toxic effects of a chemical in the laboratory is established, extrapolation can subsequently be based on the data from that species. Furthermore, the functioning of any ecosystem in which that species exists is protected, provided the structure is not sufficiently distorted to cause an imbalance. It is accepted that protection of the most sensitive species should protect the structure.

With regard to the assessment of impacts that affect ecosystems, E–R or damage functions have been developed for acidic deposition on natural and semi-natural terrestrial ecosystems. That means it accounts for the impact corresponding to the sub-area of protection (see Chapter 3) biodiversity and natural landscapes. Currently, the most widely applicable approach for the analysis of pollutant effects on terrestrial ecosystems is the critical load/ level approach. The following definitions of the terms critical load and levels were given by UN-ECE.

> *Critical load*: The highest deposition of acidifying compounds that will not cause chemical changes leading to long-term harmful effects on ecosystems structure and function according to the present knowledge.
> *Critical levels*: The concentration of pollutants in the atmosphere above which direct adverse effects on receptors, such as plants, ecosystems or material may occur, according to the present knowledge.

Critical loads have been defined for several pollutants and ecosystems. However, they cannot be used directly to assess damages per se; rather, they simply identify the areas where damages are likely to occur. In the present study, the relative exceedance weighted (REW) ecosystem area approach is used to assess the environmental impact. The contribution of a specific source of pollutants to the exceedance of critical loads for ecosystems is analyzed by taking into account predefined background conditions. The relative exceedance factor fRE is the contribution of the concentration increase Δc due to the emission to the height of exceedance of the critical load divided by the critical load C_{CL} itself (see Equation 4.2). The REW ecosystem area indicator is expressed in km^2 and is obtained by the multiplication of f_{RE} by the ecosystem exceeded area A_{EE} where the critical load is exceeded by the pollutant (see Equation 4.4) (IER 1998).

$$f_{RE} = \Delta c / C_{CL} \qquad\qquad (4.3)$$

$$REW = f_{RE} * A_{EE} \qquad\qquad (4.4)$$

For example, the additional REW by NO$_x$ are calculated in the following way for imaginative values of $\Delta c/$, C_{CL}, and A$_{EE}$:

$$f_{RE} = 0.00082 \ (mg/m^3)/ \ FU/100 \ mg/m^3 = 0.0000082/FU$$

$$REW = 0.0000082/FU * 10,000 \ km^2 = 0.082 \ km^2/FU$$

where:
$\Delta c = 0.00082$ (µg/m^3) concentration increase/functional unit (FU)
$C_{CL} = 100$ µg/m^3 (concentration corresponding to critical load)
$A_{EE} = 10,000$ km^2 (ecosystem exceeded area)

Moreover, damage functions exist for environmental damage on crops and material, as what in the present is defined as the AoP man-made environment. There are two basic pathways through which plants can be harmed by SO$_2$. The first is through foliar uptake of pollutants, and the second through effects of acid deposition on the soil. Damages to material refers to surface damage, especially in buildings, bridges and cars, due to acidic deposition. However, not included are cultural damages, e.g., to ancient cathedrals, due to their intrinsic value as mankind's patrimony.

4.7   RISK CHARACTERIZATION AND CONSEQUENCE DETERMINATION

The last stage of the risk assessment process, risk characterization, involves a prediction of the probability and severity of health and ecological impact in the exposed population and environmental damages. That is, the information from the dose–response evaluation (*What human dose or PEC is necessary to cause an effect?*) is combined with the information from the exposure assessment (*What human dose or PEC is the population or ecosystem receiving?*) to produce an estimate of the likelihood of observing an effect in the population or ecosystem being studied. An adequate characterization of risks from hazards associated with environmental contamination problems allows risk management and corrective action decisions to be better focused. To the extent feasible, risk characterization should include the distribution of risk among the target populations.

4.7.1   HUMAN HEALTH RISK

Human health risk characterization is the estimation of whether adverse effects are likely to affect humans who are exposed to certain substances. This process includes the comparison of the hazard and dose–response information, usually derived from animal experiments, or *in vitro* test systems, with data on human exposure levels, as has been explained before. For notified new substances, risk characterization should take account of each adverse effect for which the substances have been assigned a hazard classification, together with any other effect of possible concern. Workers, consumers and humans exposed indirectly via the environment must be considered.

The health risk for exposed populations to environmental pollutants is characterized by the calculation of noncarcinogenic hazard quotient and carcinogenic risks. These parameters can then be compared with benchmark criteria or standards in order to arrive at risk decisions about an environmental pollution problem.

4.7.1.1   Carcinogenic Risk to Human Health

Cancer risk expresses the likelihood of suffering cancer due to a definite daily dose of a pollutant. In this way, carcinogen risk is defined by the incremental probability of an individual developing cancer over a lifetime as a result of exposure to carcinogenic substances. It can be described by a dose–response estimate. Cancer risk is nonthreshold; this means that even the lowest doses have a small, or finite, probability of generating a carcinogenic response. Although risk decreases with the dose, it does not become zero until the dose becomes zero. For the characterization of cancer risk, the specific exposure is compared with a corresponding health benchmark for the relevant contaminant. The results correspond to the probability that cancer occurs, e.g., a value of 10^{-6} means that the probability exists that one person in a million may get cancer due to the study exposition over a lifetime.

In this way, cancer risk is estimated as excess risk (ER). ER does not express total cancer. It is an incremental risk due to exposure to the considered pollutant. In general, risks associated with the inhalation and noninhalation pathways (oral pathways) may be considered separately, and therefore different human health benchmarks are required. They can be estimated in accordance with the following generic relationships:

$$\text{Inhalation risk} = \text{exposure concentration } (\mu g/m^3) \times \text{inhalation unit risk } (m^3/\mu g)^{-1} \tag{4.5}$$

$$\text{Noninhalation risk} = \text{dose } (\mu g/kg\text{-day}) \times \text{potency slope factor } (\mu g/kg\text{-day})^{-1}$$

Therefore, different human health benchmarks are required. Inhalation risk of carcinogenic substances is assessed using the inhalation unit risk factor, while the noninhalation risk is the oral cancer slope factor. Those carcinogenic factors do not represent a safe exposure level, but relate the exposure with the probability of causing carcinogenic effects. Since cancer risk describes the probability of developing cancer over a lifetime, the entire duration of exposure must be considered for risk assessment.

4.7.1.2   Noncarcinogenic Risk to Human Health

Depending on the exposure level, adverse health effects other than cancer can be associated with all chemical substances. Therefore, a noncancer risk characterization is always a dose–response analysis that determines whether the actual human exposure exceeds a defined exposure level. This critical exposure level represents the threshold below which adverse effects are assumed to be unlikely, and it is determined in a toxicity assessment.

The human health benchmarks most widely used are the RfD for oral exposure and the RfC for inhalation of contaminants. RfD and RfC are lower-bound estimates of the NOAEL (no observed adverse effects level) of a pollutant, expressed for the different types of human exposure. These human health benchmarks are established for chronic exposure and do not account for acute toxicity of a pollutant. Noncancer risks are expressed by the hazard quotient (HQ), which relates the oral exposure to the RfD (EPA 2002) or the exposure concentration to the Reference Concentration for inhalation (EPA 2009). HQ refers only to the potential for some individuals to be affected and cannot address the absolute level of risk. If HQ is higher than 1, this does not necessarily indicate a potential health risk. In consequence, noncarcinogenic quantitative estimates only identify the exposure level below which adverse effects are unlikely but say nothing about incremental risk for higher exposure. Although cancer risks are expressed as an increased probability of the occurrence of carcinogenic effects due to additional exposure, noncancer risks are assessed for the total exposure to a pollutant.

4.7.2   Ecological Risk

Having conducted the exposure assessment and the dose (concentration)–response (effect) for all environmental compartments, risk characterization is carried out by comparing the PEC with the PNEC. This is done separately for each of the protection goals identified before, for instance: aquatic ecosystem, terrestrial ecosystem, atmosphere, top predator, microorganisms in sewage treatment plants, etc. For the risk characterization of the aquatic and terrestrial ecosystems, a direct comparison of the PEC and PNEC values must be carried out. If the PEC/PNEC ratio is greater than one, the substance is "of concern" and further action must be taken. For the air compartment, only a qualitative assessment of abiotic effects is carried out. If there are indicators that one or more of these effects occur for a given substance (for example, for ozone depletion substances), expert knowledge and consulting such as that provided by the responsible body in the United Nations Environment Programme (UNEP) will be necessary.

4.8   IMPACT PATHWAY ANALYSIS

Impact pathway analysis (IPA) was introduced in the nineties as a simplified way to assess the environmental fate and exposure of emissions to air; it allows the expression of effects in physical impact parameters, such as cancer cases and restricted activity days that can be evaluated in monetary terms. IPA takes into account damages on a regional level due to pollutants with a long residence time and last-used exposure–response functions based on epidemiological studies and also to dose–response functions based on toxicological tests.

IPA is also known as a damage function or "bottom-up" approach that traces the passage of pollutants from the place where they are emitted to the endpoint, i.e., the receptor that is affected by them. The approach provides a logical and transparent way of quantifying environmental damages, i.e., externalities. This methodology was developed through the ExternE-Project (Externalities of Energy) in a collaborative study between the European Commission and the U.S. Department of Energy.

The difference between IPA and the earlier used "top-down" damage assessment methodologies (Hohmeyer 1992; Friedrich and Voss 1993) is that in IPA, specific emission data are used for individual locations. These data are computed with pollution dispersion models, detailed

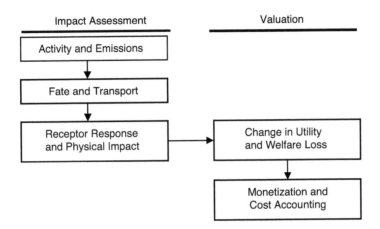

FIGURE 4.6   Illustration of the main steps of the IPA. (From EC, *ExternE—Externalities of Energy*, Vol. 6, EUR 16520-162525. DG XII Science, Research and Development, Brussels, 1995.)

information about receptors and dose–response and exposure–response functions in order to calculate the physical impact of increasing emissions. Finally, the impacts are valuated economically.

The principal steps are illustrated in Figure 4.6 and described as follows:

- Activity and emission: characterization of the relevant technologies and the environmental burdens they cause (e.g., mg of SO_2 per Nm^3 emitted by the considered process)
- Fate and transport: calculation of the increasing concentration in the affected regions via atmospheric dispersion models and chemical reactions (e.g., SO_2 transport and transformation into sulfates)
- Receptor response and physical impact: characterization of the receptors exposed to the incremental pollution, identification of suitable dose–response and exposure–response functions and their linkage to given estimated physical impacts (e.g., number of asthma cases due to increase of sulfates)
- Monetization and cost accounting: economic valuation of the mentioned impacts, determination of external costs that have not been internalized by governmental regulations (e.g., multiplication of the monetary value with the asthma incidents gives the damage costs).

Although the IPA is a complex approach, its application is quite easy thanks to the support of integrated impact assessment models like EcoSense, developed by Krewitt et al. (1995), or PathWays (Rabl et al. 1998). In this study, the EcoSense model was applied; therefore, further details are given about this model.

EcoSense stems from the experiences learned in the ExternE project (EC 1995, 1999) to support the assessment of priority impacts resulting from exposure to airborne pollutants, namely, impacts on health, crops, building materials, forests, and ecosystems. Although global warming is certainly among the priority impacts related to air pollution, EcoSense does not cover this impact category because of the very different mechanism and the global nature of impact. Priority impacts, like occupational or public accidents, are not included either because the quantification of impacts is based on the evaluation of statistics rather than on modeling. Version 3.0 of EcoSense covers 13 pollutants, including the "classical" pollutants SO_2, NO_x, particulate matter and CO, and photochemical ozone creation, as well as some of the most important heavy metals and hydrocarbons, but does not include impacts from radioactive nuclides.

In view of increased understanding of the major importance of long-range transboundary transport of airborne pollutants, also in the context of external costs from electricity generation, there was an obvious need for a harmonized European-wide database supporting the assessment of environmental impacts from air pollution. In the beginning of the ExternE project, work was focused on the assessment of local-scale impacts and teams from different countries made use of the data sources available in each country. Although many teams spent a considerable amount of time compiling data on population distribution, land use etc., it was realized that country-specific data sources and grid systems were hardly compatible when the analysis had to be on a European scale. Thus, it was logical to set up a common European-wide database by using official sources like EUROSTAT and make it available to all ExternE teams. Once there was a common database, the consequent next step was to establish a link between the database and all the models required for assessment of external costs to guarantee a harmonized and standardized implementation of the theoretical methodological framework (EC 1995).

Taking into account this background, the further objectives for the development of the EcoSense model were:

- To provide a tool supporting a standardized calculation of fuel cycle externalities
- To integrate relevant models into a single system
- To provide a comprehensive set of relevant input data for the whole of Europe
- To enable the transparent presentation of intermediate and final results
- To support easy modification of assumptions for sensitivity analysis

Because health and environmental impact assessment is a field of large uncertainties and incomplete but rapidly growing understanding of the physical, chemical and biological mechanisms of action, it was a crucial requirement for the development of the EcoSense system to allow easy integration of new scientific findings into the system. As a consequence, all the calculation models (except for the integrated ISCST model) were designed so that they were model interpreters rather than models. Model specifications, e.g., chemical equations, exposure–response functions or monetary values, are stored in the database (Paradox format) and can be modified by the user. This concept should allow easy modification of model parameters; at the same time, the model should not necessarily appear as a black box because the user can trace back what the system is actually doing.

Figure 4.7 shows the modular structure of the EcoSense model. All data —input data, intermediate and final results —are stored in a relational database system. The two air quality models integrated

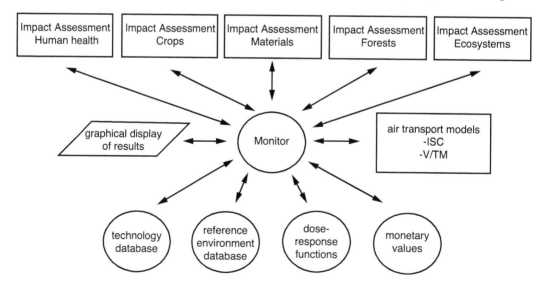

FIGURE 4.7   Structure of the EcoSense model. (From EC, *ExternE—Externalities of Energy*, Vol. 6, EUR 16520-162525. DG XII Science, Research and Development, Brussels, 1995.)

in EcoSense (ISCST-2 and the Windrose trajectory model—WTM) are stand-alone models linked to the system by pre- and postprocessors.

There are individual executable programs for each of the impact pathways, which make use of common libraries. The following sections give a more detailed description of the different EcoSense modules (IER 1998).

4.8.1 Reference Technology Database

The reference technology database holds a small set of technical data describing the emission source that are mainly related to air quality modeling, e.g., emission factors, flue gas characteristics, stack geometry and the geographic coordinates of the site.

4.8.2 Reference Environment Database

The reference environment database is the core element of the EcoSense database, providing data on the distribution of receptors and meteorology, as well as a European-wide emission inventory. All geographical information is organized using the EUROGRID coordinate system, which defines equal-area projection grid cells of 10,000 and 100 km^2 (Bonnefous and Despres 1989), covering all EU and European non-EU countries. Data on population distribution and crop production are taken from the EUROSTAT REGIO database and, in some few cases, have been updated using information from national statistics. The material inventories are quantified in terms of the exposed material area from estimates of representative buildings. Critical load maps for nitrogen deposition are available for nine classes of different ecosystems, ranging from Mediterranean scrub over alpine meadows to tundra areas. To simplify access to the receptor data, an interface presents all data according to administrative units (e.g., country, state) following the EUROSTAT NUTS classification scheme. Meteorological data (precipitation, wind speed and wind direction) and a European-wide emission inventory for SO_2, NO_x and NH_3 from EMEP (Sandnes and Styve 1992), transferred to the EUROGRID-format, are also included for the long-range pollutant transport model.

4.8.3 Dose–Response and Exposure–Response Functions

Using an interactive interface, the user can define any exposure effect model as a mathematical equation. The user-defined function is stored as a string in the database, which is interpreted by the respective impact assessment module at runtime. All dose–response and exposure–response functions compiled by the various experts of the ExternE project are stored in the database. Examples are given in Section 4.6.

4.8.4 Monetary Values

The database provides monetary values for most of the impact categories following the recommendations of the ExternE economic valuation task group according to guidelines from the European Commission, explained in Chapter 3.

4.8.5 Impact Assessment Modules

The impact assessment modules calculate the physical impacts and, as far as possible, the resulting damage costs by applying the dose–response and exposure–response functions selected by the user to each individual grid cell, taking into account the information on receptor distribution and concentration levels of air pollutants from the reference environment database. The assessment modules support the detailed step-by-step analysis for a single endpoint, as well as a more automated analysis including a range of pre-specified impact categories.

4.8.6 PRESENTATION OF RESULTS

Input data as well as intermediate results can be presented on several steps of the IPA in numerical or graphical format. Geographical information like population distribution or concentration of pollutants can be presented as maps. EcoSense generates a formatted report with a detailed documentation of the final results that can be imported into a spreadsheet program.

4.8.7 AIR QUALITY MODELS

A special feature of EcoSense is the fact that air quality models are included. Apart from the local-scale ISCST-2 or -3 model, for which a set of site-specific meteorological data must be added by the user, a long-range pollutant transport model is included; both models have also been applied separately in this study for the calculation of site-dependent impact factors.

Close to the plant, i.e., at distances of some 10–100 km from the plant, chemical reactions in the atmosphere have little influence on the concentrations of primary pollutants. For these reasons, the computation of ambient air concentrations of primary pollutants on a local scale is done with a model that neglects chemical reactions, but is detailed enough in the description of turbulent diffusion and vertical mixing. An often-used model that meets these requirements is the Gaussian plume model. The concentration distribution from a continuous release into the atmosphere is assumed to have a Gaussian shape (see Equation 4.6):

$$c(x,y,z) = \frac{Q}{u2\pi\sigma_y\sigma_z} \cdot \exp\left[-\frac{y^2}{2\sigma_y^2}\right] \cdot \left(\exp\left[-\frac{(z-h)^2}{2\sigma_z^2}\right] + \exp\left[-\frac{(z+h)^2}{2\sigma_z^2}\right]\right) \qquad (4.6)$$

where:

$c(x,y,z)$ = concentration of pollutant at receptor location (x,y,z)
Q = pollutant emission rate (mass per unit time)
u = mean wind speed at release height
σ_y = standard deviation of lateral concentration distribution at downwind distance
σ_z = standard deviation of vertical concentration distribution at downwind distance x
h = plume height above terrain

The assumptions embodied into this type of model include those of idealized terrain and meteorological conditions so that the plume travels with the wind in a straight line, mixing with the surrounding air, both horizontally and vertically, to produce pollutant concentrations with a normal (Gaussian) spatial distribution (Figure 4.8). Dynamic features that affect the dispersion, for example, vertical wind shear, are ignored. These assumptions generally restrict the range of validity of the application of these models to the region within some 100 km of the source. Pollution transport however, extends over much greater distances. The assumption of a straight line is justified for a statistical evaluation of a long period, where mutual changes in wind direction cancel out each other, rather than for an evaluation of short episodes.

In this study, the Industrial Source Complex Short Term model, version 2 (ISCST-2) of the U.S. EPA (1992) and version 3 (ISCST-3) of U.S. EPA (1995) in the form of BEEST (Beeline 1998) have been applied. The model calculates hourly concentration values of gases and particulate matter

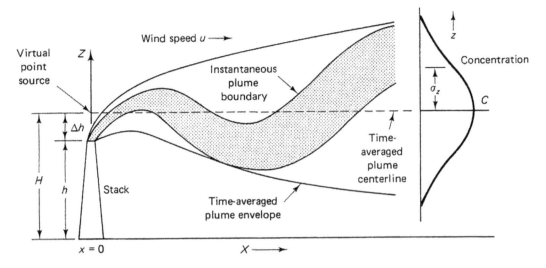

FIGURE 4.8 Gaussian plume shape (H mixing layer height, h – stack height, H – plume height assuming flat terrain, H* – plume height above terrain).

for 1 year at the center of each specified grid. Effects of chemical transformation are neglected. Annual mean values are obtained by temporal averaging of the hourly model results.

Mean terrain heights for each grid cell are necessary and it is also the responsibility of the user to provide the meteorological input data. These include wind direction, wind speed, stability class, as well as mixing height, wind profile exponent, ambient air temperature and vertical temperature gradient.

With increasing distance from the stack, the plume spreads vertically and horizontally due to atmospheric turbulence. Outside the area of the local analysis (i.e., at distances beyond 100 km from the stack), it can be assumed for most purposes that the pollutants have vertically been mixed throughout the height of the mixing layer of the atmosphere and that chemical transformations can no longer be neglected on a regional scale. The most economic way to assess annual regional scale pollution is a model with a simple representation of transport and a detailed enough representation of chemical reactions.

The Windrose trajectory model (WTM) used in EcoSense to estimate the concentration and deposition of acid species on a regional scale was originally developed at Harwell Laboratory by Derwent and Nodop (1986) for atmospheric nitrogen species, and extended to include sulfur species by Derwent et al. (1988). The model is a receptor-oriented Lagrangian plume model employing an air parcel with a constant mixing height of 800 m moving with a representative wind speed. The results are obtained at each receptor point by considering the arrival of 24 trajectories weighted by the frequency of the wind in each 15° sector. The trajectory paths are assumed to be along straight lines and are started at 96 h from the receptor point. In addition to dealing with primary pollutants, the WTM is also able to calculate concentrations of secondary sulfate and nitrate aerosols formed from emissions of SO_2 and NO_x, respectively. The chemical reaction schemes implemented in the model are shown in Figure 4.9.

As we have seen, IPA and ERA share many similarities. In Figure 4.10 the steps of both methods are compared.

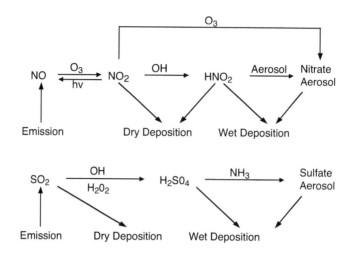

FIGURE 4.9   Chemical scheme in WTM. (From EC, *ExternE—Externalities of Energy*, Vol. 6, EUR 16520-162525. DG XII Science, Research and Development, Brussels, 1995.)

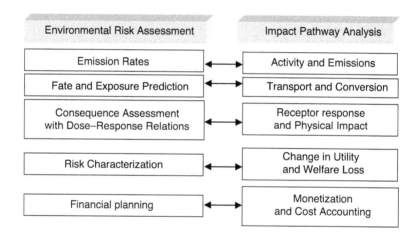

FIGURE 4.10   Steps of an IPA compared to those in a conventional ERA.

4.9   ADVANCES IN APPROACHES FOR ERA

Currently, there are many projects focused on topics aimed to refine the different steps of the ERA methodology (hazard characterization, exposure assessment, dose–response methods and models and risk characterization) by means of incorporating scientific innovations (EPA 2015). These advances in approaches are intended to address emerging challenges such as the evaluation of indoor air quality, risk of mixtures or the risks associated to emerging engineered nanomaterials, among others.

4.9.1   INDOOR AIR QUALITY

Although in the past, most ERA studies were focused on outdoor industrial environments, nowadays in industrialized countries, people spend about 80%–90% of their time indoors (home, work, etc.), therefore a range of health effects are linked to the characteristics of indoor air quality (IAQ). If only outdoor concentrations are considered, risks due to pollutant exposure may be miscalculated. In fact, personal exposure is a function of exposures in different microenvironments that are determined by the pollutant level and the time spent in each microenvironment (home, work, outside, etc).

The sources of indoor pollutants are linked to: (1) building material (e.g., paint, varnishes, preservatives), (2) equipment (e.g., TV, computers), (3) occupant's activities (e.g., cooking, heating, tobacco smoke), as well as (4) outdoor pollutant infiltration. Common indoor pollutants may be: high CO_2 concentrations as a result of insufficient ventilation, high PM due to combustion processes, VOCs emitted from paints, flame retardants such as several organic and inorganic compounds based on bromine (e.g., high brominated aromatic and cycloaliphatic compounds) or chlorine (e.g., chloroparaffins, declorane plus) (Kumar et al. 2016).

Since data regarding indoor aerosol particles are not usually available, "Indoor Aerosol Models" may be used as an alternative to estimate the amount of outdoor particles that penetrate indoors. This may be useful in the absence of indoor sources in naturally ventilated buildings where indoor concentrations show similar trends to outdoor environments.

4.9.2   RISKS OF MIXTURES

Until recently, risk assessment was mostly performed separately for each chemical, and generally considered only a single route of exposure. Therefore, in order to build more real models, there is a need to address combined exposure to mixtures of multiple chemicals. The European Food Safety Administration (EFSA) and the European Commission's Directorate-General for Health and Consumers (DG SANCO) have made considerable progress in this area and have recently reported methodology for cumulative risk assessment (CRA) for dietary exposure to structurally related pesticides. However, several issues are still open, such as: how to prioritize chemicals for inclusion and the mode of action (s) to consider in testing? How to group chemicals in the right way into cumulative assessment groups (CAGs)? How to extrapolate results of mixtures *in vitro* bioassays and *in silico* models to humans, etc.? Some research projects, such as the EU project EuroMix (European Test and Risk Assessment Strategies for Mixtures) (H2020-SFS- 633172) are developing strategies for refining cumulative and aggregated risk assessment using *in silico* and *in vitro* tools that will be verified against *in vivo* data. In turn, expanding CRA methods to integrate and evaluate impacts of chemical and non-chemical stressors on the environment and health, is another major project area of research within the Human Health Risk Assessment Strategic Research Action Plan 2016–2019 (EPA 2015).

4.9.3   ENGINEERED NANOMATERIALS

The term engineered nanomaterials (ENMs) is attributed to particles that are deliberately produced with a physical size of less than 100 nm in at least two dimensions. During the last years, the development of nanomaterials has rapidly progressed, and, currently, there is a long list of commercially produced nanoparticles (i.e., SiO_2, TiO_2, Al_2O_3, C60, and Au) due to their wide range of applications in the fields of food, agriculture, medicine, electronics, textiles, etc. However, the implications of nanoparticles with respect to health and safety have not been taken into account by regulators yet. There is a need to evaluate the distribution of ENMs and their potential toxicological effects. An ERA framework for ENMs is considered to require the same information as ERA for traditional chemicals (Koelmans 2015). However, specific characteristics of ENMs need to be addressed. For example, current technology for environmental detection and measurements of ENMs is still in a developing stage. From the exposure point of view, in addition, apart from workers in nanotechnology companies, the general population will be exposed via lots of consumer products. Studies on biopersistence, bioaccumulation and ecotoxicity have only just started. Regarding toxicity evaluation is expected to be different from that of molecules, as well as from bulk materials of the same composition, because size and shape play an important role in the toxicity mechanisms. While trying to deal with all those limitations of ENMs ERA, computational modeling approaches, such as MendNano or SimpleBOx4nano, have emerged to estimate ENMs release rates, as well as their fate in the environment (Liu and Cohen 2014). Those models, which are based on LCIA approaches, are currently used to assess the potential environmental implications of ENMs.

4.10   THE ROLE OF RISK ASSESSMENT IN ENVIRONMENTAL MANAGEMENT DECISIONS

The risk assessment proposes to give the risk management team the best possible evaluation of all available scientific data, in order to arrive at justifiable and defensible decisions on a wide range of issues. In relation to the types of risk management actions necessary to an environmental contamination problem or hazardous situation, decisions could be made based on the results of a risk assessment. In fact, risk-based decision-making will generally result in the design of better environmental management programs, because risk assessment can produce more efficient and consistent risk reduction policies; at the same time, it can also be used as a screening tool for setting priorities.

Risk management based on the risk assessment result is the process of deciding what has to be done. Given the estimates of risk established, political and social judgment is required to decide which risk is acceptable; therefore an acceptable risk must be defined. On the other hand, risk assessment has several specific applications that could affect the types of decisions to be made in relation to environmental management programs. Further general discussion of some of the more prominent applications scenarios is offered in Sections 4.10.1 to 4.10.5.

4.10.1   ENVIRONMENTAL IMPACT ASSESSMENTS

Environmental impact assessment (EIA), as described in Chapter 1, is the analysis of the likely environmental consequences associated with a determined human activity. The main goal of an EIA is to ensure that environmental aspects are incorporated into planning for decision-making and implementation of development activities. Typically, an identification of factors contributing the most to overall risks of exposures to the environmental hazards of concern may be included. It may also incorporate an analysis of baseline risks, as well as a consistent process to document potential public health and environmental threats from the relevant activity. EIAs are designed to prevent or minimize the adverse impacts of a development activity while maximizing its beneficial effects.

As an example, quantitative risk assessment is often undertaken as part of the sitting process for newly proposed facilities. In many cases, it is carried out as a regulatory requirement for EIAs or, sometimes, is used for operating facilities to evaluate implications from design changes or changes in exposure parameters. In fact, the risk assessment of stack emissions from MSWI facilities seems to be one of the most important applications for this type of evaluation. This becomes necessary because MSWIs typically release various potentially toxic compounds—some of which escape pollution control equipment and enter the outside air. These chemicals may include metals (e.g., arsenic, cadmium, and mercury), organic compounds, such as polychlorinated dibenzo(p)-dioxins and furans (PCDD/Fs) (Vilavert et al. 2014).

4.10.2   ENVIRONMENTAL ASSESSMENT AND CHARACTERIZATION

Environmental assessments are invariably a primary activity in the general processes involved in the management of environmental assessment. The main goal of environmental assessments is to determine the nature and extent of potential impacts from the release or threat of emission of hazardous substances.

First, a representative sampling from the potentially contaminated media must be collected, along with historical data and sufficient details about the likely environmental contaminants. Sampling programs can be designed to search for specific chemical pollutants that become indicator parameters for sample analysis. For instance, a multimedia approach to environmental characterization can be adopted for most environmental pollution problems so that the significance of appropriate field sampling and analysis procedures increases. The activities included are expected to provide high-quality environmental data needed to support possible corrective action response decisions. To accomplish this, samples are collected and analyzed for the pollutants of concern. Proper protocols in field sampling and laboratory analysis procedures are used to help minimize uncertainties associated with data collection and evaluation activities.

The results from these activities will be a complete analysis of the pollutants detected in the environment. Risk assessment techniques and environmental characterization are typically applied to provide the development of effective environmental characterization programs. Thus, the information obtained is used to evaluate current and potential future risks to human health and the environment. In addition to information about the nature and magnitude of potential risks associated with potential environmental contamination problems, risk assessment also affords a basis for judging the need for mitigating actions. On this basis, corrective actions are developed and implemented with the main goal of protecting public health and the environment.

4.10.3 Corrective Action

A variety of corrective action strategies may be applied in case of contaminating processes to restore sites into healthier and more ecologically sound conditions. The processes involved will generally incorporate a consideration of the complex interactions among the environment, regulatory policies and the technical feasibility of remedial methodologies. A clear understanding of the fate and behavior of the pollutants in the environment is essential for developing successful corrective action response programs, and also to ensure that the problem is not exacerbated.

The design of corrective action response programs for contaminated site problems includes various formalized steps. In general, when the existing site information has been analyzed and a conceptual understanding of a site is obtained, then remedial action objectives should be defined for all impacted media at the contaminated site. Subsequently, alternative site restoration programs can be developed to support the requisite corrective action decision. Overall, risk assessment plays a very important role in the development of remedial action objectives for contaminated sites, the identification of feasible remedies that meet the remediation objectives and the selection of an optimum remedial alternative.

Risk assessment has become particularly useful in determining the level of clean-up most appropriate for potentially contaminated sites. By utilizing methodologies that establish clean-up criteria based on risk assessment principles, corrective action programs can be conducted in a cost-effective and efficient manner.

These procedures can help determine whether a particular remedial alternative will pose unacceptable risks following implementation and to determine the specific remedial alternatives that will result in the least risk upon achieving the clean-up goals or remedial action objectives for the site. Consequently, risk assessment tools can be used as an aid in the process of selecting among remedial options for contaminated sites.

4.10.4 Ecological Risks

Often, especially in the past, only limited attention has been given to the ecosystems associated with contaminated sites, as well as to the protection of ecological resources during site remediation activities. Instead, much of the focus has been on the protection of human health and resources directly affecting public health and safety. In recent times, however, the ecological assessment of contaminated sites has gained considerable attention. This is the result of prevailing knowledge or awareness of the intricate interactions between ecological receptors and systems and contaminated site clean-up processes.

In fact, remedial actions can alter or destroy aquatic and terrestrial habitats; the consequences of ecosystem disturbances and other ecological effects must therefore be given adequate consideration during the corrective action response process. Thus, it is important to integrate ecological investigation results and general concerns into the overall site clean-up process.

4.10.5 ENVIRONMENTAL MANAGEMENT STRATEGIES

Environmental pollution problems have reached an important level in most societies globally because pollution through chemicals represents a significant portion of the overall problem of environmental protection. The effective management of environmental pollution problems has certainly become an important environmental priority that will remain a growing social concern for next future. This is mainly because of the numerous complexities and inherent uncertainties involved in the analysis of such problems.

Whatever the cause of an environmental pollution problem, the impacted media must be remedied. However, restoration or clean-up may not be economically or technically feasible. In this case, risk assessment and monitoring the situation, together with institutional control measures, may be acceptable risk management strategies in lieu of remediation.

Overall, a risk assessment will generally provide the decision maker with scientifically defensible procedures for determining whether a potential environmental pollution problem could represent a significant adverse health effect, and whether environmental pollution problems could represent a significant adverse candidate for mitigative actions. In fact, the use of health and environmental risk assessments in environmental management decisions in particular, and a corrective actions program in general, is becoming an important regulatory requirement in several places. For instance, a number of environmental regulations and laws in various jurisdictions increasingly require risk-based approaches in determining clean-up goals and related decision parameters.

4.11 CASE STUDY: APPLICATIONAL ERA TO MSWI IN TARRAGONA, SPAIN

Consider the example of estimating risk by lifetime of a person living in the surroundings of the municipal solid waste incinerator (MSWI) under study. In this example, the methodology for estimating the distribution of daily PCDD/Fs intake for the population living near a MSWI is presented. First, risk assessment requires identification of the pathways through which people will be exposed to the potential chemicals of concern, in this case PCDD/Fs. The quantitative estimation of noncancer and cancer risks due to a PCDD/F exposure has been considered to be a combination of six pathways: ingestion of soil, ingestion of vegetation from the area, inhalation of re-suspended soil particles, inhalation of air, dermal adsorption and through diet. These pathways have been classified depending on whether they are due to direct deposition of the MSWI emissions or to an indirect exposure. Ingestion of soil, ingestion of edible vegetables from the area, dermal absorption, inhalation of re-suspended particles, and air inhalation have been considered direct exposure, while the diet exposure has been considered an indirect pathway of exposure.

The concentrations of PCDD/Fs were determined in soil and vegetation samples collected near the MSWI in Tarragona, Spain (Vilavert et al. 2015). Food samples, which were randomly obtained from local markets and supermarkets, were also analyzed for PCDD/Fs (Perelló et al. 2012).

We may begin by calculating the total intake via the six different exposure pathways.

1. *Ingestion of contaminated soil (Ings)*. Humans ingest small amounts of soil indirectly (hand-to-mouth transfer) when they work outdoors or during home gardening. Although outdoor workers can be exposed during the whole year, most people have contact with soil only when they work in their gardens. Exposure to soil is a function of the pollutant concentrations in soil and the individual consumption rate. Average daily dose resulting from ingestion of contaminated soil is (Table 4.5):

TABLE 4.5

Parameter Value for Direct Exposure for the Population Living in Area Surrounding an MSWI

Parameter	Symbol	Units	Value	References
Soil ingestion rate	SIR	mg/day	3.44	LaGrega et al. (1994)
Fraction absorption ingestion of soil	AFIS	unitless	40	Nessel et al. (1991)
Vegetable ingestion rate	VIR	g/day	160	Serra-Majem et al. (2003)
Fraction absorption ingestion of vegetables	AFIV	unitless	60	Nessel et al. (1991)
Fraction vegetables from the area	IA	unitless	5	Personal communication
Re-suspended particles from the soil	RES	unitless	50	Hawley (1985)
Ventilation rate	Vr	m^3/day	11	Shin et al. (1998)
Fraction retained in the lung	RET	unitless	60	Nessel et al. (1991)
Particle concentration	Pa	µg/m^3	133	Personal communication
Fraction absorption inhalation	AFIn	unitless	100	Nessel et al. (1991)
Contact time soil–skin	CT	h/day	1.50	EPA (1990)
Exposed skin surface area	SA	cm^2	1980	EPA (1990)
Dermal absorption factor	Add	unitless	0.003	Katsumata and Kastenberg (1997)
Soil to skin adherence factor	AF	mg/cm^2	1.00	EPA (1990)
PCDD/F concentrations in soil from the area	Sc	ng/kg	0.63	Vilavert et al. (2015)
PCDD/F concentrations in vegetables from the area	Vc	ng/kg	0.07	Vilavert et al. (2015)
PCDD/F concentrations in air	Ac	fg/m^3	10.1	Vilavert et al. (2015)

$$\text{Ings} = \text{Sc} \cdot \text{SIR} \cdot \text{AFIs}$$

where:

Sc: PCDD/F concentrations in soil (ng/kg)

SIR: soil ingestion rate (mg soil/day)

AFIS: fraction absorption ingestion of soil (unitless)

SIR varies depending upon the age of the individual, amount of outdoor/indoor activity, frequency of hand-to-mouth contact and seasonal climate (U.S. EPA 1990).

2. *Ingestion of vegetables from the area (Ingv).* In general, a high fraction of consumed vegetables is not grown in the region where a person lives but is imported from other regions. Therefore, only a fraction of 5% of total vegetables ingested (equivalent to locally grown) was considered. The average daily intake of TCDD equivalents was estimated by multiplying the PCDD/F concentrations in vegetables by the daily amount of intake, by the fraction of vegetables from the area, and by the absorption factor (Table 4.5):

$$\text{Ingv} = \text{Vc} \cdot \text{VIR} \cdot \text{AFIV} \cdot \text{IA}$$

where:

Vc: PCDD/F concentrations in vegetables (ng/kg)

VIR: vegetables ingestion rate (mg vegetables/day)

AFIV: fraction absorption ingestion of vegetables (unitless)

IA: fraction vegetables from the area (unitless)

3. *Inhalation of re-suspended particles of soil (Inhp)*. Contaminants that deposit to the ground become aggregated with soil particles. Natural and mechanical disturbances, e.g., wind, construction and demolition of buildings, can lead to a resuspension of particles from soil into the atmosphere. Along with this re-suspended dust, the adhering pollutants reach the atmosphere and are subsequently inhaled by people who live or work in the region. Inhalation exposure from emissions was calculated by assuming that individuals were exposed to contaminated air and that indoor air exposure was equal to outdoor exposure (Nessel et al. 1991) (Table 4.5):

$$Inhp = Sc \cdot RES \cdot Vr \cdot RET \cdot Pa \cdot AFIn$$

where:
   Sc: PCDD/F concentrations in soil (ng/kg)
   RES: fraction of re-suspended particles from soil (unitless)
   Vr: ventilation rate (m^3/day)
   RET: fraction retained in lungs (unitless)
   Pa: particle concentration (μg/m^3)
   AFIn: fraction absorption inhalation (unitless)

4. *Air inhalation (Inh*; Table 4.5). The inhaled quantity of an airborne pollutant depends mainly on the atmospheric concentration and the individual inhalation rate. Vapor and particle-bound pollutants are taken up likewise. The total daily intake was related to the body weight in order to obtain a daily inhalation dose.

$$Inh = Ac \cdot Vr \cdot AFIn$$

where:
   Ac: PCDD/F concentrations in air (fg/m^3)
   Vr: ventilation rate (m^3/day)
   AFIn: fraction absorption inhalation (unitless)

5. *Dermal absorption exposure (Ads)*. Dermal absorption was assumed to occur only in case of direct skin contact to contaminated soil. People come in contact with soil when they work outdoors or during home gardening. Although outdoors workers are exposed during the whole year to contaminated soil, most people are affected only a limited period of the year because bad weather impedes the stay in their garden or they tend to other weekend activities. Daily dermal exposure was estimated by the following model (Table 4.5):

$$Ads = Sc \cdot SA \cdot CT \cdot AF \cdot Add$$

where:
   Sc: PCDD/F concentrations in soil (ng/kg)
   SA: exposed skin surface area (cm^2)
   CT: contact time soil to skin (h/day)
   AF: soil to skin adherence factor (mg/cm^2)
   Add: dermal absorption factor (unitless)

6. *Ingestion through diet (Ingd)*. Human daily PCDD/Fs intake from diet is calculated by multiplying the concentration of PCDD/Fs in each food group by the amount of food group consumed daily and by the absorption fraction (Nessel et al. 1991). Food groups were the following: meat, eggs, fish, milk, dairy products, oil, cereals, pulses, vegetables and fruits (Table 4.6):

TABLE 4.6

Parameter Value for Ingestion through Diet Exposure for the General Population of Tarragona, Spain

Parameter	Symbol	Units	Value	References
Intake of meat and meat products	Imne	g/day	172	Serra-Majem et al. (2003)
Intake of eggs	Ineg	g/day	31	Serra-Majem et al. (2003)
Intake of fish and seafood	Infi	g/day	68	Serra-Majem et al. (2003)
Intake of milk	Inmi	g/day	128	Serra-Majem et al. (2003)
Intake of dairy products	Indm	g/day	76	Serra-Majem et al. (2003)
Intake of oils and fats	Inol	g/day	27	Serra-Majem et al. (2003)
Intake of cereals	Ince	g/day	224	Serra-Majem et al. (2003)
Intake of pulses	Inpu	g/day	30	Serra-Majem et al. (2003)
Intake of vegetables	Inve	g/day	160	Serra-Majem et al. (2003)
Intake of fruits	Infr	g/day	194	Serra-Majem et al. (2003)
PCDD/F concentrations in meat	Mec	ng/kg	0.111	Perelló et al. (2012)
PCDD/F concentrations in eggs	Egc	ng/kg	0.032	Perelló et al. (2012)
PCDD/F concentrations in fish	Fic	ng/kg	0.120	Perelló et al. (2012)
PCDD/F concentrations in milk	Mic	ng/kg	0.005	Perelló et al. (2012)
PCDD/F concentrations in dairy products	Dmc	ng/kg	0.056	Perelló et al. (2012)
PCDD/F concentrations in oils	Olc	ng/kg	0.086	Perelló et al. (2012)
PCDD/F concentrations in cereals	Cec	ng/kg	0.007	Perelló et al. (2012)
PCDD/F concentrations in pulses	Puc	ng/kg	0.003	Perelló et al. (2012)
PCDD/F concentrations in vegetables	Vec	ng/kg	0.002	Perelló et al. (2012)
PCDD/F concentrations in fruit	Frc	ng/kg	0.003	Perelló et al. (2012)
Fraction absorption ingestion in food	AFIF	unitless	60	Nessel et al. (1991)

$$Ingd = SFc \cdot FIR \cdot AFIF$$

where:
Fc: PCDD/F concentration in a food group (ng/kg)
FIR: food ingestion rate (mg food/day)
AFIF: fraction absorption ingestion food (unitless)

The addition of the PCDD/F amount through the different pathways gives the total dose.

Table 4.7 summarizes the value of the direct exposure to PCDD/Fs by the population living in the proximities of the MSWI of Tarragona, Spain. Inhalation of air from the area and ingestion of vegetables are the pathways that contribute more to the direct exposure. The other pathways of exposure—dermal absorption, soil ingestion and inhalation—are, in fact, minimal. With regard to PCDD/F exposure through diet, Table 4.8 shows the daily intake of PCDD/F from different food groups and from total diet. The intake of fish is the exposure daily intake that contributes more to the total diet exposure. Dairy products, fat and meat have a considerable contribution to the total diet exposure. The results corresponding to the diet exposure are based on a Mediterranean diet (See Tables 4.7 and 4.8).

Once the human daily intake of PCDD/Fs through the different exposure pathways has been calculated, the noncarcinogenic and carcinogenic risks can be calculated. To determine if the contaminant poses a noncancer risk to human health, daily intake is compared with the reference dose (RfD) for chronic exposure. The carcinogenic risk is calculated by multiplying the estimated dose

TABLE 4.7
Different Types of Direct PCDD/Fs Exposure[a] of Population Living in Proximity of the MSWI

Type of Exposure	Mean
Soil ingestion	9.23×10^{-7}
Vegetable ingestion	3.40×10^{-4}
Inhalation of air	2.77×10^{-6}
Dermal absorption	9.84×10^{-7}
Total direct exposure	3.45×10^{-4}

[a] ng WHO-TEQ/day. WHO-TEQ/day stands for toxicity equivalent (TEQ) defined by the World Health Organization (WHO) per day.

TABLE 4.8
Daily Intake of PCDD/Fs[a] from the Diet

Food Group	Mean
Meat	1.27
Eggs	1.09
Fish	3.54
Milk	0.70
Dairy products	1.95
Fat	1.34
Cereals	1.41
Pulses	0.09
Vegetables	0.34
Fruits	0.47
Total diet	15.72

[a] pg WHO-TEQ/day. WHO-TEQ/day stands for toxicity equivalent (TEQ) defined by the World Health Organization (WHO) per day.

TABLE 4.9
Parameter Values for PCDD/Fs

Parameter	Symbol	Units	Value	References
Body weight	BW	kg	67.52	—
Noncarcinogenic potency factor	NCP	pg/kg/day	1–4	
Carcinogenic potency factor	CP	$(mg/kg/day)^{-1}$	34,000–56,000	Katsumata and Kastenberg (1997)

Source: PCDD toxicity values from: IRIS.

by the carcinogenic potency factor for PCDD/Fs (Table 4.9). The predicted carcinogenic risk is an upper-bound estimate of the potential risk associated with exposure. Table 4.10 shows the noncarcinogenic and carcinogenic risks from direct, indirect (= diet) and total exposure.

The total cancer risk is 7.34×10^{-5}, which means that a person living in the surroundings of the MSWI has chance of less than one in a million of developing cancer during his or her lifetime.

TABLE 4.10
Noncarcinogenic and Carcinogenic Risks by Direct Diet, Direct Risk, and Total Exposure

Noncarcinogenic		Carcinogenic	
Direct risk	3.15×10^{-3}	Direct risk	2.77×10^{-7}
Diet risk	6.79×10^{-1}	Diet risk	7.31×10^{-5}
Total exposure	6.82×10^{-1}	Total exposure	7.34×10^{-5}

The total noncancer risk is 6.82×10^{-1}, which means that the population exposure does not exceed the threshold value. Neither the emissions from the MSWI nor the indirect exposure (diet) to PCDD/Fs in the Tarragona area would mean an additional noncarcinogenic risk for health to general population living in the area.

The Integrated Risk Information System (IRIS) is a human health assessment program that evaluates quantitative and qualitative risk information on effects that may result from exposure to environmental contaminants. IRIS was initially developed for EPA staff in response to a growing demand for consistent information on substances for use in risk assessments, decision-making and regulatory activities. The California Environmental Protection Agency (CALEPA), Office of Environmental Health Hazard Assessment's (OEHHA), provides Chronic Reference Exposure Levels (RELS) from December 18, 2008, and the Cancer Potency Values (PDF) from July 21, 2009.

Another point is that, in both cases (noncarcinogenic and carcinogenic risk), the total risk is due mainly to diet. Consequently, under the present conditions, Tarragona's MSWI would not cause a substantial additional exposure to PCDD/Fs in the area under potential influence of the plant (see Table 4.9).

4.12 CASE STUDY: APPLICATION OF HHRA TO NANOPARTICLE PRODUCTION IN INDUSTRIAL APPLICATIONS

The following section presents a human health risk assessment (HHRA) on the occupational exposures that can occur during the manufacturing and handling of ENM in the workplace. The case study specifically focuses in on nano-TiO_2 and PCBM. Both of these ENMs are used in the manufacturing of OPV solar panels, introduced in Chapters 2 and 3, and was highlighted as a part of the LCIA results in Chapter 3. There were no estimations of nano-TiO_2 or PCBM emissions in the workplace and their potential impact on human health.

Occupational settings involve the manufacturing, handling and storage of ENM where workers are in close proximity to crude amounts of ENM (Aschberger et al. 2010; Bouillard and Vignes 2014; Kuempel et al. 2012; Schneider and Jensen 2009; Stebounova et al. 2012) Additionally, these situations may present ENM at uniquely elevated concentrations, small uniform sizes, with distinct shapes and structures not normally found in the environment (Buzea et al. 2007). Although the actual synthesis of ENM may occur in closed reaction chambers, exposure during handling, transporting and post-treatment are all viable possibilities (Biswas and Wu 2005; Brouwer et al. 2013; Nowack et al. 2012). Industrial-scale development will lead to large volumes of ENM being produced and utilized during OPV production. Proper emissions controls such as personal protective equipment may theoretically be in place, actual uptake of such equipment has been shown to be inconsistent and often low (Kimberyly-Clark 2012).

4.12.1 HAZARD IDENTIFICATION OF C_{60}-FULLERENES AND PCBM

The principal properties specific to the functionality of C_{60}-fullerenes, and thus PCBM, are closely related to its ability to reversibly accept upwards of six additional electrons (Taylor and Walton 1993;

Wudl 1992). C_{60}-fullerenes are also nearly insoluble; however, functionalization can produce variants of this molecule such as PCBM that are, in effect, soluble (Hummelen et al. 1995). Due to their very small size and classification as an ENM, there is an initial assumption that some of the toxicological trends seen with other spherical ENM may be relevant for C_{60}-fullerenes (Johnston et al. 2010a). On the contrary, and as a consequence of their electrophilic nature, C_{60}-fullerenes have been shown to act as free-radical sponges and have anti-inflammatory effects (Bogdanović et al. 2008; Gharbi et al. 2005; Huang et al. 2008; Injac et al. 2009; Lin et al. 2002; Liu et al. 2009; Roursgaard et al. 2008; Tykhomyrov et al. 2008). This may be the reason why some inhalation *in vivo* toxicity studies have found minimal inflammation, (Fujita et al. 2009) no inflammation (Baker et al. 2007; Morimoto et al. 2010) or even anti-inflammatory (Roursgaard et al. 2008) effects upon exposure to particles of C_{60}-fullerenes. Several *in vivo* studies do, however, demonstrate the pro-inflammatory nature of C_{60}-fullerenes as they have been associated with enhanced reactive oxygen species production, damage to cell membrane integrity and lipid peroxidation, for example (Kamat 2000; Sayes et al. 2005). Very few studies on dermal exposure have been carried out, demonstrating limited to no uptake by the skin and a lack of evidence that C_{60}-fullerenes could result in skin irritation (Aoshima et al. 2009; Huczko et al. 1999). Upon oral administration, limited uptake by the digestive tract has been demonstrated (Yamago et al. 1995) with no obvious toxicological effects taking place either in the gut or systemically (Chen et al. 1998; Mori et al. 2006).

4.12.2 Hazard Identification of Titanium Dioxide Nanoparticles

Titanium dioxide is an important inorganic metal oxide used in industry. Industrial-scale production of TiO_2 began in the early twentieth century and has steadily increased since then. Industrial use of TiO_2 is generally in the rutile or anatase forms, although the rutile form has found greater commercial use. The range of nano-TiO_2 induced *in vivo* toxicological responses in the respiratory system spans inflammation, histopathology (e.g., endothelial or epithelial changes), cytotoxicity, oxidative stress, oncogenic effects (e.g., increase rates of carcinoma) and other genotoxic-related events (Bermudez et al. 2004; Chang et al. 2013; Czajka et al. 2015; Heinrich et al. 1995; Hext et al. 2005; Iavicoli et al. 2012; Johnston et al. 2010b; National Institute for Occupational Safety and Health 2011; Noël and Truchon 2015; Shakeel et al. 2016; Shi et al. 2013; Warheit 1996; Zhang et al. 2015). In certain cases, these effects were found to be more pronounced for nano-TiO_2 compared to its bulk counterpart (Varner et al. 2010). In their 2011 report, NIOSH specifically addressed concerns over nano-TiO_2, acknowledging a potentially higher hazard potential in terms of inflammation and carcinogenicity of corresponding nanoparticles of this material as compared to bulk sizes (National Institute for Occupational Safety and Health 2011). For dermal exposure, there is limited to no evidence from *in vivo* animal studies showing nano-TiO_2 is able to penetrate the (intact) skin (Filipe et al. 2009; Furukawa et al. 2011; Monteiro-Riviere et al. 2011; Newman et al. 2009; Sadrieh et al. 2010). Penetration often remains to the superficial layers of the epidermis (Sadrieh et al. 2010), however dermal (Wu et al. 2009) and hypodermal penetration, such as in the hair follicles (Bennat and Müller-Goymann 2000; Senzui et al. 2010), has been observed. Potential toxicological effects from dermal exposure range from limited to no skin irritation (Warheit et al. 2007) to oxidative stress (Unnithan et al. 2011) and lesions in certain organ systems (Wu et al. 2009). Oral administration to animals has demonstrated their uptake particularly to the liver, but also other organs such as the kidney, brain and testes (Wang et al. 2007). Potential toxicological effects from oral exposure range from oxidative stress and neuroinflammation (Ze et al. 2014) to necropathy, (Yakubu 2012; Shrivastava et al. 2014; Wang et al. 2007) for example.

4.12.3 Methods

Thus, the lack of sufficient positively-correlated toxicological data on C_{60}-fullerenes or PCBM, particularly for *in vivo* long-term multi-dose studies, precludes its further analysis from the HHRA. This decision not to carry out a quantitative HHRA on C_{60}-fullerenes should not be seen as a formal determination that these materials pose no human health risk across the life-cycle of OPV, since as the qualitative exposure assessment discussed, there is a high likelihood that emissions will occur along the OPV life cycle. On the other hand, current human health relevant toxicological literature does not point to clear and consistent adverse responses nor is there sufficient data to carry out a quantitative assessment. There is, however, sufficient evidence in the literature that points to both adverse non-cancerous and cancerous, chronic and sub-chronic human relevant toxicity of nano-TiO$_2$. Sufficient data also exist to carry out an appropriate dose–response analysis of this material. The remainder of this chapter focuses on the quantitative HHRA of nano-TiO$_2$ in occupational exposure scenarios in the OPV life cycle.

4.12.4 DOSE–RESPONSE ASSESSMENT

The benchmark dose (BMD) method was used to quantify the exposure concentration at which the adverse health effect occurs (i.e., dose–response relationship). This concentration was estimated from the corresponding, pre-defined benchmark response (BMR). The BMD is an alternative to the no-observed-adverse-effect-level (NOAEL) and lowest-observed-adverse-effect-level (LOAEL). In this chapter, the terminology benchmark concentration (BMC) is used to clarify that the dose–response modeling used data correlating to exposure concentrations from whole body animal studies. Calculation of the BMC was completed using the Netherland's National Institute for Public Health and the Environment's (RIVM) PROAST software (www.rivm.nl). PROAST was used to first calculate a BMC$_{animal}$ (BMC$_a$) based on the *in vivo* animal data. The BMC$_a$ was extrapolated to a human effect threshold referred to as the BMC$_{human}$ (BMC$_h$), by applying the appropriate extrapolation factors (European Chemicals Agency 2012) using Equation 4.7:

$$\mathrm{BMC}_h = \frac{\mathrm{BMC}_a}{\mathrm{EF}_{\mathrm{inter}} \cdot \mathrm{EF}_{\mathrm{intra}} \cdot \mathrm{UF}_i} \tag{4.7}$$

where:
  BMC$_h$: Human equivalent benchmark concentration
  BMC$_a$: Benchmark concentration based on animal toxicological data
  EF$_{\mathrm{inter}}$: Interspecies extrapolation factor
  EF$_{\mathrm{intra}}$: Intraspecies extrapolation factor
  UF$_i$: sources of uncertainty (i)

Extrapolation factors represent differences between and among species according to anatomical differences (e.g., body size) and physiological functions (e.g., metabolism). Values are scaled with the extrapolation factors according the allocation scaling principle. The difference (i.e., increase) in breathing rates for persons exposed in the occupational setting are assumed to be elevated compared to persons at rest or not working and are captured within the EF$_{\mathrm{intra}}$.

No relevant chronic *in vivo* multi-dose studies were found in the literature, although there was one such sub-chronic study (Bermudez et al. 2004) (i.e., 90-day exposure duration) about whole body inhalation exposure to nano-TiO$_2$. Bermudez et al. exposed rats, mice and hamsters to uncoated, nano-TiO$_2$ obtained from Evonik (formerly DeGussa) with a nominal particle diameter of 21 nm as supplied by the manufacturer (Bermudez et al. 2004). Each species group was exposed to nano-TiO$_2$ at concentrations of 0.0 mg/m^3 (control), 0.5 mg/m^3, 2.0 mg/m^3 and 10 mg/m^3 for 6 hours/day, 5 days/week, for 13 weeks, while the control group received filtered air only (Table 4.11).

Bronchoalveolar lavage (BAL) was completed on the lungs of the sacrificed animals at varying post-exposure times to measure the counts of macrophages, neutrophils, eosinophils and lymphocytes. Only the effects at post exposure time zero (i.e., immediately following completion of exposure) were considered in the dose–response assessment. Bermudez et al. (2004) reported that statistically significant changes in the BAL results were limited to the macrophages, neutrophil and

TABLE 4.11

Summary of the Study (Bermudez et al. 2004) and Select Dose–Response Data Used to Characterize the Inflammatory Response upon Inhalation Exposure to Nano-TiO$_2$

Nanomaterial	Species	Exposure Type	Concentration (mg/m³)	Effects	BMR
21 nm P25 Degussa (Evonik)	Rat, Mice	13 weeks (6 h/day; 5 days/wk) whole-body inhalation	Control, 0.5, 2.0, 10	Pulmonary inflammation	20% increase in neutrophil cell counts

Control: Filtered air with corresponding concentration of 0.0 mg/m³.
BMR: Benchmark response.

lymphocyte cell types in the highest dose-groups. Furthermore, only data from the mice and rats were found to have statistically significant changes in percent cell counts over the control and thus the hamster data was not considered in the dose–response assessment.

4.12.5 Exposure Assessment

Measurements for occupational exposure to ENM are not abundant and the literature generally focuses on airborne concentrations as opposed to internal lung, internal oral, or dermal loading doses (Brouwer et al. 2016; Pietroiusti and Magrini 2014). There are currently no models to estimate dermal or oral exposure from airborne concentrations. Thus, this HHRA focuses on the inhalation exposure to nano-TiO$_2$ in the occupational setting. All exposure assessments were made with the near-field–far-field exposure assessment model in NanoSafer v.1.1. (Hristozov et al. 2016; Jensen et al. 2016; Liguori et al. 2016). The three exposure scenarios (ES) considered in this chapter (Table 4.12) were based on prompts taken from the MARINA (www.marina-fp7.eu) and NANEX (www.nanex-project.eu) databases.

MARINA was the Seventh Framework Program (2007–2013) of the European Union (EU FP7) project aimed at developing risk management methods for ENM, including development of occupational release and ES. NANEX was also an EU FP7 project which aimed to catalogue potential exposure to EMN across the life cycle including their manufacturing and industrial

TABLE 4.12

Description of the Exposure Scenarios Used for the Human Health Risk Assessment of Inhalation Exposure to Nanoparticles of Titanium Dioxide in the Occupational Workplace

ES1	Transfer (dumping) of powder from a 10 kg bag into a mixing tank over 10-minute work cycles with 10-minute pauses in-between over an 8-hour workday. Work is performed in a room with high ventilation rate. The process is assumed to be of high energy with the pouring height assumed to be 0.3–1 m corresponding to $H_i = 0.8$.
ES2	Transfer (dumping) of powder from a 560 kg container into a larger holding vessel over 10-minute work cycles with 20-minute pauses in-between over an 8-hour workday. Work is performed in a hall with low ventilation rate. The process is assumed to be of high energy with the pouring height assumed to be 0.3–1 m corresponding to $H_i = 0.8$.
ES3	Continuous filling (pouring) of bag bin with a total 250 kg of powder for 8 hours. Work is performed in a room with low ventilation rate. The process is assumed to be of high energy with the pouring height assumed to be 0.3–1 m corresponding to $H_i = 0.8$.

H_i: handling energy factor.

TABLE 4.13

Parameters Used in the NanoSafer v1.1 Exposure Assessment Model

No.	Exposure Scenario	E_i, [mg/min]	H_i	t_{wc}, [min]	p_{wc}, [min]	n_{wc}	$A_{transfer}$ [kg]	V_{tot} [m³]	AER [h⁻¹]
ES1	Manufacturer dumping into mixing tank	1.20E+01	0.8	10	10	24	10	100	8
ES2	Dumping large amount of powder in vessel	6.72E+02	0.8	10	20	16	560	75	4
ES3	Bag Bin filling	6.26E+00	0.8	480	0	1	250	70	4

H_i = handling energy factor, t_{wc} is work cycle time, p_{wc} is pause between work cycles, n_{wc} is number of work cycles, $A_{transfer}$ is amount of material transferred per transfer event within each work cycle, V_{tot} is the total volume of the work room, and AER is the general air exchange ratio in the work-room.

use. These scenarios correlated with workplace activities that take place during the production of nano-TiO$_2$. However, it was assumed that similar workplace scenarios would present themselves during OPV manufacturing activities. The NanoSafer v.1.1 model parameters used for describing the emissions and estimation of exposure concentration are given in Table 4.13.

The near-field (NF) was defined as a box with a fixed volume of 2.3 m³ that surrounds the ENM emission source. The far-field (FF) was defined as a box equal to the total volume of the work room (V_{tot}, m³) minus the NF volume. Emissions (E_i, mg/min) were quantified (Equation 4.8) as a function of the respirable powder dustiness index (DI$_{resp}$, mg/kg), determined using the rotating drum system (European Committee for Standardization 2013).

$$E_i = \left[\frac{A_{transfer}}{t_{transfer}} \right] \cdot \left[\left(DI_{resp} \right) \cdot \left(H_i \right) \right] \tag{4.8}$$

where:

E_i: Emissions

$A_{transfer}$: Amount of nano-TiO$_2$ handled during work-cycle

$t_{transfer}$: Time length of work-cycle

DI_{resp}: Dustiness index of nano-TiO$_2$

H_i: Handling energy factor of work-type

$A_{transfer}$ (kg) is the amount of powder transferred during a work cycle, of given time-length $t_{transfer}$ (min), and H_i is the handling energy factor (unit-less). H_i is based on a scale of zero to one, whereby zero is a no- energy event (e.g., no handling of the material) and one is a high energy event (e.g., crushing, dropping from greater than 2 m height) (Ramachandran 2005). In addition to those parameters used to derive source-term E_i, NanoSafer v.1.1 includes the duration of the activities and work cycle (t_{wc}), the number of work cycles (n_{wc}), the volume of work room (V_{tot}), and the general air-exchange rate (h⁻¹) in the work-room to estimate the NF and FF concentrations. Daily (8-hour) exposure potentials were calculated in both the NF and FF (Jensen et al. 2016). The DI$_{resp}$ for the TiO$_2$ as considered in this chapter was 15 mg/kg, resulting in a standard deviation of 2.5 mg/kg. The standard deviation was then applied to E_i for quantifying the exposure distribution of each ES. The final airborne exposure concentrations are reported as average 8-hour time weighted averages.

4.12.6   Risk Characterization

Risk was calculated as a RCR (Equation 4.9).

$$RCR = \frac{EXP_i}{BMC_h} \tag{4.9}$$

RCR: Risk characterization ratio
EXP_i random sample from the exposure concentration distribution for scenario i

A risky scenario was defined as RCR values ≥ 1.0 (i.e., exposure levels higher than the BMC_h) and represents the risk potential of a worker exposed to the 8-hour time weighted indoor air nano-TiO_2 concentrations over 45-years, 50 weeks/year, 5 days/week. The RCR distributions were calculated by sampling the BMC_h distributions and the three NF and three FF exposure distributions.

4.12.7   Results

4.12.7.1   Dose–Response Analysis

In general, a BMR (i.e., the relevant toxicological response of concern or safety) should approach a lower limit of reasonably measurable effects. For changes in total white blood cell counts relevant to inflammation, increases in 10% over the background response are often found to be significant (Crump 1984). However, due to uncertainty in measuring percent changes as opposed to total cell count, significant macrophage, neutrophil and lymphocyte percent changes were determined using a BMR of 20%. For both mice and rats, neutrophil and lymphocyte percent changes increased whereas macrophage changes decreased. Ultimately, the dose–response analysis was completed using the neutrophil rat data since (i) the decrease in macrophage percent change was directly correlated with the measured change in neutrophils and lymphocytes, (ii) neutrophil percent increases were much stronger and dominant compared to lymphocyte percent changes and lastly (iii) rats showed inflammatory responses at lower concentrations than did mice (Figure 4.11).

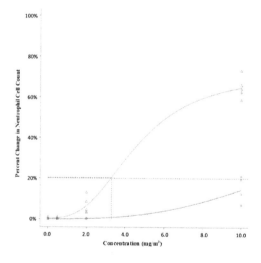

FIGURE 4.11   Fitted log-logistic models using PROAST software with reported confidence intervals to the mice and rat neutrophil percent changes upon inhalation of nano-TiO_2. The dose–response results demonstrate differences in the slope of the lines per species, with a much more sensitive response for rats. The does-response curve shown in this example was fitted with a log-likelihood of -1374. The benchmark dose for rats is shown by the lower curve corresponding to data with large circles, while the benchmark dose for mice is shown for the upper curve using corresponding triangular data points.

TABLE 4.14

Daily Averaged, Inhalation Benchmark Concentrations (mg/m³) for *In Vivo* Animal Studies and Corresponding Models Fit for a 20% Increase in Neutrophil Count in Mice

Model	Two-Stage	Log-Logistic	Weibull	Log-Probabilistic	Gamma	Exponential	Hill	Aggregation of Models
Median	12.76	12.54	13.04	12.39	12.77	11.44	11.61	**12.32**
Mean	12.84	12.62	13.15	12.47	12.87	11.48	11.66	**12.44**
Standard Deviation	1.15	1.05	1.24	1.06	1.11	0.64	0.99	**1.20**
Minimum	9.25	9.58	9.60	9.37	9.66	9.57	8.82	**8.82**
Maximum	18.20	19.54	21.01	18.39	18.85	14.66	17.78	**21.01**

For rats, seven valid models were fitted to the neutrophil data (Table 4.14). The models were aggregated into an overall daily averaged inhalation concentration (i.e., BMC$_a$) distribution that was normally distributed with a mean of 3.71 ± 0.56 mg/m³. After applying EF$_{inter}$ and EF$_{intra}$, the BMC$_h$ had a mean of 0.91 mg/m³.

4.12.7.2 Exposure Assessment

The daily NF and FF indoor air concentrations of nano-TiO$_2$ were calculated for three different exposure scenarios that ultimately involved three different amounts of nano-TiO$_2$ throughout the 8-hour workday. The applied emission rates ranged from a low of 6.26 mg/min (ES3) to a high of 672 mg/min (ES2), while the value for ES1 was 12.0 mg/min. Figure 4.12 shows the time-integrated exposure concentrations in each of the exposure scenario.

Geometric means representing 8-hour time weighted averages for NF airborne concentrations of nano-TiO$_2$ ranged from a low of 0.825 mg/m³ for ES3 and a high of 36.2 mg/m³ for ES2, while the NF concentration for ES1 was 0.93 mg/m³ (Table 4.15).

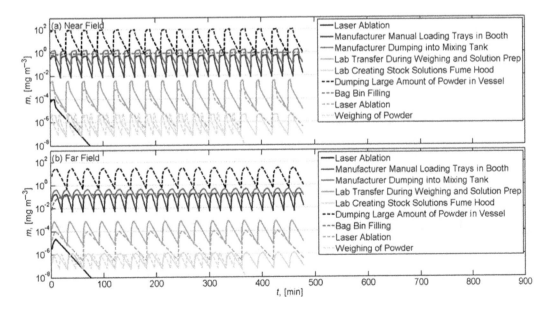

FIGURE 4.12 Potential exposure time-series in the (a) near field and (b) far field. Note: This figure shows the results for a larger range of exposure scenarios than presented in this chapter. These additional scenarios were completed as a part of a larger publication out of direct context of this chapter.

TABLE 4.15

Calculated Near-Field and Far-Field Airborne Concentrations of Nano-TiO$_2$ for the Three Separate Exposure Scenarios Considered in the Human Health Risk Assessment

No.	Exposure Scenario	C_{NF} [mg m^{-3}]	C_{FF} [mg m^{-3}]
ES1	Manufacturer dumping into mixing tank	8.93E−01	2.59E−01
ES2	Dumping large amount of powder in vessel	3.62E+01	1.18E+01
ES3	Bag Bin filling	8.25E−01	1.60E−01

Note: These are the geometric means 8-hour time weighted averages that were all defined as having a geometric standard deviation of 2.5.

C: exposure concentration.

Large variations in the NF and FF concentrations observed between the exposure scenarios were due to differences in the time-integrated substance emission, the dilution by room size, and the air exchange rate. As expected, the resulting exposure potentials varied directly with the scenario-specific integrated emission levels given by the emission rate and duration of the process resulting in a higher exposure potential for ES2 than in the other scenarios. In the case of ES2, the high emissions and small workroom volume were coupled with a high handling energy factor (i.e., dumping powder into a vessel) and lower air exchange rate (4 h^{-1}) as compared to the conditions in the other scenarios. Compared to the NF, the FF exposure concentrations were consistently lower but generally less than an order of magnitude lower. As was the case for the NF exposure scenarios, geometric means for FF exposure concentrations ranged from a low of 0.16 mg/m^3 for ES3 and a high of 11.8 mg/m^3 for ES2, while the value for ES1 was 0.259 mg/m^3.

4.12.7.3   Risk Characterization and Uncertainty

Table 4.16 displays the results of the risk characterization for the NF and FF exposure scenario, respectively, representing the risk potential of a worker exposed to an 8-hour time weighted indoor air nano-TiO$_2$ concentrations over 45-years, 50 weeks/year, 5 days/week.

The results demonstrate that there is only the potential for inflammation in workers handling nano-TiO$_2$ in ES2. The risk potential occurs both in the NF and FF; however, the RCR for the NF is nearly 3-fold larger than the FF. ES2 specifically dealt with handling and use of high volume nano-TiO$_2$ powders over 8-hour workdays. The risk to lung inflammation was calculated as a 20% increase in certain white blood cells over background levels; however, this response level only

TABLE 4.16

Summary of the Risk Characterization (Reported as Risk Characterization Ratios) Distributions for Each Near- and Far-Field Exposure Scenarios. Results Represent 10,000 Monte-Carlo Simulations.

Exposure (Near Field)	ES1	ES2	ES3
Risk characterization ratio	9.91E−02	3.99E+00	9.05E−02
Exposure (Far Field)	**ES1**	**ES2**	**ES3**
Risk characterization ratio	2.83E−02	1.29E+00	1.77E−02

indicates the onset of inflammation as opposed to the degree and/or severity of the pathology. OPV are not produced at the industrial scale and thus the exposure scenarios presented in this thesis do not represent a specific OPV industrial profile. Instead, potential and realistic exposure scenarios that involve the production and handling of nano-TiO_2 were presented, and, therefore, provide a first-tier assessment for OPV. For example, if OPV devices utilizing PCBM and/or TiO_2 come to full industrial scale and uptake to the marketplace, large volumes of these substances will be handled, loaded and dispensed either into other containers, holding tanks or vessels for mixing and/or application to the OPV panel. ES1 and ES3 involve exposure scenarios for nano-TiO_2 powder, ranging over an order of magnitude less compared with ES2. Thus, these represent production volumes that are much more moderate than ES2.

It should be noted that the fate and exposure model, NanoSafer, used in this work largely relied on advection and bulk air flow to estimate the concentration of nano-TiO_2 in the workplace. The exposure also is based on external airborne concentrations as opposed to internal doses. Particle sizes are an important characteristic influencing the effectiveness of ventilation control, personal protective equipment and ultimately where ENM retention will occur in the lung. For example, ENM might lodge deeper and more uniformly in the lung (i.e., alveoli) as compared to larger-size materials (Organization for Economic Cooperation and Development 2012). Modifying factors (e.g., use of personal protective equipment) were not considered and should be considered for future assessments. For example, Fransman et al. (2011) provides protection factors for common localized controls and personal protective equipment, although these can deviate significantly from applied protection factors for ENM (Koivisto et al. 2015). Additionally, a few occupational exposure studies have shown that particle size distributions in some cases may be similar for both ENM and conventional powders (Koivisto et al. 2012, 2014). Such considerations could be included in models that utilize size-resolved concentration data (e.g., mass median aerodynamic diameters) in their models (Bermudez et al. 2004; Heinrich et al. 1995; Oberdorster et al. 1994). Given these relative uncertainties, the exposure assessments and corresponding risk calculations in this HHRA can be considered as worst-case scenarios.

4.13   CONCLUSIONS

These findings presented in this chapter suggest that the potential human health impacts from ENM used in OPV cannot, by default, be assumed to be negligible and may, in fact, be highly probable. This issue may not necessarily be relevant to the current technology and production level of OPV, but should be considered for future scale-ups of these devices. It is important to note that the HHRA results are not scaled to the functional units defined by the LCA in Chapters 2 and 3. The difference in scopes means that the risk pertains to the handling of a mass of the ENM that is independent of exact amounts of nano-TiO_2 required to fulfill the functional unit (e.g., production of 1 Wp of OPV). Therefore, the risk cannot be directly interpreted in terms of its overall relevance to the OPV life cycle, particularly any prospective analysis of future, large-scale production of this technology. Of course, the exact exposure to these ENM used at current lab- and small-scale production per functional unit could, in some cases, be determined for the HHRA. While these results might be beneficial from a health and safety perspective for laboratory managers, for example, it may ultimately provide little relevance from an eco-design perspective where material choices and occupational workplace conditions at the industrial scale are being extrapolated from lab- or research-based scenarios and where uncertainty is thus high. Instead, potential human health impacts calculated within LCA, as opposed to absolute values of risk, might prove more useful for early design and development of emerging technologies. The next chapter presents the methods and results of calculating the potential human health impacts for ENM within the context of life-cycle impact assessment and the OPV-LCA case studies presented in earlier chapters.

4.14   QUESTIONS AND EXERCISES

1. Indicate possible and ideal target/indicator species for an ERA in the following cases: (a) dioxins emission of a MSWI; (b) acid rain; (c) indoor radiation. Explain their advantages and disadvantages.

2. Which aspects do you consider the main difficulties in the selection of target species of an ecosystem to overall risk assessment? How do you think these problems can be noticeably improved?

3. Describe the main concept of risk assessment and sum up the main fields and boundaries of application, as well as the main tools utilized in it.

4. Calculate an acceptable concentration (8-h time-weighted average) to prevent cancer effects in workers where there is working lifetime exposure to an airborne threshold toxicant. The pollutant has a potency factor of 0.002 $(mg/kg/day)^{-1}$, the absorption factor is estimated at 80%, and the exposure time is 5 days per week, 50 weeks per year over a 30-year period. The worker is assumed to breathe for 3.5 h per workday at the rate of 1.5 m^3/h and 3.5 h per workday at a moderate breathing rate of 1 m^3/h.

5. The reference dose (RfD) for arsenic is set at 3.0 $\times 10^{-4}$ mg/kg/day and the carcinogenic oral scope is 1.75 $(mg/kg/day)^{-1}$. Discuss which would be more stringent: oral concentration standard based on a carcinogenic risk or on RfD.

6. A 70-kg person is exposed to 1.33 ng/m^3 of cadmium in the air and also consumes an average of 53 g fish per day twice a week taken from a contaminated river with a cadmium concentration of 0.2 µg/g. The reference concentration is 2.0 $\times$ 10^{-2} $µg/m^3$, reference dose is 1.0 $\times$ 10^{-3} mg/kg/day and the inhalation unit factor $(mg/m^3)^{-1}$. Calculate his or her lifetime cancer and noncancer risks.

7. Suppose a 70-kg person consumes an average of 5.6 g fish per day with a concentration of DDT (oral potency factor = 0.34 $(mg/kg-day)^{-1}$) equal to 50 ppb (0.05 mg/L). Calculate the maximum lifetime cancer risk from this source.

8. Suppose that a 70-kg individual eats 7 g fish per day taken from a river contaminated by methylene chloride and that the bioconcentration factor (BCF) is 2 L of water per kilogram of fish. What concentration of methylene chloride (mg/L) in the river water would produce a lifetime risk of 6 $\times$ 10^{-7} to this individual? (Bioconcentration factor is a measure for the characterization of the accumulation of a chemical in an organism. It is defined as the concentration of a chemical in an organism—plants, microorganisms, animals—divided by the concentration in a reference compartment, e.g., food, surrounding water.)

9. The reference dose (RfD) for 1,1-dichloroethylene is set at 0.009 mg/kg/day and its oral potency factor is 0.58 $(mg/kg-day)^{-1}$. Discuss which would be more stringent: a dichloroethylene oral concentration standard based on a carcinogenic risk of 1 $\times$ 10^{-6} or a standard based on RfD.

10. Consider the issue of indoor air pollution caused by sidestream smoke emitting roughly 0.4 mg/cigarette of 1,3-butadiene. Calculate the average concentration of 1,3-butadiene producing a 1 $\times$ 10^{-6} lifetime cancer risk using standard values of inhalation. The inhalation potency factor of 1,3- butadiene is 6.1 $\times$ 10^{-1} $(mg/kg-day)^{-1}$.

11. Estimate and discuss what would be better regarding the cancer risk: drinking unchlorinated groundwater with 25 ppb of benzene, which has an oral potency factor of 2 $\times$ 10^{-2} $(mg/kg-day)^{-1}$, or switching to a surface water supply that, as a result of chlorination, has a chloroform concentration of 50 ppb (which has a potency factor of 6.1 $\times$ 10^{-3} $(mg(kg-day)^{-1})$.

REFERENCES

Aoshima, H. et al. 2009. Safety evaluation of highly purified fullerenes (HPFs): Based on screening of eye and skin damage. *The Journal of Toxicological Sciences* 34(5): 555–562.

Aschberger, K. et al. 2010. Review of carbon nanotubes toxicity and exposure—Appraisal of human health risk assessment based on open literature. *Critical Reviews in Toxicology* 40(July): 759–790.

Baker, G.L. et al. 2007. Inhalation toxicity and lung toxicokinetics of C60 fullerene nanoparticles and microparticles. *Toxicological Sciences* 101(1): 122–131.

Bakhtyari, N.G., Baderna, D., Boriani, E., Schuhmacher, M., Heise, S., and benfenati, E. 2013. Toxicological and ecotoxicological studies for additives, in *Global Risk-Based Management of Chemical Additives II*. Berlin, Germany: Springer-Verlag.

Beeline. 1998. *Beest for Windows. User's Manual*. Asheville, NC: Beeline Software.

Bennat, C. and C.C. Müller-Goymann 2000. Skin penetration and stabilization of formulations containing microfine titanium dioxide as physical UV filter. *International Journal of Cosmetic Science* 22(4): 271–283.

Bermudez, E. et al. 2004. Pulmonary responses of mice, rats, and hamsters to subchronic inhalation of ultrafine titanium dioxide particles. *Toxicological Sciences* 77(2): 347–357.

Biswas, P. and C.-Y. Wu. 2005. Nanoparticles and the environment. *Journal of the Air & Waste Management Association* 55(6): 708–746.

Bogdanović, V. et al. 2008. Fullerenol C60(OH)24 effects on antioxidative enzymes activity in irradiated human erythroleukemia cell line. *Journal of Radiation Research* 49(3): 321–327.

Bonnefous, S. and A. Despres. 1989. Evolution of the European database, IPSN/EURATOMCEA Association, Fontenay-Aux-Roses, France.

Bouillard, J.X. and A. Vignes. 2014. Nano-evaluris: An inhalation and explosion risk evaluation method for nanoparticle use. Part I: Description of the methodology. *Journal of Nanoparticle Research* 16(2): 2149.

Brouwer, D.H. et al. 2013. Workplace air measurements and likelihood of exposure to manufactured nano-objects, agglomerates, and aggregates. *Journal of Nanoparticle Research* 15(11): 2090.

Brouwer, D.H. et al. 2016. Occupational dermal exposure to nanoparticles and nano-enabled products: Part 2, exploration of exposure processes and methods of assessment. *International Journal of Hygiene and Environmental Health* 219(6): 503–512.

Buzea, C., I.I. Pacheco, and K. Robbie. 2007. Nanomaterials and nanoparticles: Sources and toxicity. *Biointerphases* 2(4): 17–71.

Chang, X., Y. Zhang, M. Tang, and B. Wang. 2013. Health effects of exposure to nano-TiO2: A meta-analysis of experimental studies. *Nanoscale Research Letters* 8(1): 51.

Chen, H.H.C. et al. 1998. Acute and subacute toxicity study of water-soluble polyalkylsulfonated C60 in rats. *Toxicologic Pathology* 26(1): 143–151.

Crump, K.S. 1984. A new method for determining allowable daily intakes. *Fundamental and Applied Toxicology* 871(5): 854–871.

Czajka, M. et al. 2015. Toxicity of titanium dioxide nanoparticles in central nervous system. *Toxicology in Vitro* 29(5): 1042–1052.

Derwent, R.G. and K. Nodop. 1986. Long-range transport and deposition of acidic nitrogen species in northwest Europe. *Nature* 324: 356–358.

Derwent, R.G., G.J. Dollard, and S.E. Metcalfe. 1988. On the nitrogen budget for the United Kingdom and northwest Europe. *Quarterly Journal of the Royal Meteorological Society* 114: 1127–1152.

EC (European Commission). 1995. ExternE—Externalities of Energy, Vol. 6, EUR 16520-162525. DG XII Science, Research and Development, Brussels.

EC (European Commission). 1999. ExternE—Externalities of Energy, Vol. 7, EUR 19083. DG XII Science, Research and Development, Brussels.

EC (European Commission). 2000. Externalities of fuel cycles—Volume 7: Methodology update, Report of the JOULE research project.

EC (European Commission). 2003. Technical Guidance Document (TGD) in support of the commission directive 93/67/EEC on risk assessment for new notified substances and the commission regulation (EC) 1488/94 on risk assessment for existing substances, Parts I–IV, JointResearch Centre, Italy.

EPA. 1990. Methodology for assessing health risks associated with indirected exposure to combustor emissions, Office of Health and Environmental Assessment, Cincinnati, OH, EPA/600/6-90/003.

EPA. 1992. User's guide for the industrial source complex (ISC2)—Dispersion models volumes I–III. EPA-450/4-92-008a, EPA-450/4-92-008b, EPA-450/4-92-008c. Triangle Park, NC.

EPA. 2002. A brief description of the AERMOD. Available online at: http://www.epa.gov/scram001/tt26.htm#aermod

EPA. 2002. A review of the reference dose and reference concentration processes. EPA/630/P-02/002F. Prepared for the Risk Assessment Forum U.S. Environmental Protection Agency, Washington, DC.

EPA. 2005. Guidelines for carcinogen risk assessment. EPA/630/P-03/001F. Risk Assessment Forum U.S. Environmental Protection Agency, Washington, DC.

EPA. 2009. Risk assessment guidance for superfund volume I: Human health evaluation manual (Part F, Supplemental Guidance for Inhalation Risk Assessment). EPA-540-R-070-002. Office of Superfund Remediation and Technology Innovation Environmental Protection Agency. Washington, DC.

EPA. 2015. Human health risk assessment. Strategic research action plan 2016–2019. EPA 601/K-15/002.

European Chemicals Agency. 2012. Guidance on information requirements and chemical safety assessment Chapter R.8: Characterisation of Dose[concentration]-Response for Human Health. Helsinki.

European Committee for Standardization (CEN). 2013. EN 15051 workplace exposure. Measurement of the dustiness of bulk materials. requirements and choice of test methods.

Filipe, P. et al. 2009. Stratum corneum is an effective barrier to TiO2 and ZnO nanoparticle percutaneous absorption. *Skin Pharmacology and Physiology* 22(5): 266–275.

Fransman, W. et al. 2011. Advanced reach tool (ART): Development of the mechanistic model. *Annals of Occupational Hygiene* 55(9): 957–979.

Friedrich, R. and A. Voss. 1993. External costs of electricity generation. *Energy Policy* 21: 114–122.

Fujita, K. et al. 2009. Gene expression profiles in rat lung after inhalation exposure to C60 fullerene particles. *Toxicology* 258(1): 47–55.

Furukawa, F. et al. 2011. Lack of skin carcinogenicity of topically applied titanium dioxide nanoparticles in the mouse. *Food and Chemical Toxicology* 49(4): 744–749.

Gharbi, N. et al. 2005. [60]Fullerene is a powerful antioxidant in vivo with no acute or subacute toxicity. *Nano Letters* 5(12): 2578–2585.

Hawley, J.K. 1985. Assessment of health risk from exposure to contaminated soil. *Risk Analysis* 5: 78–85.

Heinrich, U. et al. 1995. Chronic inhalation exposure of Wistar rats and two different strains of mice to diesel engine exhaust, carbon black, and titanium dioxide. *Inhalation Toxicology* 7(4): 533.

Hext, P.M, J.A. Tomenson, and P. Thompson. 2005. Titanium dioxide: Inhalation toxicology and epidemiology. *The Annals of Occupational Hygiene* 49(6): 461–472.

Hohmeyer, O. 1992. Renewables and the full costs of energy. *Energy Policy* 20: 365–375.

Hristozov, D. et al. 2016. Demonstration of a modelling-based multi-criteria decision analysis procedure for prioritisation of occupational risks from manufactured nanomaterials. *Nanotoxicology* 1–14.

Huang, S.-T., J.-S. Liao, H.-W. Fang, and C.-M. Lin. 2008. Synthesis and anti-inflammation evaluation of new C60 fulleropyrrolidines bearing biologically active xanthine. *Bioorganic & Medicinal Chemistry Letters* 18(1): 99–103.

Huczko, A., H. Lange, and E. Calko. 1999. Short communication: Fullerenes: Experimental evidence for a null risk of skin irritation and allergy. *Fullerene Science and Technology* 7(5): 935–939.

Hummelen, J.C. et al. 1995. Preparation and characterization of fulleroid and methanofullerene derivatives. *The Journal of Organic Chemistry* 60(3): 532–538.

Iavicoli, I., V. Leso, and A. Bergamaschi. 2012. Toxicological effects of titanium dioxide nanoparticles: A review of in vivo studies. *Journal of Nanomaterials* 2012: 1–36.

IER (Institut für Energiewirtschaft und Rationale Energieanwendung). 1998. EcoSense 2.1 Software, University of Stuttgart, Germany.

Injac, R. et al. 2009. Protective effects of fullerenol C60(OH)24 against doxorubicin-induced cardiotoxicity and hepatotoxicity in rats with colorectal cancer. *Biomaterials* 30(6): 1184–1196.

Jensen, K.A. et al. 2016. NanoSafer Version 1.1: A web-based precautionary risk assessment and management tool for manufactured nanomaterials using first order modeling. unpublished work.

Johnston, H.J. et al. 2010a. The biological mechanisms and physicochemical characteristics responsible for driving fullerene toxicity. *Toxicological Sciences* 114(2): 162–182.

Johnston, H.J. et al. 2010b. A review of the in vivo and in vitro toxicity of silver and gold particulates: Particle attributes and biological mechanisms responsible for the observed toxicity. *Critical Reviews in Toxicology* 40(4): 328–346.

Kamat, J. 2000. Reactive oxygen species mediated membrane damage induced by fullerene derivatives and its possible biological implications. *Toxicology* 155(1–3): 55–61.

Katsumata, P.T. and W.E. Kastenberg. 1997. On the impact of future land use assumptions on risk analysis for superfund sites. *Air & Waste Management Association* 47: 881–889.

Kimberly-Clark. 2012. Alarming number of workers fail to wear required protective equipment (NYSE:KMB). http://investor.kimberly-clark.com/releasedetail.cfm?ReleaseID=712258 (May 1, 2015).

Koelmans, A.A., N.J. Diepens, I. Velzeboer, E. Besseling, J.T.K. Quik, and D. vand de Meent. 2015. Guidance for the prognostic risk assessment of nanomaterials in aquatic ecosystems. *Science of the Total Environment* 535: 141–149.

Koivisto, A.J. et al. 2012. Industrial worker exposure to airborne particles during the packing of pigment and nanoscale titanium dioxide. *Inhalation Toxicology* 24(12): 839–849.

Koivisto, A.J. et al. 2014. Range-finding risk assessment of inhalation exposure to nanodiamonds in a laboratory environment. *International Journal of Environmental Research and Public Health* 11(5): 5382–5402.

Koivisto, A.J. et al. 2015. Workplace performance of a loose-fitting powered air purifying respirator during nanoparticle synthesis. *Journal of Nanoparticle Research* 17(4): 177.

Krewitt, W., A. Trukenmueller, P. Mayerhofer, and R. Freidrich. 1995. EcoSense—An integrated tool for environmental impact analysis, in Kremers, H. and Pillmann, W., Eds., *Space and Time in Environmental Information Systems*, Umwelt-Informatik aktuell, Band 7. Metropolis–Verlag, Marburg, Germany. (SETAC).

Kuempel, E.D., C.L. Geraci, and P.A. Schulte. 2012. Risk assessment and risk management of nanomaterials in the workplace: Translating research to practice. *The Annals of Occupational Hygiene* 56(5): 491–505.

Kumar, P. et al. 2016. Review: Real-time sensors for indoor air monitoring and challenges ahead in deploying them to urban buildings. *Science of the Total Environment* 560–561: 150–159.

LaGrega, M.D., P.L. Buckingham, and J.C. Evans. 1994. *Hazardous Waste Management*. New York: McGraw-Hill.

Liguori, B. et al. 2016. Sensitivity analysis of the exposure assessment module in NanoSafer Version 1.1: Ranking of determining parameters and uncertainty. unpublished work.

Lin, A.M.-Y. et al. 2002. Local carboxyfullerene protects cortical infarction in rat brain. *Neuroscience Research* 43(4): 317–321.

Liu, Y. et al. 2009. Immunostimulatory properties and enhanced TNF-α mediated cellular immunity for tumor therapy by C60(OH)20 nanoparticles. *Nanotechnology* 20(41): 415102.

Liu, H.H. and Y. Cohen. 2014. Multimedia environmental distribution of engineered nanomaterials. *Environmental Science & Technology* 48: 3281–3292.

Monteiro-Riviere, N.A. et al. 2011. Safety evaluation of sunscreen formulations containing titanium dioxide and zinc oxide nanoparticles in UVB sunburned skin: An in vitro and in vivo study. *Toxicological Sciences* 123(1): 264–280.

Mori, T. et al. 2006. Preclinical studies on safety of fullerene upon acute oral administration and evaluation for no mutagenesis. *Toxicology* 225(1): 48–54.

Morimoto, Y. et al. 2010. Inflammogenic effect of well-characterized fullerenes in inhalation and intratracheal instillation studies. *Particle and Fibre Toxicology* 7(1): 4.

Nadal, M., F. Fàbrega, M. Schuhmacher, and J.L. Domingo. 2013. PCDD/Fs in plasma of individuals living near a hazardous waste incinerator. A comparison of measured levels and estimated concentrations by PBPK modeling. *Environmental Science & Technology* 47(11): 5971–5978.

National Institute for Occupational Safety and Health. 2011. Current intelligence bulletin 63: Occupational exposure to titanium dioxide.

Nessel, C.S. et al. 1991. Evaluation of the relative contribution of exposure routes in a health risk assessment of dioxin emissions from a municipal waste incinerator. *Journal of Exposure Analysis and Environmental Epidemiology* 1(3): 283–307.

Newman, M.D., M. Stotland, and J.I. Ellis. 2009. The safety of nanosized particles in titanium dioxide– and zinc oxide–based sunscreens. *Journal of the American Academy of Dermatology* 61(4): 685–692.

Noël, A. and G. Truchon. 2015. Inhaled titanium dioxide nanoparticles: A review of their pulmonary responses with particular focus on the agglomeration state. *Nano LIFE* 5(1): 1450008.

Nowack, B. et al. 2012. Analysis of the occupational, consumer and environmental exposure to engineered nanomaterials used in 10 technology sectors. *Nanotoxicology* 7(February): 1–5.

Oberdorster, G., J. Ferin, and B.E. Lehnert. 1994. Correlation between particle size, in vivo particle persistence, and lung injury. *Environmental Health Perspectives* 102: 173–179.

Olsen, S.I. et al. 2000. Indicators for human toxicity in life-cycle impact assessment, position paper for SETAC-Europe WIA-2 Task Group on Human Toxicity, intermediate draft.

Organization for Economic Cooperation and Development. 2012. Inhalation toxicity testing: Expert meeting on potential revisions to OECD test guidelines and guidance document. Paris, France.

Perelló, G. et al. 2012. Assessment of the temporal trend of the dietary exposure to PCDD/Fs and PCBs in Catalonia, over Spain: Health risks. *Food and Chemical Toxicology* 50: 399–408.

Pietroiusti, A. and A. Magrini. 2014. Engineered nanoparticles at the workplace: Current knowledge about workers' risk. *Occupational Medicine* 64(5): 319–330.

Pilkington, A., F. Hurley, and P. Donnan. 1997. Health effects in the ExternE transport—Assessment and exposure–response functions, technical report, Institute of Occupational Medicine, Edinburgh, Scotland.

Rabl, A., J.V. Spadaro, and P.D. McGavran. 1998. Health risks of air pollution from incinerators—A perspective. *Waste Management & Research* 16(4): 368–388.

Ramachandran, G. 2005. *Occupational Exposure Assessment for Air Contaminants*. Boca Raton, FL: CRC Press.

Rovira, J. et al. 2013. A revision of current models foe environmental and human health impact and risk assessment for the application to emerging chemicals, in *Global Risk-Based Management of Chemical Additives II*. Berlin, Germany: Springer-Verlag.

Roursgaard, M. et al. 2008. Polyhydroxylated C 60 fullerene (Fullerenol) attenuates neutrophilic lung inflammation in mice. *Basic & Clinical Pharmacology & Toxicology* 103(4): 386–388.

Royal Society. 1992. *Risk Analysis, Perception and Management*. London, UK: The Royal Society.

Sadrieh, N. et al. 2010. Lack of significant dermal penetration of titanium dioxide from sunscreen formulations containing nano- and submicron-size TiO2 particles. *Toxicological Sciences* 115(1): 156–166.

Sandnes, H. and H. Styve. 1992. Calculated budgets for airborne acidifying components in Europe 1985, 1987, 1988, 1989, 1990 and 1991. MSC-W Report 1/92. Technical Report no. 97, EMEP. Oslo, Norway: The Norwegian Meteorological Institute.

Sayes, C.M. et al. 2005. Nano-C60 cytotoxicity is due to lipid peroxidation. *Biomaterials* 26(36): 7587–7595.

Schneider, T. and K. Jensen. 2009. Relevance of aerosol dynamics and dustiness for personal exposure to manufactured nanoparticles. *Journal of Nanoparticle Research* 11(7): 1637–1650.

Senzui, M. et al. 2010. Study on penetration of titanium dioxide (TiO2) nanoparticles into intact and damaged skin in vitro. *The Journal of Toxicological Sciences* 35(1): 107–113.

Serra-Majem, L. et al. 2003. Ngo Avaluació de l'estat nutricional de la població catalana 2002–2003. Evolució dels hàbits alimentaris i del consum d'aliments i nutrients a Catalunya (1992–2003). Direcció General de Salut Pública, Departament de Sanitat i Seguretat Social, Generalitat de Catalunya, Barcelona, Spain (in Catalan).

Shakeel, M. et al. 2016. Toxicity of nano-titanium dioxide (TiO2-NP) through various routes of exposure: A review. *Biological Trace Element Research* 172(1): 1–36.

Shi, H., R. Magaye, V. Castranova, and J. Zhao. 2013. Titanium dioxide nanoparticles: A review of current toxicological data. *Particle and Fibre Toxicology* 10(1): 15.

Shin, D., J. Lee, J. Yang, and Y. Yu. 1998. Estimation of air emission for dioxin using a mathematical model in two large cities of Korea. *Organohalogen Compounds* 36: 449–453.

Shrivastava, R. et al. 2014. Effects of sub-acute exposure to TiO2, ZnO and Al2O3 nanoparticles on oxidative stress and histological changes in mouse liver and brain. *Drug and Chemical Toxicology* 37(3): 336–347.

Stebounova, L.V., H. Morgan, V.H. Grassian, and S. Brenner. 2012. Health and safety implications of occupational exposure to engineered nanomaterials. *Wiley Interdisciplinary Reviews: Nanomedicine and Nanobiotechnology* 4(3): 310–321.

Taylor, R. and D.R.M. Walton. 1993. The chemistry of fullerenes. *Nature* 363(6431): 685–693.

Tykhomyrov, A.A., V.S. Nedzvetsky, V.K. Klochkov, and G.V. Andrievsky. 2008. Nanostructures of hydrated C60 fullerene (C60HyFn) protect rat brain against alcohol impact and attenuate behavioral impairments of alcoholized animals. *Toxicology* 246(2–3): 158–165.

U.S. EPA. 1995. User's guide for the industrial source complex (ISC3)—Dispersion models volumes I–II. EPA-454/B-95-003a, EPA-454/B-95-003b. Triangle Park, NC.

Unnithan, J., M.U. Rehman, F.J. Ahmad, and M. Samim. 2011. Aqueous synthesis and concentration-dependent dermal toxicity of TiO2 nanoparticles in Wistar rats. *Biological Trace Element Research* 143(3): 1682–1694.

Varner, K.E., K. Rindfusz, A. Gaglione, and E. Viveiros. 2010. *Nano Titanium Dioxide Environmental Matters: State of the Science Literature Review*. Washington, DC. http://cfpub.epa.gov/si/si_public_file_download.cfm?p_download_id=498019.

Vilavert, L., M. Nadal, M. Schuhmacher, and J.L. Domingo. 2014. Seasonal surveillance of airborne PCDD/Fs, PCBs and PCNs using passive samplers to assess human health risks. *Science of the Total Environment* 466–467: 733–740.

Vilavert, L., M. Nadal, M. Schuhmacher, and J.L. Domingo. 2015. Two decades of environmental surveillance in the vicinity of a waste incinerator: Human Health Risks Associated with metals and PCDD/Fs. *Archives of Environmental Contamination and Toxicology* 69: 241–253.

Wang, J. et al. 2007. Acute toxicity and biodistribution of different sized titanium dioxide particles in mice after oral administration. *Toxicology Letters* 168(2): 176–185.

Warheit, D. 1996. Subchronic inhalation of high concentrations of low toxicity, low solubility participates produces sustained pulmonary inflammation and cellular proliferation. *Toxicology Letters* 88(1–3): 249–253.

Warheit, D. et al. 2007. Development of a base set of toxicity tests using ultrafine TiO2 particles as a component of nanoparticle risk management. *Toxicology Letters* 171(3): 99–110.

Wild, C.P. 2005. Complementing the genome with an "exposome": The outstanding challenge of environmental exposure measurement in molecular epidemiology. *Cancer Epidemiology, Biomarkers & Prevention* 14: 1847–1850.

Wu, J. et al. 2009. Toxicity and penetration of TiO2 nanoparticles in hairless mice and porcine skin after subchronic dermal exposure. *Toxicology Letters* 191(1): 1–8.

Wudl, F. 1992. The chemical properties of Buckminsterfullerene (C60) and the birth and infancy of fulleroids. *Accounts of Chemical Research* 25(3): 157–161.

Yakubu, A. 2012. Determination of Lindane and its metabolites by HPLC-UV-Vis and MALDI-TOF. *Journal of Clinical Toxicology* 1(S1). http://www.omicsonline.org/2161-0495/2161-0495_Toxicology-2012_AcceptedAbstracts.digital/2161-0495_Toxicology-2012_AcceptedAbstracts.html.

Yamago, S. et al. 1995. In vivo biological behavior of a water-miscible fullerene: 14C labeling, absorption, distribution, excretion and acute toxicity. *Chemistry & Biology* 2(6): 385–389.

Ze, Y. et al. 2014. TiO2 nanoparticles induced hippocampal neuroinflammation in mice. *PLoS One* 9(3): e92230.

Zhang, X., W. Li, and Z. Yang. 2015. Toxicology of nanosized titanium dioxide: An update. *Archives of Toxicology* 89(12): 2207–2217.

5 Criticality Assessment of Natural Resources

Dimitra Ioannidou and Guido Sonnemann

CONTENTS

5.1 INTRODUCTION

The steep rise in the extraction of fuel and non-fuel resources in the last 50–60 years in combination with population growth and the ensuing high levels of consumption has stirred interest on the assessment of natural resources and their availability for future generations. At the same time, the awareness of the finite boundaries of our system (Meadows et al., 1972) and of the importance of natural resources for the economy and society has led to the development of "solid and comprehensive systems which measure human resource use through appropriate indicators" (Giljum et al., 2011).

> "Indicators arise from values (we measure what we care about) and they create values (we care about what we measure)" (Meadows, 1998).

> "What is not measured often gets ignored in policy processes" (Giljum et al., 2011).

In the field of natural resources, an increasing number of research endeavors is currently focusing on developing methodologies and indicators to evaluate whether supply can meet the current and anticipated demand. Environmental impact assessment methods such as Life-Cycle Assessment (LCA) use different indicators to evaluate resource consumption and abiotic depletion (Goedkoop and Spriesma, 2001; Guinée et al., 2002; Jolliet et al., 2003). These indicators focus mainly on the geological availability of the resource or the impact of its extraction and ignore the importance of the resource to the economy and the society. A resource in small concentrations on the earth's crust would, hence, have a high risk, even if the demand for this resource is relatively low. To this end, new methodologies were developed towards complementing the existing abiotic depletion potential indicators. One of the main concepts introduced to address additional aspects related to resources is the criticality concept. Criticality is gaining popularity in the industry, as a critical raw resource can negatively impact a firm through price surges and can lead companies to change their policies and increase recycling or substitution to decrease the use of the specific resource (Alonso et al., 2008).

At this point, it would be useful to provide a definition for natural resources. According to Sonderegger et al. (2017):

> "Natural resources are material and non-material assets occurring in nature that are at some point in time deemed useful for humans. Natural resources include minerals and metals, air components, fossil fuels, renewable energy sources, water, land and water surface, soil, and biotic natural resources such as wild flora and fauna."

This chapter contains a short summary of the existing abiotic depletion indicators. Furthermore, it presents the concept of criticality and analyzes three of the main criticality methodologies developed: the methodology of the United States National Research Council, the methodology developed by the European Commission, as it was updated in 2017, and the methodology of Graedel and his co-authors at Yale University. With respect to the European Commission Criticality Assessment, the most recent list of critical materials (published in 2017) will be presented and discussed. The chapter closes with an overview of the latest research in this field, namely studies on dynamic criticality and extension of the criticality methodologies to other resources.

5.2   ABIOTIC DEPLETION INDICATORS

The early assessments of the depletion of natural resources were based on estimates of their reserve and reserve base. A definition of the terms resource, reserve base and reserve was provided by USGS (2009):

> *Resource*: A concentration of naturally occurring solid, liquid, or gaseous material in or on the Earth's crust in such form and amount that economic extraction of a commodity from the concentration is currently or potentially feasible.
> *Reserve base*: That part of an identified resource that meets specified minimum physical and chemical criteria related to current mining and production practices, including those for grade, quality, thickness, and depth. The reserve base includes those resources that are currently economic (reserves), marginally economic (marginal reserves), and some of those that are currently subeconomic (subeconomic resources).
> *Reserves*: That part of the reserve base which could be economically extracted or produced at the time of determination.

A graphical representation of the above definitions is given in Figure 5.1.

One of the first Life-Cycle Assessment (LCA) methods, CML, developed by the homonymous institute CML (Centrum voor Milieuwetenschappen Leiden), includes the abiotic depletion indicator,

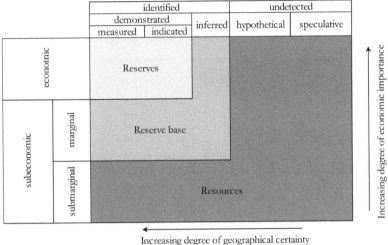

FIGURE 5.1 Reserves, reserve base, and resources as defined by USGS (2009). (Adapted from Achzet, B. and Helbig, C., *Resour. Polic.*, 38, 435–447, 2013.)

which expresses depletion in relation to the reference resource antimony (Guinée et al., 2002). The Abiotic Depletion Potential of resource *I* (ADPi) is defined as the ratio of the extraction rate for resource *i* over the ultimate reserve for this resource (Equation 5.1):

$$ADP_i = \frac{DR_i}{\left(R_i\right)^2} * \frac{\left(R_{Sb}\right)^2}{DR_{Sb}} \tag{5.1}$$

where:

DR_i and DR_{Sb} are the extraction rates (kg/year) for resource *i* and antimony, respectively
R_i and R_{Sb} are their ultimate reserves (kg)

In Eco-Indicator 99, the calculation of the resource depletion is based on the assumption that a certain extraction leads to an additional energy requirement for further mining of this resource in the future, due to lower resource concentrations or other unfavorable characteristics of the remaining reserves (Goedkoop and Spriensma, 2001).

In the method of Environmental Priority Strategies in product development (EPS), the impact of a resource is related to the processes associated with its extraction and processing, accounting for the amount of materials or processes used (Steen, 1999). The environmental impact is expressed in Environmental Load Units (ELUs). For example, the environmental impact of natural gravel is calculated through its substitution by crushed rock (Habert et al., 2010). Following a series of assumptions (an energy consumption of 4 MJ per ton, a capital cost of $86000, an operating time of 4000 hours per year, a lifetime of 5 years and a demand of manpower of 1 person-hour/operating hour) and taking into account extra equipment, like conveyor belts and necessary vehicles, the cost of crushing rock to gravel is about $2/ton. This corresponds to a resource value of 0.002ELU/kg.

In Impact2002+ (Jolliet et al., 2003), the impact of the resource consumption is the sum of the midpoint categories of nonrenewable energy consumption and mineral extraction (MJ additional energy/kg extracted). It is expressed in MJ primary nonrenewable energy, according to Eco-Indicator 99.

Table 5.1 summarizes the main methodologies to evaluate resource consumption in Life-Cycle Assessment.

TABLE 5.1

Methodologies Developed to Evaluate Resource Consumption

	Methodology	Indicator	Type of Resource	Calculation
Midpoint methods	CML 2001	Abiotic depletion	Energy, material	Abiotic depletion potential: $$ADP_i = \frac{DR_i}{(R_i)^2} * \frac{(R_{Sb})^2}{DR_{Sb}},$$ DR: extraction rate (kg·year^{-1}), R: ultimate reserve (kg)
	EDIP	Resource	Energy, material	Flow (kg)
	Ecopoint 97	Resource	Energy	Flow (kg)
Endpoint methods	Eco 99	Fossil fuels	Energy	MJ surplus/kg extracted (Increase of energy demand due to the lowering of ore grades)
	Eco 99	Minerals	Minerals	MJ surplus/kg extracted (Increase of energy demand due to the lowering of ore grades)
	EPS	Resources	Resources	Environmental Load Unit
	Impact 2002+	Fossil fuels	Energy	MJ surplus/kg extracted (Increase of energy demand due to the lowering of ore grades)
	Impact 2002+	Minerals	Minerals	MJ surplus/kg extracted (Increase of energy demand due to the lowering of ore grades)

Source: Habert, G. et al., *Resour. Conserv. Recyc.*, 54, 364–376, 2010.

Outside the LCA framework, the production and depletion of fossil fuels have been extensively assessed using the Hubbert peak approach (Hubbert, 1956, 1962); the methodology has also been applied to non-fuel minerals (Calvo et al., 2017). Otherwise, Singer and Menzie (2010) followed a three-step assessment in order to provide estimates of undiscovered mineral resources, based on the geologic map of a region and on mineral deposit (grade-and-tonnage) models.

However, the aforementioned indicators mainly focus on the supply of the resource and do not account for additional parameters that can threaten the equilibrium between supply and demand. Various attempts have thus been directed lately towards complementing the existing abiotic depletion potential indicators (Schneider et al., 2011, 2014) using methods of industrial ecology such as Material and Energy Flow Accounting (MEFA). The MEFA method is used to evaluate the industrial metabolism of a territory (Ayres, 1989) and focuses on the interaction between economic activities and the environment (Haberl et al., 2004). In a new indicator they developed, Habert et al. (2010) used MEFA analysis to study the temporal evolution of the importation over consumption ratio in order to reach conclusions on the state of resource depletion in a specific area. Both parameters, Imports (I) and Domestic Material Consumption (DMC), are calculated following a MEFA analysis. If I/DMC is increasing with time, this indicates an increasing difficulty to have access to the resource, therefore it is assumed that the risk of depletion of the resource in this area is high. On the contrary, a decrease in the ratio of I/DMC is related to the fact that more consumption needs can be supported by local production (Habert et al., 2010).

Another shortcoming of the depletion indicators is that they do not study the issue of accessibility to resources, which can pose a barrier to their exploitation (Bridge, 2000; Topp et al., 2008; Giurco et al., 2010; Prior et al., 2012). Topp et al. (2008) explained that the initial focus is on high-quality, readily accessible deposits, since they are the most profitable ones. When these resources become depleted, then the remaining deposits "of lower grade, in more remote locations, deeper in the ground, mixed with greater impurities, require more difficult extraction techniques." Therefore, more "effort," in terms of capital and labor, is required to extract a unit of a resource, when the latter is not easily accessible (Topp et al., 2008). At the same time, Sonnemann et al. (2015) and Lloyd et al. (2012) built on the idea of a holistic assessment of the availability of resources and underlined the importance of a criticality assessment of resources, where the accessibility to a specific resource would be taken into account.

5.3 CRITICALITY CONCEPT

Over the last decade, with the emergence of many new, innovative technologies and the vast use of resources that did not have a widespread use until some years ago, the concept of criticality has gained ground. Criticality refers both to a high potential impact of shortage of a resource (when the resource is particularly important for the value chain and has few or no substitutes) and to a comparatively high probability of such a shortage (National Research Council, 2008; European Commission, 2011; Buijs et al., 2012; Jin et al., 2016). The methodologies of criticality assessment of minerals integrate environmental, socio-economic, and geopolitical aspects related to the availability and use of minerals (Sonnemann et al., 2015; Drielsma et al., 2016).

Criticality stems from the classical risk theory framework. In the work of Glöser et al. (2015), risk is defined as the product of probability of occurrence of a specific scenario and the consequences caused by this scenario. In the same study, Glöser et al. (2015) define raw material criticality as the product of the likelihood of supply disruptions and their economic consequences (Equation 5.2) and highlight that the representation of criticality in a criticality matrix is an adaptation of the classical risk matrix:

$$\text{Raw material criticality} = \text{Supply Risk} * \text{Vulnerability}$$
$$= \text{Likelihood of supply disruptions} * \text{Economic consequences}$$

(5.2)

Criticality assumes a risk assessment perspective in estimating the impacts of several external factors (such as environmental and socio-economic) on a company's production supply chain and its decision making procedure (Cimprich et al., 2017). Cimprich et al. (2017) define criticality as an "outside-in" view of the system, where the worst case scenario is studied. On the other hand, Life-Cycle Assessment provides an "inside-out" approach of the impacts of the internal operation of a company on the society and the environment (Cimprich et al., 2017). LCA studies the average impacts corresponding to a functional unit. Nevertheless, Frenzel et al. (2017) note that criticality is considered an abstraction of the classical risk theory, since by strict terms it does not abide by the algebraic formulas of calculation of risk.

Looking at the evolution of criticality and the different methodologies developed, one of the first studies on criticality was performed by the United States National Research Council in 2008 (National Research Council, 2008). The purpose of this study was the assessment of the criticality of non-fuel minerals for preventive policy in the field of mineral resources. In this study, criticality was defined as a combination of supply risk and importance in use. Two years afterwards, the European Commission (2010) published a study on the criticality of raw materials based on two components, supply risk and economic importance (equivalent to the importance in use of the study of the National Research Council). Since then, there have been numerous studies dealing with the topic of criticality. One of the most widely discussed criticality methodologies integrating the above two dimensions (Supply Risk and Vulnerability to Supply Restriction) plus a third one regarding the environmental implications was developed at Yale University by Graedel et al. (2012). Nevertheless, even though the criticality studies adopt a more or less common approach with respect to the criticality dimensions, there is significant divergence with respect to the indicators that are aggregated to provide the axis values. The most common of these are the degree of country concentration in either the reserves or current production, the geopolitical risks of those countries, and the level of reserves relevant to current consumption (Achzet and Helbig, 2013; Graedel and Reck, 2016). All of these indicators evaluate the Supply Risk dimension.

5.4   UNITED STATES NATIONAL RESEARCH COUNCIL CRITICALITY ASSESSMENT

In 2008, the Committee on Critical Mineral Impacts on the U.S. Economy, appointed by the United States National Research Council (NRC), published a study where they identified and reviewed nonfuel minerals that are "critical" for the domestic industry and emerging technologies (National Research Council, 2008). The evaluation of the criticality of minerals was based on whether a particular industry sector would be adversely affected by a restriction on the supply of that mineral. The criticality of minerals was contrasted to the criticality of fossil fuels in that the former can be recycled and therefore continue to be part of the material flow analysis (as secondary resources) even after the initial use. Moreover, a supply restriction (an imbalance between supply and demand) of any critical mineral would have more limited consequences than would a restriction in the supply of oil.

In this study, criticality depends on two factors: importance in use and availability. Importance in use is directly linked to the mineral's physical and chemical properties and to its degree of substitutability, i.e. using another mineral to provide the same function and performance as the initial mineral. When minerals have few or no substitutes, the impact of a possible restriction in their supply will be significant. The importance in use is calculated for each of the end uses of a mineral and an average score provides the overall impact of supply restriction, considering the weighted impact of every end use in the US economy.

On the other hand, availability (or supply risk) refers to the process of mining and mineral processing and accounts also for the recycling potential. Regarding the primary available reserves, estimates are continuously updated with the advancement in mineral production technologies. In addition, availability considers past inflows and outflows of the stock of the specific mineral that is available for recycling. To estimate the availability of a mineral, the methodology considers environmental and social, geologic, political, technical, and economic aspects. Each of these aspects is evaluated for a certain mineral and the overall availability is defined as the highest of these evaluation scores. When availability is less than the demand, this can lead either to physical unavailability of the resource, or to higher prices, resulting in a reallocation of the available supply toward those users who are willing to pay more for this mineral; this provides incentives to alternative resources or products to enter the market.

The overall criticality in this methodology is given by the criticality matrix, a risk assessment matrix, which has two dimensions: Supply Risk (related to availability) and Impact of Supply Restriction (related to importance). The position of a mineral in both axes is estimated through a semi-quantitative scoring. As critical are considered the minerals that are situated in the upper right-hand corner of the matrix.

In this study, the US NRC applied the criticality matrix to 11 minerals or mineral groups: copper, gallium, indium, lithium, manganese, niobium, PGMs, REs, tantalum, titanium, and vanadium. According to this assessment, among the minerals studied, Rare Earths, Platinum Group Metals, indium, manganese, niobium, and, potentially, gallium were deemed as critical for the US economy.

Figure 5.2 presents graphically the criticality matrix, with its two dimensions: Supply Risk and Impact of Supply Restriction. Supply Risk accounts for the available deposits (through the reserve-to-production ratio) and the concentration of production, while Impact of Supply Restriction considers mainly substitution. The larger the distance of a point from the origin of the graph, the higher its criticality. For example, in the graph, mineral A is more critical than mineral B.

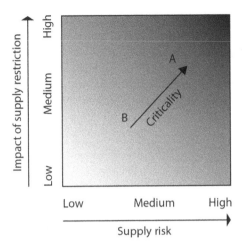

FIGURE 5.2 Graphical representation of the criticality matrix. (From National Research Council, *Minerals, Critical Minerals, and the U.S. Economy*, National Academies Press, Washington, DC, 2008.)

5.5 EUROPEAN COMMISSION CRITICALITY ASSESSMENT

Shortly after the US NRC published the study on critical minerals, the European Commission created the Raw Materials Initiative in 2008 to address challenges related to the access to raw materials at the EU level and secure a sustainable supply of raw materials for the EU (Nuss and Blengini, 2018). The European Commission (EC) developed a methodology to identify critical raw materials and published a list of 14 critical raw materials (European Commission, 2010). Since then, it has published two updated lists; in 2014, containing 20 critical raw materials and in 2017, featuring 27 critical raw materials for the EU. The raw materials that are evaluated by the EC are non-energy and non-agricultural.

The EC criticality assessment provides "a snapshot in time of the current situation, based on the recent past" (Blengini et al., 2017). A critical raw material, according to this approach, has a high economic importance for the EU and is vulnerable to supply disruption, i.e. presents a high risk of not being able to have an adequate supply to meet the EU demand. The two dimensions are equivalent to the axis of the criticality matrix of the US NRC. What follows is the explanation of the two components and their calculation steps.

5.5.1 Economic Importance

The Economic Importance expresses the importance of a raw material for the EU economy and is an indicator of the potential consequences in the case of inadequate supply. The Economic Importance is therefore related to the end use applications of a resource and to the corresponding economic sectors that are affected by its use. It is calculated by Equation 5.3:

$$EI = \sum_s \left(A_s * Q_s \right) * SI_{EI} \tag{5.3}$$

where:
A_s is the share of end use of a raw material in a sector s
Q_s is the sector's value added
SI_{EI} is the substitution index of the raw material

The substitution index considers only "proven substitutes that are readily available today and able to reduce the consequences of a disruption and/or influence the risk of a disruption" (Blengini et al., 2017). Substitutability and potential future substitutes are not taken into account here. The substitution index for the economic importance is calculated via the Substitution Cost Performance matrix, which has two dimensions: substitute material technical performance (if a substitute can replace the function of the main resource) and substitute material cost (comparing the cost of the substitute to the main resource). This matrix is applied for each substitute material within a given application.

5.5.2 Supply Risk

The Supply Risk of a raw material evaluates the risk of not having an adequate supply of the material to meet the demand. The calculation of the Supply Risk depends on the concentration of the primary supply of the resource and the governance situations in the corresponding countries producing the resource (Equation 5.4):

$$\text{SR} = \left[\left(HHI_{WGI-t} \right)_{GS} * \frac{IR}{2} + \left(HHI_{WGI-t} \right)_{EU_{sourcing}} * \left(1 - \frac{IR}{2} \right) \right] * \left(1 - EoL_{RIR} \right) * SI_{SR} \qquad (5.4)$$

where:

    HHI is the Herfindahl-Hirschman Index (used as a proxy for country concentration)

    WGI is the scaled World Governance Index (used as a proxy for country governance)

    t is the trade adjustment (of WGI) for Global Supply (GS) and actual EU supply ($EU_{sourcing}$), since the initial indicator of WGI does not account for export restrictions and international trade agreements; t depends on the export tax or physical quota imposed or the export prohibition introduced for a certain raw material

    IR refers to the Import Reliance, which is given by the ratio Net Import/Apparent Consumption

    EoL_{RIR} is the End-of-Life Recycling Input Rate, i.e. the input of secondary material to the EU from old scrap to the total input of material (primary and secondary)

    SI_{SR} is the Substitution Index related to the Supply Risk dimension

The Substitution Index for Supply Risk is dependent on the production volume, criticality and degree of co-/by-production of the substitute, as per Equation 5.5:

$$SI_{SR} = \sum_i \left[\left(SP_i * SCr_i * SCo_i \right)^{1/3} * \sum_a \left(\text{Subshare}_{i,a} * \text{Share}_a \right) \right] \qquad (5.5)$$

where:

    SP_i refers to the global production of the substitute in relationship to the main material, indicating whether the substitute can provide an adequate volume to substitute the raw material in question i

    SCr_i is the criticality of the substitute

    SCo_i refers to the production of the substitute as a co- or by-product

    a denotes the application of the main material i

With respect to the World Governance Indicators, this is a set of indicators providing information about the governance situation in a country (Kaufmann and Kraay, 2014). These indicators include: Voice and Accountability, Political Stability and Absence of Violence/Terrorism, Government Effectiveness, Regulatory Quality, Rule of Law, and Control of Corruption. In Equation 5.4, the average of the values of the six WGI dimensions is used.

Finally, the modified Herfindahl-Hirschman Index of Equation 5.4, which accounts for export restrictions, is calculated as follows:

$$\left(HHI_{WGI-t}\right)_{GS\ or\ EU_{sourcing}} = \sum_{C}\left(S_C\right)^2 *WGI_C *t_C \qquad (5.6)$$

where:

HHI_{WGI-t} is the HHI_{WGI} for the concentration of the Global Supplier country or EU adjusted
 by t
S_C is the share of country c in the global supply (or EU28 sourcing) of the raw material considered
WGI_C is the rescaled score in the World Governance Indicators (WGI) of country c
t_C is the value of the trade-related variable t of country c for the raw material considered

5.5.3 Updated List of Critical Raw Materials

In 2017, the European Commission published a new list of critical raw materials for the EU that present a high risk of supply shortage and have a significant impact on the economy. The EC evaluated 78 raw materials and deemed that 27 among them belong to the high criticality region, namely antimony, baryte, beryllium, bismuth, borate, cobalt, fluorspar, gallium, germanium, hafnium, helium, indium, magnesium, natural graphite, natural rubber, niobium, phosphate rock, phosphorus, scandium, silicon metal, tantalum, tungsten, vanadium, the Platinum Group Metals, and the Heavy and Light Rare Earth Elements (European Commission, 2017). The EC also included in the above list coking coal; this is not considered critical anymore, but was included for reasons of complete and conservative assessment, since its criticality is at the borderline and it was part of the previous criticality list of 2014. The new materials on the aforementioned list are baryte, bismuth, hafnium, helium, natural rubber, phosphorus, scandium, tantalum and vanadium. These are mainly materials with low domestic production (apart from hafnium), imported from countries outside the EU. It is also interesting to note that since 2014, the EC includes in the criticality assessment biotic materials as well, such as natural rubber (critical), natural cork, natural teak wood and sapele wood. Criticality in these resources is associated with responsible and sustainable sourcing and management of these resources, land use competition with other uses (such as agriculture), variety of end-uses that the supply should cover, geographical concentration of the supply sources, by-production dynamics related to the various products resulting from the same raw resource and frequency of occurrence of natural disasters (European Commission, 2017).

5.6 METHODOLOGY OF METAL CRITICALITY DETERMINATION

Another methodology extensively used for assessing criticality is the "Methodology for Metal Criticality Determination," developed by Graedel and his co-authors (2012) at Yale University. The goal of this methodology was to prescribe criticality as a snapshot in time by incorporating data from different sources, such as geology, technology, economics, human behavior, and expert assessment. It adopted a more systematic and transparent method of assessment of criticality by defining various indicators and following a specific weighing procedure. Criticality in this methodology is comprised of three dimensions: Supply Risk (SR), Environmental Implications (EI), and Vulnerability to Supply Restriction (VSR). The evaluation of SR and VSR is performed based on three components and each component consists of different indicators.

5.6.1 Supply Risk

Graedel et al. (2012) assumed two different time scales in order to adequately serve the complete spectrum of interested parties: a medium-term scale (5–10 years) and a long-term scale (a few

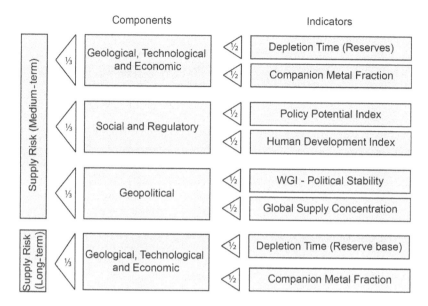

FIGURE 5.3 Components and Indicators of the Supply Risk dimension for the medium and long-term perspective. (Reprinted with permission from Graedel, T.E. et al., *Environ. Sci. Technol.*, 46, 1063–1070, 2012. Copyright 2012 American Chemical Society.)

decades). The former is intended for corporations and for governments, while the latter is more oriented to "planners, futurists, and the community of scholars dealing with sustainability." The methodology evaluates the supply risk on the basis of three components: (1) geological, technological, and economic, (2) social and regulatory, and (3) geopolitical. The first component assesses the potential availability of a metal's supply, including both primary and secondary (recycled) sources, and whether it is technologically feasible and economically profitable to extract the metal. The second one attempts to integrate regulations and social attitudes that can support or pose a threat to mineral extraction. Finally, the geopolitical component refers to governmental policies and stabilizing actions that affect the supply of mineral resources. Each component is evaluated on the basis of two indicators, as shown in Figure 5.3. All indicators are scored on a common 0–100 scale, with higher values suggesting a higher level of risk. When aggregated, the scores of these components yield a metal's overall SR score.

5.6.1.1 Geological, Technological, and Economic Component

The geological, technological, and economic component is comprised of two equally weighted indicators: Depletion Time and Companion Metal Fraction.

Depletion Time (DT) serves to evaluate the relative availability of a metal. Depending on the medium- or long-term outlook of the methodology, DT is based on the metal's reserves or reserve base. Furthermore, for the calculation of the depletion time, the methodology takes into account, apart from the mining production, the losses to tailings and slag as well as the end-of-life recycling rate. Depletion Time is not used to calculate how long it will take until the metal is depleted, but is rather used as a relative indicator of the contemporary balance between supply and demand for the metal in question.

The second indicator of the geological, technological, and economic component is the Companion Metal Fraction. It indicates the percentage of the production of a mineral for which it is not the main reason for the quarrying activity, but is instead extracted as a result of the principal, "host" mineral. The reason is that some metals do not form usable deposits on their own, but occur in the ores of metals with similar physical and chemical properties. The former are termed "companion metals" while the latter "host metals." Companion Metal Fraction ranges from 0 to 100. A value of 100 means that 100% of the production of a metal results from mines, where it is mined as a companion metal.

5.6.1.2 Social and Regulatory Component

This component aims to account for governmental and non-governmental factors that can impact the investment of mining companies. It is comprised of two indicators, Policy Potential Index (PPI) and Human Development Index (HDI).

The Policy Potential Index is derived from the Fraser Institute Annual Survey (Jackson and Green, 2014). It is a metric of how much a country's policies and regulatory environment encourage the investments of mining companies. Some of the factors evaluated in this index are current regulations, the legal system, labor regulations, political stability, trade barriers, quality of the geological database, security, and labor and skills availability.

The Human Development Index is published by the United Nations Development Program (UNDP, 2014) and evaluates the countries with respect to life expectancy and health, education, and standard of living. Its inclusion in the supply risk evaluation is based on the rationale that the better a country performs in the three aforementioned fields, the stronger emphasis it places on human quality of life over industrial development (Graedel et al., 2012).

5.6.1.3 Geopolitical Component

This component assesses the political and regulatory stability of the producing countries, which can affect the mining operations. It is built on two indicators: the Worldwide Governance Indicator – Political Stability & Absence of Violence/Terrorism and the Global Supply Concentration.

The Worldwide Governance Indicator – Political Stability (WGI-PS) evaluates the political situation in a country (Kaufmann and Kraay, 2014). It is part of a group of six indicators that evaluate the governance situation in a country, i.e. the traditions and institutions that define the political scene, as presented in paragraph 5.5.2.

The second indicator is founded on the idea that the more concentrated the mineral deposits, the higher the risk of supply restriction. In the case of supply of a material from various sources, a company or country can better cope with a potential cease of supply of the resource from one of these countries due to geological or political factors. The Global Supply Concentration employs a metric, the Herfindahl-Hirschman Index, which takes into account the distribution of the mineral production in various countries. The Herfindahl-Hirschman Index (HHI) is calculated by summing the squares of each country's annual production share. The Global Supply Concentration (GSC) is then calculated based on Equation 5.7:

$$GSC = 17.5 * \ln(HHI) - 61.18 \qquad (5.7)$$

5.6.2 ENVIRONMENTAL IMPLICATIONS

The second dimension in the criticality methodology is called Environmental Implications. This dimension aims to reflect the environmental impact of the production of a metal in order "to indicate to designers, governmental officials, and nongovernmental agencies the potential environmental implications of utilizing a particular metal" (Graedel et al., 2012). The calculation of the environmental implications of using a specific metal relies on the Ecoinvent database (Kellenberger et al., 2007) and the ReCiPe endpoint method (Goedkoop et al., 2009), with "world" normalization and "hierarchist" weighting. The endpoint categories "Human Health" and "Ecosystems" are weighted by 42.9% and 57.1% respectively, while the Resource Availability damage category is omitted, since it was already taken into account in the calculation of the

Supply Risk. The first two damage categories are summed and transformed to Environmental Implications (EI) according to Equation 5.8:

$$\text{Environmental Implications} = \log_{10}\left(ReCiPe \text{ points} + 1\right) * 20 \qquad (5.8)$$

5.6.3 VULNERABILITY TO SUPPLY RESTRICTION

This is the third dimension in the criticality methodology and is comprised of three components: Importance, Substitutability and Susceptibility. The aim of quantifying the Vulnerability to Supply Restriction is to have an indication of the extent of impact that a potential cease of operation of the minerals production chain can have on society. In this direction, the three components assess how indispensable is a mineral or metal for the economy and society. The indicators used in this dimension vary, depending on the organizational level of assessing the criticality of a metal (corporate, national, or global). Figure 5.4 shows the indicators taken into account at the national level of evaluation of criticality.

5.6.3.1 Importance

The importance component aims to quantify how important the element being evaluated is to a given population. It is broken down to two indicators, the National Economic Importance (NEI) and the Material Assets (MA), which replaced the initial indicator Percentage of Population Utilizing.

The National Economic Importance is a measure of the relative importance of a metal to a nation's economy and is expressed as the ratio of the value of the metal that is used in a country in a year over the country's gross domestic product (GDP) for the specified year.

The Percentage of Population Utilizing of the initial methodology (Graedel et al., 2012) was superseded by Material Assets in the work of Harper et al. (2015) because the choice of proxies

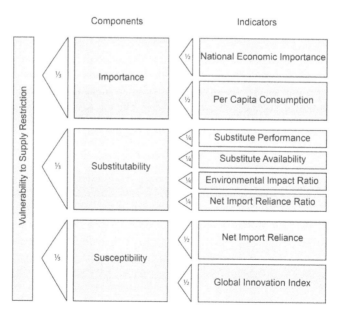

FIGURE 5.4 Components and Indicators of the Vulnerability to Supply Restriction dimension. (Reprinted with permission from Graedel, T.E. et al., *Environ. Sci. Technol.*, 46, 1063–1070, 2012. Copyright 2012 American Chemical Society.)

related to the calculation of the initial indicator was fairly subjective and problematic. Material Assets determines the relative amount of the material analyzed which is used to sustain current lifestyles in the geographic region of focus (Harper et al., 2015).

5.6.3.2 Substitutability

This component aims to quantify the ease of substitution of a metal with another one in a given economy, as well as the repercussions of such a substitution (technical, economic, environmental). Assessing the substitutability of a mineral involves studying four indicators: Substitute Performance, Substitute Availability, Environmental Impact Ratio and Net Import Reliance Ratio.

Substitute Performance (SP) is assessed by examining the past, concurrent, and future use of the substitutes and whether there is a significant differentiation in the final product when using the substitute.

Substitute Availability (SA) is defined as the supply risk of the metal that is used to substitute the metal of focus.

The Environmental Impact Ratio (EIR) serves to compare the environmental impact of a metal to its substitute. It is given by Equation 5.9:

$$EIR = 50 * \frac{EI_{substitute}}{EI_{metal\ of\ focus}} \qquad (5.9)$$

Finally, the last indicator in the substitutability component is the Net Import Reliance Ratio (NIRR). The Net Import Reliance aims to capture the reliance of a nation on imports from other countries (Equation 5.10). For the Net Import Reliance Ratio, the Net Import Reliance of the substitute is compared to the Import Reliance of the metal of focus.

5.6.3.3 Susceptibility

The Susceptibility component is comprised of two indicators: the Net Import Reliance and the Global Innovation Index.

As mentioned above, the Net Import Reliance (NIR) expresses the net imports of a country in comparison to the mineral consumption in that country (Equation 5.10):

Net Import Reliance

$$= \frac{Imports - Exports + adjustments\ for\ government\ and\ industry\ stock\ changes}{Apparent\ Consumption} * 100 \qquad (5.10)$$

The second indicator, Global Innovation Index (GII), was developed by INSEAD in collaboration with the Confederation of Indian Industry (Cornell University, INSEAD, WIPO, 2014). The index accounts for a series of variables in order to give a measure of the innovation level of every country. The rationale is that more innovative nations can more quickly overcome a supply restriction.

5.6.4 CRITICALITY SCORE

Graedel et al. (2012) define the overall criticality of a metal as a combination of its three dimensions. The criticality vector joins the origin to a metal's location in the criticality space. Its magnitude $\|C\|$ is normalized on a 0–100 scale and is given by Equation 5.11:

$$\|C\| = \frac{\sqrt{SR^2 + EI^2 + VSR^2}}{\sqrt{3}} \qquad (5.11)$$

5.7   CURRENT RESEARCH IN THE FIELD

The criticality methodologies presented above adopt a backward-looking approach, i.e. they evaluate criticality based on current information on reserves and the existing market conditions and technologies. Most criticality methodologies adopt a simplified approach to systems thinking and provide a snapshot of the current situation. Nevertheless, this static approach has some limitations and does not consider the dynamic nature of many indicators. As underlined in the report of the EU (European Commission, 2014), "the dynamic character of the global market and the technology developments can change the picture of raw materials in the criticality map and materials which are currently not identified as critical can become critical over time." For example, changes in technology can entail material substitution in some products and the use of raw resources that were more expensive until now or more difficult to extract. Therefore, they can cause changes in the supply chain and accentuate the need of updating the list of critical materials.

Furthermore, according to Bradshaw et al. (2013), a mineral cannot be critical by itself; criticality is rather a condition of the system, triggered by some attribute or property, which is said to become critical. Habib and Wenzel (2016) also argue that criticality is a dynamic rather than static phenomenon, since the nature of the indicators used to assess criticality is dynamic as well.

To date, few studies have dealt with a forecast of future criticality. Roelich et al. (2014) were one of the few to develop a dynamic criticality methodology. Criticality here is assumed as the product of supply disruption potential (P) by the exposure to disruption (E). Both indices are produced as a forecast time series, which estimates criticality over time and identifies trends of increasing (or decreasing) criticality. Knoeri et al. (2013) also suggested an approach for dynamic criticality by developing a framework that couples an agent-based behavior model with a dynamic material flow model. Similarly, Riddle et al. (2015) used an agent-based behavior model to introduce the future changes in the criticality assessments. Moreover, the Joint Research Centre of the European Commission published a study where it performs a detailed assessment of the criticality of various metals and uses a bottom-up approach (Moss et al., 2013). The research team compiled an inventory of all metals used in the low-carbon technologies and estimated the projected diffusion of each technology under different scenarios, accounting for market and geopolitical factors. Finally, Habib and Wenzel (2016) assess the dynamic criticality of rare earth metals in wind turbine technology using a product design tree approach and assuming dynamic reserves.

Until recently, criticality had a very restricted field of application, mainly with respect to metals used in low-carbon technologies. However, the criticality concept slowly expands to include more raw, as well as biotic materials, in addition to the metals. As discussed above, the European Commission has included biotic materials in its criticality evaluations and identified natural rubber as a critical resource (European Commission, 2017). Ongoing research is focusing on applying criticality to other biotic resources, such as wood and biomass. The criticality methodology developed by Graedel et al. (2012) has been used to evaluate the depletion risk of various metals (Nassar et al., 2012; Panousi et al., 2015; Harper et al., 2015) and has also found its application in other resources, such as water (Sonderegger et al., 2015) and gravel (Ioannidou et al., 2017). In addition, Meylan et al. (2017) suggested a modified framework for the criticality of nutrient flows of agricultural livelihood systems.

5.8   QUESTIONS AND EXERCISES

1. Provide a definition for the terms resource, reserve base, and reserve.
2. What is the difference between the criticality concept to Life-Cycle Assessment and the abiotic depletion indicators?
3. Explain the main dimensions used in criticality assessment methodologies.
4. Explain the different components and indicators of the methodology for metal criticality determination developed by Graedel et al. (2012).

5. What is the field of application of the criticality concept?
6. Calculate the supply risk score for iron, given the following data (data modified from Nassar et al., 2012):
   - Assume that the following countries are the suppliers of iron:

Countries	% of Production
China	52
Japan	15
Russia	12
India	8
Brazil	7
United States	6

   - Iron is a main metal and not a by-product of the mining process.

REFERENCES

Achzet, B., and Helbig, C. (2013). How to evaluate raw material supply risks—An overview. *Resources Policy* 38, 435–447.

Alonso, E., Field, F.R., and Kirchain, R.E. (2008). *The Dynamics of the Availability of Platinum Group Metals for Electronics Manufacturers*. IEEE, San Francisco, CA.

Ayres, R. (1989). Industrial metabolism. In: Ausubel J, (Ed.), *Technology and Environment*. National Academy Press, Washington, DC.

Blengini, G.A., Blagoeva, D., Dewulf, J., Torres de Matos, C., Nita, V., Vidal-Legaz, B. et al. (2017). *Assessment of the Methodology for Establishing the EU List of Critical Raw Materials*. Publications Office of the European Union, Luxembourg, GD Luxembourg. doi:10.2760/73303, JRC106997.

Bradshaw, A.M., Reuter, B., and Hamacher, T. (2013). The potential scarcity of rare elements for the Energiewende. *Green* 3(2), 93–111.

Bridge, G. (2000). The social regulation of resource access and environmental impact: Production, nature and contradiction in the US copper industry. *Geoforum* 31, 237–256.

Buijs, B., Sievers, H., and Espinoza, L.A.T. (2012). Limits to the critical raw materials approach. *Proceedings of the ICE-Waste and Resource Management* 165(4), 201–208.

Calvo, G., Valero, A., and Valero, A. (2017). Assessing maximum production peak and resource availability of non-fuel mineral resources: Analyzing the influence of extractable global resources. *Resources, Conservation and Recycling* 125, 208–217.

Cimprich, A., Young, S.B., Helbig, C., Gemechu, E.D., Thorenz, A., Tuma, A., and Sonnemann, G. (2017). Extension of geopolitical supply risk methodology: Characterization model applied to conventional and electric vehicles. *Journal of Cleaner Production* 162, 754–763.

Cornell University, INSEAD, and WIPO. (2014). *The Global Innovation Index 2014*. The Human Factor in Innovation, Switzerland.

Drielsma, J.A., Allington, R., Brady, T., Guinée, J., Hammarstrom, J., Hummen, T. et al. (2016). Abiotic raw-materials in life cycle impact assessments: An emerging consensus across disciplines. *Resources* 5, 12.

European Commission. (2010). Critical raw materials for the EU. *Report of the Ad-hoc Working Group on defining critical raw materials*. European Commission. Enterprise and Industry. Available from: https://ec.europa.eu/growth/tools-databases/eip-raw-materials/en/community/document/critical-raw-materials-eu-report-ad-hoc-working-group-defining-critical-raw. Accessed on November 20, 2017.

European Commission. (2011). *Tackling the Challenges in Commodity Markets and on Raw Materials*. European Commission, Brussels, Belgium.

European Commission. (2014). Report on critical raw materials for the EU. *Report of the Ad hoc Working Group on Defining Critical Raw Materials*. May 2014. Brussels, Belgium.

European Commission. (2017). *Communication from the Commission to the European Parliament, the Council, the European Economic and Social Committee and the Committee of the Regions on the 2017 list of Critical Raw Materials for the EU*. COM (2017) 490 final. Brussels, Belgium.

Frenzel, M., Kullik, J., Reuter, M.A., and Gutzmer, J. (2017). Raw material "criticality"—Sense or nonsense? *Journal of Physics D: Applied Physics* 50(12), 123002.

Giljum, S., Burger, E., Hinterberger, F., Lutter, S., and Bruckner, M. (2011). A comprehensive set of resource use indicators from the micro to the macro level. *Resources, Conservation and Recycling* 55(3), 300–308.

Giurco, D., Prior, T., Mudd, G.M., Mason, L., and Behrisch, J. (2010). *Peak Minerals in Australia: A Review of Changing Impacts and Benefits.* Institute for Sustainable Futures, Sydney, Australia.

Glöser, S., Espinoza, L.T., Gandenberger, C., and Faulstich, M. (2015). Raw material criticality in the context of classical risk assessment. *Resources Policy* 44, 35–46.

Goedkoop, M., and Spriensma, R. (2001). The eco-indicator 99, a damage oriented method for life cycle impact assessment, methodology report. 3rd ed., PRé Consultants B.V., June 22, 132 p. Available from: http://www.pre-sustainability.com/download/EI99_methodology_v3.pdf. Accessed on March 20, 2015.

Goedkoop, M., Heijungs, R., Huijbregts, M., De Schryver, A., Struijs, J., and van Zelm, R. (2009). ReCiPe 2008. Main Report, Part 1: Characterization, 1st ed.; Ministerie van Volkshuisvesting, Ruimtelijke Ordening en Milieubeheer (VROM), The Hague, the Netherlands.

Graedel, T.E., and Reck, B.K. (2016). Six years of criticality assessments: What have we learned so far? *Journal of Industrial Ecology* 20(4), 692–699.

Graedel, T.E., Barr, R., Chandler, C., Chase, T., Choi, J., Christoffersen, L. et al. (2012). Methodology of metal criticality determination. *Environmental Science & Technology* 46, 1063–1070.

Guinée, J.B., Gorrée, M., Heijungs, R., Huppes, G., Kleijn, R., van Oers L. et al. (2002). *Handbook on Life Cycle Assessment. Operational Guide to the ISO Standards.* I: LCA in perspective. IIa: Guide. IIb: Operational annex. III: Scientific background. Kluwer Academic Publishers, Dordrecht, the Netherlands, 692 p.

Haberl, H., Fischer-Kowalski, M., Krausmann, F., Weisz, H., and Winiwarter, V. (2004). Progress towards sustainability? What the conceptual framework of material and energy flow accounting (MEFA) can offer. *Land Use Policy* 21(3), 199–213.

Habert, G., Bouzidi, Y., Chen, C., and Jullien, A. (2010). Development of a depletion indicator for natural resources used in concrete. *Resources Conservation and Recycling* 54, 364–376.

Habib, K., and Wenzel, H. (2016). Reviewing resource criticality assessment from a dynamic and technology specific perspective–using the case of direct-drive wind turbines. *Journal of Cleaner Production* 112, 3852–3863.

Harper, E.M., Kavlak, G., Burmeister, L., Eckelman, M.J., Erbis, S., Sebastian Espinoza, V., Nuss, P., and Graedel, T.E. (2015). Criticality of the geological zinc, tin, and lead family. *Journal of Industrial Ecology* 19, 628–644.

Hubbert, M.K. (1956). *Nuclear Energy and the Fossil Fuels.* Shell Development Company. Publication No. 95, Houston, TX, 40 p.

Hubbert, M.K. (1962). *Energy Resources: A Report to the Committee on Natural Resources.* National Academy of Sciences, Washington, DC.

Ioannidou, D., Meylan, G., Sonnemann, G., and Habert, G. (2017). Is gravel becoming scarce? Evaluating the local criticality of construction aggregates. *Resources Conservation and Recycling* 126, 25–33.

Jackson, T., and Green, K.P. (2015). *Fraser Institute Annual Survey of Mining Companies*, 2014. Fraser Institute. http://www.fraserinstitute.org. Accessed on November 20, 2017.

Jin, Y., Kim, J., and Guillaume, B. (2016). Review of critical material studies. *Resources Conservation and Recycling* 113, 77–87.

Jolliet, O., Margni, M., Charles, R., Humbert, S., Payet, J., Rebitzer, G. et al. (2003). IMPACT 2002+: A new life cycle impact assessment methodology. *The International Journal of Life Cycle Assessment* 47, 356–374.

Kaufmann, D., and Kraay, A. (2014). The worldwide governance indicators, 2014 update. Aggregate Indicators of Governance 1996–2013. Available from: http://info.worldbank.org/governance/wgi/index.aspx#home. Accessed on November 20, 2017.

Kellenberger, D., Althaus, H.J., Künniger, T., Lehmann, M., Jungbluth, N., and Thalmann, P. (2007). Life cycle inventories of building products, Data v2.0.ecoinvent report No. 7. Swiss Center for Life Cycle Inventories, Dubendorf, Switzerland.

Knoeri, C., Wäger, P.A., Stamp, A., Althaus, H.J., and Weil, M. (2013). Towards a dynamic assessment of raw materials criticality: Linking agent-based demand—With material flow supply modelling approaches. *Science of the Total Environment* 461, 808–812.

Lloyd, S., Lee, J., Clifton, A., Elghali, L., and France, C. (2012). Recommendations for assessing materials criticality. *Proceedings of the Institution of Civil Engineers-Waste and Resource Management* 165, 191–200.

Meadows, D.H. (1998). Indicators and information systems for sustainable development. A report to the Balaton Group, The Sustainability Institute, Hartland, CA, September 1998.

Meadows, D.H., Meadows, D.L., Randers, J., and Behrens, W.W. (1972). *The Limits to Growth.* Universe Books, New York, Vol. 102, p. 27.

Meylan, G., Thiombiano, B.A., and Le, Q.B. (2017). *Nutrient Flow Scenarios for Sustainable Smallholder Farming Systems in Southwestern Burkina Faso.* Research Partnership between USYS TdLab/Swiss Federal Institute of Technology (ETH) Zurich and CGIAR Research Program on Dryland Systems (CRP-DS)/International Center for Agricultural Research in Dry Areas (ICARDA), Zurich, Switzerland, 44 p.

Moss, R.L., Tzimas, E., Willis, P., Arendorf, J., Thompson, P., Chapman, A. et al. (2013). Critical metals in the path towards the decarbonisation of the EU energy sector—Assessing rare metals as supply-chain bottlenecks in low-carbon energy technologies. JRC Scientific and Policy Reports. Report EUR 25994 EN, Luxembourg, GD Luxembourg.

Nassar, N.T., Barr, R., Browning, M., Diao, Z., Friedlander, E., Harper, E.M. et al. (2012). Criticality of the geological copper family. *Environmental Science & Technology* 46, 1071–1078.

National Research Council. (2008). *Minerals, Critical Minerals, and the U.S. Economy.* National Academies Press, Washington, DC.

Nuss, P., and Blengini, G.A. (2018). Towards better monitoring of technology critical elements in Europe: Coupling of natural and anthropogenic cycles. *Science of the Total Environment* 613, 569–578.

Panousi, S., Harper, E.M., Nuss, P., Eckelman, M.J., Hakimian, A., and Graedel, T.E. (2015). Criticality of seven specialty metals. *Journal of Industrial Ecology* 20(4), 837–853.

Prior, T., Giurco, D., Mudd, G., Mason, L., and Behrisch, J. (2012). Resource depletion, peak minerals and the implications for sustainable resource management. *Global Environmental Change* 22, 577–587.

Riddle, M., Macal, C.M., Conzelmann, G., Combs, T.E., Bauer, D., and Fields, F. (2015). Global critical materials markets: An agent-based modeling approach. *Resources Policy* 45, 307–321.

Roelich, K., Dawson, D.A., Purnell, P., Knoeri, C., Revell, R., Busch, J., and Steinberger, J.K. (2014). Assessing the dynamic material criticality of infrastructure transitions: A case of low carbon electricity. *Applied Energy* 123, 378–386.

Schneider, L., Berger, M., Schüler-Hainsch, E., Knöfel, S., Ruhland, K., Mosig, J., Bach, V., and Finkbeiner, M. (2014). The economic resource scarcity potential (ESP) for evaluating resource use based on life cycle assessment. *The International Journal of Life Cycle Assessment* 19, 601–610.

Schneider, L., Markus, B., and Finkbeiner, M. (2011). The anthropogenic stock extended abiotic depletion potential (AADP) as a new parameterisation to model the depletion of abiotic resources. *The International Journal of Life Cycle Assessment* 16, 929–36.

Singer, S.A., and Menzie, W.D. (2010). *Quantitative Mineral Resource Assessments. An Integrated Approach.* 1st ed. Oxford University Press, Oxford, UK, 219 p.

Sonderegger, T., Dewulf, J., Fantke, P., de Souza, D.M., Pfister, S., Stoessel, F. et al. (2017). Towards harmonizing natural resources as an area of protection in life cycle impact assessment. *The International Journal of Life Cycle Assessment* 22, 1912–1927.

Sonderegger, T., Pfister, S., and Hellweg, S. (2015). Criticality of water: Aligning water and mineral resources assessment. *Environmental Science & Technology* 49, 12315–12323.

Sonnemann, G., Gemechu, E.D., Adibi, N., De Bruille, V., and Bulle, C. (2015). From a critical review to a conceptual framework for integrating the criticality of resources into life cycle sustainability assessment. *The Journal of Cleaner Production* 94, 20–34.

Steen, B. (1999). *A Systematic Approach to Environmental Priority Strategies in Product Development (EPS): Version 2000-general System Characteristics* (p. 67). Centre for Environmental Assessment of Products and Material Systems, Gothenburg, Sweden.

Topp, V., Soames, L., Parham, D., and Bloch, H. (2008). Productivity in the mining industry: measurement and interpretation. *Productivity Commission Staff Working Paper*, Canberra, Australia.

UNDP. (2014). Human Development Report 2014. Sustaining Human Progress: Reducing Vulnerabilities and Building Resilience. New York.

USGS. (2009). *Mineral commodity summaries.* Available from: http://minerals.usgs.gov/minerals/pubs/mcs/2009/mcs2009.pdf. Accessed on November 20, 2017.

6 Combining and Integrating Life-Cycle and Risk Assessment

Michael Tsang and Guido Sonnemann

CONTENTS

6.1 MOTIVATIONS FOR USING LIFE-CYCLE ASSESSMENT AND RISK ASSESSMENT

Life-cycle assessment (LCA) is a broadly scoped environmental management tool defined by ISO 14040:2006 and ISO 14044:2006 (International Organization for Standardization 2006a, 2006b). Although LCA is founded on the concept of environmental management, it is a tool capable of measuring and communicating the relative environmental impacts of a product or process. LCA tabulates the resources used, energy consumed, emissions generated, and resulting impacts across the life cycle of a "function," which in turn represents a product, process, or service.

The strengths of LCA are that it is a holistic tool, modeling and estimating multiple environmental and human health metrics throughout the entire supply chain of a function. Due to this nature as a broad scoping tool, LCA integrates emissions and resource uses over space and time. One consequence of this is that the reported environmental and human health metrics represent relative impact values as opposed to absolute values. Hence, it is often used to make comparative evaluations to guide eco-design strategies for, quantify the resource efficiency of, or communicate profiles, such as in Environmental Product Declarations (ISO 14025), for products, as opposed to making judgments of regulatory compliance or determining levels of safety. LCA is particularly useful in stage-gate product development approaches, where LCA studies can be built very early on in product development (i.e., early-stage development) to provide hot-spot analysis and guidance for sustainable production and consumption.

Risk Assessment (RA) is a tool used to calculate the eco- or human-toxicological risks posed by chemicals and pollutants emitted into the environment and can be used on a case-by-case basis to identify whether potentially toxic substances pose such risks. Specifically, RA involves analyzing pollutant emissions, fate, and exposure for specific exposure settings. The result is that RA directly measures or estimates toxicological impacts in absolute values at the individual level for human health, for example, in focused exposure scenarios. This is in contrast to LCA, which calculates relative values of toxicity, as described above. Absolute values are those that can be directly compared to established regulatory reference values (e.g., acceptable daily intake, ADI; threshold limit value, TLV) to know whether an individual's exposure exceeds the safety threshold, while relative values cannot be used for such purposes. Consequently, and due to the existing regulatory framework governing chemicals in commerce, RA is a fundamental tool for many scientists, regulators and commercial enterprises that act to ensure the safety of workplaces and products in the consumer marketplace. RA, therefore, would have little relevance during the early-stage development of products or communicating the environmental preference of a product since absolute values of exposures at this level would give a premature indication of risk due to the uncertainty associated with moving from early-stage development to full-scale production.

Thus, while LCA's strengths lie in comparative analyses of resource efficiencies and global hot-spot identification of certain environmental and human health impacts of products, processes, or services, RA's strengths are in making more definitive statements about the potential hazards and risks posed to the environment and human health from individual pollutant emissions. In the next two sections of this chapter, we first outline the general reasons driving the isolated, complementary, and integrated uses of LCA and RA. Then, this is explored in greater detail using the topic of nanotechnology in order to illustrate how evaluation of such products benefits from each particular approach.

6.2 COMBINING LIFE-CYCLE AND RISK ASSESSMENT

6.2.1 Isolated and Complementary Use of Life-Cycle Assessment and Risk Assessment

In current practice, LCA and RA are often used in isolation of one another, according to decision-makers' conceptions regarding the value of, the preferences for, or familiarity of using one tool over another. This is what we refer to as *Isolated Use* of either tool. *Isolated Use* serves many types of purposes and many different stakeholders, whereby certain stakeholders may need to determine a measure of safety (i.e., regulatory compliance) while others may want to create a lower carbon footprint technology than their competitor. For example, a project design team might be interested in knowing how to forecast which materials they should choose to lower their overall toxicological impacts across the product's life cycle. The team is likely to run through several hypothetical or pilot-scale scenarios in which multiple feedstocks, manufacturing routes, and final product designs are considered. In this case, LCA would prove useful as it can give global averages for these impacts that, while the assessment does not provide absolute values of safety, it does constructively add to the decision-making process. On the other hand, a health and safety manager at the company might be concerned with the occupational exposure to volatile organic carbons emitted during the manufacturing stages of existing full-scale product lines. In this case, an RA would allow the manager to identify if exposure levels are exceeding specific values of safety. In other cases, decision-makers may wish to evaluate both of these considerations for the same project or at the same time, for example, if the project team wants to do their due diligence on a final design choice and forecast whether occupational exposure levels are safe at large-scale versions of that technology. This could be done by combining the results independently derived by each tool. We refer to this approach as the *Complementary Use* of both LCA and RA as separate tools and then combining their results (Harder et al. 2015; Shatkin 2008). This could be done, for example, using multi-criteria decision analysis (MCDA) to convert and weigh the results for comparison (Linkov and Seager 2011).

Whatever the approach, *Complementary Use* can serve as an obvious and straightforward way to include both types of evaluations.

Because of its incorporation of RA, *Complementary Use* should be viewed as an appropriate method for evaluating fully scaled and mature technologies or when it is necessary to comply with regulatory standards or ensure levels of ecological or human health safety (Barberio et al. 2014; Kikuchi and Hirao 2008; Walser et al. 2014). To put it another way, the level of detail involved during the RA process is not warranted in early-stage development common to emerging technologies given the uncertainty regarding the operating parameters and conditions of early-stage development compared to full-scale production processes. In such cases, it may be more prudent to lead off with a hot-spot analysis of the pollutant emissions using LCA and then follow up with an RA, if potential impacts are observed (i.e., *Isolated Use*).

Secondarily, reconciling the different scopes of each tool may prove difficult for the average user. While LCA scales its impacts to a "functional" unit (e.g., production of an average 1 kWh of energy), RA is defined by a relevant exposure scenario (e.g., the human health risk posed by indoor air emissions of a pollutant in the occupational setting). For instance, an RA might provide the health risk of workers in a certain exposure scenario, a, given an amount of work performed, b, resulting in certain emissions, c, over a specified length of time, d. However, parameters a, b, c, and d, for example, are not likely to correlate with the amount of work performed, emissions involved, and length of time of the LCA's functional unit, thus, making it difficult to interpret and reconcile results of the RA and LCA. The RA would have to be performed such that parameters a, b, c, and d correspond with the LCA, or vice versa, clearly challenging the study designs of each approach and efforts of the decision maker. Thus, if such things are not reconciled, *Complementary Use* misses the opportunity to express how proportionate the resulting chemical-specific human health impacts are compared to the other toxicological and environmental hazards estimated across the life cycle using LCA.

6.2.2 INTEGRATION OF LIFE-CYCLE ASSESSMENT AND RISK ASSESSMENT

Integration refers to combining the methods of LCA and RA as opposed to combining their individual results (i.e., providing estimations of human health impacts within LCA as opposed to using RA as a separate tool to determine this) (Harder et al. 2015). One advantage of this approach is that it provides a way to directly compare and interpret ecological and human health hazards against the resource efficiencies across the life cycle of a product, process, or service. Integration effectively would provide relative values of human health impacts that could be used in comparative assessments for eco-design or making environmental product declarations, for example.

We distinguish between three different approaches, which differ by their extent of integration (see Figure 6.1). These three approaches are categorized as *Site-Generic*, *Site-Dependent*, and *Site-Specific* integration, whose terminology is consistent with existing literature related to the general subject of combining LCA and RA (Harder et al. 2015; Potting and Hauschild 2006). The first level of integration, referred to as *Site-Generic*, estimates ecological or human health impacts through models that calculate the fate and exposure of chemicals in the environment using steady-state assumptions (i.e., low temporal resolution) for generalized environments and settings (i.e., low special resolution). The second level of integration, referred to as *Site-Dependent*, removes the steady-state assumptions and applies dynamic fate and exposure models to the estimation of ecological or human health impacts. However, model parameters are defined in several classes such that they can be estimated per class, as opposed to exact values for each specific scenario under analysis. Lastly, the third level of integration, referred to as *Site-Specific*, refers to the use of parameters in a dynamic model that are defined specifically for the exact scenario under analysis, on a case-by-case basis.

In the following section, we use the example of engineered nanomaterials (ENMs) and nanotechnologies to illustrate how LCA and RA can be used in such evaluations and how different approaches to integration may assist in certain cases.

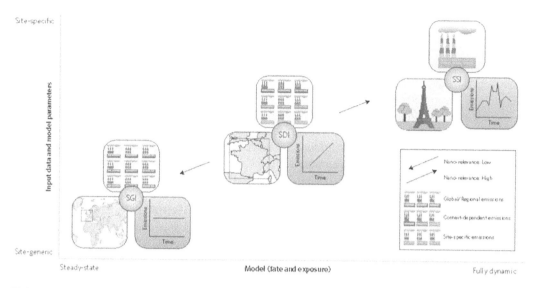

FIGURE 6.1 Illustration depicting the integration of risk assessment into life-cycle assessment at the methodological level. Integration is measured on its level of spatial and temporal specificity, as well as model complexity or model type. Site-Generic Integration uses steady-state modeling that does not consider specific emission or exposure scenarios. At the steady-state level, global (that is, highly aggregated) and temporally undefined (that is constant) inputs and parameters to the model are used. Site-Dependent Integration assesses the fate and exposure of emissions for broad "classes" of emission and exposure scenarios. Site-Specific Integration nears the estimation of absolute exposure values through the derivation of empirically determined changes in single-source emissions data measured in an unconditional timeframe in a fully dynamic model. SGI: Site-Generic Integration; SDI: Site-Dependent Integration; and SSI: Site-Specific Integration.

6.3 ILLUSTRATIVE EXAMPLE: ENGINEERED NANOMATERIALS AND NANOTECHNOLOGY

As a "key enabling technology" of economic growth and competitiveness (Savolainen et al. 2013), there is tremendous optimism for the role that ENMs will play in revolutionizing whole industries. ENMs are broadly defined as those that have one dimension at or less than 100 nm in size (Som et al. 2010). Their novel characteristics and properties can be used in a diverse set of emerging technologies, such as newer generation solar photovoltaics, for example, to create renewable sources of electricity. This was the case demonstrated in Chapters 2 and 3, where the organic photovoltaic (OPV) panels had employed nanoparticles of titanium dioxide (nano-titania) to enhance the energy conversion efficiency of the device. As of 2015, there were more than 1,800 such products on the market, according to the crowdsourced inventory of commercial nanotechnologies called the Project of Emerging Nanotechnologies (PEN) (Vance et al. 2015). However, these same novel characteristics could pose adverse impacts on the environment or human health which need to be evaluated at early stages of development to avoid unforeseen repercussions from adaptation of these technologies (Busnaina et al. 2013; Greßler and Nentwich 2012b).

6.3.1 Life-Cycle Assessment and the Resource Opportunities of ENMs

LCA, in theory, can be applied to ENM-related systems (Miseljic and Olsen 2014). The first formal promulgation of using LCA for nano-related systems came during a 2006 workshop titled *Nanotechnology and Life Cycle Assessment* (Klöpffer et al. 2007), where it was acknowledged that "[LCA] is fully suitable to nanomaterials and nano-products." In practice, however, Hischier and Walser point out that there have been few *complete* nano-related LCAs, particularly those whose system boundaries are cradle-to-grave (Hischier and Walser 2012). This is, in part, due to difficulties obtaining relevant nano-production data from manufacturers (Greßler and Nentwich 2012a; Miseljic and Olsen 2014; Tsang et al. 2014a). Companies may be less willing to provide such proprietary information due to fear of patent infringement, competition, or otherwise. This limitation can be categorized as an *accessibility gap*, where information is technically attainable yet often inaccessible (Tsang et al. 2014a). These data gaps often concern data that are generally not nano-specific (i.e., characterization of the material), but instead, data regarding the production and manufacturing conditions such as energy use, material consumption, synthesis parameters, and waste generation.

Additionally, current knowledge about the characterization, release, behavior, and effects of ENMs over the life cycle is also limited (Gavankar et al. 2012). This limitation can be categorized as an *evaluation gap*, where the generation of completely new data is needed to complete the assessment (Tsang et al. 2014a). Such data might be partially an accessibility issue, for instance, if a production facility knows the environmental release of their ENMs. However, this information might simply be unknown, requiring considerable amounts of resources and time to acquire (Hristozov et al. 2012). Adding to this problem is the inability to properly measure and identify particular ENM emissions across the life cycle (Miseljic and Olsen 2014), for instance, trying to determine the fraction of ENM in an airborne emission.

A third gap, defined as *methodological*, concerns the ability of life cycle impact assessment methods to characterize and create impact scores directly for the ENM themselves (Gavankar et al. 2012; Greßler and Nentwich 2012a; Hischier and Walser 2012; Miseljic and Olsen 2014; Tsang et al. 2014a). This is particularly relevant for human health impacts. The impact factors currently used in impact assessment methods such as ReCiPe, Uniform System for the Evaluation of Substances adapted for LCA (USES-LCA), and USEtox do not model fate, transport, and exposure of ENMs. Current LCIA methods rely on characterization factors that are dependent on fugacity-based models that employ criteria such as vapor pressures and octanol-water partition coefficients. Such criteria describe fate, transport, and exposure of organics appropriately in many cases (Salieri et al. 2015), but this approach either only describes a subset of nano-related fate criteria or do not include them at all (Eckelman et al. 2012).

In contrast to organic molecules, ENMs behave like colloidal substances that exist in their own phase when released into a specific medium (Quik et al. 2011). Therefore, ENMs are not thermodynamically stable, even if some may be kinetically stable for long periods of time (Praetorius et al. 2014). In other words, ENM concentration ratios between two mediums cannot be predicted from a previously measured partition coefficient after the further addition or removal of ENMs from that system (Praetorius et al. 2014). Instead, the behavior of ENMs in the environment is concentration dependent. Consequently, their behavior in the environment can be estimated kinetically using dynamic models that describes the rates of the processes that control their behavior (Praetorius et al. 2014; Salieri et al. 2015), such as homo- and hetero-aggregation (Meesters et al. 2014) or even clearance and retention from an organism (Braakhuis et al. 2014; Hodas et al. 2015; Li et al. 2015). Evaluating such concentration-dependent manifestations requires non-steady-state models. However, the use of dynamic models is a challenge for most LCA practitioners, as precise emissions data and detailed modeling procedures may not be readily available, much less warranted for these types of evaluations. A framework for this approach is presented later in this chapter on integration, and a case study defining the model is presented in Chapter 7.

6.4 RISK ASSESSMENT OF ENGINEERED NANOMATERIALS

Risk is defined by the United States Environmental Protection Agency (U.S. EPA) as the chance that harmful effects to human health or ecological systems results from exposure to an environmental stressor, such as a chemical or other material substance (U.S. EPA 2015). In the context of human health, a stressor could be any physical, biological, or chemical entity, while exposure is related to the dose and duration one is subjected to. To quantify the risk to human health, an RA must be conducted. In contrast to LCA, the aim of RA is to quantify *absolute* values of risk and to minimize such risk by designating thresholds of "acceptable" exposure levels (Barberio et al. 2014; MacPhail et al. 2013).

RA has been applied to quantify human health impacts of chemicals for developing strategy of chemicals management at the national level for decades. The U.S., E.U., and the Japanese governments, for example, all employ human health risk assessment as a part of their chemical regulatory policy. Specific RA of ENM requirements do not exist for these governments, but the U.S. EPA and the E.U. European Chemicals Agency, for example, can demand and gather data to assist in the RA procedure for ENMs if they find it necessary to do so (European Chemicals Agency n.d.; U.S. Environmental Protection Agency 2016). In Japan, research projects, including field studies at manufacturing sites, have been conducted in several national institutes (National Institute of Advanced Industrial Science and Technology [AIST] 2011), (National Institute of Occupational Safety and Health, Japan n.d.) for developing guidelines for RA and management of ENMs.

Although the methodological framework of RA is applicable to ENMs, further efforts are needed for data accumulation of a wider range of ENMs, especially for the establishment of thresholds and the measurement of workplace exposure. In order to systematically cover ENMs of different physical, chemical, and physicochemical properties, the harmonization of strategies and conditions in toxicity tests and exposure monitoring is needed.

6.5 FRAMEWORK FOR EVALUATING OPPORTUNITIES AND RISKS OF NANOTECHNOLOGIES: INTEGRATION OF LIFE-CYCLE ASSESSMENT AND RISK ASSESSMENT FOR NANOTECHNOLOGIES

As the ENM sector continues to grow, scientists and policy makers have highlighted the need to identify a holistic evaluation across the life cycle that takes into account competing opportunities and risks for nanomaterials (Beaudrie et al. 2013; Busnaina et al. 2013; Meyer and Upadhyayula 2014; Wong and Karn 2012), as has been proposed for regular chemicals (Sonnemann and Schuhmacher 2004). Steinfeldt et al. set out to evaluate both opportunities and risks of ENMs in their 2007 book *Nanotechnology, Hazards and Resource Efficiency*, using LCA to evaluate different ENM case studies (Steinfeldt and Petschow 2007). They noted that LCA should be coupled with ENM hazard and risk data for a more thorough evaluation of the technology. As discussed earlier in this chapter, LCA is not adequately equipped to assess the potential human health and/or environmental impacts or risks arising from the ENMs themselves. Since LCA can still provide useful results evaluating the comparative impacts and thus opportunities across systems, many experts in the field have discussed the tandem use LCA with RA to produce a more complete assessment of the system (Breedveld 2013; Canis et al. 2010; Grieger et al. 2012; Jahnel et al. 2013; Meyer and Upadhyayula 2014; Rousk et al. 2012; Seager and Linkov 2008; Som et al. 2010; Steinfeldt and Petschow 2007).

6.5.1 COMPLIMENTARY USE OF LIFE-CYCLE ASSESSMENT AND RISK ASSESSMENT FOR ENGINEERED NANOMATERIALS

The *Complimentary Use* of LCA and RA for evaluating the risks posed by ENMs was discussed in the review article by Grieger et al. (2012). Based on their review, the authors identified and distinguished between two types of LCA and RA integration presented in the literature. The first approach, described as "LC-based RA," involves RA conducted at various stages of a chemical's life cycle (Wardak et al. 2008), which does not actually involve an LCA, but instead revolves around the concept of life cycle thinking. The second approach, described as "RA-complemented LCA," is an LCA complimented with an RA, which is to say that an LCA is performed and, additionally, an RA would be performed where deemed appropriate (Barberio et al. 2014). This latter approach is one of a result-oriented integration (i.e., integration post-assessment), where the decision maker is left with considering how to weigh and balance the individual results of the assessment (Linkov and Seager 2011; Tsang et al. 2014b).

However, another approach of bridging LCA and RA together to evaluate ENM is to integrate them at the methodological level, as introduced in the sections above. In this approach, data, methods, and/or the scope of RA are integrated into LCA (Barberio et al. 2014; Eckelman et al. 2012; Miseljic and Olsen 2014; Salieri et al. 2015) such that human health impacts of ENMs can be considered alongside LCA impacts using a single tool and single dataset. This is not unlike the approach already used in LCA for calculating the human health impacts of regular chemicals as discussed further below.

6.5.2 Integration of Life-Cycle Assessment and Risk Assessment for Engineered Nanomaterials

As introduced above, the first level of integration is referred to as *Site-Generic* integration. This approach estimates ecological or human health impacts of ENMs through models that calculate the fate and exposure of ENMs in the environment using steady-state assumptions (i.e., low temporal resolution) for generalized environments and settings (i.e., low special resolution). Steady-state assumptions assume that substances come into thermodynamic equilibrium with their environments and that a final distribution between various mediums can be estimated using partition coefficients. *Site-Generic* integration, for example, has been demonstrated in a few studies using the USEtox steady-state toxicity model by incorporating ENM-specific partition coefficients and kinetic descriptors of behavior into the model (Eckelman et al. 2012; Miseljic and Olsen 2014; Salieri et al. 2015).

In each case, the authors modified existing characterization factors in USEtox to make them nano-specific. The USEtox method describes toxicity across a cause and effect chain involving three principal factors: fate, exposure, and effect. When calculating ecotoxicity characterization factors, Eckelman et al. (2012) used existing USEtox variables for the fate and exposure factors, but with values found in the literature specifically for carbon nanotubes (CNT) in order to simulate their behavior in water. Their conclusions acknowledged the difficulty deriving realistic fate and exposure factors when using nano-specific properties in a fugacity-type fate and transport model governing their behavior. Salieri et al. derived ecotoxicity characterization factors using experimental fate factors for nano-silver based on colloidal chemistry (e.g., aggregation and settling), albeit in a steady-state model, and used a simplified exposure factor equal to 1 (i.e., all material was considered bio-accessible), observing that a lack of knowledge on bioavailability prohibited deriving a reasonable exposure factor (Salieri et al. 2015). Additionally, (Miseljic and Olsen 2014) derived ecotoxicity characterization factors for both nano-silver and nano-titania by setting both the fate and exposure factor to 1 for an extremely simplified description of ENM behavior in aquatic systems, reinforcing the difficulty to model fate and transport of ENMs with existing LCA methods. Pini et al. defined human health characterization factors for nano-titania using steady-state fate and transport models that estimated the amount of airborne ENM and the multiple-pathway particle dosimetry (MPPD) model to estimate the exposure of these airborne ENMs.

Such an approach allows for a relatively straightforward, yet rudimentary estimation of the fate, exposure, and ultimately human health effects of EMNs. This approach is agreeable with a comparative evaluation of potential impacts, for which LCA is often used, but the approach could be improved based on several outstanding issues. However, ENMs are assumed to behave similarly to colloidal materials in the environment (Darlington et al. 2009; Klaine et al. 2008; Lin et al. 2010; Meesters et al. 2014), inasmuch that they do not come into thermodynamic equilibrium with their surrounding media, even if some may demonstrate kinetic stability over prolonged periods of time. One implication of this is that, in certain cases, the behavior of ENMs in the environment may be highly concentration dependent and best described dynamically over time using kinetically defined modeling parameters (Meesters et al. 2014). Thus, changes in ENM concentration over time should be considered when possible and where necessary, for example, by consideration of variable emissions that are expected in occupational indoor settings (Walser et al. 2015), from ENM-specific transformations such as homo- and hetero-aggregation, or through pharmacokinetic and physiological processes in the human body or other organisms (Li et al. 2015).

We therefore designate a second level of integration, *Site-Dependent* Integration, that incorporates these considerations using dynamic fate and exposure models. While dynamic models present a more relevant modeling environment for ENMs, it also presents the opportunity to introduce fate and exposure parameters that are unique to a certain emission scenario (i.e., context). To avoid reproducing fate and exposure parameters on a case-by-case basis, *Site-Dependent* Integration relies on some level of generalization of these parameters by creating parameter "classes" (Potting and Hauschild 2006; Sonnemann and Schuhmacher 2004). For example, emissions of ENMs could be defined by a series of magnitude (e.g., emission rates) classes that result in statistically significant differences in the fate and exposure of ENMs between classes but not within classes. This has been done previously in LCA for chemicals, for example, where statistically determined classes of environmental conditions (e.g., meteorological data), operating parameters (e.g., stack height of emissions source) and receptor densities (e.g., number of persons exposed in a given area) of chemical pollutants were determined for waste disposal sites across Spain (Sonnemann and Schuhmacher 2004). In this way, the spatial distribution around an emission point could be considered for estimating the concentration increment per receptor (e.g., person) in a geographically distinguishable environment. This type of classification fits well with dynamic models which make use of spatially and temporally specific data inputs. *Site-Dependent* Integration presents a particularly effective way to support early-stage development and eco-design, by allowing one to create classes that forecast prospective changes and developments of the product. For nanotechnologies, human health impacts from indoor air emissions of and exposure to ENMs could prove an extremely practical first case study for the *Site-Dependent* Integration approach, given that indoor environmental conditions are more constrained than ambient ones (i.e., influenced by fewer parameters) (Hinds 1999; Walser et al. 2015). These conditions could allow for the relatively simplified classification of indoor, occupational model parameters, similar to what has been done for regular chemicals in industrial processes (Sonnemann and Schuhmacher 2004) and consumer products (Nigge 2000).

While *Site-Dependent* Integration provides a greater level of relevance for estimating the human health impacts from emissions of ENMs, the use of classes implies that some level of generalization is still occurring. The implication of this is that the results represent potential instead of absolute human health impacts. Absolute impacts could be achieved within the integration framework but would require a case-by-case modeling approach, whereby the emissions and values of each fate and exposure parameter are defined for a specific emission setting as opposed to generalized classes. This is what we refer to as the third level of *Site-Specific* Integration. Due to the inclusiveness of the data and modeling requirements, this approach is best used for full-scale development and/or the need to evaluate absolute values of exposure with regulatory or safety thresholds. Mature nanotechnology market products, such as certain paints and pigments, have well-established manufacturing routes, production volumes, and intended uses that pose potential exposures across the life cycle. In such real-world scenarios, absolute damage and evaluations of risk are necessary to ensure human health safety, local ecosystem integrity, and to ensure regulatory compliance. While *Site-Specific* Integration, in theory, also supports comparative assessments for early-stage development, eco-design, and environmental product declarations, for example, the level of data and model detail required of *Site-Specific* Integration, and, similar to RA, would be unnecessary in cases of early product and process design of nanotechnologies given the uncertainty associated with extrapolating the results from early-stage development to full-scale production scenarios. Furthermore, as the number of ENM-emitting processes along the life cycle become large (e.g., >20), data collection may become untenable and the option of *Site-Specific* Integration impractical.

6.5.3   A STRATEGIC-GUIDANCE DIAGRAM FOR USING LIFE-CYCLE ASSESSMENT AND RISK ASSESSMENT TO EVALUATE NANOTECHNOLOGIES

A Strategic Guidance Diagram (SGD) (see Figure 6.2) is presented to guide decision-makers using both LCA and RA either in isolation, complementary to one another (i.e., combining the results of both tools) and in an integrated way at the methodological level. LCA and RA are found as isolated options, respectively, when either looking at environmental and human health impacts not related to ENM emissions, referred to as resource efficiencies of the nanotechnology, or when specifically looking at the ecological or human health risks from ENM emissions. Complementary and integrated uses of LCA and RA are introduced when those two sets of criteria come together, whereby integration is further delineated by the resolution (i.e., specificity) of the emissions and modeling data used in the approach. To illustrate with an example, 3rd-generation solar photovoltaic devices, introduced in Chapters 2 and 3, are a mostly lab-scale technology that exploit the photo-electric properties of ENMs to generate electricity. Use of LCA as a single tool has demonstrated the resource efficiency of these devices, such as potential to reduce embodied energy and greenhouse gas emissions of photovoltaics (Tsang et al. 2016a, 2016b). However, these studies lack any

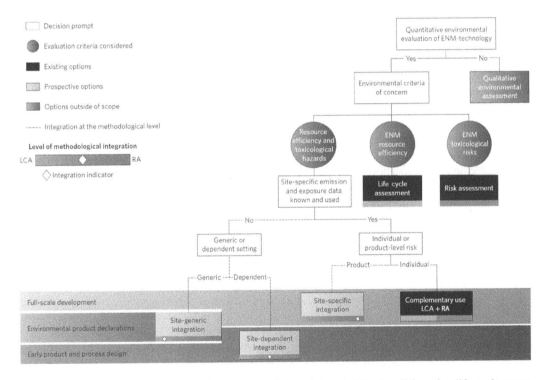

FIGURE 6.2   Two separate, one complementary, and three integrated "options" for using life-cycle assessment and risk assessment to evaluate nanotechnologies. Options are presented as a guidance diagram, taking into consideration the type of assessment (i.e., quantitative versus qualitative), the ENM-specific criteria of concern (i.e., resource efficiencies versus toxicological risks), and the scale of the technology (e.g., early-stage design versus full-scale development). The dotted lines indicate when LCA and RA should be integrated at the methodological level based on (1) the specificity of the emissions and exposure data and (2) whether the human health impacts are estimated at the product or individual-substance (e.g., single chemical) level. Integration is denoted by an indicator that reflects the degree to which the fate and exposure are incorporated into LCA. The indicator rests closer to LCA when steady-state assumptions for estimating the fate and exposure are used, while the indicator is closer to RA when fate and exposure use dynamic models that are individually defined for case-by-case scenarios. *ENM*: engineered nanomaterial. LCA: life-cycle assessment. RA: risk assessment.

estimation of ENM toxicological hazards, as was pointed out in Chapter 4. According to the SGD, a decision-maker at early-stage development (e.g., lab- or pilot-scale) concerned with addressing both the resource efficiency and the toxicological hazards, would find *Site-Generic* and/or *Site-Dependent* Integration most suits their needs (i.e., effective hot-spot analysis to identify where in the life cycle toxicological impacts and, eventually, risks may occur). This is because at the current scale of 3rd-generation photovoltaic development, there is too much uncertainty in production volumes or operating conditions, for example, and thus absolute damage (i.e., using *Site-Specific* Integration) or risk (i.e., using *Complementary Use*) would have little to no basis.

6.6   CONCLUSIONS

This chapter introduced and expanded upon the role that LCA and RA generally play for certain decision-makers and discussed the ways in which they might be strategically used together to make a more holistic evaluation of certain products, processes, and services. This was elaborated on and hypothesized specifically in the context of nanotechnologies. As an emerging technology, the development and integration of ENMs into industrial processes and consumer goods is a fast-transforming and evolving area of specialization. Nanotechnologies are demonstrating novel properties and creative uses, while also introducing potentially new and varied environmental and human health hazards. The integration of LCA and RA at the methodological level can provide a tool to support and promote nanotechnologies through early-stage evaluations, eco-design, environmental product declarations, and other comparative assessments. Specifically, *Site-Generic* Integration involves a moderate level of effort to achieve and can be considered an immediate and near-term (i.e., circa 5 years) objective for most ENMs. *Site-Dependent* Integration, on the other hand, will require greater coordination among the scientific community to identify the relevant data and appropriate emission, fate, and exposure "classes" most pertinent to current and anticipated nanotechnology development and uses.

When regulatory compliance is an objective of the evaluation process, *Complementary Use* is the only suitable option. This is because LCA delivers relative impacts and not absolute impacts, the latter being required in RA. However, if the objective is to evaluate product improvement and sustainable development, then comparing relative impacts would be the most appropriate. Moreover, purely risk-based evaluations often include political, economic, and risk management engineering variables which move beyond strict evaluation of science and environmental performance which may be less informative during technology evaluation phases (Hellweg et al. 2005; McKone et al. 2006).

More importantly, producing an RA for each instance of ENM exposure may not be feasible or simply not required during early-development stages of a product, process, or service. An important point to realize here is that conducting an RA at each instance of ENM exposure takes for granted the knowledge of where such exposure would occur in the first place. However, such *a priori* knowledge is not likely to be anticipated 100% of the time by the developer or evaluator of a technology, and, thus, the potential to screen potential human health impacts would be missed by relying on RA exclusively. Using LCA with integrated methods to evaluate relative human health impacts of ENMs would allow for hot-spot identification where further RA might be warranted.

Lastly, ENMs are not used in isolation, but are integrated into processes and products that affect a greater scope of activities. Human health impacts, therefore, will not be an exclusive ENM-dependent phenomenon. The use of ENMs may change the entire system under study, including energy production, extraction, and modification of secondary materials, and generation of waste, for example, that will also have human health impacts in other life cycle stages (Hellweg et al. 2005). Being able to track and avoid such "burden-shifting" is a primary function of LCA and an integrated methodology would allow one to evaluate this type of outcome (Eckelman et al. 2012).

It is the opinion of the authors that methodological integration of LCA and RA is a necessary research action for ENMs. USEtox provides an appropriate method to work with as it is a consensus model built on formerly used risk assessment methods such as Californian EPA Multimedia Total Exposure Model for Hazardous-Waste Sites (CalTOX). USEtox integrates fate and transport processes of chemicals to estimate an exposure dose is done in risk assessment, but on much broader spatial and temporal scales (Rosenbaum et al. 2015). As the fate, transport, and exposure to ENMs relies particularly on localized environmental conditions, such adoption of USEtox could lead to instances where fate and exposure are inappropriately estimated. A reasonable research goal would be to adapt the discreet, workplace environment nested into the USES-LCA framework by Hellweg et al., but specific for ENMs in a tool such as USEtox. Chapter 7 introduces such a *Site-Dependent* model for indoor air emissions of ENMs.

REFERENCES

Barberio, G. et al. 2014. Combining life cycle assessment and qualitative risk assessment: The case study of alumina nanofluid production. *The Science of the Total Environment* 496: 122–131.

Beaudrie, C. E. H. et al. 2013. From cradle-to-grave at the nanoscale: Gaps in U.S. regulatory oversight along the nanomaterial life cycle. *Environmental Science & Technology* 47(11): 5524–5534.

Braakhuis, H. M. et al. 2014. Physicochemical characteristics of nanomaterials that affect pulmonary inflammation. *Particle and Fibre Toxicology* 11(1): 18.

Breedveld, L. 2013. Combining LCA and RA for the integrated risk management of emerging technologies. *Journal of Risk Research* 16(3–4): 459–468.

Busnaina, A. A. et al. 2013. Nanomanufacturing and sustainability: Opportunities and challenges. *Journal of Nanoparticle Research* 15(10): 1–6.

Canis, L. et al. 2010. Application of stochastic multiattribute analysis to assessment of single walled carbon nanotube synthesis processes. *Environmental Science & Technology* 44(22): 8704–8711.

Darlington, T. K. et al. 2009. Nanoparticle characteristics affecting environmental fate and transport through soil. *Environmental Toxicology and Chemistry/SETAC* 28(6): 1191–1199.

Eckelman, M. J. et al. 2012. New perspectives on nanomaterial aquatic ecotoxicity: Production impacts exceed direct exposure impacts for carbon nanotoubes. *Environmental Science & Technology* 46(5): 2902–2910.

European Chemicals Agency. 2016. Nanomaterials. https://echa.europa.eu/regulations/nanomaterials (accessed September 15, 2016).

Gavankar, S. et al. 2012. Life cycle assessment at nanoscale: Review and recommendations. *The International Journal of Life cycle Assessment* 17(3): 295–303.

Greßler, S., and M. Nentwich. 2012a. Nano and environment–Part II: Hazard potentials and risks. *Nano Trust Dossiers* 27: 1–6.

Greßler, S., and M. Nentwich. 2012b. Nano and the environment–Part I: Potential environmental benefits and sustainability effects. *Nano Trust Dossiers* 26: 1–4.

Grieger, K. D. et al. 2012. Analysis of current research addressing complementary use of life cycle assessment and risk assessment for engineered nanomaterials: Have lessons been learned from previous experience with chemicals? *Journal of Nanoparticle Research* 14(7): 1–23.

Harder, R. et al. 2015. Review of environmental assessment case studies blending elements of risk assessment and life cycle assessment. *Environmental Science & Technology* 49: 13083–13093.

Hellweg, S. et al. 2005. Confronting workplace exposure to chemicals with LCA: Examples of trichloroethylene and perchloroethylene in metal degreasing and dry cleaning. *Environmental Science & Technology* 39(19): 7741–7748.

Hinds, W. C. 1999. *Aerosol Technology: Properties, Behavior, and Measurement of Airborne Particles.* 2nd ed. New York: Wiley-Interscience.

Hischier, R., and T. Walser. 2012. Life cycle assessment of engineered nanomaterials: State of the art and strategies to overcome existing gaps. *Science of the Total Environment* 425: 271–282.

Hodas, N. et al. 2015. Indoor inhalation intake fractions of fine particulate matter: Review of influencing factors. *Indoor Air.* doi:10.1111/ina.12268.

Hristozov, D. R. et al. 2012. Risk assessment of engineered nanomaterials: A review of available data and approaches from a regulatory perspective. *Nanotoxicology* 6: 1–19.

International Organization for Standardization. 2006a. *ISO 14040—Environmental Management—Life cycle Assessment—Principles and Framework.* The International Organization for Standardization (ISO), Geneva, Switzerland.

International Organization for Standardization. 2006b. *ISO 14044—Environmental Management—Life cycle Assessment—Requirements and Guidelines.* The International Organization for Standardization (ISO), Geneva, Switzerland.

Jahnel, J. et al. 2013. Risk assessment of nanomaterials and nanoproducts—Adaptation of traditional approaches. *Journal of Physics: Conference Series* 429: 1–10.

Kikuchi, Y., and M. Hirao. 2008. Practical method of assessing local and global impacts for risk-based decision making: A case study of metal degreasing processes. *Environmental Science & Technology* 42(12): 4527–4533.

Klaine, S. J. et al. 2008. Nanomaterials in the environment: Behavior, fate, bioavailability, and effects. *Environmental Toxicology and Chemistry/SETAC* 27(9): 1825–1851.

Klöpffer, W. et al. 2007. Nanotechnology and life cycle assessment. *A Systems Approach to Nanotechnology and the environment: Synthesis of Results Obtained at a Workshop*, Washington, DC, October 2–3, 2006. European Commission, DG Research, jointly with the Woodrow Wilson International Center for Scholars.

Li, D. et al. 2015. *In vivo* biodistribution and physiologically based pharmacokinetic modeling of inhaled fresh and aged cerium oxide nanoparticles in rats. *Particle and Fibre Toxicology* 13(1): 45.

Lin, D. et al. 2010. Fate and transport of engineered nanomaterials in the environment. *Journal of Environmental Quality* 39(6): 1896–1908.

Linkov, I., and T. P. Seager. 2011. Coupling multi-criteria decision analysis, life cycle assessment, and risk assessment for emerging threats. *Environmental Science & Technology* 45(12): 5068–5074.

MacPhail, R. C. et al. 2013. Assessing nanoparticle risk poses prodigious challenges. *Wiley Interdisciplinary Reviews: Nanomedicine and Nanobiotechnology* 5(4): 374–387.

McKone, T. E. et al. 2006. Dose-response modeling for life cycle impact assessment—Findings of the portland review workshop. *The International Journal of Life cycle Assessment* 11(2): 137–140.

Meesters, J. A. J. et al. 2014. Multimedia modeling of engineered nanoparticles with simpleBox4nano: Model definition and evaluation. *Environmental Science & Technology* 48(10): 5726–5736.

Meyer, D. E., and V. K. K. Upadhyayula. 2014. The use of life cycle tools to support decision making for sustainable nanotechnologies. *Clean Technologies and Environmental Policy* 16(4): 757–772.

Miseljic, M., and S. I. Olsen. 2014. Life cycle assessment of engineered nanomaterials: A literature review of assessment status. *Journal of Nanoparticle Research* 16(6): 1–33.

National Institute of Advanced Industrial Science and Technology (AIST). 2011. *Risk Assessment of Manufactured Nanomaterials—Overview of Approaches and Results.* Retrieved February 18, 2015, from http://www.aist-riss.jp/projects/nedo-nanorisk/nano_rad2/docs/ES_20130222e.zip.

National Institute of Occupational Safety and Health, Japan. 2015. Information *provision about handling of nanomaterials (in Japanese).* http://www.jniosh.go.jp/publication/nanomaterial.html (accessed February 18, 2015).

Nigge, K.-M. 2000. *Life cycle Assessment of Natural Gas Vehicles—Development and Application of Site-Dependent Impact Indicators.* Berlin, Germany: Springer.

Potting, J., and Hauschild, M. 2006. Spatial differentiation in life cycle impact assessment: A decade of method development to increase the environmental realism of LCIA. *The International Journal of Life cycle Assessment* 11(S1): 11–13.

Praetorius, A. et al. 2014. The road to nowhere: Equilibrium partition coefficients for nanoparticles. *Environmental Science: Nano* 1(4): 317.

Quik, J. T. K. et al. 2011. How to assess exposure of aquatic organisms to manufactured nanoparticles? *Environment International* 37(6): 1068–1077.

Rosenbaum, R. K. et al. 2015. Indoor air pollutant exposure for life cycle assessment: Regional health impact factors for households. *Environmental Science & Technology* 49: 12823–12831.

Rousk, J. et al. 2012. Comparative toxicity of nanoparticulate CuO and ZnO to soil bacterial communities. *PLoS One* 7(3): e34197.

Salieri, B. et al. 2015. Freshwater ecotoxicity characterisation factor for metal oxide nanoparticles: A case study on titanium dioxide nanoparticle. *The Science of the Total Environment* 505: 494–502.

Savolainen, K. et al. 2013. *Nanosafety in Europe 2015–2025: Towards Safe and Sustainable Nanomaterials and Nanotechnology Innovations.* Helsinki, Finland: Finnish Institute of Occupational Health.

Seager, T. P., and I. Linkov. 2008. Coupling multicriteria decision analysis and life cycle assessment for nanomaterials. *Journal of Industrial Ecology* 12(3): 282–285.

Shatkin, J. A. 2008. Informing environmental decision making by combining life cycle assessment and risk analysis. *Journal of Industrial Ecology* 12(3): 278–281.

Som, C. et al. 2010. The importance of life cycle concepts for the development of safe nanoproducts. *Toxicology* 269(2–3): 160–169.

Sonnemann, G., and M. Schuhmacher. 2004. *Integrated Life cycle and Risk Assessment for Industrial Processes*. Boca Raton, FL: CRC.

Steinfeldt, M., and U. Petschow. 2007. *Nanotechnologies, Hazards and Resource Efficiency*. Berlin, Germany: Springer-Verlag.

Tsang, M. et al. 2014a. Life cycle assessment for emerging materials: Case study of a garden bed constructed from lumber produced with three different copper treatments. *The International Journal of Life cycle Assessment* 19(6): 1345–1355.

Tsang, M. P. et al. 2014b. Benefits and risks of emerging technologies: Integrating life cycle assessment and decision analysis to assess lumber treatment alternatives. *Environmental Science & Technology* 48(19): 11543–11550.

Tsang, M. P. et al. 2016a. A comparative human health, ecotoxicity, and product environmental assessment on the production of organic and silicon solar cells. *Progress in Photovoltaics: Research and Applications* 24(5): 645–655.

Tsang, M. P. et al. 2016b. Life cycle Assessment of Cradle-to-Grave Opportunities and Environmental Impacts of Organic Photovoltaic Solar Panels Compared to Conventional Technologies. *Solar Energy Materials and Solar Cells* 156: 37–48.

U.S. EPA. 2015. Risk Assessment. http://www.epa.gov/risk (accessed February 18, 2015).

U.S. EPA. 2016. *Control of nanoscale materials under the toxic substances control act.* https://www.epa.gov/reviewing-new-chemicals-under-toxic-substances-control-act-tsca/control-nanoscale-materials-under (accessed November 20, 2016).

Vance, M. E. et al. 2015. Nanotechnology in the real world: Redeveloping the nanomaterial consumer products inventory. *Beilstein Journal of Nanotechnology* 6: 1769–1780.

Walser, T. et al. 2015. Life cycle assessment framework for indoor emissions of synthetic nanoparticles. *Journal of Nanoparticle Research* 17(245): 1–18.

Walser, T. et al. 2014. Indoor exposure to toluene from printed matter matters: Complementary views from life cycle assessment and risk assessment. *Environmental Science & Technology* 48(1): 689–697.

Wardak, A. et al. 2008. Identification of risks in the life cycle of nanotechnology-based products. *Journal of Industrial Ecology* 12(3): 435–448.

Wong, S., and B. Karn. 2012. Ensuring sustainability with green nanotechnology. *Nanotechnology* 23(29): 1–2.

7 Life-Cycle Toxicity Assessment

Michael Tsang and Guido Sonnemann

CONTENTS

7.1 BACKGROUND ON THE DEVELOPMENT OF THE LIFE-CYCLE IMPACT ASSESSMENT METHODS

According to ISO 14040:2006, the life-cycle impact assessment (LCIA) is the third of four total stages in a life-cycle assessment (LCA) study. Specifically, LCIA refers to the analysis of the LCI according to its environmental significance and, more specifically, to the conversion of the life-cycle inventory flows (e.g., tons of carbon dioxide emissions into the air) into environmental and human health impacts (International Organization for Standardization 2006; United Nations Environmental Programme 2003). As discussed in Chapter 3, this is done in two (mandatory) steps, (1) classification and (2) characterization (Figure 3.1).

Classification refers to the assignment of inventory flows into specific impact categories (e.g., climate change potential, human health toxicity, smog formation potential), while characterization refers to conversion of those flows into quantified impacts. There are a wide set of both environmental and human health impacts one can consider in an LCA, and, accordingly, this means a wide set of models needed to derive the characterization factor (CF) values. These models are specific to the nature of the inventory flow in terms of the substance (i.e., material, energy, chemical), place and scenario (e.g., where an emission occurs), and impact category particular to each inventory flow scenario (i.e., place of emission). Often, these can be chosen from pre-defined LCIA methodologies such as ReCiPe (latest LCIA method from The Netherlands) (Goedkoop et al. 2013) (www.lcia-recipe.net), European Commission's International Reference Life Cycle Data System (ILCD) (European Commission 2011), Tool for the Reduction and Assessment of Chemical and Other Environmental Impacts (TRACI) (US EPA), LC-Impact (www.lc-impact.eu), Institute of Environmental Sciences (CML), (Leiden University, The Netherlands), Cumulative Energy Demand (CED), or USEtox (www.usetox.org), among others. Table 7.1 lists a number of commonly employed LCIA methods and their corresponding midpoint impact categories. Many, but not all, of these methods further consolidate these impacts from midpoint to endpoint as discussed in Chapter 3. These LCIA methodologies include a compiled database of substance- and flow-specific CFs that automatically classify and then subsequently characterize flows into their respective environmental and human health impacts.

These LCIA methods differ on the types of impact categories included, the spatial representativeness of the methods, and their modeling choices (e.g., fate and transport of substances in the environment), among other things. Although Table 7.1 is not a comprehensive account of all possible impact categories, it is representative of 14 commonly used and important environmental and human health impacts. Some of these impacts, for instance, represent multiple impact criteria such as ecotoxicity, which can be subdivided into freshwater or marine water ecotoxicity. Other impact categories, which are nevertheless important, such as impacts from noise pollution, are not commonly included in these methods, lack formal reporting within LCI databases, and are, thus, not included in the Table 7.1 (Cucurachi et al. 2012). The majority of the LCIA methodologies are specific for European conditions and impacts. These include CML, Eco-indicator99, EDIP, ILCD and ReCiPe. However, there have been others developed over time for other regions. TRACI, for example, was developed specifically for estimating the impacts within North America, while Life-cycle Impact assessment Method based on Endpoint modeling (LIME), Japan, was originally developed for Japan but has since expanded its focus. Other impact methodologies are narrower in scope, such as Eco-Scarcity, Ecoinvent, and USEtox, which focus on resource depletion, CED, and toxicity, respectively. These can have the added benefit of providing greater detail and impact pathway modeling resolution.

7.2   LIFE-CYCLE TOXICITY ASSESSMENT

The advent of LCA and LCIA, where it was originally referred to as a Resource and Environmental Profile Analysis (REPA) in its earliest days, was mainly focused on energy and waste flows, providing an approach to resource and solid waste management.

REPA's approach and methods gradually evolved towards what we now refer to as LCA, but in a non-centralized and non-uniform way. It was not until the early 1990s, upon the release of "Environmental Life Cycle Assessment of Product—Guide and Backgrounds" by Heijungs et al. (1992),[1] that a standardized approach to this discipline emerged.

The Heijungs et al. approach included both human (health) toxicity and ecotoxicity (i.e., both aquatic and terrestrial ecotoxicity) as "pollution"-related problems. Up until that time, risk assessment (RA) had been the conventional way of addressing the toxicity of pollutants in the environment. Thus, it is not surprising that Heijungs et al. (1992) converted emissions into toxicity impacts using *human toxicological* and/or *ecotoxicological classification factors,* which themselves were defined by regulatory and/or safety-related values such as Acceptable Daily Intake (ADI) values. The use of regulatory and/or safety-related values is common practice in RA and regulatory circles but it is one of several points of difference within the scientific community regarding how to

---

[1] A collaborative effort led by Leiden University's Centre of Environmental Sciences (CML) and including the Netherlands Organization for Applied Scientific Research (TNO) and the Fuels and Raw Materials Bureau.

TABLE 7.1
Commonly Applied Life-Cycle Impact Assessment Methodologies and Their Commonly Defined Midpoint Impact Categories

Methods	Acidification	Climate Change	Cumulative Energy Demand[a]	Resource Depletion	Ecotoxicity	Eutrophication	Human Toxicity	Ionizing Radiation	Land Use	Odor	Ozone Layer Depletion	Particulate Matter	Photo-Chemical Oxidation	Water Use[b]
CML, 2001 (Baseline)	●	●	○	●	●	●	●		○	○	●	○	●	○
CML, 2001 (Non-Baseline)	●	●	○	●	●	●	●	○	●	●	●	○	●	○
Ecoinvent	○	○	●	○	○	○	○	●	○	○	○	○	○	○
Eco-Indicator 99	●	●	○	●	●	●	○	●	●	○	●	○	●	●
Eco-Scarcity 2006	○	○	○	●	○	○	○	○	○	○	○	○	○	●
EDIP 2003	●	●	○	○	●	●	●	○	○	○	●	●	●	●
ILCD 2011 (Midpoint)	●	●	○	●	●	●	●	●	●	○	●	●	●	●
IMPACT World+	●	●	○	●	●	●	●	●	●	○	●	●	●	●
LIME	●	○	○	●	●	●	●	○	●	○	●	●	●	●
ReCiPe	●	●	○	●	●	●	●	○	●	○	●	●	●	●
TRACI 2.1	●	●	○	●	●	●	●	○	○	○	●	○	●	○
USEtox	○	○	○	○	●	○	●	○	○	○	○	○	○	○

Source: Acero, A.P. et al., *LCIA Methods—Impact Assessment Methods in Life Cycle Assessment and their Impact Categories*, GreenDelta GmbH, Berlin, Germany, 2014.

[a] Cumulative energy demand is not an impact category in the strict definition of the term; instead, it represents an approach for calculating an energy-based inventory, and it is often reported by LCA studies as a proxy-impact indicator.

[b] Water use methods vary between inventory-based calculations, akin to cumulative energy demand, but can also represent actual impacts estimated across the cause and effect chain.

characterize and estimate toxicity from a life-cycle perspective. The use of safety-related values such as ADI, without the consideration of fate and transport of the pollutants, carried over into subsequent proposals and models throughout the 1990s. In 1993, Guinee and Heijungs expanded on the Heijungs et al. method by proposing a model in which the fate and transport models would be used to describe the behavior of toxic pollutants in the environment, using multimedia fugacity models (Guinée and Heijungs 1993; Mackay 2001; Mackay et al. 1992). The quantified outcome of this new approach was introduced as a "Human Toxicity Potential" (HTP), which was taken up and consistently modified through the 1990s and all the way to the present day.

Leading up to the release of the Heijungs et al. publication, the Society of Environmental Toxicology and Chemistry (SETAC) had become heavily involved with the development of LCA's methods, so much as to develop a "Code of Practice," which was the forerunner to the ISO 14040 series (de Bruijn et al. 2002). Later on, as more methodologies came into existence, SETAC, along with the United Nations Environmental Programme (UNEP), jointly sponsored the Life Cycle Initiative, which put some of its efforts towards enhancing and building global consensus on life-cycle impact assessment methods (Sonnemann and Valdivia 2014).

Furthermore, the idea of consensus models had been borne out of many discussions dating back from the mid- to late-1990s and summarized in the 2003 United Nations Environmental Programme Report (UNEP) on *Evaluation of Environmental Impacts in Life Cycle Impact Assessment* (United Nations Environmental Programme 2003). By that time, there was some agreement that a standardized or consensus approach would benefit the scientific community by providing comparability between studies, since too many methods might undermine the credibility of LCIA (United Nations Environmental Programme 2003). However, there was considerable debate as to which perspectives should be included and put forth in such a model (United Nations Environmental Programme 2003).

Toxicological modeling, from the point of emission to the ultimate uptake and damage at a receptor (i.e., human organ), is, in itself, a quite rigorous exercise involving many different disciplines such as chemistry, biology, geosciences, and toxicology. Given the number of different potentially toxic substances and their physical chemical properties, as well as the variety of environmental conditions pollutants interact with and biological species that pollutants interact with, there are endless considerations to be made in modeling toxicity. As a result, the history of life-cycle toxicity modeling has always been about the trade-off between competing specializations, the accuracy of the resulting impact values, and the feasibility of implementing the amount of details required in accounting for all relevant parameters.

Eventually, USEtox emerged out of the Life-Cycle Initiative as the consensus model for incorporating ecological and human health toxicological impacts into LCA. USEtox quantifies impacts for freshwater ecotoxicity as well as non-cancerous and cancerous human health toxicity (Rosenbaum et al. 2008a). It was developed, in part, from authors of formerly and widely used environmental impact models such as Californian EPA Multimedia Total Exposure Model for Hazardous-Waste Sites (CalTOX) (McKone et al. 2001), IMPACT 2002+ (Jolliet et al. 2003), Berkeley-Trent North American contaminant fate model (BETR) (MacLeod et al. 2001), Environmental Design of Industrial Products (EDIP) (Wenzel et al. 1997), WAter and SOil environmental fate and exposure model of noxious substances at the European scale (WATSON) (Bachmann 2006), the Uniform System for the Evaluation of Substances adapted for LCA purposes (USES-LCA) (Huijbregts et al. 2000), and EcoSense (European Commission 2005), and employs multimedia, fugacity models, and Human Toxicity Potential (HTP) values that were described previously.

7.3   THE CONSENSUS MODEL

7.3.1   ECOTOXICITY

The LCIA methodology embodied by USEtox considers the potential eco-toxicological impacts that result from exposure to organic chemicals and certain non-organic metals in freshwater. Multi-compartmental models are used to trace the fate and transport of a chemical from the source of its emission in one medium to the final concentration in another medium. Steady-state equilibrium models are employed, which are largely dependent on partition (equilibrium) coefficients for characterizing these behaviors (Rosenbaum et al. 2008a). Exposure is estimated using steady-state bioaccumulation models using the same types of partition coefficients mentioned above (Rosenbaum et al. 2008a). This estimated level of exposure (e.g., bioavailable mass) is then combined with an estimation of the concentration at which 50% of the population has the observed impact, referred to as the effective concentration at 50% (EC50) (Rosenbaum et al. 2008a). The EC50 value is calculated from *in vivo* toxicological studies for at least three different species/trophic levels (Rosenbaum et al. 2008a). The midpoint impact is defined here as a non-reference unit expressed as the comparative toxic unit for ecosystems (CTU$_e$) (Rosenbaum et al. 2008a). The endpoint may be combined with a damage factor that estimates the number of potentially disappeared fraction of species.

7.3.2 Human Toxicity

The LCIA methodology embodied by USEtox considers the potential human health impacts that result from exposure to organic chemicals and certain non-organic metals. Multi-compartmental models are used to trace the fate and transport of a chemical from the source of its emission in one medium to the final concentration of another medium. Steady-state equilibrium models are employed, largely dependent on partition (equilibrium) coefficients for characterizing these behaviors (Equation 7.1) (Rosenbaum et al. 2008a).

$$CF_{i,j} = FF_i \cdot XF_i \cdot EF_{i,j} = (iF_i) \cdot EF_{i,j} \qquad (7.1)$$

where:
   FF_i is the fate factor (days) for substance i
   XF_i is the exposure factor (days^{-1}) for substance i
   iF_i is the intake fraction for substance i
   EF_i is the effect factor (cases/kg) for substance i and pathology j

The FF represents the various transformation and removal mechanisms from one compartment to another and is reported in days. Exposure is limited to inhalation and ingestion, without consideration of dermal exposure and is expressed as the XF in units of 1/days. It is assessed on a population-based level as opposed to an individual level, the latter being more conventional for Human Health Risk Assessment (HHRA) methods. The results of the fate and exposure calculations provide a lifetime intake fraction (iF), which is defined as the ingested or inhaled amount per mass of emitted substance. This intake fraction is then combined with an estimation of the human-equivalent dose at which 50% of the population has the observed impact. This is referred to as the effective dose at 50% (ED50) and is represented as the EF with units of cases of cancerous or non-cancerous disease per kg of emitted substance (Rosenbaum et al. 2008a). This dose is generally quantified using *in vivo* animal dose–response toxicological data at which 50% of the test subjects gave a response. Both carcinogenic and noncarcinogenic doses may be defined depending on toxicity and health impact data. Midpoint impacts are expressed in a non-equivalent unit defined as the comparative toxic unit for humans (CTU$_h$) (Rosenbaum et al. 2008a). Endpoints may be presented in terms of the disability-adjusted life years (DALY), where one DALY is defined as one year of lost life (YLL) due to mortality and/or due to disease and disabilities (YLD) (WHO 2014).

7.3.3 Expanding on the Consensus Model for Engineered Nanomaterials

LCA is meant to be an iterative process, and as such, the level of detail in an LCA and subsequent LCIA can go from purely qualitative to quantitative. This is referred to as the "level of sophistication" of an LCIA, whereby the greater the level of detail incorporated in an LCIA model, the more sophisticated it is said to become (United Nations Environmental Programme 2003). Moreover, the quantitative approach itself can be a general, low-sophisticated approach (i.e., time-generic and spatially generic) or it can be a more specified, high-sophisticated approach (i.e., temporally acute and spatially acute).

As such, consensus models, such as USEtox, are by their nature somewhat low-sophisticated approaches in which background concentrations and the spatial and temporal influences of the impact pathway are ignored, while simultaneously aggregating all relevant inventory flows per impact category (United Nations Environmental Programme 2003).

This is very beneficial to the scientific community, because it allows for a feasible and scientifically justifiable way to calculate potential toxicity, facilitating the comparability between LCA studies, and also demonstrating the state-of-the-science on the topic. That being said, the state-of-the-science has arguably shifted in the last decade. In particular, LCA is being used on a wider array of technologies and industries in which the underlying LCIA models are challenged. One particular example concerns the evaluation of engineered nanomaterials (ENMs).

While ENMs provide benefits across many sectors (Vance et al. 2015), they also raise concerns regarding potential environmental and human health hazards (Greßler and Nentwich 2012; Stone 2009; Yokel and MacPhail 2011). Nanotechnologies can be evaluated with life-cycle assessment (LCA) to determine their potential resource efficiencies and environmental and human health impacts, although LCA does not evaluate direct environmental and human health hazards posed by ENM emissions themselves. ENMs may be released into the environment at any stage of their life cycle (Gottschalk and Nowack 2011; Keller et al. 2013). Although it is possible to build life-cycle inventories (LCI) that estimate and quantify ENM emissions, (Gottschalk et al. 2009, 2013; Keller et al. 2014; Sun et al. 2014) currently available life-cycle impact assessment (LCIA) methodologies do not cover nano-specific characterization factors (CFs) that are necessary for quantifying the fate of, exposure to, and impacts from those emissions in an LCA (Gavankar et al. 2012). A few studies have made first approximations to define nano-specific CF, (Eckelman et al. 2012; Miseljic and Olsen 2014; Pini et al. 2016; Salieri et al. 2015), but otherwise, the direct impacts from ENM have not been addressed by past LCA studies on ENMs (Hischier and Walser 2012; Miseljic and Olsen 2014). Additionally, ambient (i.e., outdoor) emissions have historically been the focus of LCI (Hellweg et al. 2005), and is reflected in the scope of nano-specific CF published to date. However, indoor emissions, particularly in occupational settings, can be important contributors to the overall LCA results (Hellweg et al. 2009, 2005; Kikuchi et al. 2012; Meijer et al. 2005; Walser et al. 2014). Neglecting such emissions and their potential impacts in an LCA may result in burden shifting from the environment to workers.

Currently, characterization models take advantage of a number of assumptions that describe the fate and transport of small organic molecules and metals well, but these methods are not appropriate for ENMs (Praetorius et al. 2014). Traditional characterization models often assume steady-state or thermodynamic equilibrium conditions (Goedkoop et al. 2013; Rosenbaum et al. 2008b) ignoring the changes in concentration of the pollutant over time. Therefore, these models often rely on (equilibrium) partition coefficients (e.g., octanol-water partition coefficient, K_{ow}) for estimating the fate of pollutants.

However, ENMs behave, in effect, like colloidal substances that, when released into a specific medium, exist in their own phase. Therefore, ENMs are not thermodynamically stable, even if certain ENMs may be kinetically stable for long periods of time (Praetorius et al. 2014). In other words, ENM concentration ratios between two mediums cannot be predicted from a previously measured partition coefficient after the further addition or removal of ENMs from that system (Praetorius et al. 2014). Instead, the behavior of ENMs in the environment is heavily concentration dependent and their behavior in the environment (Praetorius et al. 2014; Salieri et al. 2015) and could otherwise be estimated kinetically using a dynamic model that describes the rates of the processes that control their behavior, such as homo- and hetero-aggregation (Meesters et al. 2014) or even clearance and retention from an organism (Braakhuis et al. 2014; Hodas et al. 2015; Li et al. 2015). While recent adaptations to existing LCIA models have been made to estimate both ecotoxicity (Eckelman et al. 2012; Miseljic and Olsen 2014; Salieri et al. 2015) and human health toxicity (Pini et al. 2016; Walser et al. 2015) impacts, steady-state assumptions were used in all cases.

7.3.4 Characterization Factor Development

We present a dynamic LCIA model for calculating the human health impacts from occupational, indoor air emissions of metallic nanoparticles, with a focus on nano-TiO_2. Specifically, we define a CF (Equation 7.2) for use in LCA (Rosenbaum et al. 2015; Wenger et al. 2012):

$$CF_i = iF_i \cdot EF_i \qquad (7.2)$$

The CF_i is based on the concept proposed in USEtox for calculating the human health life-cycle impacts resulting from the emission of a substance (i), given its intake fraction (iF) and its effect factor (EF). The iF represents the mass of substance, i, to which one is exposed per mass of that emitted substance. The EF is defined by the toxicological dose–response relationship of the substance. However, unlike in USEtox, the iF in this paper is not estimated using (equilibrium) partition coefficients of steady-state models. Instead, we defined and used the *Retained-intake fraction* (*RiF*) derived from a dynamic fate and exposure model using kinetically defined fate and exposure parameters.

7.4 EXPOSURE SCENARIOS AND OCCUPATIONAL, INDOOR AIR EMISSIONS OF NANO-TIO₂

A total of six exposure scenarios were modeled based on a previous occupational RA involving the handling of nano-TiO_2 (Tsang et al. 2017). These scenarios were six variations of a single representative workplace activity involving the handling of nano-TiO_2 (Table 7.2).

The description of the exposure scenario was adapted from the Development of Exposure Scenarios for Manufactured Nanomaterials (NANEX) (www.nanex-project.eu) database and describes a situation where pre-fabricated ENMs were *handled* as opposed to scenarios that estimate exposure *during* the production of raw-ENMs. The latter scenarios were not included due to the assumption that production of raw-ENMs is more likely to occur under automated, enclosed settings where fugitive (i.e., accidental) exposures were near zero. Specifically, each exposure scenario involves the transfer of nano-TiO_2 powders into an open vessel. Emissions rates (E_i) were estimated as a function of the total amount (kg) of nano-TiO_2 handled ($A_{handled}$), the dustiness index (DI) of the ENM and the handling energy factor (H, *unit-less*) of the work-related activity (Equation 7.3) (Hristozov et al. 2016).

$$E_i = \frac{A_{handled}}{t_{wc}} \cdot (DI) \cdot (H_i), \qquad (7.3)$$

TABLE 7.2
Parameters Used in the Fate and Transport Model Describing the Exposure Scenarios Involved with Transferring Large Amounts of Powder to a Vessel

No.	Exposure Scenario	E_i [mg/min]	H_i	t_{wc} [min]	p_{wc} [min]	n_{wc}	$A_{handled}$ [kg]	V_{tot} [m³]	AER [h⁻¹]
ES1	e-high, *f-short*	6.72E+02	0.80	10	20	16	5.60E+02	100	8
ES2	e-high, *f-long*	6.72E+02	0.80	60	60	4	3.36E+03	100	8
ES3	e-high, *f-daily*	6.72E+02	0.80	480	0	1	2.69E+04	100	8
ES4	e-high, *f-single pulse*	6.72E+02	0.80	1	0	1	5.60E+01	100	8
ES5	e-medium, *f-short*	6.72E+00	0.80	10	20	16	5.60E+00	100	8
ES6	e-low, *f-short*	6.72E-02	0.80	10	20	16	5.60E-02	100	8

Note: The exposure scenarios differ based on the magnitude of the emission, *e*, per minute and the frequency of the work-cycle activity, *f*. h_i is the handling energy factor, t_{wc} is work cycle time (i.e., duration of the work-related activity), p_{wc} is pause between work cycles, n_{wc} is number of work cycles, $a_{handled}$ is amount of material transferred per transfer event within each work cycle, v_{tot} is the total volume of the work room, and *AER* is the air exchange rate of the work room.

---

t_{wc} is the duration of the work activity and DI was estimated as 15 mg/kg (Tsang et al. 2017). H is based on a scale of 0–1, whereby 0 is a no-energy event (e.g., no handling of the material) and 1 is a high-energy event (e.g., dropping from greater than a height of 2 m) (Ramachandran 2005).

All exposure scenarios were characterized with a high handling energy factor and high air exchange rates. ES1–ES6 (Table 7.2) represent variations of the same handling activity, but are based on differences in (a) the emission rate, e, per minute and (b) the frequency of the work-cycle activity, f (i.e., a function of both duration and length of emission). ES2, ES3, and ES4 all had the same emission rates, but at varying frequencies of 60 min emission events (i.e., work-cycle) with 60 minute pauses in between, 480 min all-day emission events with no pauses, and 1 min emission events with an indefinite pause the remainder of the workday, respectively. Compared to ES1, whose frequency of 10 min was defined as short, ES2, ES3, and ES4 represent long, daily (i.e., non-interrupted), and single-pulse frequencies, respectively. Conversely, ES5 and ES6 had the same emission frequencies as ES1, but modified emission rates that were 2- and 4-orders of magnitude smaller than ES1, respectively. Compared to ES1, whose emission rate was considered high, ES5 and ES6 were considered medium and low rates, respectively.

7.4.1 FATE AND TRANSPORT OF AIRBORNE NANO-TiO₂

In this paper, a two-zone, dynamic fate and transport model is presented for use with indoor, occupational ENM airborne emissions. We assume that there is only one emission source fully located inside a near-field (NF) zone. The remaining indoor air room volume was defined as far-field (FF) and should not be confused with outdoor air volumes. The NF is defined as the volume of a hemisphere with a radius of 0.8 m. This radius corresponds to being an arm's length away from the source of emission (Ramachandran 2005) and is relevant given the workplace activities described in ES1–ES6. Both NF and FF zones were modeled as well-mixed compartments connected by advective air flow (Nicas 1996). Existing LCA indoor air impact assessment methodologies utilize a one-box model under the assumption that there is only one emission source in a well-mixed room (Demou et al. 2009) (Equation 7.4):

$$V \cdot \frac{dC_i}{dt} = S - \left(C_i - 1 \cdot Q \right)$$

(7.4)

where:

  S is the emissions rate (mg/hr)
  V is the volume of the compartment (i.e., the one-box)
  C_i is the concentration at a given time-step (µg/m³)
  C_{i-1} is the concentration at the previous time-step
  Q is the ventilation rate (m³/hr)

In the case of indoor air emissions from single point sources, it can be anticipated that large concentration gradients will exist between the point of emission and points further away from the source (Jr. et al. 1996). To accommodate for imperfect mixing, LCIA methods can use a mixing factor, m (Equation 7.5) (Demou et al. 2009; Hellweg et al. 2009; Rosenbaum et al. 2015; Walser et al. 2015).

$$V \cdot \frac{dC_i}{dt} = S - \left(C_i - 1 \cdot Q \cdot m \right)$$

(7.5)

However, the use of m can still result in underestimations of NF exposure concentrations upwards of 50% compared to the results of a two-zone model (Jr. et al. 1996; Nicas 1996). Thus, in this paper, a two-zone model was used to address the difference between NF (Equation 7.6) and FF (Equation 7.7) concentrations of airborne nano-TiO₂.

$$V_{NF} \cdot \frac{dC_{NF}}{dt} = S + \left(\beta \cdot C_{FF} - 1 \right) - \left(C_{NF} - 1 \cdot Q \right) - \left(\beta \cdot C_{NF} - 1 \right)$$

(7.6)

$$V_{FF} \cdot \frac{dC_{FF}}{dt} = \left(\beta \cdot C_{NF} - 1\right) - \left(C_{FF} - 1 \cdot Q\right) - \left(\left(\beta + Q\right) \cdot C_{FF} - 1\right) \tag{7.7}$$

β is the inter-zonal advective exchange rate connecting the *NF* and *FF* air volumes. It was defined as the volume of air entering and leaving through half of the curved surface area of a hemisphere, whose radius, *r*, was equal to that of the *NF*, and the average air speed between the *NF* and *FF*, *s* (Equation 7.8) (Nicas 1996)

$$\beta = s \cdot \pi \cdot r^2 \tag{7.8}$$

ENMs settle out of the air at a relatively slow rate (Phalen and Phalen 2011); thus, when indoor ENM emissions rates are sufficiently low and ventilation rates are high, steady-state ventilation models may adequately describe particle loss and estimate indoor ENM concentrations (Walser et al. 2015). However, these conditions may not always be met, as has been demonstrated for organic chemicals (Wenger et al. 2012). For ENMs, this would mean other non-ventilation sources of particle loss, such as homo- and hetero-aggregation and gravitational settling, should be considered (Walser et al. 2015). The introduction of non-ventilation sources of particle loss (*k*) in a two-zone, dynamic model is represented by Equations (7.9) and (7.10).

$$V_{NF} \cdot \frac{dC_{NF}}{dt} = S + \left(\beta \cdot C_{FF} - 1\right) - \left(C_{NF} - 1 \cdot V_{NF} \cdot \sum k_i\right) - \left(\beta \cdot C_{NF} - 1\right) \tag{7.9}$$

$$V_{FF} \cdot \frac{dC_{FF}}{dt} = \left(\beta \cdot C_{NF} - 1\right) - \left(C_{FF} - 1 V_{FF} \cdot \sum k_i\right) - \left(\left(\beta + Q\right) \cdot C_{FF} - 1\right) \tag{7.10}$$

where:
    k_i is any process (*i*) of non-ventilation removal
    V_{NF} is the near-field volume
    V_{FF} is the far-field volume

Non-ventilation particle losses considered in this model are (1) homo- and hetero-aggregation, herein referred to as aggregation, and (2) gravitational settling. Aggregation was estimated using a first order rate constant (k_h) of 9.4E–5 min^{-1} (Meesters et al. 2014). The removal, k_{set}, due to gravitational settling, based on Stokes' Law, is defined by Equation (7.11):

$$k_{set} = \frac{v_{set}}{h} = \left[\frac{\rho_p \cdot d_e^2 \cdot g \cdot C_s}{18 \cdot n \cdot x} \right] \cdot \left[\frac{1}{h} \right] \tag{7.11}$$

where:
    v_{set} is the settling velocity
    ρ_p is the particle density
    d_e^2 is the equivalent volume diameter
    g is the gravitational force
    C_s is the Cunningham slip correction factor (Allen and Raabe 1985)
    n is the viscosity of the medium
    x is the dynamic shape factor (i.e., perfectly spherical materials have a value of 1)
    h is the height of the emission source

Gravitational settling is likely to be more important for ENMs and aggregates that are ≥100 nm, while Brownian motion might be more important for ENMs below 100 nm, although this distinction is not absolute and may differ based on the properties of the ENMs under consideration (Hodas et al. 2015; Kleinstreuer et al. 2008; Walser et al. 2015). Particles that deposit onto surfaces could potentially be re-suspended, attach to the skin upon contact from a surface, or be removed by cleaning (e.g., sweeping). It was assumed that there was no direct contact of contaminated surfaces to the skin and that all of the nano-TiO$_2$ deposited onto the ground was cleaned each day and not allowed to accumulate. All the parameters and their corresponding values that were used to run the fate and transport model are listed in Table 7.3. The model was constructed in MatLab 9.0 (MathWorks, USA).

TABLE 7.3

Parameters and Their Values Used in the Fate and Transport Model That Describes the Occupational Workplace Settings for ES1–ES6

Parameter	Description	Value	Units	Additional Information
r	Radius of near-field	0.80	m	Average arm's length from emission source (Ramachandran 2005)
V_{NF}	Volume of the near-field	1.07	m³	Volume of a hemisphere with radius, r
V_{FF}	Volume of the far-field	98.9	m³	Defined as $V_{tot}-V_{NF}$, where V_{tot} is 100 m³
β	Inter-zonal air flow	21.9	m³/min	Equation 7.8
s	Air flow between near- and far-fields	0.18	m/s	Calculated from reported (measured) indoor air speeds at occupational workplaces dealing with powder mixers and packers, excluding the outlier (*see Supporting Information*) (Baldwin and Maynard 1998)
$k_{h,a}$	Aggregation rate constant	9.4E–5	min⁻¹	First-order rate constant in air (Meesters et al. 2014) assumed for both homo- and hetero-aggregation.
v_{set}	Gravitational settling based on Stoke's law	N/A	N/A	Equation 7.11
ρ_p	ENM particle density	3900	kg/m³	(Garner et al. 2017)
d_e	Diameter of ENM	21.0	nm	Equivalent volume diameter
g	Gravitational acceleration	9.8	m/s²	
C_s	Cunningham slip correction factor	1.0	Unit-less	(*see Supporting Information*) (Clift et al. 1978)
n	Viscosity of air	1.1E–3	kg/m-min	(Garner et al. 2017)
x	Dynamic shape factor	1.0	Unit-less	Values of 1 for perfectly spherical particles
h_w	Height of the workplace	2.50	m	(Hristozov et al. 2016)
h_e	Height of the emission source	1.6	m	Interpreted from the handling energy factor of 0.8, whereby a value of 1 means a drop height greater than 2 meters. (Hristozov et al. 2016)
T	Time scale	10,080	min	Number of minutes in a week
t	Time-step	1	min	Time resolution at which the model was integrated
Q (k$_{ex}$)	Air-exchange rate	8.0	hr⁻¹	In a two-zone model, this represents the air exchange rate of the far-field room volume (Hristozov et al. 2016)

7.4.2   Exposure to Occupational Indoor Air Concentrations of Nano-TiO$_2$

In regards to human exposure to ENMs, emissions may be taken up by inhalation, through direct contact with the skin, or ingestion. While all routes of exposure are a cause for concern, this study focuses on the inhalation of airborne particulates to workers in the occupational setting. Exposure during work-related activities was assumed to occur only in the NF, while workers were assumed to be exposed in the FF during non-work cycles (i.e., pause cycles). The exposure, therefore, represents the influence from both the NF and FF. A physiologically based pharmacokinetic (PBPK) model was adapted from Li et al. (2015) to establish the deposition and retention of nano-TiO$_2$ in the lungs over time. This model was originally built for estimating the exposure of inhaled cerium oxide nanoparticles in rats, and was adapted for human-based PBPK modeling upon inhalation of nano-TiO$_2$ using human-relevant parameter values for human, adult, male lungs (*see Supporting Information*) (Brown et al. 1997; Carlander et al. 2016; Fox 2010; Molina and DiMaio 2012a, 2012b; Salmasi and Iskandrian 1993). The reader is referred to Li et al. (2015) for complete details on the model, but a brief explanation follows.

The overall exposure model does not assume any use of personal protective equipment (PPE) and assumes direct interaction between the airways and indoor air. The MPPD v3.01 dosimetry model was first used to derive the regional (fractional) deposition of ENMs in the lungs (*see Supporting Information*) (Centers for Health Research (CII) and the Netherlands' National Institute for Public Health and the Environment RIVM 2015). Because the current MPPD model does not allow for modeling deposition and clearance under variable exposure conditions, inhalation, and/or kinetically defined biological processes, the regional deposition fraction MPPD data was fed into the PBPK model to estimate bio-distribution in the lung. The PBPK model then estimated the final retention after considering (1) mucociliary clearance, (2) phagocytosis, and (3) translocation of nano-TiO$_2$ into systemic circulation (Li et al. 2015). The model estimates deposition in (1) the head or upper airways, (2) the tracheobronchial region, and (3) the pulmonary regions of the lung (i.e., air exchange occurs at the alveoli). Flow- and diffusion-limited processes, defined by their permeability and partition coefficients (not to be confused by partition coefficients (PCs) previously discussed for steady-state fate and transport modeling), governed the exchange of ENMs with blood and tissues (*see Supporting Information*). PCs had organ-specific saturation levels that limited their rate of ENM sequestration as PCs reached saturation. Finally, the mucociliary clearance rate was defined as a constant, irrespective of ENM loading in the lung. It has been shown that pulmonary clearance of particulates found in lavages is 10 times faster in rats compared with humans and might be partially explained by the greater mucociliary clearance rate in rats compared to humans (Kuempel et al. 2015). Thus, the transport factor governing translocation of loaded-PCs to the tracheobronchial region was estimated as 1.44E–6 min^{-1}, which is one order of magnitude slower than reported by Li et al. (2015). Exposure was expressed as total wet lung burden and reported as an internal mass dose of nano-TiO$_2$ per mass of wet lung. The wet lung was defined as the pulmonary and interstitial regions of the lung and the corresponding blood and PCs found in those compartments yet excluding the upper airway and trachea-bronchial regions. The PBPK model was implemented in Berkeley Madonna™ version 8.3.23 (Berkeley, CA).

7.4.3   Retained-Intake Fraction of Nano-TiO$_2$ Emissions to Occupational Indoor Air

The emissions and resulting exposures were combined into an overall RiF, which represents the averaged internal wet lung mass dose of TiO$_2$ (EXP$_{int}$) per averaged lifetime-emitted mass of TiO$_2$ (E$_{life}$). This is different from traditional inhalation iF, which is the inhaled amount per emitted amount (Bennett et al. 2002). Two different time horizons were used to represent limited and chronic exposures, defined using a (1) 1-year and (2) 45-year (i.e., lifetime) work period, respectively. The time-weighted average wet lung burden was calculated over an assumed life expectancy of 70 years, defined by periods of (1) non-work years from 1-20, (2) work years from 21 to 65, and (3) non-work in their retirement years from 66 to 70. Thus, the model assumed emissions that persisted throughout the complete set of working years of 20–65 (E_{life}) irrespective of whether the person was working or not the entire time. Therefore, the 1-year time-weighted lung burden assumed one year of exposure given 70-years of emissions, while the lifetime time-weighted lung burden assumed 45 years of exposure given 70-years of emissions. These wet lung burdens were then divided through by the 70-year life expectancy to obtain a total retained wet lung burden over lifetime. Finally, the 1-year and lifetime ratios of lifetime wet lung dose and lifetime of emissions were scaled by the number (population) of workers (POP) inside of the exposure zones (Equation 7.12).

$$RiF = \frac{EXP_{int}}{E_{life}} \cdot POP \tag{7.12}$$

The number of exposed workers was estimated from Walser et al. (2015) and defined as a lognormal distribution with a geometric mean of 8.7 and geometric standard deviation of 2.8 (*see Supporting Information*). Three different POP scenarios were defined as low, average, and high. These three scenarios represented the 5% confidence interval, geometric mean, and 95%-CI of the distribution, respectively.

7.4.4   DOSE–RESPONSE RELATIONSHIP AND EFFECT FACTOR FOR NANO-TiO$_2$

The effect factor (EF) represents the underlying dose–response relationship of a substance (Rosenbaum et al. 2008a). The EF was estimated based on the USEtox approach (Rosenbaum et al. 2008a) defined according to Equation (7.13):

$$EF = \frac{0.5}{ED50, h, int} \tag{7.13}$$

where $ED50, h, int$ is defined as the human-equivalent (h) dose at which 50% of population experiences a carcinogenic or noncarcinogenic impact upon inhalation exposures, and where 0.5 represents the fraction of the worker population that experiences the adverse human health impact (Rosenbaum et al. 2008a). Unlike USEtox, which is concerned with external exposure concentrations, the ED_{50} considered here (Equation 7.14) is reported as an internal dose per g of wet lung in order to be in agreement with the RiF that is represented as a lifetime-averaged internal wet lung dose.

$$ED50, h, int = \frac{ED50a, t, int}{AF_a \cdot AF_t} \tag{7.14}$$

$ED50a,t,int$ (mg/g-wet lung) is the animal (*a*) internal dose (*int*) of nano-TiO$_2$ per gram of lung that results in a 50% response rate, *t* is the duration of the study (e.g., chronic), AF_a is an interspecies extrapolation factor (i.e., between animals to humans), and AF_t is a study-time conversion factor (i.e., between acute to chronic studies) (Huijbregts et al. 2005, 2010; Rosenbaum et al. 2008a). Dose–response data reported for dry lung samples were converted to wet lung with the conversion factor of 0.11 based on previous results showing that nearly 89% of the lung-weight is lost after drying (Levine et al. 2003). Interspecies extrapolation factors account for the difference in physiology, such as breathing and body weight, between species, however, a factor of 1 was applied in this study, under the assumption that effects based on internal doses were equivalent in animals and humans (Huijbregts et al. 2005). Furthermore, the $ED50a,t,int$ assumed a 1:1 correlation of inflammation, reported as *extra risk*, and disease probability. Extra risk is the fraction of animals that respond to a dose, among animals who do not respond, effectively taking into account the background response rate (Crump 1984). An AF_t of 2 was used to convert sub-chronic dose–response data to chronic-equivalent values (Huijbregts et al. 2005; Rosenbaum et al. 2008a).

A noncarcinogenic $ED50a,t,int$ was defined using the benchmark dose (BMD) approach (Crump 1995) and estimated using the Netherland's National Institute for Public Health and the Environment's (RIVM) PROAST software (www.rivm.nl). Based on previously published data, (Bermudez et al. 2004) the relationship between internal nano-TiO$_2$ mass loadings per gram of lung and lung inflammation measured from a sub-chronic animal study (*see Supporting Information*). A carcinogenic $ED50a,t,int$ was defined using the same approach as defined above and derived from the relationship between internal nano-TiO$_2$ mass loadings per gram of lung and an incidence of lung tumors measured from three chronic animal studies (*see Supporting Information*).

7.5 RESULTS

7.5.1 Exposure Scenarios and Occupational, Indoor Air Emissions of Nano-TiO$_2$

Although the emission rates for ES1–ES4 were the same (6.72E+02 mg/min), their total daily emissions (Table 7.4) were not equal due to changes in the frequency of the emission events.

Total daily emissions for ES1–ES3 differed by less than one-order of magnitude, even though emission frequencies were noticeably different. Particularly, ES2 was characterized by 60-minute emission cycles with 60-minute pauses between, resulting in total daily nano-TiO$_2$ emissions of 1.61E+05 mg. This was only 35% larger than ES1, even though emissions lasted only 10 minutes with 20-minute breaks between. ES3 had a constant all-day emission event that resulted in a total daily nano-TiO$_2$ emission of 3.23E+05 mg, roughly 3 times the daily emission of ES1. ES4 was characterized by a single-pulse emission event that resulted in a total maximum daily nano-TiO$_2$ emission of 6.72E+02 mg. The total daily emissions for ES5 and ES6 were roughly 2- and 4-orders of magnitude smaller, respectively, than ES1. This was in direct correlation with their 2- and 4-orders of magnitude decrease in emission rate.

TABLE 7.4

Emission Rates, Total Daily Emissions, Average Daily, and Maximum Near-Field and Far-Field Airborne Concentrations

No.	Exposure Scenario	E_i, [mg/min]	Total Daily Emissions (mg)	C^a_{NF} [mg m^{-3}]	C^a_{FF} [mg m^{-3}]	C^b_{NF} [mg m^{-3}]	C^b_{FF} [mg m^{-3}]
ES1	e-high, *f-short*	6.72E+02	1.08E+05	6.50E+01	3.45E01	2.49E01	1.48E01
ES2	e-high, *f-long*	6.72E+02	1.61E+05	8.15E+01	5.09E01	4.07E01	2.54E01
ES3	e-high, *f-all day*	6.72E+02	3.23E+05	8.15E+01	5.09E01	8.08E01	5.02E01
ES4	e-high, *f-single pulse*	6.72E+02	6.72E+02	3.63E+01	6.01E00	0.170E+00	0.106E+00
ES5	e-medium, *f-short*	6.72E+00	1.08E+03	0.677E+00	0.371E+00	0.271E+00	0.169E+00
ES6	e-low, *f-short*	6.72E-02	1.08E+01	6.77E-03	3.71E-03	2.71E-03	1.69E-03

Note: C: exposure concentration.

[a] Average daily concentrations during working hours.

[b] Maximum daily concentrations during work hours.

7.5.2   FATE AND TRANSPORT OF AIRBORNE NANO-TiO$_2$

The results of the fate model are reported as two different NF and FF airborne concentrations. Figure 7.1 shows the trends in NF and FF concentrations during the 8-hour workday.

ES2 and ES3 both reached a maximum NF airborne concentration of 8.2E+04 µg/m³ shortly after the work-cycles began, even though emissions were ongoing throughout the remainder of the workday. Maximum NF daily airborne concentrations in ES1, ES5, and ES6 reached 6.5E+04, 6.8E+02, and 6.8E+00 µg/m³, respectively, but were still increasing at the time the work-cycle ended and emissions stopped. The maximum NF airborne concentration for ES4 of 3.6E+04 µg/m³ coincided with the exact time it was released at the beginning of the workday upon the initiation of the first and only work-cycle. The trends in FF concentrations were similar to the NF, however these concentrations were, on average, 50% lower than their respective NF values.

Average daily airborne concentrations in both the NF and FF during working hours were 45% lower than their corresponding maximum daily values. However, this correlation was not linear, particularly for ES1, ES2, and ES4 (Figure 7.1). While there were distinct differences in the NF and FF concentrations during the emission events, their respective concentrations eventually

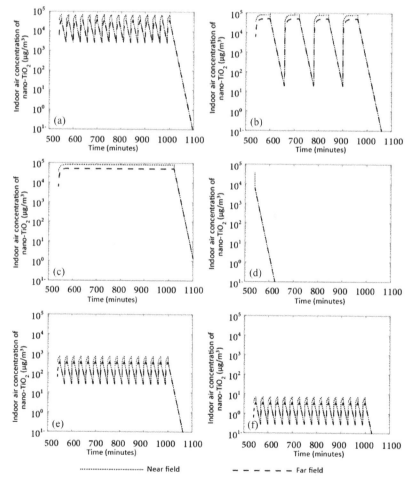

FIGURE 7.1   Results of the fate and transport model showing changes in the near- and far-field airborne concentrations of nano-TiO$_2$ over the course of the 8-hour workday. Note that the beginning of the workday begins at minute 500 and ends just after minute 980. Parts (a) through (f) refer to ES1 through ES6, respectively.

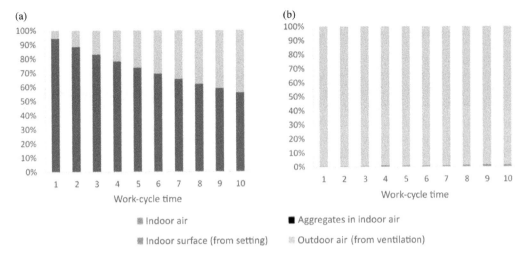

FIGURE 7.2   Mass fraction of total nano-TiO$_2$ emitted in the day, per "compartment" during (a) the first 10 minutes of the first emission cycle of ES1 and (b) the final 10-minute emission cycle of ES1 at the end of the workday.

equilibrated during non-emission events. In all cases, airborne concentrations effectively reached zero before the next workday, and, thus, concentrations were not cumulative from day to day (*see Supporting Information*).

The dominant mechanism driving the fate and transport of nano-TiO$_2$ was the air exchange rate between the indoor and outdoor air compartments and, to a lesser extent, the inter-zonal air flow between the NF and FF. In ES1, during the first minute of the first work-cycle, nearly 94% of the nano-TiO$_2$ remained in the indoor air (i.e., combined NF and FF), while 5.7% had been transferred to outside air (Figure 7.2).

An additional 0.0025% of the emissions in ES1 had been transformed into larger aggregates, and 0.0021% had settled to the surface due to gravitational settling. By the end of the first 10 minutes of that emission cycle, the amount of total emissions remaining in the indoor air was only 55.84%, while 44% had been transferred to the outdoor air. The proportion of the emissions that had been removed by aggregation slightly increased to 0.0030% by the second and third minute and then fell back to 0.25% by the end of the emission event. The proportion of emissions removed by gravitational settling onto surfaces increased by over an order of magnitude to 0.03%. These trends continued through the workday whereby the exchange of indoor with outdoor air contributed to nearly 99% removal of the total daily emissions (Figure 7.2). The contributions from gravitational settling and aggregation remained minimal and accounted for roughly 0.07% and 0.00005% of nano-TiO$_2$ removal from the indoor air. By the end of the final minute of the last emission event of the workday, nearly 1.0% of the total daily emissions remained in the indoor air compartment, as non-agglomerated nano-TiO$_2$.

In ES2, the overall fate and transport patterns for ES2, ES3, and ES4 were similar to ES1. However, they differed by the rates at which the indoor air concentrations of nano-TiO$_2$ decreased. For example, after only 40 minutes, the amount of nano-TiO$_2$ emissions that remained in the indoor air were 25%, 19%, 19%, and 0.49% for ES1, ES2, ES3, and ES4, respectively. The relative amount of nano-TiO$_2$ emissions that remained in the indoor air decreased at the fastest rate for ES4 since it was only a single pulse event without further addition of nano-TiO$_2$ over time during the workday. Thus, the effect of the removal mechanisms, mainly driven by the overall air exchange rate of the room, was more apparent in this scenario. Although the emission rates of ES5 and ES6 were lower than ES1, they had the same relative fate and transport pattern as ES1.

7.5.3 EXPOSURE TO AND RETAINED-INTAKE FRACTION OF OCCUPATIONAL, INDOOR AIR EMISSIONS OF NANO-TiO₂

The results of the MPPD model showed that 75.2% of the airborne nano-TiO_2 deposited into the lung, irrespective of the exposure scenario or concentration of particles in the air (*see Supporting Information*). The regional deposition was 9.8%, 24.6%, and 40.8% in the upper airway, tracheobronchial region, and the pulmonary (i.e., alveolar) regions, respectively. The results of the PBPK model, which estimated the clearance and ultimate retention of that deposited fraction, are reported as mass of nano-TiO_2 per mass of wet lung as well as its corresponding 1-year and lifetime RiF (Table 7.5).

In general, greater total yearly emissions resulted in greater total lung burdens (e.g., ES1-3 versus ES4-6). Over the course of each workday, the wet lung burden increased with all emission events. Between emissions, internal lung doses continued to rise, and between workdays or workweeks (i.e., over the weekends), the lung burden did not decrease sufficiently to clear the lung of its total nano-TiO_2 load (Figure 7.3). For example, the maximum exposure by the end of the first workweek in ES1 was 333 μg/g-wet lung, while the remaining lung burden at the beginning of the second workweek was 238 μg/g-wet lung. This trend continued until the 5th workweek, after which maximum weekly accumulations slowed considerably, having already reached 453 μg/g-wet lung, which was 95% of the maximum lung burden of 478 μg/g-wet lung observed at the end of the year (Figure 7.3). Similar trends were seen in the other exposure scenarios, except for ES6, whose total lung burden continually increased at a steady rate compared to the plateaus seen within a few weeks of the other ES (*see Supporting Information*).

Additionally, internal lung doses did not clear effectively between work years (i.e., two weeks of vacation), as was demonstrated by the 1-year time-weighted lung burden for ES1 over a lifetime of 5.72 μg/g-wet lung, and the lifetime lung-burden was 273 μg/g-wet lung. The 1-year time-weighted lung burdens ranged from a high of 25 μg/g-wet lung for ES3 to a low of 0.017 μg/g-wet lung for ES6 (Table 7.5). Similarly, the lifetime-weighted lung burdens ranged from a high of 119 μg/g-wet lung for ES3 to a low of 4.0 μg/g-wet lung for ES6 (Table 7.5).

For ES1, total lung burden was mainly due to the retention factor of nano-TiO_2 in the interstitial tissue, where it represented up to 80% of the retained wet lung mass by the end of the first work year (Figure 7.4). In contrast, the pulmonary region had cleared itself of all deposited nano-TiO_2 by the beginning of each subsequent work week. Even so, the pulmonary region contributed up to 17% of the total *maximum* retention observed each day by the end of the first work year. The PC located in the interstitial and pulmonary regions reached maximum retentions of 3202 and 4644 μg, respectively, very quickly within the first workday, and did not accumulate more over the duration of the exposure period. This represented roughly 0.7% and 1.0% of the total lung burden by the end of the work year. The trachea-bronchial, or upper airway, regions were not defined as part of the wet lung, and, thus, do not directly contribute to the overall lung burden under consideration (*see Supporting Information*).

TABLE 7.5
Results for the Internal Wet Lung Burden (μg/g-wet Lung) Reported as Either a Lifetime or 1-Year Value

No.	Exposure Scenario	Lung Burden (Lifetime)	Lung Burden (1-year)
ES1	e-high, *f-short*	2.73E+02	5.72E+00
ES2	e-high, *f-long*	5.53E+02	1.16E+01
ES3	e-high, *f-daily*	1.19E+03	2.50E+01
ES4	e-high, *f-single pulse*	7.04E+00	1.44E−01
ES5	e-medium, *f-short*	8.10E+00	1.67E−01
ES6	e-low, *f-short*	3.99E+00	1.70E−02

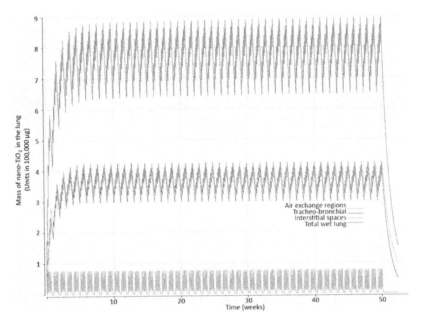

FIGURE 7.3   Retention of nano-TiO$_2$ in the lung estimated over one full work year for ES1, whereby work continued five days a week (i.e., no work on the weekends) for 50 weeks with a 2-week holiday at the end of the year during weeks 51 and 52.

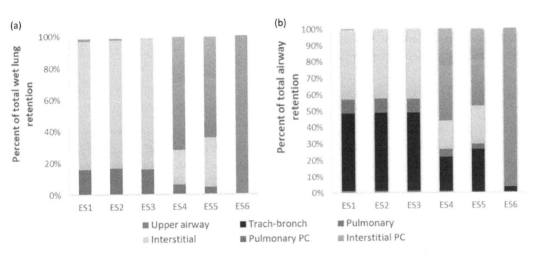

FIGURE 7.4   Relative retention of nano-TiO$_2$ in the (a) wet lung and (b) total airway system per region of the lung, for each of the six exposure scenarios. Results represent the relative retention for the maximum exposure obtained after one year of work.

The physiologic pattern of retention was very similar for ES1–3, whose total yearly emissions and retained lung burdens were much greater than ES4–6. As total yearly emissions, and thus lung burdens, decreased, the pattern of retention shifted towards greater number of particles captured by the pulmonary and interstitial PCs. Subsequently, in the scenario with the lowest emissions and lowest lung burden, ES6, nearly 100% of the particles were captured by pulmonary-PCs.

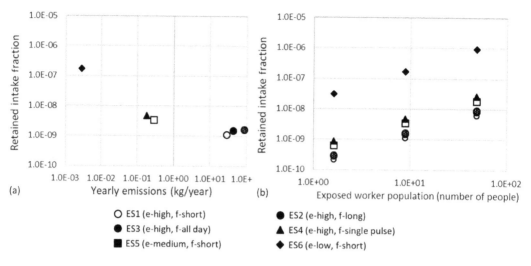

FIGURE 7.5 Lifetime-retained intake fractions as a function of (a) total yearly emissions and (b) worker population size. All axes are shown in log-scale.

The corresponding 1-year RiF values range from a high of 7.81E–10 for ES6 to a low of 2.62E–11 for ES1. Similarly, the lifetime RiF values ranged from a high of 2.21E–7 for ES6 to a low of 1.25E–9 for ES1. Consequently, an inverse relationship was seen between the total yearly emissions and the RiF (Figure 7.5).

As expected, the RiF increases linearly with the increasing number of workers. Thus, the RiF for the highest worker population was 1.5-orders of magnitude greater than the RiF in the lowest worker population scenario. These results corresponded directly to the 1.5-order of magnitude difference in their respective worker populations.

7.5.4 Dose–Response Relationship and Effect Factor for Nano-TiO_2

The carcinogenic dose–response analysis is shown in Figure 7.6. The corresponding $ED50_a$ (i.e., BMD) was 1.43 m^2/g-dry lung based on the excess risk of 50% over background cancer rates, with the dose being explicitly expressed as the TiO_2 surface area concentration per gram of dry lung. After conversion to a mass-based dose-metric, assuming a value of 48 m^2/g-TiO_2, the $ED50_a$ was 2.98E+04 μg/g-dry lung. After converting the dry lung doses to wet lung, the $ED50_a$ became 3.16E+03 μg/g-wet lung. Applying the relevant extrapolation factors (Equation 7.14), the resulting $ED50_h$ value was 1.58E+03 μg TiO_2 (per g-wet lung) and the final EF was 3.17E–04 cases/μg nano-TiO_2 (per g-wet lung).

The results of the noncarcinogenic dose–response analysis are shown in Figure 7.7. BMD values were interpreted from the results of a sub-chronic whole-body inhalation study measuring the changes in BAL fluids upon exposure to nano-TiO_2. The reported benchmark doses were 27352 μg/g-dry lung for mice and 7807 μg/g-dry lung for rats based on the excess risk of 50% over background inflammation rates. Covariation, based on species type, showed that there were distinct dose–response slopes for both mice and rats, with rats having lower BMD values (i.e., more sensitive). Based on the approach put forth in USEtox (Rosenbaum et al. 2008a), the BMD for rats was used to estimate the EF, because it was the most sensitive species. This $ED50_a$ was 8.27E+02 μg/g-wet lung. The resulting $ED50_h$ value was 4.14E+02 μg TiO (per g-wet lung) and the final EF was 1.21E–03 cases/μg (internal) TiO_2 dose.

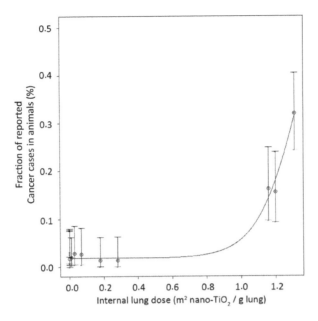

FIGURE 7.6 Benchmark dose results for cancerous impacts in rats upon inhalation of nano-TiO$_2$ and reported as fraction of cancer cases per internal lung dose.

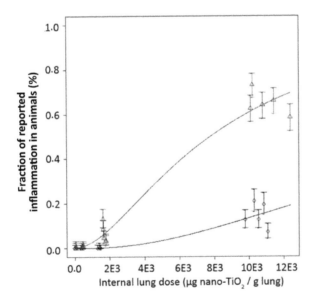

FIGURE 7.7 Benchmark dose results for noncancerous impacts to both mice (circles) and rats (triangles).

7.5.5 CHARACTERIZATION FACTORS FOR HUMAN HEALTH IMPACTS FROM OCCUPATIONAL, INDOOR AIR EMISSIONS OF NANO-TiO$_2$

For each exposure scenario, a 1-year and lifetime CF are reported for nano-TiO$_2$ occupational, airborne emissions (Table 7.6). One-year CF$_C$ values ranged from a low of 3.96E−4 cases of cancer per kg of nano-TiO$_2$ emissions in ES1 to a high of 5.80E−2 cases of cancer per kg of nano-TiO$_2$ emissions in ES6. The 1-year, *noncarcinogenic* CF values were slightly larger and ranged from a low

TABLE 7.6
Effect Factors (EF), Retained-Intake Fractions (RiF), Carcinogenic Characterization Factors (CF$_C$), and Noncarcinogenic Characterization Factors (CF$_{NC}$) for the Six Emission and Exposure Scenarios, Worker Population, as Well as Timeframe of the Exposure

Reference	EF$_C$	EF$_{NC}$	Lifetime[c]			1-Year[c]			Low Population[d]			High Population[d]		
			RiF	CF$_C$	CF$_{NC}$	RiF	CF$_C$	CF$_{NC}$	RiF	CF$_C$	CF$_{NC}$	RiF	CF$_C$	CF$_{NC}$
ES1 (e-high, *f-short*)	3.17E+05	1.21E+06	1.25E−09	3.96E−04	1.51E−03	2.62E−11	8.31E−06	3.17E−05	2.29E−10	7.25E−05	2.77E−04	6.83E−09	2.16E−03	8.25E−03
ES2 (e-high, *f-long*)			1.69E−09	5.35E−04	2.04E−03	3.55E−11	1.12E−05	4.29E−05	3.09E−10	9.79E−05	3.74E−04	9.22E−09	2.92E−03	1.11E−02
ES3 (e-high, *f-all day*)			1.82E−09	5.75E−04	2.19E−03	3.82E−11	1.21E−05	4.62E−05	3.33E−10	1.05E−04	4.02E−04	9.92E−09	3.14E−03	1.20E−02
ES4 (e-high, *f-single pulse*)			5.16E−09	1.63E−03	6.24E−03	1.05E−10	3.34E−05	1.27E−04	9.46E−10	2.99E−04	1.14E−03	2.82E−08	8.92E−03	3.40E−02
ES5 (e-medium, *f-short*)			3.71E−09	1.18E−03	4.48E−03	7.64E−11	2.42E−05	9.23E−05	6.80E−10	2.15E−04	8.21E−04	2.03E−08	6.42E−03	2.45E−02
ES6 (e-low, *f-short*)			1.83E−07	5.80E−02	2.21E−01	7.81E−10	2.47E−04	9.44E−04	3.36E−08	1.06E−02	4.05E−02	1.00E−06	3.17E−01	1.21E+00
Pini et al.	4.19E+02[a]	1.72E−02	3.41E−05	1.43E−02	5.85E−07	—	—	—	—	—	—	—	—	—
USEtox	—	—	1.19E−04[b]	—	—	—	—	—	—	—	—	—	—	—

[a] Reported as an ED4: 4% increase in inflammation rate.
[b] Reported for generic "organics" and "inorganics" and not specific for nanoparticles or titanium dioxide.
[c] Reported for the average number of workers.
[d] Reported for the lifetime exposure scenarios.

of 1.51E-3 cases of lung inflammation per kg of nano-TiO_2 emissions to a high of 2.21E-1 cases of lung inflammation per kg of nano-TiO_2 emissions. Compared to 1-year exposure periods, lifetime CF values were generally 1-2 orders of magnitude larger.

An inverse relationship is then noticed between the total emissions and the corresponding CF (*see Supporting Information*). Consequently, ES6 had the largest CF_C and CF_{NC} values, even though it had the lowest total yearly emissions, while ES1-3 had the lowest CF values, but the largest emissions.

7.6 DISCUSSION

Currently, the estimation of ENM emissions, fate, exposure, and human health impacts from indoor air exposure within LCA have been largely unaddressed in the scientific literature. Steady-state fate and exposure methodologies (Eckelman et al. 2012; Miseljic and Olsen 2014; Pini et al. 2016; Salieri et al. 2015) provide a straightforward first approximation to understanding ENM behavior in the environment; however, dynamic models can provide additional insight into the behavior and exposure to ENMs that are not captured by steady-state models. However, dynamic models require more data, such as reported or estimated emissions over time. In this paper, we utilized an approach for estimating indoor airborne emissions of spherical, metal oxide ENM powders. Hence, this approach may have limited application for estimating the effects of producing or handling ENMs whose physical characteristics are not spherical or are in solution (i.e., whose emissions are directly into soil and/or water).

The difference between steady-state and dynamic models can be striking. For example, if we apply a steady-state approximation (see *Supporting Information*) (Nicas 1996) to ES1, the resulting near-field airborne concentration would be 3.52 mg/m³. The result is approximately 95% smaller than the maximum near-field concentration predicted by the dynamic model, which demonstrated increases in ENM concentration over the course of the emissions. Removal by aggregation and gravitational settling was minimal, which might be expected given that the settling velocity of even 1 um sized particles is circa 12.5 cm/hr. (Phalen and Phalen 2011). These removal mechanisms will likely become much more important if air exchange rates are kept low. While air exchange rates in the workplace are likely to be kept high for safety issues, some studies have reported much lower air exchange rates for ENM manufacturing conditions, (Demou et al. 2008) not to mention other non-industrial workplaces (Sundell et al. 2006).

The model used to estimate the exposure to airborne ENM emissions departed from conventional LCIA methods that assume inhalation exposures are linearly related to an individual's respiration rate. Walser et al. (2015) recently proposed, and Pini et al. (2016) estimated, final retention values based on steady-state deposition and clearance using the MPPD model. Whereas our lifetime RiF was over 2-orders of magnitude less than what was reported in Pini et al. (Table 7.6), our results are similar to those from previously reported *in vivo* studies which were on the order of hundreds of days in the air exchange and interstitial regions of the lung. (Braakhuis et al. 2014; Kuempel et al. 2015; Oberdorster et al. 1994). Of the three main pathways in the model that governed ENM retention in the lung, translocation of loaded-PCs from the pulmonary (i.e., alveolar) to the tracheo-bronchial region was the limiting factor determining clearance of ENM from the lung. This value was modified to reflect the slower clearance rate of particulates in the human lung compared with the rat lung (Kuempel et al. 2015). In the highest exposure scenario, PCs were easily saturated and overburdened, leading to accumulation in many of the lung regions. The unlimited accumulation found in the air-exchange and interstitial regions for some of the exposure scenarios may be occurring due to mucociliary clearance becoming overburdened, which cannot remove particles at a fast-enough rate to reach zero lung burden even at times of low- or no-exposure. However, these results may also be limited by the model's capability to adjust for the increase or decrease of mucociliary clearance activity depending on the mass loading of ENM in the airway. The current version of the PBPK model uses a constant transfer factor to describe this mechanism instead of one that might change based on the internal load. Additionally, the high concentration of nano-TiO_2 found in the tracheobronchial region was expected given the primary particle sizes upon exposure were assumed to be 21 nm. Particles ≤ 100 nm have a greater chance of reaching the alveolar region, however particles ≤ 30 nm, particularly those below 10 nm (Kuempel et al. 2015) will show greater deposition and retention in the tracheobronchial region due to airflow dynamics of smaller diameter particles (Braakhuis et al. 2014).

The RiF, as defined in this study, showed an inverse relationship between the rate of nano-TiO_2 emitted into the occupational indoor air workplace and the retained amount of nano-TiO_2 in the lung. This finding is counterintuitive given that conventional-iF values in conventional LCIA methods are independent of the emission rates or total emission amounts. This can be explained by the saturation of PCs in the lungs and the relative contribution of PC-sequestered ENMs to the overall wet lung burden. Neither of these scales linearly with emission rate, thus bearing the inverse relationship discovered in this study. As shown in the results for the exposure, after saturation, the relative contribution of PCs sequestered ENMs to overall wet lung burden decreases as emission rates increased. The amounts in the pulmonary region and interstitium of the lungs scale proportionally with the emission rate after the PCs are saturated. The lifetime RiF, for example, differed by a maximum two-orders of magnitude while emissions per year increased up to four-orders of magnitude. This is because PCs still play a major role at lower emission rates. This highlights the importance of the emission rate of the RiF when considering the target organ's physiology.

It should be noted that the RiF and exposure results should be interpreted with caution. Although adaptations were made to the original rat-based model developed by Li et al. (2015) to estimate biodistribution in humans, the results of this model have not been validated by experimental evidence. In addition, the original PBPK model published by Li et al. (2015) was built based on exposure concentrations that were 2-orders of magnitude lower than the highest exposure scenario presented in this study, which is, in effect, the worst-case scenario. The results of applying the PBPK-rat model to human exposure scenarios should be considered as potential impacts as opposed to absolute values. The resulting RiF calculated in this paper did not include the effects and use of personal-protection-equipment (PPE) in the work environment. At the time of this study, it was not evident which occupational scenarios would require PPE, at what rate PPE would be used, and the effectiveness of PPE to filtering ENMs. However, it is expected that some level of PPE would apply, resulting in lower amounts of exposure than predicted in this paper. Therefore, if patterns of PPE usage are known, this should be included in the exposure estimation.

The RiF was also significantly influenced by the exposure timeframe assumed in the model, whereby 1-year RiF and their corresponding CF were one- to two-orders of magnitude smaller than their lifetime counterparts. The relative decrease in retained-intake and, therefore, cases of human health impact per emitted mass of nano-TiO_2 was expected given the significantly shorter amount of exposure time in the 1-year exposure scenarios. In this way, the CF can be used in a manner that is compatible with LCIA methodologies that contain different "perspectives" based on the decision-maker's preferences for short-term versus long-term priorities and objectives (Goedkoop et al. 2013; Hofstetter et al. 2000).

Additionally, the CF presented in this study were calculated for inhalation exposures in the occupational setting, since inhalation is the primary intake route in the workplace scenario (Walser et al. 2015). While the fate and transport model predicts indoor surface concentrations that could be used for approximating dermal exposures, dermal exposure and, thus, toxicity is expected to be low, unless in contact with broken skin (Filipe et al. 2009; Furukawa et al. 2011; Monteiro-Riviere et al. 2011; Newman et al. 2009; Sadrieh et al. 2010). Furthermore, apart from direct impacts to the lungs and lung-related injuries, ENMs that are deposited in the lung may translocate to other regions of the body after inhalation (Borm et al. 2006; Fertsch-Gapp et al. 2011; Geiser et al. 2005). Although the exposure model presented in this approach allows for the estimation of ENMs in 10 other organs of the body, these values were not yet used to address systemic human health impacts upon inhalation to nano-TiO$_2$. Future work in this area should address the compounded effects from multisystem organ toxicity upon inhalation to ENM.

The EF$_C$ reported in this study was almost 1-order of magnitude smaller than the EF$_{NC}$. It should be noted that the noncarcinogenic endpoint was pulmonary inflammation, a precursor to the carcinogenic endpoint of pulmonary tumors. Thus, these results intuitively make sense. It would be expected that many more cases of noncancerous diseases occur compared with more advanced pathologies such as cancer. This is in contrast to the results presented by the only other study in the literature that has reported a human health EF for nano-TiO$_2$ (Pini et al. 2016). Pini et al. (2016) report an EF$_C$ that is 4-orders of magnitude greater than their reported EF$_{NC}$, thus implying that there are many more expected cases of cancer compared with inflammation. However, a likely explanation regarding this discrepancy is the fact that Pini et al. (2016) used an inflammation study (Bermudez et al. 2004) as a proxy for the carcinogenic endpoint. Therefore, a comparison of their EF$_C$ to our EF$_{NC}$ is more appropriate. This comparison shows that our EF$_{NC}$ was still over 3-orders of magnitude greater than their EF$_C$ value. This may be partially explained by the fact that their EF$_C$ value was calculated as an ED$_4$ (i.e., a 4% increase over background response), and at a response rate of only 4% over the background the slope of the dose–response curve—the slope of the dose-response curve is, in essence, the EF—could be nearly flat if it is still near or under the point of departure. Comparatively, the slope of the dose–response curve at a response rate of 50% will be significantly larger, leading to a significant deviation between our EF and that previously reported by Pini et al.

7.6.1   CF Application in OPV Thesis

7.6.1.1   Life-Cycle Assessment of Organic Photovoltaics with ENM-Specific Characterization Factors

Life-cycle inventories do not allow for time integrated emission-flows nor are they categorized by time-scale (e.g., kg emissions per hour). If there is exact information regarding the rate of production for the reference-flow (i.e., product), then the reported emissions for that reference-flow could be defined as an emission rate for use in the fate and transport model. Often, this level of life-cycle inventory detail might not be attainable. In such a case, assumptions regarding the emission rate would have to be made. As explained above, the results of the CF$_{NS}$ calculations were most sensitive to the magnitude of the emission rate as opposed to emission frequency. Thus, static inventory flows should be interpreted as the magnitude of an emission rate for best correlation. Based on the previous set of results, it was assumed that the emissions reported in the OPV inventory correspond with the highest emission rate. This is because the highest emission rate was characteristic of an industrial-scale scenario, and since the purpose of this thesis is to forecast the potential environmental and human health risks for a prospective, large-scale OPV device, this assumption is warranted. Furthermore, the highest emission rate was correlated with the highest level of risk based on the HHRA as discussed in Chapter 4. It was useful to explore whether the LCA results would echo this human health hazard. Therefore, the highest emission rate was used as an initial proof of concept to estimate the human health hazards of nano-TiO$_2$ use in OPV. Within this category of emission rates, there was a range of half an order of magnitude based on differences in emission frequencies. Since the original frequency was originally described for an actual occupational exposure scenario at the industrial scale, this value was chosen as a proof of concept due to its relevance in forecasting human health impacts at the industrial scale. Both the long-term (lifetime) and the short-term (1-year) CF$_{NS}$ were tested for their influence on the LCA results.

Thus, lifetime CF$_{NS}$ for ES1 was used as a worst-case scenario proof-of-concept for a first approximation of the human health impacts from nano-TiO$_2$ handled and used at the industrial scale. Emissions across the life cycle of the OPV panel were considered only during production of the nano-TiO$_2$ power itself, and during production of the OPV panel. Emissions during nano-TiO_2 production were estimated for the production route outlined by the sulfate production route. Hischier et al. (2015) assumed there were no emissions to air during production, however this neglects the issues of handling, transferring, and packing the powder once it is created. Thus, emissions were estimated using an average of lower-estimate and higher-estimate emission scenarios as described by the material flow analysis of various ENMs reported by Gottschalk and Nowack (2011). This resulted in an estimated 0.25% of the total production rate being emitted to air, and, thus, per kg of nano-TiO_2 produced and handled, there was 0.32 mg of this material that ended up as occupational, indoor air emissions. Additionally, there were no known nano-TiO_2 emissions during the production of the OPV panel, and the same modeling assumptions stated above applied for OPV production. It was further assumed that this emission was to indoor occupational air, as opposed to outdoor environmental emissions. The reported emission (e.g., 0.32 mg) was then combined with both the noncarcinogenic and carcinogenic lifetime CF_{NS} for ES1 to calculate the potential human health impacts. This implicit assumption is that the CF_{NS} for ES1 represents an industrial-scale scenario, and that the inventory, including the emissions, built for the nano-TiO_2 production and OPV manufacturing, represent emissions of a similar magnitude over time at the industrial scale. This is a significant caveat because there is no relatable time-scale on the inventory.

7.6.2 Preliminary Results

Preliminary cradle-to-gate results show there was no noticeable increase in the carcinogenic and non-carcinogenic human health impacts after including the human health impacts of nano-TiO_2 either using the short-term (1-year) or the lifetime CF_{NS}. The application of the long-term (lifetime) CF_{NS} resulted in greater impacts compared to the 1-year CF_{NS}, as was expected given the latter is a one-year time-weighted average exposure over an entire lifetime worth of emissions. After applying the lifetime CF_{NS}, the human health impacts only slightly increased at around 1% increase overall (Table 7.7).

TABLE 7.7
Human Health Impacts per WattPeak of OPV Cell Production without a ENM-Specific Characterization Factor for Nano-TiO_2 (Left Columns) and with a ENM-Specific Characterization Factor (Right Columns)

Percent Contribution by Life-Cycle Stage	No ENM-Specific Characterization Factor		With ENM-Specific Characterization Factor (Lifetime Average)	
	Human Health (Noncarcinogenic)	Human Health (Carcinogenic)	Human Health (Noncarcinogenic)	Human Health (Carcinogenic)
Annealing	32.08%	29.55%	31.95%	29.47%
HTL	24.66%	2.46%	24.56%	2.45%
PCBM	14.47%	18.02%	14.41%	17.97%
Lamination	12.31%	14.56%	12.26%	14.51%
FTO Substrate	9.61%	10.44%	9.57%	10.41%
P3HT	0.80%	17.82%	0.80%	17.77%
Printing	3.16%	3.14%	3.14%	3.13%
ETL	1.59%	1.38%	1.58%	1.38%
Aluminum Electrode	0.63%	1.96%	0.62%	1.95%
Nano-TiO_2 Spacer	0.05%	0.04%	0.25%	0.18%
Other	0.64%	0.63%	0.86%	0.78%
Total Impacts (CTU)	9.35E−09	3.55E−09	9.38E−09	3.56E−09
Total contribution from nano-TiO_2	N/A	N/A	0.41%	0.28%

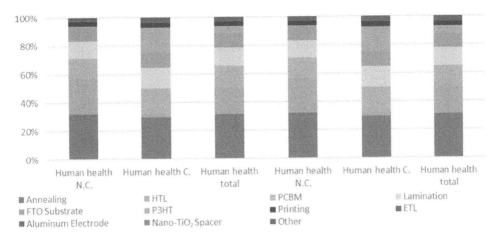

FIGURE 7.8 Contribution to human health impacts by life-cycle stage.

These nano-TiO_2 emissions to the indoor, occupational air resulted in a 0.28% and 0.41% contribution to the carcinogenic and noncarcinogenic impacts, respectively. Roughly 50% of these emissions came from handling nano-TiO_2 during OPV-manufacturing and the other half during nano-TiO_2 production. Thus, if a decision-maker wanted to reduce the human health impact of OPV, there is not much support for targeting the nano-TiO_2 emissions, for example (Figure 7.8).

Such results must be met with caution as several assumptions were made in during the calculation. First, emissions from the production route of both the nano-TiO_2 and the OPV panels were not known. These were estimated from generic release distributions previously reported in the literature for the production of ENM-related technologies and, thus, may not adequately describe the true emissions occurring in these industries (Gottschalk and Nowack 2011). It is likely that emissions to indoor air from powder handling and transfer are not adequately accounted for in these distributions; however, this assumption is not clarified in those previous publications (Gottschalk and Nowack 2011).

The nano-TiO_2 CF_{NS} were calculated using models that were emission-specific (i.e., the emission amount changed the resulting CF_{NS}). The CF_{NS} ultimately used to calculate the human health impacts from nano-TiO_2 emissions was an industrial scale emission rate for a representative industrial scale emission scenario and number of workers exposed. While this CF_{NS} might represent a valid industrial-scale scenario for handling nano-TiO_2, it is not specific to the actual emissions scenarios represented by the life-cycle inventory for OPV, and, thus, it was used to forecast impacts at the representative, prospective industrial scale of OPV production.

These preliminary results indicate that the human health impacts from occupational, indoor air emissions of ENMs across the OPV life cycle might be marginal, at best. These results are only valid for nano-TiO_2 and do not include the human health impacts of fullerenes. The magnitude of fullerene emissions during production, handling, and use were not known; however, this was qualitatively discussed in Chapter 4. While some indication of inflammation, for instance, have been demonstrated, fullerenes have also been shown to have anti-inflammatory effects. Thus, its potential to significantly contribute to the human health impacts of OPV due to emissions in occupational indoor air is not anticipated. Furthermore, these are just two material types used in OPV that have been explicitly called out as ENM. Since OPV are constructed on the order of hundreds of nanometers in thickness, this implies that all the layers would necessarily have to be composed of materials in the nanometer size range. While this is not usually explicitly stated, this technically implies that the use of ENMs comprises the entire panel (less the plastic-substrate). Consequently, the cumulative contribution of ENM emissions along the life cycle of OPV could potentially be more influential than this preliminary assessment suggests. Lastly, while human health impacts from occupational indoor air emissions seemed to be the most relevant potential impact pathway, emissions during the use phase and end-of-life phase cannot be ruled out. Such emissions will also be important for the determination of potential environmental and ecological impacts.

REFERENCES

Acero, A.P., C. Rodriguez, and A. Ciroth. 2014. *LCIA Methods—Impact Assessment Methods in Life Cycle Assessment and their Impact Categories*. Berlin, Germany: GreenDelta GmbH.
Allen, M.D., and O.G. Raabe. 1985. Slip correction measurements of spherical solid aerosol particles in an improved millikan apparatus. *Aerosol Science and Technology* 4(3): 269–286.
Bachmann, T.M. 2006. *Hazardous Substances and Human Health: Exposure, Impact and External Cost Assessment at the European Scale. Trace Metals and Other Contaminants in the Environment*. Amsterdam, the Netherlands: Elsevier.
Baldwin, P.E.J., and A.D. Maynard. 1998. A survey of wind speeds in indoor workplaces. *Annals of Occupational Hygiene* 42(5): 303–313.
Bennett, D.H. et al. 2002. Peer reviewed: Defining intake fraction. *Environmental Science & Technology* 36(9): 206A–211A.
Bermudez, E. et al. 2004. Pulmonary responses of mice, rats, and hamsters to subchronic inhalation of ultra-fine titanium dioxide particles. *Toxicological Sciences* 77(2): 347–357.
Borm, P.J.A. et al. 2006. The potential risks of nanomaterials: A review carried out for ECETOC. *Particle and Fibre Toxicology* 3:11.
Braakhuis, H.M. et al. 2014. Physicochemical characteristics of nanomaterials that affect pulmonary inflammation. *Particle and Fibre Toxicology* 11(1): 18.
Brown, R.P. et al. 1997. Physiological parameter values for physiologically based pharmacokinetic models. Toxicology and Industrial Health 13(4): 407–484.
Carlander, U. et al. 2016. Toward a general physiologically-based pharmacokinetic model for intravenously injected nanoparticles. *International Journal of Nanomedicine* 11: 625.
Clift, R., J.R. Grace, and M.E. Weber. 1978. *Bubbles, Drops, and Particles*. New York: Academic Press.
Crump, K. S. 1984. A new method for determining allowable daily intakes. *Fundamental and Applied Toxicology* 871(5): 854–871.
Crump, K. S. 1995. Calculation of benchmark doses from continuous data. *Risk Analysis* 15(1): 79–89.
Cucurachi, S., R. Heijungs, and K. Ohlau. 2012. Towards a general framework for icluding noise impacts in LCA. *The International Journal of Life Cycle Assessment* 17(4): 471–487.
de Bruijn, H., R. van Duin, and M.A.J. Huijbregts. 2002. Main characteristics of LCA. In *Handbook on Life Cycle Assessment*. B. Guinee et al. (Eds.). Dordrecht, the Netherlands: Springer.
Demou, E. et al. 2009. Evaluating indoor Exposure modeling alternatives for LCA: A case study in the vehicle repair industry. *Environmental Science & Technology* 43(15): 5804–5810.
Demou, E., P. Peter, and S. Hellweg. 2008. Exposure to manufactured nanostructured particles in an industrial pilot plant. *Annals of Occupational Hygiene* 52(8): 695–706.
Eckelman, M.J., M.S. Mauter, J.A. Isaacs, and M. Elimelech. 2012. New perspectives on nanomaterial aquatic ecotoxicity: Production impacts exceed direct exposure impacts for carbon nanotoubes. *Environmental Science & Technology* 46(5): 2902–2910.
European Commission. 2005. ExternE Externalities of Energy ExternE Externalities of Energy. http://www.externe.info/ (accessed December 15, 2017).
European Commission. 2011. *Recommendations for Life Cycle Impact Assessment in the European context*. 1st ed. Ispra, Italy: Joint Research Centre.
Fertsch-Gapp, S., M. Semmler-Behnke, A. Wenk, and W.G. Kreyling. 2011. Binding of polystyrene and carbon black nanoparticles to blood serum proteins. *Inhalation Toxicology* 23(8): 468–475.
Filipe, P. et al. 2009. Stratum corneum is an effective barrier to TiO_2 and ZnO nanoparticle percutaneous absorption. *Skin Pharmacology and Physiology* 22(5): 266–275.
Fox, S. 2010. *Human Physiology*. 12th ed. New York: McGraw-Hill.
Furtaw, E.J. Jr., M.D. Pandian, D.R. Nelson, and J.V. Behar. 1996. Modeling indoor air concentrations near emission sources in imperfectly mixed rooms. *Journal of the Air & Waste Management Association* 46(9): 861–868.
Furukawa, F. et al. 2011. Lack of skin carcinogenicity of topically applied titanium dioxide nanoparticles in the mouse. *Food and Chemical Toxicology* 49(4): 744–749.
Garner, K., S. Suh, and A. Keller. 2017. The risk of engineered nanomaterials in the environment: Development and application of the nanoFate model. *Environmental Science & Technology* 51(10): 5541–5551.

Gavankar, S., S. Suh, and A.F. Keller. 2012. Life cycle assessment at nanoscale: Review and recommendations. *The International Journal of Life Cycle Assessment* 17(3): 295–303.

Geiser, M. et al. 2005. Ultrafine particles cross cellular membranes by nonphagocytic mechanisms in lungs and in cultured cells. *Environmental Health Perspectives* 113(11): 1555–1560.

Goedkoop, M., R. Heijungs, M. Huijbregts, A. De Schryver, J. Struijs, and R. Rosalie van Zelm. 2013. *ReCiPe 2008 - A Life Cycle Impact Assessment Method Which Comprises Harmonized Category Indicators at the Midpoint and the Endpoint Level.* Amersfoort, the Netherlands: PRé Consultants.

Gottschalk, F., and B. Nowack. 2011. The release of engineered nanomaterials to the environment. *Journal of Environmental Monitoring* 13(5): 1145.

Gottschalk, F., T. Sonderer, R.W. Scholz, and B. Nowack. 2009. Modeled environmental concentrations of engineered nanomaterials (TiO_2, ZnO, Ag, CNT, Fullerenes) for different regions. *Environmental Science & Technology* 43(24): 9216–9222.

Gottschalk, F., T.Y. Sun, and B. Nowack. 2013. Environmental concentrations of engineered nanomaterials: Review of modeling and analytical studies. *Environmental Pollution* 181: 287–300.

Greßler, S., and M. Nentwich. 2012. Nano and environment–Part II: Hazard potentials and risks. *Nano Trust Dossiers* 27: 1–6.

Guinée, J., and R. Heijungs. 1993. A proposal for the classification of toxic substances within the framework of life cycle assessment of products. *Chemosphere* 26(10): 1925–1944.

Heijungs, R. et al. 1992. *Environmental Life Cycle Assessment of Products—Guide and Backgrounds.* Leiden, the Netherlands.

Hellweg, S. et al. 2005. Confronting workplace exposure to chemicals with LCA: Examples of trichloroethylene and perchloroethylene in metal degreasing and dry cleaning. *Environmental Science & Technology* 39(19): 7741–77448.

Hellweg, S. et al. 2009. Integrating human indoor air pollutant exposure within life cycle impact assessment. *Environmental Science & Technology* 43(6): 1670–1679.

Hischier, R. et al. 2015. Life cycle assessment of façade coating systems containing manufactured nanomaterials. *Journal of Nanoparticle Research* 17(2): 68.

Hischier, R., and T. Walser. 2012. Life cycle assessment of engineered nanomaterials: State of the art and strategies to overcome existing gaps. *Science of the Total Environment* 425: 271–282.

Hodas, N. et al. 2015. Indoor inhalation intake fractions of fine particulate matter: Review of influencing factors. *Indoor Air* 26(6):836–856.

Hofstetter, P., T. Baumgartner, and R.W. Scholz. 2000. Modelling the valuesphere and the ecosphere: Integrating the decision makers' perspectives into LCA. *The International Journal of Life Cycle Assessment* 5(3): 161–175.

Hristozov, D. et al. 2016. Demonstration of a modelling-based multi-criteria decision analysis procedure for prioritisation of occupational risks from manufactured nanomaterials. *Nanotoxicology* 1–14. doi:10.31 09/17435390.2016.1144827.

Huijbregts, M., M. Hauschild, O. Jolliet, M. Margni, T. McKone, R.K. Rosenbaum, and D. van de Meent. 2010. *USEtox User Manual v1.01,* usetox@usetox.org (accessed December 15, 2017).

Huijbregts, M.A.J. et al. 2000. Priority assessment of toxic substances in life cycle assessment. Part I: Calculation of toxicity potentials for 181 substances with the nested multi-media fate, exposure and effects model USES–LCA. *Chemosphere* 41(4): 541–573.

Huijbregts, M.A.J., L.J.A. Rombouts, A.M.J. Ragas, and D. van de Meent. 2005. Human-toxicological effect and damage factors of carcinogenic and noncarcinogenic chemicals for life cycle impact assessment. *Integrated Environmental Assessment and Management* 1(3): 181.

International Organization for Standardization. 2006. *ISO 14040—Environmental Management—Life Cycle Assessment—Principles and Framework.* Geneva, Switzerland.

Jolliet, O. et al. 2003. IMPACT 2002+: A new life cycle impact assessment methodology. *The International Journal of Life Cycle Assessment* 8(6): 324–330.

Keller, A.A., S. McFerran, A. Lazareva, and S. Suh. 2013. Global life cycle releases of engineered nanomaterials. *Journal of Nanoparticle Research* 15(6): 1–17.

Keller, A.A., W. Vosti, H. Wang, and A. Lazareva. 2014. Release of engineered nanomaterials from personal care products throughout their life cycle. *Journal of Nanoparticle Research* 16(7): 2489.

Kikuchi, E., Y. Kikuchi, and M. Hirao. 2012. Monitoring and analysis of solvent emissions from metal cleaning processes for practical process improvement. *Annals of Occupational Hygiene* 56(7): 829–842.

Kleinstreuer, C., Z. Zhang, and Z. Li. 2008. Modeling airflow and particle transport/deposition in pulmonary airways. *Respiratory Physiology & Neurobiology* 163(1–3): 128–138.

Kuempel, E.D., L.M. Sweeney, J.B. Morris, and A.M. Jarabek. 2015. Advances in inhalation dosimetry models and methods for occupational RA and exposure limit derivation. *Journal of Occupational and Environmental Hygiene* 12(sup1): S18–S40.

Levine, K.E. et al. 2003. Development and validation of a high-throughput method for the determination of titanium dioxide in rodent lung and lung-associated lymph node tissues. *Analytical Letters* 36(3): 563–576.

Li, D. et al. 2015. *In vivo* biodistribution and physiologically based pharmacokinetic modeling of inhaled fresh and aged cerium oxide nanoparticles in rats. *Particle and Fibre Toxicology* 13(1): 45.

Mackay, D. 2001. *Multimedia Environmental Models: The Fugacity Approach, Second Edition*. Boca Raton, FL: CRC Press.

Mackay, D., S. Paterson, and W.Y. Shiu. 1992. Generic models for evaluating the regional fate of chemicals. *Chemosphere* 24(6): 695–717.

MacLeod, M. et al. 2001. BETR North America: A regionally segmented multimedia contaminant fate model for North America. *Environmental Science and Pollution Research* 8(3): 156.

McKone, T., D. Bennett, and R. Maddalena. 2001. *CalTOX 4.0 Technical Support Document*. Berkeley, CA.

Meesters, J.A.J. et al. 2014. Multimedia modeling of engineered nanoparticles with simpleBox4nano: Model definition and evaluation. *Environmental Science & Technology* 48(10): 5726–5736.

Meijer, A., M. Huijbregts, and L. Reijnders. 2005. Human health damages due to indoor sources of organic compounds and radioactivity in life cycle impact assessment of dwellings—Part 1: Characterisation factors (8 Pp). *The International Journal of Life Cycle Assessment* 10(5): 309–316.

Miseljic, M., and S.I. Olsen. 2014. LCA of engineered nanomaterials: A literature review of assessment status. *Journal of Nanoparticle Research* 16(6): 1–33.

Molina, D.K., and V.J.M. DiMaio. 2012a. Normal organ weights in men. *The American Journal of Forensic Medicine and Pathology* 33(4): 368–372.

Molina, D.K., and V.J.M. DiMaio. 2012b. Normal organ weights in men. *The American Journal of Forensic Medicine and Pathology* 33(4): 362–367.

Monteiro-Riviere, N.A. et al. 2011. Safety evaluation of sunscreen formulations containing titanium dioxide and zinc oxide nanoparticles in UVB sunburned skin: An *in vitro* and *in vivo* study. *Toxicological Sciences* 123(1): 264–280.

National Institute for Public Health and the Environment (RIVM). 2015. *Multiple-Path Particle Dosimetry (MPPD v3.01): A Model for Human and Rat Airway Particle Dosimetry*. Bilthoven, the Netherlands: Centre for Health Protection.

Newman, M.D., M. Stotland, and J.I. Ellis. 2009. The safety of nanosized particles in titanium dioxide–and zinc oxide–based sunscreens. *Journal of the American Academy of Dermatology* 61(4): 685–692.

Nicas, M. 1996. Estimating exposure intensity in an imperfectly mixed room. *American Industrial Hygiene Association Journal* 57(6): 542–550.

Oberdorster, G., J. Ferin, and B.E. Lehnert. 1994. Correlation between particle size, *in vivo* particle persistence, and lung injury. *Environmental Health Perspectives* 102(SUPPL. 5): 173–179.

Phalen, R.F., and R.N. Phalen. 2011. Important properties of air pollutants. In *Introduction to Air Pollution Science*. 1st ed. M. Gartside (Ed.). Burlington, NJ: Jones & Bartlett Learning.

Pini, M. et al. 2016. Human health characterization factors of nano-TiO_2 for indoor and outdoor environments. *The International Journal of Life Cycle Assessment*. doi:10.1007/s11367-016-1115-8.

Praetorius, A. et al. 2014. The road to nowhere: Equilibrium partition coefficients for nanoparticles. *Environmental Science: Nano* 1(4): 317.

Ramachandran, G.Y. 2005. *Occupational Exposure Assessment for Air Contaminants*. Boca Raton, FL: CRC Press.

Rosenbaum, R.K. et al. 2008a. USEtox—the UNEP-SETAC toxicity model: Recommended characterisation factors for human toxicity and freshwater ecotoxicity in life cycle impact assessment. *The International Journal of Life Cycle Assessment* 13(7): 532–546.

Rosenbaum, R.K. et al. 2008b. USEtox—the UNEP-SETAC toxicity model: Recommended characterisation factors for human toxicity and freshwater ecotoxicity in life cycle impact assessment. *The International Journal of Life Cycle Assessment* 13(7): 532–546.

Rosenbaum, R.K. et al. 2015. Indoor air pollutant exposure for life cycle assessment: Regional health impact factors for households. *Environmental Science & Technology* 49: 12823–12831.

Sadrieh, N. et al. 2010. Lack of significant dermal penetration of titanium dioxide from sunscreen formulations containing nano- and submicron-size TiO_2 particles. *Toxicological Sciences* 115(1): 156–166.

Salieri, B., S. Righi, A. Pasteris, and S.I. Olsen. 2015. Freshwater ecotoxicity characterisation factor for metal oxide nanoparticles: A case study on titanium dioxide nanoparticle. *The Science of the Total Environment* 505: 494–502.

Salmasi, A.-M., and A.S. Iskandrian 1993. Cardiac output: Physiological concepts. In *Cardiac Output and Regional Flow in Health and Disease*. A.-M. Salmasi and A.S. Iskandrian (Eds.). Dordrecht, the Netherlands: Springer.

Sonnemann, G., and S. Valdivia. 2014. The UNEP/SETAC life cycle initiative. In Springer, Dordrecht, the Netherlands, pp. 107–144. doi:10.1007/978-94-017-8697-3_4.

Stone, V. 2009. *Engineered Nanoparticles: Review of Health and Environmental Safety*. Edinburgh, UK: ENRHES Coordination and Support Action, Edinburgh Napier University.

Sun, T. Y., F. Gottschalk, K. Hungerbühler, and B. Nowack. 2014. Comprehensive probabilistic modelling of environmental emissions of engineered nanomaterials. *Environmental Pollution* 185: 69–76.

Sundell, J., H. Levin, and D. Novosel. 2006. *Ventilation Rates and Health: Report of an Interdisciplinary Review of the Scientific Literature*. Alexandria, Egypt.

Tsang, M.P. et al. 2017. Probabilistic RA of emerging materials: Case study of titanium dioxide nanoparticles. *Nanotoxicology* 11(4): 558–568.

United Nations Environmental Programme. 2003. *Evaluation of Environmental Impacts in Life Cycle Assessment*. Paris, France.

Vance, M.E. et al. 2015. Nanotechnology in the real world: Redeveloping the nanomaterial consumer products inventory. *Beilstein Journal of Nanotechnology* 6: 1769–1780.

Walser, T. et al. 2015. LCA framework for indoor emissions of synthetic nanoparticles. *Journal of Nanoparticle Research* 17(245): 1–18.

Walser, T., R. Juraske, E. Demou, and S. Hellweg. 2014. Indoor exposure to toluene from printed matter matters: Complementary views from life cycle assessment and RA. *Environmental Science & Technology* 48(1): 689–697.

Wenger, Y., D. Li, and O. Jolliet. 2012. Indoor intake fraction considering surface sorption of air organic compounds for life cycle assessment. *The International Journal of Life Cycle Assessment* 17(7): 919–931.

Wenzel, H., M.Z. Hauschild, and L. Alting. 1997. *Environmental Assessment of Products*. London, UK: Springer U.S.

WHO. 2014. *WHO|Metrics: Disability-Adjusted Life Year (DALY)*. Geneva, Switzerland.

Yokel, R.A., and R.C. MacPhail. 2011. Engineered nanomaterials: Exposures, hazards, and risk prevention. *Journal of Occupational Medicine and Toxicology* 6(1): 7.

8 Life-Cycle Criticality Assessment

Eskinder Demisse Gemechu and Guido Sonnemann

CONTENTS

8.1 INTRODUCTION

The use of industrial minerals has increased by factor of 27, from the early 1900s to 2005 (Krausmann et al. 2009). The number of metals utilized in different technologies also expanded from a few to almost the full range of elements in the periodic table. If we look at the development of a computer chip in the last two decades, the spectrum of metal used has increased from 12 to 70 (National Research Council 2008). Such a remarkable increase in the dynamism of metal utilization, together with their potential supply restriction, results in price fluctuation. These situations have recently drawn the attention of material scientists, environmental practitioners and researchers, decision-makers at different organizational levels and the general public to the issue of the criticality assessment of raw materials (Erdmann and Graedel 2011), as introduced in Chapter 5.

Since its development, the direct impact from the use of natural resources, referred as to *Resource Depletion*, has been an integral part of life-cycle assessment (LCA) (Jolliet et al. 2004). A number of life-cycle impact assessment (LCIA) methods have been developed to assess resources depletion as an impact category. However, the way how each method addresses the impacts of resources in general, and mineral metals use in particular, has been one of the most controversial issues of the LCA community (Weidema et al. 2005). A lack of methodological consistency has hampered the development of widely acceptable indicators for resource use (Stewart and Weidema 2005; Wäger and Classen 2006; Yellishetty et al. 2009; Emanuelsson et al. 2013). This lack of consensus on how resource depletion should be handled urges for the development of a harmonized LCIA method for resource use (Eldh and Johansson 2006). The missing alignment among different LCIA methods for resource use impact comes not only from the differences in the nature of modeling but also from the differences in definitions and understandings of what the resource depletion is, what limits the access to resources and why there is a need to consider resources as an Area of Protection (AoP). Here is a strong paradox, as, in theory, all agree that what has to be protected is the access to a functional value of the resources. That means the services provided by the resources are what the society has to protect, not the resource for the sole value of its existence. However, in practice, most LCIA methods are only based on the geological availability of resources without any consideration of their functionality or of the multiple barriers for their access.

Recently, a strong interest has been raised on how to overcome the resource assessment issue through integrating the concept of criticality (see Chapter 5) within LCA under the life-cycle sustainability assessment (LCSA) framework (European Commission 2012; Peña 2013; Sonnemann 2013). This chapter explains why there is a need in LCA to integrate resource criticality and how to proceed to such integration. Furthermore, the chapter illustrates the practical aspect of the method with an electric vehicle case study.

8.2   RESOURCES IN LIFE-CYCLE IMPACT ASSESSMENT

In LCIA (introduced in Chapter 3), impacts associated with a product's life-cycle are modeled either as midpoint or endpoint indicators. Midpoints are problem-oriented indicators that are defined through developing an environmental mechanism to reflect the relative importance of emissions or resource extractions, e.g. acidification, global warming potentials, ozone depletion potential, resource depletion potential. Endpoints are damage-oriented indicators which are defined by linking midpoint indicators through a cause-effect chain to impacts at the endpoint level, which are referred to as areas of protection (AoP). AoP are the defining classes of endpoint category indicators society wants to protect, namely: Human Health, Natural Environment and Natural Resources (Udo de Haes et al. 1999; Guinée et al. 2002; Udo de Haes and Lindeijer 2002; Finnveden et al. 2009). They ensure a link between damages due to environmental interventions with societal values.

While impacts on human health and ecosystem service are well defined in LCA, the way how direct impacts, due to the use of functions of natural resources, are quantified and linked with societal values are controversial (Drielsma et al. 2016). This is mainly due to the fact that Natural Resources as an AoP is not clearly defined in LCA. Dewulf et al. (2015) provided a new conceptual framework to elaborate the definition of the Natural Resources AoP and to highlight how the direct impacts from the use of natural resources within LCA or other socioeconomic assessment tools are used to assess the sustainability implications of resource use under the LCSA framework. Dewulf et al.'s (2015) framework provides five perspectives that define safeguard subjects for the Natural Resources AoP in the LCIA: asset, global functions, provisioning capacity, supply chain and human welfare. The Natural Resources AoP represents elements that are physically extracted from nature and enter the economy for human use. Here, the term natural resources refers to abiotic and biotic resources, land and freshwater. Therefore, the damage on the Natural Resources AoP focuses on measuring the potential depletion in the future from their current use in supporting the economy. It has an anthropological view. Natural resources, such as minerals, have functional value to humans, but not existence value, therefore, their societal value is solely due to the possible function they provide to humans (could be linked to welfare or health) (Stewart and Weidema 2005). Hence, the direct impact from the use of natural resources then have to be modeled as the potential impact of their physical depletion or limited accessibility due to social and geopolitical related factors on human welfare (Finnveden 2005). Currently, direct impacts from the use of resources could be handled as socioeconomic impacts or environmental impacts; however, there are not yet clear boundaries between the environment and socioeconomic systems, and there is a need to clarify the boundaries between the ecosphere and socioeconomic systems. The current LCIA methods for impacts from natural resource use, hence, have several shortcomings in complying with the Natural Resources AoP. In most cases, their intrinsic value is considered, but fails to address the scarcity issue. The impact on resource is measured as a resource depletion, which is entirely based on estimating the limit to future geological availability as a result of current consumption trend. But the limitations on their accessibility due to economic, geopolitical, social and other circumstances are ignored. Therefore, there is a desperate need of a new perspective that integrates both the concept of criticality and LCA to give emphasis on a broader dimension of resource "availability." In this context, the next section discusses the methodological aspects of LCA and resource criticality integration to include other aspects, such as geopolitical and social components, to the environmental indicators to extend it to LCSA context.

8.3 INTEGRATION OF LIFE-CYCLE ASSESSMENT AND CRITICALITY ASSESSMENT

There is a growing concern over the sustainable supply of natural resources; therefore, it is among the top priorities in the political agendas of developed and economically emerging countries. We are living in the world where science has reached its greatest discoveries and high technological advancements, in most cases with the intent to reduce the environmental impact associated with their use and increase human welfare. However, with regard to natural resource use, the advancement of modern theology increases the spectrum of materials used from very few in the past to almost the full range of the periodic table. Advanced technologies are highly dependent on specific functional materials, for example, manganese for wind turbines and rare earth elements for electronic, optical and magnetic application technologies, are highly dependent on specific functional materials (Moss et al. 2013). Most of the functional materials are not substitutable in their main applications (Graedel et al. 2015). As a result of technological advancement and its high dependency on functional materials, there is an imbalance supply and demand of resources, which is usually reflected by their price fluctuation (Graedel et al. 2015). This has motivated the development of a criticality assessment as a systematic analytical tool to assess the risk incorporated with the raw material usage in different technologies. A detailed introduction on the criticality assessment of raw materials is given in Chapter 5. This section mainly focuses on the methodological integration of criticality assessment within LCA under the LCSA framework.

Resource criticality assessment aims at a qualitative assessment and displays an aggregation of different risks associated with their use at different economic levels (Graedel et al. 2012). It incorporates supply risks and the potential consequences of risks, considering all three sustainability aspects: economic, environmental and social, into account. Resource criticality assessment is based on material flow analysis. On the other hand, LCA has been an internationally recognized assessment tool to measure and evaluate the environmental burdens associated with the life-cycle of a product (Sonnemann et al. 2003, ISO 2006a, 2006b). It is based on the material and energy accounting principle at unit process level and linking the emissions output to a number of environmental problems through developing environmental mechanisms. The life-cycle inventory (LCI) data are finally translated to quantify the potential impact on human health, ecosystem service and natural resources. There are a number of well-established impact assessment methods and databases to estimate the damage on human health and the ecosystem (Stewart and Weidema 2005; Wäger and Classen 2006; Yellishetty et al. 2009 Emanuelsson et al. 2013); however, addressing the direct impact from the use of natural resources has been a controversial and debatable issue within the life-cycle community. However, there is an interest from both LCA practitioners and material scientists for the potential of an integrative approach of LCA and resource criticality under the LCA framework to meaningfully address the resource depletion issue.

The conceptual framework on the integration of resource criticality assessment with LCA under the LCSA framework is presented in Figure 8.1. A resource criticality method proposed by Graedel et al. (2012) has been used to demonstrate the framework. Some aspects in the criticality assessments are already covered by the current LCA framework. The components that are not well covered in LCA are identified and linked in the framework. From the environmental dimension, the LCI from unit process modeling that are linked with damage to human health and ecosystem quality could be handled by using a well-established method such as ReCiPe, whereas impacts on human resource area of protection from their direct use can be captured in the socioeconomic dimension as resource scarcity and accessibility (Sonnemann et al. 2015; Gemechu et al. 2016).

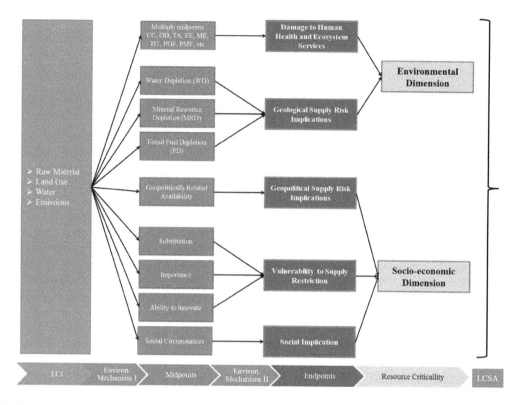

FIGURE 8.1 An LCSA framework for the integration of resource criticality and LCA. (Based on Sonnemann, G. et al., *J. Clean. Prod.*, 94, 20–34, 2015; Gemechu, E.D. et al., *J. Ind. Ecol.*, 20, 154–165, 2016.). The criticality assessment indicators. (Adapted from Graedel, T.E. et al., *Environ. Sci. Technol.*, 46, 1063–1070, 2012.)

8.4 GEOPOLITICALLY RELATED AVAILABILITY

From the resource criticality features integrated in the LCSA framework presented in Figure 8.1, the methodological aspect of geopolitical resource availability component is elaborated in this section, later followed by a case study applied to an electric vehicle. The geopolitical resource supply risk is a midpoint indicator that measures the potential supply disruption due to geopolitical constraints. The mathematical model to calculate the geopolitical-related supply risk, hereinafter referred to as *GeoPolRisk*, is described in Equation 8.1, derived in detail by Gemechu et al. (2016).

$$SR_{c,i} = \left[\left(\sum_{k=1}^{n} S_k^2 \right) * \left(\sum_{k=1}^{n} g_k * f_{i,k} \right) \right] \tag{8.1}$$

where:

$S_{c,i}$ expresses the supply risk of country i concerning the commodity of interest c

S_k is the global production share of country k in the global production (mining or refining) of the commodity c

g_k refers to the political instability indicator of country k derived from Worldwide Governance Indicators (WGI) of the World Bank

$f_{i,k}$ is an import share of country k in the supply chain of country i

The *GeoPolRisks* are then expressed as a socioeconomic risk-oriented midpoint indicator with values between 0 and 1, which can be interpreted as a share of the commodity supply being at risk, assuming a linear relation between WGI score and supply disruption probability.

The model has two basic components, the supply concentration and political stability as a WGI. The WGI scores are weighted by the import share so that they can reflect the regional particularities through the consideration of the typical supply chain patterns of each country. This allows attribution of the geopolitical risk depending on the instability of the most relevant import partners in addition to the market concentration. The latter reflects more the overall market ability to restructure trade flows in order to compensate decreased or disrupted sourcing from more instable countries. It is used as a multiplicative risk mitigation factor as described in the equation (Gemechu et al. 2016).

The main database consulted to calculate the *GeoPolRisk* are: global production distribution data from US Geological Survey database (USGS 2013), import share data from the UN Comtrade database (UN 2013) and the WGI data from the World Bank (2013).

8.5 ELECTRIC VEHICLE CASE STUDY

The main aim of the electric vehicle (EV) case study is to elaborate, with examples, about how the scope of conventional LCA can be broadened from use as an environmental assessment tool to include resource criticality as an additional dimension. The study shows the importance and practicability of integrating both environmental and criticality concepts within the LCSA framework with a common understanding of the Natural Resource AoP and the importance of social and economic dimensions as complementary to the environmental aspects. The case study is based on the structure and inventory data for a European representative first-generation battery small EV published by Hawkins et al. (2012), which provides a current, transparent and detailed LCI on this technology. The LCI data includes the energy and material requirements and associated emissions through the entire life-cycle of an EV per the functional unit, one EV. From the bill of materials, 14 important metal resources utilized in the production of an EV are considered both for the environmental and *GeoPolRisk* assessment. Their relative mass contribution is presented in Table 8.1.

TABLE 8.1
Main Resources Used in the Production of One EV

	Weight (kg)	Weight (%)
Steel and Iron (Fe)	839.124	71.45
Aluminum (Al)	206.222	17.56
copper (Cu)	125.338	10.67
Neodymium (Nd)	1.667	0.14
Other metals	0.800	0.07
Zinc (Zn)	0.462	0.04
Lead (Pb)	0.308	0.03
Magnesium (Mg)	0.240	0.02
Brass	0.231	0.02
Boron (B)	0.062	0.01
Bronze	0.014	0.00
Tin (Sn)	0.013	0.00
Nickel (Ni)	0.001	0.00
Chromium (Cr)	0.000	0.00
Gold (Au)	0.000	0.00
Silver (Ag)	0.000	0.00
Barium (Ba)	0.000	0.00
Platinum Group Metals (PGMs)	0.000	0.00

Source: Hawkins, T.R. et al., *Int. J. Life Cycle Assess.*, 17, 997–1014.

The environmental LCIA was conducted to identify the main contribution of each resource to selected midpoint impact categories using the ReCiPe midpoint method with Hierarchist perspective. They are global warming potential (GWP), metal depletion potential (MDP), human toxicity potential (HTP), and freshwater ecotoxicity potential (FETP). For the background system, ecoinvent database (ecoinvent v3.1) is consulted.

For the GeoPolRisk assessment, the method is applied to the following regions and countries: Australia, Canada, China, EU, France, Germany, Greece, India, Italy, Japan, Norway, UK and USA to determine the supply risk characterization factor specific to each country. These countries represent a reasonable diverse set of regions that include primary resource producers, manufacturers of EVs, and end-use economies of EVs.

The GeoPolRisk results for the 14 metal resources for each country are displayed in Figure 8.2 and Table 8.2 in line with Gemechu et al. (2017). Each country has a specific risk factor, which depends on the instability of important trade partners and the global supply concentration of the resource. For example, if we look at neodymium, the supply risk for Australia is relatively low due its high import share from USA, which accounts for more than 75%, since USA is a country with, comparatively, high political stability. However, the GeoPolRisks for Canada, India and Norway are high as their import is highly dependent on China, which is responsible for more than 95% of import share for each country. Here, it is worth mentioning that the GeoPolRisk calculation is exclusively applied to the imported volumes only, and it does not account for the domestically produced and consumed commodities, as they are not affected by geopolitical factors. However, it is important to consider them in assessing the overall supply risk and the potential consequence of supply restriction in a given economy, which is beyond the scope of this study.

Mg and Nd display the highest geopolitically related supply risk among all the 14 resources considered in this study. Both metals are among those 20 materials that were identified as critical from the list of 54 candidate materials by the recent EU report on critical materials (European Commission 2014). Nd is a widely used type of rare earth metal, which is applied as a permanent

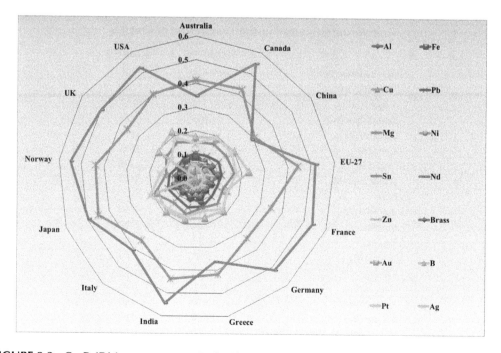

FIGURE 8.2 GeoPolRisk assessment results for 14 metal resources relevant to EVs. (From Gemechu, E. D. et al., *Int. J. Life Cycle Assess.*, 22, 31–39, 2017. With permission.)

TABLE 8.2
WGI Weighted GeoPolRisk Characterization Factor

Resources	AUS	CAN	CHN	EU-27	FRA	DEU	GRC	IND	ITA	JPN	NOR	UK	USA	Maz	Min
Al	0.10	0.08	0.09	0.08	0.08	0.07	0.11	0.09	0.08	0.09	0.08	0.07	0.08	0.11	0.07
Fe	0.09	0.07	0.07	0.09	0.07	0.06	0.08	0.08	0.07	0.09	0.06	0.07	0.08	0.09	0.06
Cu	0.06	0.05	0.06	0.07	0.05	0.05	0.07	0.07	0.06	0.07	0.04	0.06	0.06	0.07	0.04
Pb	0.10	0.11	0.12	0.11	0.10	0.10	0.12	0.13	0.11	0.13	0.11	0.09	0.10	0.13	0.09
Mg	0.41	0.43	0.31	0.44	0.35	0.34	0.42	0.44	0.35	0.45	0.44	0.36	0.40	0.45	0.31
Ni	0.02	0.04	0.05	0.05	0.04	0.05	0.04	0.06	0.04	0.06	0.04	0.04	0.04	0.06	0.02
Sn	0.15	0.16	0.16	0.17	0.13	0.14	0.13	0.16	0.14	0.19	0.13	0.15	0.17	0.19	0.13
Nd	0.35	0.55	0.29	0.52	0.54	0.52	0.36	0.54	0.41	0.50	0.55	0.50	0.53	0.55	0.29
Zn	0.07	0.07	0.07	0.07	0.07	0.06	0.06	0.07	0.07	0.08	0.05	0.05	0.06	0.08	0.05
Brass	0.04	0.05	0.08	0.10	0.07	0.07	0.06	0.06	0.08	0.07	0.05	0.06	0.05	0.10	0.04
Au	0.03	0.03	0.00	0.02	0.02	0.01	0.02	0.02	0.02	0.03	0.02	0.02	0.03	0.03	0.00
B	0.17	0.19	0.20	0.23	0.14	0.18	0.17	0.18	0.19	0.17	0.00	0.20	0.22	0.23	0.00
Pt	0.15	0.20	0.19	0.18	0.15	0.18	0.14	0.17	0.17	0.21	0.15	0.13	0.19	0.21	0.13
Ag	0.06	0.04	0.04	0.04	0.03	0.04	0.04	0.04	0.04	0.05	0.03	0.04	0.05	0.06	0.03

magnet, with high-performance magnetic properties. It is also used in electronic devices, such as computers, headphones, loudspeakers, wind turbines, and magnetic resonance imaging (MRI) equipment. Mg is used in the interior of vehicle instrument panels, where both strength and light weight are very essential. It is known for its high specific strength (strength-to-weight ratio) that makes it a very crucial resource in the automotive and aerospace industries. Other resources such as Fe, Al and Cu constitute the largest quantity of metals by mass, but do not exhibit supply risks. This is because their production is relatively widely distributed around the globe, and includes numerous low-risk country sources as measured by the WGI.

Figure 8.3 presents both the environmental and GeoPolRisk results per functional unit an EV (Gemechu et al. 2017). The results reflect only the production phase impacts, without taking into account the environmental burden from the use and end-of-life phases. The environmental impacts from the metals' primary production are due to mining, extraction and refining processes. Impacts from Cu, Al and Fe are high across different categories. The GeoPolRisk results in the last column display the relative criticalness among multiple resources by assessing them at a time. It can be seen: there is a clear difference between the GeoPolRisk and environmental impacts for different resources. While global impacts per functional unit are dominated by Cu, Al and Fe, Mn and Nd are most relevant to the GeoPolRisks. Due to the low mass requirements of Mn and Nd, their environmental impact contributions are not significant. They are utilized in very small amounts, 240 and 2 kg, respectively. More importantly, their contributions to resource impact are very low compared with other elements. The resource impact assessment in most existing LCIA methods is based on geological availability. Particularly in the ReCiPe method, the metal depletion at midpoint level is estimated as a marginal lower grade of a deposit, which results from a marginal increase in yield, caused by an extraction of the deposit. The metal depletion at endpoint, which is a damage measure on Natural Resource AoP, is defined as the additional costs society has to pay as a result of extraction (Goedkoop et al. 2009). The cost increase results from the need of additional effort to extract a certain amount of a resource in the future due to a reduction in its ore grade. The midpoint metal depletion results in Figure 8.3 (4th column), which is a traditional LCA indicator, shows that Cu and Fe (Steel) have the highest impact per functional unit. Metal depletion in the LCA also seems to follow mass, like the environmental measures; however, looking outside LCA, global resource concerns are dominated by PGMs, Ag, gold and Rh—and these do not appear in the LCA results due to their low mass contribution. This is because resource availability is not entirely determined by geological parameters, but also by other factors, such as socioeconomic, political, environmental regulation and others. The GeoPolRisk results (6th column) reflect the importance of considering

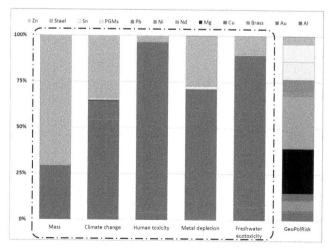

FIGURE 8.3 Comparison of the environmental versus the GeoPolRisk of an EV. (From Gemechu, E. D. et al., *Int. J. Life Cycle Assess.*, 22, 31–39, 2017. With permission.)

such other factors to meaningfully address the issue of resource accessibility. The metal depletion results are based on global availability, and attempt to assess resource scarcity for mid- and long-term time perspective without regarding short-term availability (accessibility). The dominancy of neodymium and magnesium in the GeoPolRisk indicator, on the other hand, points to the higher criticality of these resources with regard to their accessibility compared with other resources.

The case study advances the use of the GeoPolRisk assessment method. Resource challenges in traditional LCA are usually addressed from long-term time perspective. Short-term time perspective can be better addressed by including social and geopolitical factors, in addition to the conventional indicators which are based on their geological availability. This is more significant for modern technologies, such as electronic devices, in which critical resources contribute to important components. The EV case study elaborates the important contribution of the GeoPolRisk assessment as it provides a broader socioeconomic dimension, which could be added to conventional environmental LCA. While all environmental indicators tend to point to the same hotspots that arise from considerable usage of resources in vehicle production, the GeoPolRisk indicator highlights the importance of key elements that are used in very small quantities, but are crucial to the overall LCSA. Such an integrated assessment provides a more comprehensive result as it allows magnification of metals, such as Nd, Mn, and PGMs, that provide special functions, but are utilized in small amounts, so that they are usually left out of the traditional LCA assessment due to their low environmental significance. Such a wide-ranging presentation of environmental and GeoPolRisk indicators can provide more information that is helpful for both businesses and policy-makers to make an informed decisions.

8.6 QUESTIONS AND EXERCISES

1. What are the areas of protections? Explain how they have been addressed in the traditional LCA.
2. Why is the natural resource issue so important nowadays?
3. What is LCIA? Why is the direct impact from the use of natural resources controversial in LCA?
4. Explain the difference between geological and geopolitical resource availability.
5. Discuss how a supply disruption of a resource could affect the deployment of advanced technology?
6. What is geopolitical-related supply risk?
7. What are the main factors affecting GeoPolRisk?
8. Explain the complementariness of environmental impacts and resource criticality.
9. Why are the environmental impacts of Nd and Mg small, whereas their GeoPolRisks are high?
10. List and discuss the geopolitics of critical resources in your country or continent.

REFERENCES

Dewulf, J. et al. (2015). Rethinking the area of protection "natural resources" in life cycle assessment. *Environmental Science & Technology* 49(9): 5310–5317.
Drielsma, J. A. et al. (2016). Mineral resources in life cycle impact assessment defining the path forward. *The International Journal of Life Cycle Assessment* 21(1): 85–105.
Eldh, P. and J. Johansson. (2006). Weighting in LCA based on ecotaxes—Development of a mid-point method and experiences from case studies. *The International Journal of Life Cycle Assessment* 11(1): 81–88.
Emanuelsson, A. et al. (2013). Recommended assessment framework, method and characterisation and normalisation factors for resource use impacts: Phase 1, LC-IMPACT.
Erdmann, L. and T. E. Graedel. (2011). Criticality of non-fuel minerals: A review of major approaches and analyses. *Environmental Science & Technology* 45(18): 7620–7630.
European Commission. (2012). *Security of Supply and Scarcity of Raw Materials: Towards a Methodological Framework for Sustainability Assessment*. Luxembourg, Joint European Centre—Institute for Environment and Sustainability.

European Commission. (2014). Report on critical raw materials for the EU—Report on the Ad hoc Working Group on defining critical raw materials. Brussels, Belgium, EU.

Finnveden, G. (2005). The resource debate needs to continue. *The International Journal of Life Cycle Assessment* 10(5): 372–372.

Finnveden, G. et al. (2009). Recent developments in life cycle assessment. *Journal of Environmental Management* 91(1): 1–21.

Gemechu, E. D. et al. (2016). Import-based indicator for the geopolitical supply risk of raw materials in life cycle sustainability assessments. *Journal of Industrial Ecology* 20: 154–165.

Gemechu, E. D. et al. (2017). Geopolitical-related supply risk assessment as a complement to environmental impact assessment: The case of electric vehicles. *International Journal of Life Cycle Assessment* 22(1): 31–39.

Goedkoop, M. et al. (2009). ReCiPe 2008: A life cycle impact assessment method which comprises harmonised category indicators at the midpoint and the endpoint level. First ed., *Report I. Characterisation Ministry of Housing, Spatial Planning and Environment*, The Hague, the Netherlands.

Graedel, T. E. et al. (2012). Methodology of metal criticality determination. *Environmental Science & Technology* 46(2): 1063–1070.

Graedel, T. E. et al. (2015). On the materials basis of modern society. *Proceedings of the National Academy of Sciences* 112(20): 6295–6300.

Guinée, J. B. et al. (2002). *Handbook on Life Cycle Assessment: Operational Guide to the ISO Standards.* London, UK, Kluwer Academic Publishers.

Hawkins, T. R. et al. (2012). Environmental impacts of hybrid and electric vehicles—A review. *The International Journal of Life Cycle Assessment* 17(8): 997–1014.

ISO. (2006a). ISO 14040 international standard. In: *Environmental Management—Life Cycle Assessment—Principles and Framework*. Geneva, Switzerland, International Organisation.

ISO. (2006b). ISO 14040 international standard. In: *Environmental Management—Life Cycle Assessment—Requirements and Guidelines*. Geneva, Switzerland, International Organisation.

Jolliet, O. et al. (2004). The LCIA midpoint-damage framework of the UNEP/SETAC life cycle initiative. *The International Journal of Life Cycle Assessment* 9(6): 394–404.

Krausmann, F. et al. (2009). Growth in global materials use, GDP and population during the 20th century. *Ecological Economics* 68(10): 2696–2705.

Moss, R. L. et al. (2013). The potential risks from metals bottlenecks to the deployment of Strategic Energy Technologies. *Energy Policy* 55(Supplement C): 556–564.

National Research Council. (2008). *Minerals, Critical Minerals, and the U.S. Economy.* Washington, DC, The National Academies Press.

Peña, C. (2013). A geopolitical model for the implementation of Life Cycle Thinking based methodology in Latin America. *6th International Conference on Life Cycle Management—LCM 2013*. Gothenburg, Sweden.

Sonnemann, G. (2013). Geopolitical implications of life cycle assessment. *SETAC Europe Annual Meeting*. Glasgow, UK, SETAC.

Sonnemann, G. et al. (2003). *Integrated Life-Cycle and Risk Assessment for Industrial Processes*. Boca Raton, FL, Lewish Publishers.

Sonnemann, G. et al. (2015). From a critical review to a conceptual framework for integrating the criticality of resources into Life Cycle Sustainability Assessment. *Journal of Cleaner Production* 94(Supplement C): 20–34.

Stewart, M. and B. P. Weidema. (2005). A consistent framework for assessing the impacts from resource use—A focus on resource functionality. *The International Journal of Life Cycle Assessment* 10(4): 240–247.

The World Bank. (2013). Worldwide governance indicators. Retrieved from http://info.worldbank.org/gover nance/wgi/index.aspx#home (accessed December 1, 2014).

Udo de Haes, H. A. and E. Lindeijer. (2002). The conceptual structure of life-cycle impact assessment. *Life-Cycle Impact Assessment: Striving Towards Best Practice*. H. A. Udo de Haes, G. Finnveden, M. Goedkoop et al. (Eds.), Pensacola, FL, The Society of Environmental Toxicology and Chemistry (SETAC), pp. 209–225.

Udo de Haes, H. A. et al. (1999). Best available practice regarding impact categories and category indicators in life cycle impact assessment. *The International Journal of Life Cycle Assessment* 4(3): 167–174.

UN. (2013). Commodity trade statistics database. Retrieved from http://comtrade.un.org/db/default.aspx (accessed December 1, 2014).

USGS. (2013). *Mineral Commodity Summaries 2013*. Washington, DC, U.S. Geological Survey.

Wäger, P. and M. Classen. (2006). Metal availability and supply: The many facets of scarcity. *1st International Symposium on Material, Minerals, & Metal Ecology (MMME 06)*. Cape Town, South Africa.

Weidema, B. P. et al. (2005). Impacts from resource use—A common position paper. *The International Journal of Life Cycle Assessment* 10(6): 382–382.

Yellishetty, M. et al. (2009). Life cycle assessment in the minerals and metals sector: A critical review of selected issues and challenges. *The International Journal of Life Cycle Assessment* 14(3): 257–267.

9 Uncertainty Assessment by Monte Carlo Simulation[1]

*Guido Sonnemann, Luiz Kulay, Yolanda Pla,
Michael Tsang, and Marta Schuhmacher*

CONTENTS

---

[1] Extracts of this chapter are reprinted from *Environment International*, 28, Sonnemann, G.W., Pla, Y., Schuhmacher, M., and Castells, F., pp. 9–18, 2002a; and *Journal of Cleaner Production* 11, Sonnemann, G.W., Schuhmacher, M., and Castells, F., pp. 279–292, 2002b. Copyright 2002 with permission from Elsevier.

9.1   INTRODUCTION

The high uncertainty present in the implementation of life-cycle and environmental risk assessment studies introduces a crucial limitation when interpreting the environmental impact and damage estimations provided by these methodologies. Within this particular context, this chapter presents a strategic procedure to better deal with an uncertainty assessment based on the stochastic model of Monte Carlo (MC) simulation. Initially, we will present an overview about the importance of implementing uncertainty assessments in studies of life-cycle assessment (LCA) and impact pathway analysis (IPA). Then, the basic statistical concepts related to MC simulation are introduced. This section also compares some of the best known commercial software packages that apply MC simulation to uncertainty evaluation. The third section describes the core of the uncertainty assessment strategic procedures. Finally, in the last section, we will use the municipal solid waste incinerator (MSWI) process in Tarragona, Spain, as the practical case study of risk assessment for explaining the procedure better.

9.2   TYPES OF UNCERTAINTIES IN ENVIRONMENTAL IMPACT ANALYSIS

The various sources of uncertainty in environmental impact analysis can be systematically classified in accordance with the following categories (Huijbregts, 1998).

9.2.1   PARAMETER UNCERTAINTY

A life-cycle inventory (LCI) analysis and the models that calculate fate, exposure and effect within an impact and risk assessment usually need a large amount of data. Uncertainty of these parameters reflects directly on the outcome of any environmental impact method. Empirical inaccuracy (imprecise measurements), unrepresentative data (incomplete or outdated measurements) and lack of data (no measurements) are common sources of parameter uncertainty. Weidema and Wesnæs (1996) describe a comprehensive procedure for estimating combined inaccurate and unrepresentative LCI data qualitatively and quantitatively. Although this procedure may substantially improve the credibility of LCA outcomes, uncertainty analyses are generally complicated by a lack of knowledge of uncertainty distributions and correlations among parameters.

9.2.2   MODEL UNCERTAINTY

According to conclusions presented by various authors, the predicted values for environmental impact and risk generally respond in a linear manner to the amount of emitted pollutant. Moreover, in life-cycle impact assessment (LCIA) and IPA, thresholds for environmental interventions are disregarded. Additionally, in LCIA the derivation of characterization factors causes model uncertainty because these are calculated with the aid of simplified environmental models without considering spatial and temporal characteristics.

9.2.3   UNCERTAINTY DUE TO CHOICES

In many cases, performing choices is unavoidable in environmental impact analysis. Considering the step of LCI from LCA, examples of choices leading to uncertainty include the selection of the functional unit (or definition of the allocation procedure for multioutput processes), multiwaste processes and open-loop recycling. Moreover, the socioeconomic evaluation step in LCA and IPA is an area in which choices play a crucial role. Although experts from the social sciences have suggested many different weighting schemes, only a few are operational and no general agreement exists as to which one should be preferred.

9.2.4   SPATIAL AND TEMPORAL VARIABILITY

In most LCAs, environmental interventions are summed up regardless of their spatial context, thus introducing model uncertainty. Temporal variations, in turn, are present in LCI and other impact assessment methods. In general, variations of environmental interventions over a relatively short time period, such as differences in industrial emissions on weekdays vs. weekends or even short disastrous emissions, are not taken into account.

9.2.5   VARIABILITY AMONG SOURCES AND OBJECTS

In LCA or in other impact assessment methods, variability among sources and objects may influence the outcome of a study. This means, for example, that some variability in LCIs may result from differences in inputs and emissions of comparable processes within a product system (due to the use of different technologies in factories producing the same material). Furthermore, variability among objects exists in weighting environmental problems during impact assessment due to variability in human preferences. For instance, when a method such as the willingness to pay (WTP) method is used to determine the external environmental cost due to a specific damage, differences related to individual preferences cause inherent variation in the final result.

9.3   WAYS TO DEAL WITH DIFFERENT TYPES OF UNCERTAINTY

Huijbregts et al. (2000) have offered solutions on how to deal with the issues of uncertainty previously discussed. The tools available to address different types of uncertainty and variability in LCAs include probabilistic simulation, correlation and regression analysis, additional measurements, scenario modeling, standardization, expert judgment or peer review, and nonlinear modeling. Scenario modeling (Pesonen et al., 2000) should be especially useful in cases in which uncertainty about choices and temporal variability is present.

When a model suffers from large uncertainties, the results of a parameter uncertainty analysis may be misleading. In most cases, the consequence of decreasing model uncertainty will be the implementation of more parameters in the calculation, thereby increasing the importance of operationalizing parameter uncertainty in the model. In the following two sections, we present an overview of previous efforts to assess uncertainties in LCA.

9.3.1 EXPERIENCES TO ASSESS UNCERTAINTY IN LIFE-CYCLE ASSESSMENT

So far the influence of data quality on final results of LCA studies has rarely been analyzed. In spite of the lack of published case studies, several approaches to carry out this kind of evaluation have been proposed during recent years. Nevertheless, from a general point of view, the existing methods can be classified in qualitative and quantitative assessments.

Qualitative assessment means describing the data used by characterizing its quality. Weidema and Wesnaes (1996) and Weidema (1998) proposed using data quality indicators depending on categories like reliability, completeness, temporal correlation, etc. In turn, Finnveden and Lindfors (1998) suggested ranges for various inventory parameters as rules of thumb. Quantitative assessment means to quantify all inherent uncertainties and variations in an LCA. In order to perform this task, many different analytical procedures have been applied. For uncertainty analysis of LCI, Hanssen and Asbjornsen (1996) used statistical analysis, Ros (1998) proved the fuzzy logic, and Maurice et al. (2000), as well as Meier (1997), decided in favor of the stochastic methods. Regarding uncertainty assessment within the impact assessment stage of LCA, Meier (1997), Hofstetter (1998) and Huijbregts and Seppälä (2000) have reported results achieved using similar techniques. Even when it is strongly effective, quantitative assessment is continuously confronted with the problem that it is hardly possible to analyze all types of uncertainties.

9.3.2 FORMER UNCERTAINTY ASSESSMENT IN IMPACT PATHWAY ANALYSIS

The IPA is a quite complex approach and, hence, risks lack of reliability in the final results. In the same way as with other environmental analysis methods, uncertainty is the key problem that makes it difficult to convince decision-makers based on the outcomes of a study.

One of the most interesting experiences is that reported by Rabl and Spadaro (1999), in which they evaluated the uncertainty and variability of damages and costs of air pollution by means of analytical statistical methods. In this case, the authors observed that the equation for the total damage is largely multiplicative, even though it involves a sum over receptors at different sites. This conclusion comes from the principle of conservation of matter, which implies that overprediction of the dispersion model at one site is compensated for by underprediction at another; the net error of the total damage arises mostly from uncertainties in the rate at which the pollutant disappears from the environment.

In the same reference, the authors discuss the typical error distributions related to the factors in the equation for the total damage, in particular, those related to two key parameters: the deposition velocity of atmospheric dispersion models and the value of statistical life; according to Rabl and Spadaro (1999), these are close to log-normal. They conclude that a log-normal distribution for the total damage appears very plausible whenever the dose–response or exposure–response function is positive everywhere. As an illustration, they show results for several types of air pollution damage: health damage due to particles and carcinogens, damage to buildings due to SO_2 and crop losses due to O_3, in which the geometric standard deviation is in the range of 3–5. Results and conclusions such as those presented by Rabl and Spadaro illustrate the necessity of dealing with uncertainty assessment in IPA in spite of its high level of complexity.

9.4 INTRODUCTION TO MONTE CARLO SIMULATION

Based on the previously mentioned experiences regarding uncertainty analysis in LCA and ERA studies, especially to IPA, it seems that the use of a stochastic model helps to characterize the uncertainties better, rather than a pure analytical mathematical approach. This can be justified because the relevant parameters follow a different frequency distribution. In this case, one of the most widespread stochastic model uncertainty analyses is the Monte Carlo (MC) simulation. In a wide approach to perform an MC simulation, the parameters under evaluation must be specified as uncertainty distributions. The method makes all the parameters vary at random, because the variation is restricted by the given uncertainty distribution for each parameter. The randomly selected values from all the parameter uncertainty distributions are inserted in the output equation. Repeated calculations produce a distribution of the predicted output values reflecting the combined parameter uncertainties. According to LaGrega et al. (1994), MC simulation can be considered the most effective quantification method for uncertainties and variability among the environmental system analysis tools available.

The term simulation can be understood as an analytical method meant to imitate a real-life system, especially when other analyses are too mathematically complex or too difficult to reproduce. Without the aid of simulation, a spreadsheet model would only reveal a single outcome, generally the most likely or average scenario. Spreadsheet uncertainty analysis uses a spreadsheet model and simulation to analyze the effect of varying inputs or outputs of the modeled system automatically. The random behavior of how MC simulation selects variable values to simulate a model is similar to that employed by games of chance. When a player rolls a die, he or she knows that a 1, 2, 3, 4, 5 or 6 will come up, but does not know which will occur in any particular roll. It is the same with the variables that have a known range of values, but an uncertain value for any particular time or event (Decisioneering, 1996).

9.4.1 USE OF SOFTWARE PACKAGES TO PROPAGATE UNCERTAINTY BY PERFORMING MONTE CARLO SIMULATION

In order to simplify the task of determining the uncertainty of a parameter by MC simulation, various commercial software packages are available. Among them, Crystal Ball[2], @Risk[3], Analytica[4], Stella II[5], PRISM[6] and Susa-PC[7] can be highlighted. Table 9.1 summarizes the main information about each of these software packages.

According to Metzger et al. (1998), @Risk was originally designed for business applications, but it has found wide use in human health and ecological risk assessment. This package includes uncertainty in estimates to generate results that show all possible outcomes. Because of its standardized spreadsheet backbone, @Risk is easy to use without a need for extensive statistical knowledge, modeling capability or programming ability.

Analytica and Stella II are two standalone programs designed for a wide variety of applications. Analytica is a model-building program that attempts to simplify sophisticated systems with the use

TABLE 9.1
Uncertainty Analysis Software Commercial Packages

Software Package	Version	Producer	Year	URL Address
Crystal Ball	4.0	Decisioneering	1996	www.decisioneering.com
@Risk	3.1	Palisade Corporation	1996	www.palisade.com
Analytica	1.0 Beta 1	Lumina Decision Systems	1996	www.decisioneering.com
Stella II	3.0.6	High Performance System	1994	www.hps-inc.com
PRISM	November 1992	Gardner et al.	1992	www.senes.com
Susa-PC	1.0	Hofner et al.; GRS–Garching	1992	—

Source: Metzger, J.N. et al., *Hum. Ecol. Risk Assess.*, 4(2), 263–290, 1998. With permission.

[2] Registered trademark of Decisioneering, Inc., Denver, CO.
[3] Registered trademark by Palisade Corporation, Newfield, NY.
[4] Registered trademark of Decisioneering, Denver, CO.
[5] Registered trademark of High Performance Systems, Lebanon, NH.
[6] Registered trademark of SENES Oak Ridge, Oak Ridge, TN.
[7] Registered trademark of Gesellschaft für Anlagenund Reaktorsicherheit (GRS) mbH, Köln, Deutschland.

of multilevel influence diagrams. Stella II is a multilevel hierarchical environment for constructing and interacting with models. These two programs are designed to simplify complex problems and, therefore, require additional effort in learning the software and building the models. However, because these models are displayed in diagrams, their interpretation for other users tends to be easier than for any other software package.

PRISM and Susa-PC are Fortran-based codes, designed specifically for use in risk analysis. PRISM is a simple free-access program that builds distributions for input into any model, and then analyzes the output of that model. Susa-PC is a more involved package that offers advanced statistical analysis specific to risk analysis. It is Excel-based and uses macros for everything except the actual results from the model. Because of their relatively wide focus, PRISM and Susa-PC are more difficult to use than spreadsheets and require some knowledge of Fortran and statistics.

Finally, the software Crystal Ball Version 4.0 from Decisioneering (1996) is a simulation program that helps analyze the uncertainties associated with Microsoft Excel spreadsheet models by MC simulation. Crystal Ball adds probability to the best, worst and most likely case versions of the same model—which only predict the range of outcomes—and automates the what-if process. Crystal Ball is an add-on to Excel; the user does not need to leave Excel to undertake such forecasting. The program works with models existing already, so the calculations do not need to be recreated. As a fully integrated Excel add-on program with its own toolbar and menus, Crystal Ball picks up where spreadsheets end by allowing the user to perform the MC analysis. The user must define a range for each uncertain value in the spreadsheet, and Crystal Ball uses this information to perform thousands of simulations. Another function of this software is the sensitivity analysis. Sensitivity charts show how much influence each assumption has on the results, allowing the user to focus further analytical effort on the most important factors. The results are dynamically summarized in forecast charts that show all the outcomes and their likelihood. An example of the activated cells of an Excel sheet, the result of a sensitivity analysis, and the final provision are shown in the Crystal Ball screenshot (Figure 9.1).

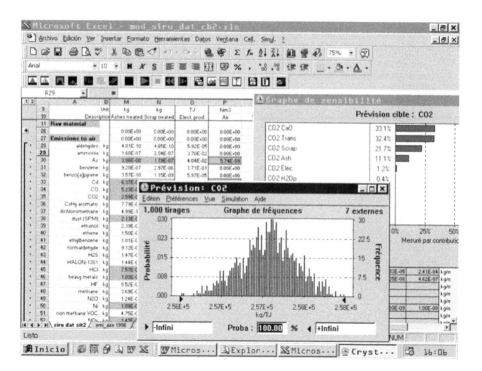

FIGURE 9.1 Crystal Ball screenshot (activated cells of an Excel spreadsheet, the result of a sensitivity analysis and the final provision).

9.4.2 FRAMEWORK FOR UNCERTAINTY ASSESSMENT BY MONTE CARLO SIMULATION IN ENVIRONMENTAL IMPACT ANALYSIS

Based on the information about previous studies on uncertainty evaluation in environmental impact analysis methods and the knowledge of the MC simulation technique, the following general strategy for the assessment of uncertainties in impact assessment studies may be established:

- Classification of data (extensively available, based on little information and data that can be ignored)
- Identification of probability distribution for considered data
- Monte Carlo simulation
- Sensitivity analysis
- Analysis and discussion of results

As mentioned briefly earlier, by means of a sensitivity analysis, it is possible to show which parameters are most relevant for the final result. If small modifications of one parameter characterized by a probability distribution strongly influence the final result, it can be concluded that the sensitivity of the considered variable is elevated for the relation between parameter and final result. This information is crucial for decision-makers in order to understand the variables to be acted upon. Also, it would be very useful to know the parameters that might be neglected, especially if it is difficult to get detailed information about them. Sensitivity can be analyzed by an approach that displays sensitivity as a percentage of contribution from each parameter to the variance of the final result. Crystal Ball Version 4.0, the software package selected to perform an example of application for the use of MC simulation, approximates this approach by lifting to square the correlation coefficients of ranks and normalizing them to 100%.

9.5 UNCERTAINTY ASSESSMENT IN DIFFERENT ENVIRONMENTAL IMPACT ANALYSIS TOOLS

In this section, methods of uncertainty assessment in different environmental impact analysis tools are presented.

9.5.1 UNCERTAINTY ASSESSMENT IN LIFE-CYCLE INVENTORY

Figure 9.2 presents an adaptation of the procedure for uncertainty analysis in LCI as it is reported in the literature (Meier, 1997; Maurice et al., 2000). The first step refers to the compilation of LCI data.

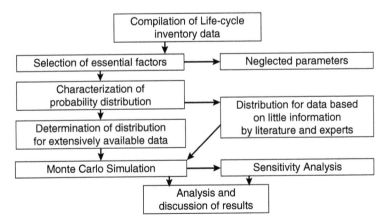

FIGURE 9.2 Procedure for the uncertainty and variability assessment in the life-cycle inventory. (Reprinted from *J. Cleaner Prod.*, 11, Sonnemann, G.W. et al., pp. 279–292, Copyright 2002b with permission from Elsevier.)

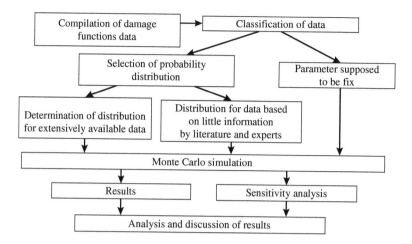

FIGURE 9.3   Framework for the assessment of uncertainty and variability in the impact pathway analysis.

If all the parameters that might have repercussions on the final result were considered, an exhaustive study would need to be carried out; however, not all these data are relevant. Hence, only the most relevant factors must be selected and, for some parameters, can be assumed to have fixed values. Once the essential factors have been selected, a characterization of the probability distributions is carried out. Therefore, the data are classified into two groups: extensively available data, for which average and standard deviation can be calculated, and data based on little information, for which literature and expert estimations must be considered. All these parameters feed the MC simulation, which gives the results in the form of a probability distribution around a mean value and allows a detailed sensitivity analysis to be carried out.

9.5.2   Uncertainty Assessment in Impact Pathway Analysis

Figure 9.3 presents the framework for uncertainty assessment in the IPA. The first step of the so-called framework is the compilation of damage function data, in which an exhaustive study must be carried out on all the parameters that have a repercussion on the final result. Although the model is processing an enormous quantity of data that are not all relevant, only fundamental facts need really be considered. Thus, a classification must be made among the most significant parameters, for which probability distributions should be defined. In addition, the parameters are supposed to be invariant and are called point estimates. Significant data are further classified into the above-mentioned two groups for advanced evaluation: extensively available data and data based on little information. In the same way as in the uncertainty assessment for LCI, these parameters feed the MC simulation that gives the results in the form of a probability distribution around a mean value, and allows a detailed sensitivity analysis to be carried out. The last step of the framework consists of the analysis and discussion of the achieved results.

9.5.3   Risk Characterization and Uncertainty Analysis

Risk assessment uses a wide array of information sources and models. Even when actual exposure-related measurements exist, assumptions or inferences will still be required. Most likely, data will not be available for all aspects of the exposure assessment and may be of questionable or unknown quality. In these situations, the exposure assessor will depend on a combination of professional judgment, inferences based on analogy with similar chemicals and conditions, estimation techniques and the like. The net result is that the exposure assessment will be based on a number of assumptions with varying degrees of uncertainty (U.S. Environmental Protection Agency, 1992).

Decision analysis literature has focused on the importance of explicitly incorporating and quantifying scientific uncertainty in risk assessment (Roseberry and Burmaster, 1991).

Several reasons lead to uncertainties concerning the validity and entirety of the results of a risk assessment. These uncertainties can be regarded in different manners and degrees depending on the methodology applied in the risk assessment process. One source of high uncertainties is the application of models that simulate the behavior of a pollutant in the environment and the uptake into the human body. Computer models that attempt to describe natural processes are always simplifications of a complex reality. They require the exclusion of some variables that, in fact, influence the results, but cannot be regarded because of increased complexity or lack of data. Moreover, many natural processes can only be approximated, but not exactly explained with mathematical correlations. Hence, a model is always affected with uncertainties and gives only an imperfect description of the reality. Different models for the same issue consider different uncertainties but disregard also different sources of uncertainty.

On the other hand, because many parameters in a model cannot be treated as fixed-point values, a range of values better represents them. This uncertainty of input parameters can result from real variability, measurement and extrapolation errors as well as the lack of knowledge regarding biological, chemical and physical processes. Uncertainties that are related to lack of knowledge or measurement and extrapolation errors can be reduced or eliminated with additional research and information. However, real parameter variability, e.g., spatial and temporal variation in environmental conditions or life-style differences, occurs always and cannot be eliminated. It leads to a persisting uncertainty of the modeling results.

Risk assessment is subject to uncertainty and variability. Specifically, uncertainty represents a lack of knowledge about factors affecting exposure or risk, whereas variability arises from true heterogeneity across people, places and time. In other words, uncertainty can lead to inaccurate or biased estimates, whereas variability can affect the precision of the estimates and the degree to which they can be generalized.

Now, let us consider a situation that relates to exposure, such as estimating the average daily dose by one exposure route—inhalation of contaminated air. Suppose that it is possible to measure an individual's daily air inhalation consumption (and concentration of the contaminant) exactly, thereby eliminating uncertainty in the measured daily dose. The daily dose still has an inherent day-to-day variability because of changes in the individual's daily air inhalation or concentration of the contaminants in air.

Clearly, it is impractical to measure the individual's dose every day. For this reason, the exposure assessor may estimate the average daily inhalation based on a finite number of measurements, in an attempt to "average out" the day-to-day variability. The individual has a true (but unknown) average daily dose, which has not been estimated based on a sample of measurements. Because the individual's true average is unknown, it is uncertain how close the estimate is to the true value. Thus, variability across daily doses has been translated into uncertainty in the parameter. Although the individual's true value has no uncertainty, the estimate of the value has some variability (U.S. Environmental Protection Agency, 1992).

The preceding discussion pertains to the air inhalation for one person. Now consider a distribution of air inhalation across individuals in a defined population (e.g., the general U.S. population). In this case, variability refers to the range and distribution of air inhalation across individuals in the population. Otherwise, uncertainty refers to the exposure assessor's state of knowledge about that distribution, or about parameters describing the distribution (e.g., mean, standard deviation, general shape, various percentiles).

As noted by the National Research Council (1994), the realms of variability and uncertainty have fundamentally different ramifications for science and judgment. For example, uncertainty may force decision-makers to judge how probable it is that exposures have been overestimated or underestimated for every member of the exposed population, whereas variability forces them to cope with a certainty that different individuals are subject to exposures above and below any of the exposure levels chosen as a reference point (U.S. Environmental Protection Agency, 1992).

To account for the uncertainty in ERA, process probabilistic models are used. These techniques generate distributions that describe the uncertainty associated with the risk estimate (resultant doses). The predicted dose for every 5th percentile to the 95th percentile of the exposed population and the true mean are calculated. Using these models, the assessor is not forced to rely solely on a single exposure parameter or the repeated use of conservative assumptions to identify the plausible dose and risk estimates. Instead, the full range of possible values and their likelihood of occurrence are incorporated into the analysis to produce the range and probability of expected exposure levels.

In addition to establishing exposure and risk distributions, probabilistic analysis can also identify variables with the greatest impact on the estimates and illuminated uncertainties associated with exposure variables through sensitivity analysis. This provides some insight into the confidence that resides in exposure and risk estimates and has two important results. First, it identifies the inputs that would benefit most from additional research to reduce uncertainty and improve risk estimates. Second, assuming that a thorough assessment has been conducted, it is possible to phrase the results in more accessible terms, such as, the risk assessment of polychlorinated biphenyls (PCBs) in small-mouth bass is based on a large amount of high-quality reliable data, and we have high confidence in the risk estimates derived. The analysis has determined that 90% of the increased cancer risk could be eliminated through a ban on carp and catfish, but there is no appreciable reduction in risk from extending such a ban to bass and trout.

9.6   TYPES OF PROBABILITY DISTRIBUTIONS USED

In the MC simulation, new values of the random variables are selected at least 10,000 times, and a new estimate of the final damage is foreseen. The results of the calculations are summarized in a single histogram of damage values; mathematical operations such as multiplication, exponential functions, matrix calculations, etc. can be managed. Among the wide range of statistical distributions (normal, log-normal, uniform, etc.) found in MC simulation, we will refer only to the most common types of probability distributions, which are:

1. *Normal distribution* (Figure 9.4). Normal distribution is appropriate to describe the uncertainties of large samples that constitute stochastic events and are symmetrically distributed around the mean. The mean and the standard deviation will define the probability density function. The normal distribution is especially appropriate if data uncertainties are given as a percentage of the standard deviation with respect to the mean, i.e., the coefficient of variation (CV).
2. *Log-normal probability distribution* (Figure 9.5). This type of distribution can be used if large numbers of quantities must be presented, no negative values are possible and the variance is characterized by a factor rather than a percentage.
3. The 50th percentile of a log-normal distribution is related to the mean of its corresponding normal distribution. The log-normal distribution is calculated assuming that the logarithm of the variable has a normal distribution. Many environmental impacts follow the log-normal model. The geometric mean, μg, and the geometric standard deviation, σ_g, of the samples are very practical and correspond to the mean and coefficient of variation for the normal distribution. Moreover, they provide multiplicative confidence intervals such as:
   $[\mu_g/\sigma_g, \mu_g.\sigma_g]$ for a confidence interval of 68%
   $[\mu/\sigma_g^2, \mu_g.\sigma_g^2]$ for a confidence interval of 95%

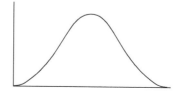

FIGURE 9.4 Normal probability distribution profile.

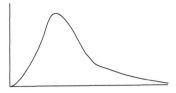

FIGURE 9.5   Log-normal probability distribution profile.

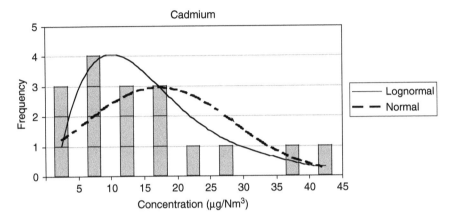

FIGURE 9.6   Log-normal distribution of cadmium emissions by adjustment of sample points in a density function. (Reprinted for *Environ. Int.*, 28, Sonnemann, G.W. et al., pp. 9–18, Copyright 2002a with permission from Elsevier.)

The way to calculate the probability distribution from an enormous amount of experimental data will be explained here with a brief example:

The Crystal Ball software facilitates the adjustment of sample points in a density function. As an example, Figure 9.6 shows the variation of such sample points for different measurements of cadmium emissions. In the diagram, 17 measurements are classified according to their range of concentration. The most frequent value is the 10 μg/Nm^3 that appears four times. However, two samples have more than 35 μg/Nm^3. The presented variation has been adjusted with normal and log-normal distribution. The different curves make evident that the log-normal distributions fit the variation of the measurements much better.

As previously explained, in the case study from Tarragona's MSWI, the variation of the pollutant concentrations in the incinerator emissions is enormous due to the heterogeneity of the incinerated waste. As can be expected, its elementary composition varies strongly at each moment. Often cadmium concentrations are low, but sometimes these increase due to the elevated cadmium amount in the waste. Thus the measured emissions are not constant over time, following a log-normal distribution.

9.7   EXAMPLE: RISK ASSESSMENT TO PCDD/Fs IN TARRAGONA, SPAIN, USING THE MONTE CARLO APPROACH

Returning to the case of the MSWI and its application of the risk assessment seen in Chapter 4, now, the same case—not as point estimation, but considering its distribution—will be shown as an example of uncertainty assessment. Applying the uncertainty analysis to the risk assessment of the population living in its vicinity will be demonstrated. Only the "direct risk" due to air emissions will be considered. To account for variability and uncertainty, the Monte Carlo simulation will be applied to estimate the set of risk estimates. In this example, model and data uncertainty will be considered. The sensitivity analysis will show how much each predictor variable contributed to the uncertainty or variability of the predictions.

9.7.1   DETERMINATION OF DISTRIBUTION FUNCTIONS

The determination of which form of distribution function to assign to each parameter depends on site-specific data and judgment based on statistical analysis. The distribution employed in this example is assembled from site-specific data, data existing in the most current literature and professional judgment; they are considered to be the most up-to-date description of the parameter (Katsumata and Kastenberg, 1997). In the vegetation production in the area of study (Tarragona), only the adult population was considered. In this example, only exposure through the air, soil and 10% of the consumed vegetation will be considered as a direct means of exposure. Because the MSWI is located close to a city, no impact to foods like meat, fish or dairy products will be considered.

Tables 9.2 and 9.3 show a description of the Monte Carlo parameter distribution for risk assessment evaluation due to direct exposure for people living in the area surrounding the MSWI

TABLE 9.2
Monte Carlo Parameter Distributions for Direct Exposure for Population Living in Areas Surrounding the MSWI

Parameter	Symbol	Units	Type	Distribution[a]	Reference
Soil ingestion rate	SIR	mg/day	Log-normal	3.44 ± 0.80	LaGrega (1994)
Fraction absorption ingestion of soils	AFIS	Unit-less	Point	40	Nessel et al. (1991)
Vegetable ingestion rate	VIR	g/day	Log-normal	99 ± 80	Arija et al. (1996)
Fraction absorption ingestion of vegetables	AFIV	Unit-less	Point	60	Nessel et al. (1991)
Fraction vegetables from the area	IA	Unit-less	Uniform	1–10	Generalitat Catalunya (Statistical Dept.)[b]
Re-suspended particles from the soil	RES	Unit-less	Point	50	Hawley (1985)
Ventilation rate	Vr	m³/day	Log-normal	2–20	Shin (1998)
Fraction retained in the lungs	RET	Unit-less	Uniform	60	Nessel et al. (1991)
Particle concentration	Pa	µg/m³	Point	133	Generalitat Catalonia (Environmental Dept.)[b]
Fraction absorption inhalation	AFIn	Unit-less	Point	100	Nessel et al. (1991)
Contact time soil–skin	CT	h/day	Uniform	1–2	EPA (1990)
Exposed skin surface area	SA	cm²	Triangular	1980 (910–2940)	EPA (1990)
Dermal absorption factor	Add	Unit-less	Triangular	0.003 (0–0.03)	Katsumata and Kastenberg (1997)
Soil to skin adherence factor	AF	mg/cm²	Uniform	0.75–1.25	EPA (1990)
PCDD/Fs soil concentration from the area	Sc	ng/Kg	Triangular	1.17 (0.10–3.88)	Schuhmacher et al. (1998a)
PCDD/Fs vegetable concentration from the area	Vc	ng/Kg	Triangular	0.197 (0.06–0.50)	Schuhmacher et al. (1998b)
PCDD/Fs air concentration	Ac	pg/m³	Triangular	0.07 (0.01–0.22)	Generalitat Catalonia (Environmental Department)[b]

[a] Distribution: means and standard deviations are used for log-normal distributions, the low and high for uniform distributions, and the mean, the low and high for triangular distributions.

[b] Personal communication.

TABLE 9.3

Monte Carlo Parameter Distributions for PCDD/Fs

Parameter	Symbol	Units	Type	Distribution[a]	Reference
Body weight	BW	kg	Log-normal	67.52 ± 12.22	Arija et al. (1996)[b]
Noncancer potency factor	NCP	pg/kg day	Uniform	1–4	Rolaf and Younes (1998)
Cancer potency factor	CP	(mg/kg day)$^{-1}$	Uniform	34,000–56,000	Katsumata and Kastenberg (1997)

[a] Distribution: mean and standard deviation are used for log-normal distributions, the low and high for uniform distributions.

[b] Personal communication.

of Tarragona, Spain. After characterizing the uncertainty and/or variability associated with each parameter, the uncertainty in the risk can be estimated. For the risk assessment presented here, the commercially available software package Crystal Ball (Version 4.0) was used. For analyzing the results, the mean and median values and the percentiles 50% and 90% were extracted and presented. In this example, ranges of exposure, rather than single-point estimates, are developed in order to account for the natural variability among members of a population and for uncertainties in the input variables. Even if it were possible to eliminate the uncertainty associated with the input variables, a probability density function would still be required because of natural variability.

Figure 9.7 shows the distribution of the different variables of direct exposure due to the incinerator emissions for the population living in the area surrounding the plant. The distribution of total direct exposure is also depicted. Figure 9.8 shows the sensitivity analysis for total direct exposure from the different exposure pathways due to MSWI emissions. This figure shows that inhalation of air from the area contributes to 50.4% of the variance and vegetation ingestion to 49.5%. The other pathways—exposure, dermal absorption, soil ingestion and inhalation of re-suspended particles—are irrelevant.

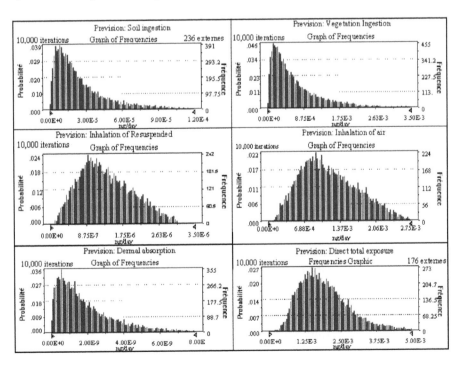

FIGURE 9.7   Distribution of the different variables of direct exposure due to MSWI emissions.

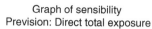
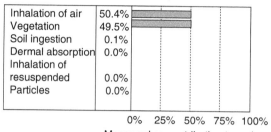

FIGURE 9.8   Distribution of total direct exposure.

Table 9.4 summarizes the exposure to polychlorinated dioxins and furans (PCDD/Fs) by the population living in the proximity of the MSWI of Tarragona, Spain, and Table 9.5 shows the PCDD/Fs dose (ng I-TEQ/day/kg) for the population living around the incinerator. The tolerable average intake levels of PCDD/Fs recently established by the WHO is between 1 and 4 pg I-TEQ/kg/day for lifetime exposure (Rolaf and Younes, 1998). Therefore, the current total exposures of 2.87×10^{-2} pg I-TEQ/day/kg for the 50th percentile and 5.37×10^{-2} for the 90th percentile (Table 9.5) are within this tolerable intake.

9.7.2   RISK EVALUATION

The noncancer and cancer risks from direct exposure are shown in Tables 9.6 and 9.7, respectively. The results show the 10th percentile, the central tendency estimates of risk (50th percentile) and the reasonable

TABLE 9.4

PCDD/Fs Exposure[a] by Different Ways of Direct Exposure of the Population Living in Proximity of the MSWI

	Mean	SD	Percentiles		
			10th	50th	90th
Soil ingestion	3.01×10^{-5}	3.37×10^{-5}	5.49×10^{-6}	1.98×10^{-5}	6.69×10^{-5}
Vegetable ingestion	7.95×10^{-4}	9.09×10^{-4}	1.36×10^{-4}	5.11×10^{-4}	1.76×10^{-3}
Inhalation of re-suspended particles	1.37×10^{-6}	6.53×10^{-7}	5.73×10^{-7}	1.28×10^{-6}	2.31×10^{-6}
Inhalation of air	1.19×10^{-3}	5.53×10^{-4}	5.20×10^{-4}	1.11×10^{-3}	1.98×10^{-3}
Dermal absorption	2.15×10^{-9}	2.00×10^{-9}	3.78×10^{-10}	1.53×10^{-9}	4.70×10^{-9}
Total direct exposure	2.02×10^{-3}	1.07×10^{-3}	6.62×10^{-4}	1.64×10^{-3}	3.81×10^{-3}

[a] ng I-TEQ/day.

TABLE 9.5

PCDD/Fs Dose[a] for Population Living around the Incinerator

	Mean	SD	Percentiles		
			10th	50th	90th
Direct dose	3.24×10^{-5}	1.80×10^{-5}	1.46×10^{-5}	2.87×10^{-5}	5.37×10^{-5}

[a] ng I-TEQ/day/kg.

TABLE 9.6
Noncancer Risk: Mean, Standard Deviation, and 10th, 50th, and 90th Percentiles

	Mean	SD	Percentiles 10th	50th	90th
Direct risk	1.44×10^{-2}	1.06×10^{-2}	5.18×10^{-3}	1.15×10^{-2}	2.65×10^{-2}

TABLE 9.7
Cancer Risk: Mean, Standard Deviation, and 10th, 50th, and 90th Percentiles

	Mean	SD	Percentiles 10th	50th	90th
Direct risk	1.42×10^{-6}	8.28×10^{-6}	6.20×10^{-7}	1.25×10^{-6}	2.36×10^{-6}

maximum exposure (RME; 90th percentile). It can be seen that the median (50th percentile) of noncancer risk due to PCDD/Fs in the population living in the area surrounding the MSWI of Tarragona is 0.015. The results reveal that the uncertainty of the risk estimated as defined by the ratio of the 90th to the 10th percentile is 9.1 (Table 9.6). With respect to total cancer risk, the median increment in individual lifetime is 1.25×10^{-6}, and the ratio between the 90th percentile and 10th percentile is about 3.8 (Table 9.7).

It can be concluded that the exposure to PCDD/Fs due to the MSWI in the Tarragona area is not producing health risks for the general population.

9.8 CASE STUDY: UNCERTAINTY ASSESSMENT BY MONTE CARLO SIMULATION FOR LCI AND IPA APPLIED TO MSWI IN TARRAGONA, SPAIN

The frameworks for uncertainty assessment by MC simulation for LCI and IPA described in Section 9.5 are applied to the formerly introduced case study of the MSWI in Tarragona, Spain.

9.8.1 APPLICATION OF THE FRAMEWORK TO LIFE-CYCLE INVENTORY OF ELECTRICITY PRODUCED BY A WASTE INCINERATOR

This section presents an application of the framework for uncertainty assessment corresponding to the case of LCIs mentioned earlier. For this purpose, the industrial process of electricity produced by the MSWI, discussed in Chapters 1 through 4, was once more selected. The following goals were proposed for the sake of a more didactical and practical example:

1. Assigning probability distributions to the parameters considered in the study
2. Assessing the uncertainties and variations in the calculation of the LCI table
3. Determining the most relevant parameters in such LCI by sensitivity analysis

9.8.1.1 Assigning Probability Distributions to Considered Parameters
The predominant pollutants identified and quantified during the implementation of the LCI for the MSWI study were selected by a combined quantitative and qualitative approach. The quantitative selection consisted of a dominance analysis performed on the basis of the results in the impact

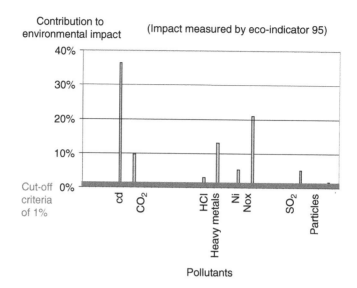

FIGURE 9.9   Selection of essential pollutants by dominance analysis. (Reprinted from *J. Cleaner Prod.*, 11, Sonnemann, G.W. et al., pp. 279–292, Copyright 2002b with permission from Elsevier.)

assessment carried out using the eco-indicator 95 method (see Chapter 3). Figure 9.9 presents the contribution of the considered pollutants to the total environmental potential impact measured by the eco-indicator 95. As a selection criterion, only the emissions with a contribution to the total environmental impact higher than 1% will be selected for the uncertainty assessment. The results of the quantitative selection established that the atmospheric emission of cadmium (Cd), carbon dioxide (CO_2), hydrochloric acid (HCl), nickel (Ni), sulfur dioxide (SO_2), other heavy metals (HMs) and particulate matter (PM) would be taken into account. Moreover, because of their carcinogenicity and consideration as primary air pollutants in the ExternE project (EC, 1995, 2000), arsenic (As), carbon monoxide (CO) and PCDD/Fs were also to be considered (Figure 9.9).

A proper determination of the probability distribution is possible if data are extensively available, as in the case of measured emissions, electricity production, working hours and flow gas volume. Here, the probability distributions were calculated from experimental data provided by the LCA study (STQ, 1998) and by the MSWI director by means of a report (Nadal, 1999), or personally. Based on a relevant number of measurements and their inherent variations, the normal or log-normal distribution was selected as the best-fitting probability density function for the respective types of data. The quality of the fitting was assessed by the Kolmogorov–Smirnov test for parameters with less than 30 measurements and by the Chi^2 test for parameters with more than 30 measurements.

The software Crystal Ball allowed the carrying out of this fitting of probability distributions. The variation of the emissions in the study was enormous due to the constant variation in the waste's incinerated composition. The concentrations of the incineration process emissions, together with their distribution type and deviations, are presented in Table 9.8. Except for PCDD/Fs, the variations of the pollutant concentration emissions were fitted from experimental data by log-normal distributions with a geometric standard deviation (σ_g) between 1.5 and 3.4. It can be seen that the variation of the measurements of nickel and other HM concentrations in the emissions is, in general, much higher than that of the macropollutants SO_2, NO_x and CO. The heterogeneous feature of the waste used in the process can explain the enormous variation once more (Table 9.8).

The probability distributions for the PCDD/Fs concentrations in the emissions were considered to be log-normal with a geometric standard deviation of 2.0, according to the estimations published by Rabl and Spadaro (1999). The consideration was done due to a lack of sufficient experimental data on this substance. Also, in the case of the life-cycle data taken from Frischknecht et al. (1996), site-specific data on transport processes, as well as the inputs and outputs for local production and the waste treatment

TABLE 9.8
Site-Specific Data: Concentrations in Emissions of the Incineration Plant[a]

Parameter	Unit	Distribution	Former Situation[b]	Current Situation[c]	Reference
		Type	Mean value (σ_g)	Mean value (σ_g)	
As	mg/Nm³	Log-normal	2.00×10^{-2} (3.4)	5.60×10^{-3} (3.4)	STQ (1998)
Cd	mg/Nm³	Log-normal	2.00×10^{-2} (1.7)	6.60×10^{-3} (1.7)	STQ (1998)
CO	mg/Nm³	Log-normal	4.00×10^{1} (1.5)	4.00×10^{1} (1.5)	STQ (1998)
PCDD/Fs	pg/Nm³	Log-normal	2.00×10^{1} (2.0)	2.00 (2.0)	STQ (1998); Rabl and Spadaro (1999)
HCl	mg/Nm³	Log-normal	5.16×10^{2} (1.6)	3.28×10^{1} (1.6)	STQ (1998)
Heavy metals	mg/Nm³	Log-normal	4.50×10^{-1} (2.5)	9.10×10^{-2} (2.5)	STQ (1998)
Ni	mg/Nm³	Log-normal	3.00×10^{-2} (2.2)	8.40×10^{-3} (2.2)	STQ (1998)
NO$_x$	mg/Nm³	Log-normal	1.91×10^{2} (1.5)	1.91×10^{2} (1.5)	STQ (1998)
Particulate matter	mg/Nm³	Log-normal	2.74×10^{1} (2.1)	4.80 (2.1)	STQ (1998)
SO$_2$	mg/Nm³	Log-normal	8.09×10^{1} (1.5)	3.02×10^{1} (1.5)	STQ (1998)

Note: σ_g = geometric standard deviation.

[a] CO_2 emissions are determined stochiometrically.

[b] Without new filters.

[c] With new filters.

process, count on little information about data quality. Consequently, the uncertainty estimations have been made according to the literature (Meier, 1997; Weidema and Wesnaes, 1996). Table 9.9 shows the technical site-specific data for the whole system, which embraces the operation of waste treatment, electricity production and consumption and transportation, as well as its inherent inflows and outflows (see Chapter 2). All these technical data show a normal distribution. The variation in the data on the annual amounts of waste treated and electricity produced are described by their normal standard deviations because these statistical factors would be calculated from a sufficient amount of data.

For the annual working hours and gas volume flow, a 5% coefficient of variation was estimated according to the information given by the technical staff of the MSWI. The probability distributions for other pieces of technical data had to be derived from the literature (Meier, 1997; Weidema and Wesnaes, 1996). Thus, a normal distribution with a CV of 10% was assumed for site-specific inflows and outflows, while for transportation, a normal distribution with a CV of 20% was chosen due to the large uncertainty in the exact description of the waste transport. An enormous amount of the data used in the LCI is not directly related to the incineration process, but to the cycles of associated inputs, outputs and transport processes, as can be seen in Table 9.9. The data of the system under study were not obtained in a site-specific manner, but from the ETH database (Frischknecht et al., 1996). These data have been collected from a Swiss perspective on a European scale. It is evident that the transfer of data to the Spanish situations definitely caused an uncertainty that, according to Meier (1997), differs depending on the considered pollutant. For information taken from databases, Meier (1997) proposed to assume classes of normal probability distributions with the following CVs:

- For data obtained by stochiometric determination, a CV of 2% needs to be considered.
- For actual emission measurements or data computable in well-known process simulation, a CV of 10% is expected.
- For well-defined substances or summed parameters, a CV of 20% can be assumed.
- For data taken from specific compounds by an elaborated analytical method, a CV of 30% is expected.

TABLE 9.9

Site-Specific Data: Technical Variables of the Incineration Plant

Parameter	Unit	Distribution	Former Situation[a]	Current Situation[b]	Reference
		Type	Mean value (CV)	Mean value (CV)	
Electricity production	TJ	Normal	158.56 (8.08)	149.55 (6.94)	Nadal (1999)
Waste treated	t	Normal	153,467 (5,024)	148,450 (5,024)	Nadal (1999)
Yearly working hours	h	Normal	8,280 (0.05)	8,280 (0.05)	Nadal (1999)
Flue gas volume	Nm³/h	Normal	90,000 (0.05)	90,000 (0.05)	Nadal (1999)
Transport	tkm	Normal	4,100,000 (0.2)	4,100,000 (0.2)	Weidema and Wesnaes (1996); STQ (1998)
Plastic proportion	%	Normal	13 (0.1)	13 (0.1)	Weidema and Wesnaes (1996); STQ (1998)
Electricity consumption	TJ	Normal	1.66 (0.1)	1.66 (0.1)	Weidema and Wesnaes (1996); STQ (1998)
Diesel	t	Normal	148.8 (0.1)	148.8 (0.1)	Weidema and Wesnaes (1996); STQ (1998)
Lubricant oil	t	Normal	2.3 (0.1)	2.3 (0.1)	Weidema and Wesnaes (1996); STQ (1998)
Lime (CaO)	kg	Normal	0 (0)	921,000 (0.1)	Weidema and Wesnaes (1996); STQ (1998)
Water deionized	m³	Normal	19,665 (0.1)	19,665 (0.1)	Weidema and Wesnaes (1996); STQ (1998)
Water refrigeration	m³	Normal	5,175 (0.1)	5,175 (0.1)	Weidema and Wesnaes (1996); STQ (1998)
Water purified	m³	Normal	7,360 (0.1)	7,360 (0.1)	Weidema and Wesnaes (1996); STQ (1998)
Unspecified water	m³	Normal	8,122 (0.1)	33,120 (0.1)	Weidema and Wesnaes (1996); STQ (1998)
Ashes treated	kg	Normal	590,000 (0.1)	3,450,000 (0.1)	Weidema and Wesnaes (1996); STQ (1998)
Scrap treated	kg	Normal	2,740,000 (0.1)	2,740,000 (0.1)	Weidema and Wesnaes (1996); STQ (1998)
Slag	t	Normal	42,208 (0.1)	42,208 (0.1)	Weidema and Wesnaes (1996); STQ (1998)

Note: CV = coefficient of variation with the exception of electricity production and waste treated, which are expressed as normal standard deviation.

[a] Without new filters.

[b] With new filters.

For a better understanding of these estimates, Table 9.10 specifies which pollutant emission corresponds to which class under study. In addition, an example of the life-cycle data for the input flow of energy consumption in Spain is presented. According to this scheme, CO_2 is the only environmental load that has been determined stochiometrically for all life-cycle data. As a result of using this relatively certain assessment method, CO_2 received a CV of 2%. CO, NO_x, and SO_2 were considered to be obtained by actual emission measurements or to be computed in well-known process simulations, depending on multiple parameters. Because of more possibilities for errors, a CV of 10% was assumed for all of these compounds. For well-defined substances or summed parameters such as

TABLE 9.10

Uncertainties in Measurements of Emissions in ETH Process Modules[a]

Parameter Types	Distribution Type	Uncertainty (CV)
Substances determined stochiometrically (CO_2)	Normal	0.02
Actual emission measurements or emissions of well-known processes depending on multiple parameters (CO, NO_x, SO_2)	Normal	0.10
Well-defined substances or sum parameters (As, Cd, HCl, HMs, Ni, PCDD/Fs)	Normal	0.20
Specific compounds with elaborated analytical methods (PM)	Normal	0.30

Note: CV = coefficient of variation.

[a] According to Meier, M., *Eco-Efficiency Evaluation of Waste Gas Purification Systems in the Chemical Industry*, LCA Documents, Vol. 2, Ecomed Publishers, Landsberg, Germany, 1997, and Frischknecht, R. et al., Ökoinventare von Energiesystemen—Grundlagen für den ökologischen Vergleich von Energiesystemen und den Einbezug von Energiesystemen in Ökobilanzen für die Schweiz. 3rd ed., ETH Zürich: Gruppe Energie-Stoffe-Umwelt, PSI Villigen: Sektion Ganzheitliche Systemanalysen, 1996.

HM, PCDD/Fs and HCl, a CV of 20% was established. Finally, in the class for specific compounds with elaborated analytical methods, the uncertainty level was considered the highest with a CV of 30%. In the present case study, this CV was assumed for the life-cycle database information on particulate matter.

9.8.1.2 Assessing Uncertainties and Variations in the Calculation of Life-Cycle Inventory

Following the procedure related to uncertainty assessment for the proposed LCI, a Monte Carlo simulation was run for each situation with the probability distribution described previously. Its final result consisted of a set of histograms—one per selected pollutant—corresponding to the two scenarios proposed in the study: (1) Scenario 1: former situation and (2) Scenario 2: current situation.

Current situation (Scenario 2) refers to the incineration process carried out with an advanced acid gas treatment system (AGTS). On the other hand, in Scenario 1, the atmospheric emissions of the same unit without AGTS were evaluated. Each simulation has been made in one separate run per scenario. Because of the inherent variability of the MC model, it is not possible to affirm that the set of values relative to the input variables used in the run of the current situation are going to be the same as those in the former situation. The reason for this is that every run occurs in different ways due to the generation of random numbers. In order to verify the importance of this variability on the final outcome, both simulations were also carried out in one run, and the results obtained showed negligible variations.

Figure 9.10 presents the results of the substances considered in this study in the current situation with advanced AGTS. In the x-axis, it is possible to observe the amount of pollutant emission per energy produced. The y-axis shows the probability of each value of the life-cycle emissions. As mentioned before, 10,000 iterations were carried out with the software Crystal Ball. The mean atmospheric emissions of the heavy metals per 1 TJ of electricity produced by the incinerator were 7.50×10^{-1} kg with a normal standard deviation of 5.50×10^{-1} kg/TJ. Because the density distribution of the results is best adjusted by log-normal density function, a geometric mean of 6.10×10^{-1} kg for HMs/TJ and a geometric standard of 1.92 were calculated. On this basis, a 68% confidence interval from 3.18×10^{-1} –1.17 kg/TJ was obtained.

Because the other pollutants can be adjusted well by a log-normal distribution, all the atmospheric emissions were treated in the same way, with their mean in the 68% confidence interval. In this framework, the calculated values of μ_g and σ_g for As, Cd and CO were, respectively, 3.37×10^{-2} kg/TJ (2.30), 2.62×10^{-2} kg/TJ (1.67) and 2.43×10^2 kg/TJ (1.37).

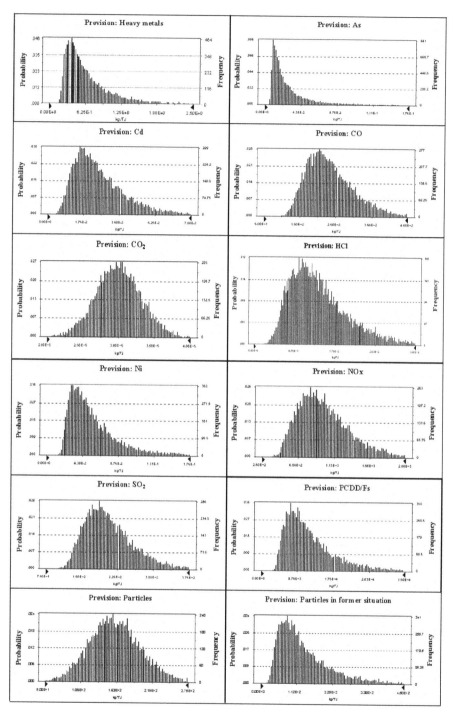

FIGURE 9.10   Monte Carlo simulation LCI results for the different substances. (Adapted from Sonnemann, G.W., et al., *J. Cleaner Prod.*, 11, 279–292b, 2002.)

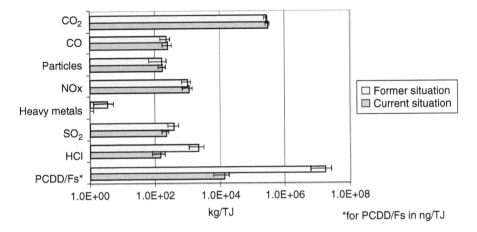

FIGURE 9.11   LCI results with confidence interval of 68%. (Reprinted from *J. Cleaner Prod.*, 11, Sonnemann, G.W. et al., pp. 279–292, Copyright 2002b with permission from Elsevier.)

The CO_2 emissions of the incineration process were not determined by the measurements assumed to have a log-normal distribution. In this case, the total amount of waste treated was multiplied by the percentage fraction of plastics present on it, because this material is the only one in the mixture component that originally comes from fossil fuels. Taking into account that 1 kg of plastic burned produces approximately 2.0 kg of CO_2, the total CO_2 produced by the waste incineration was divided by the gas volume emitted to the atmosphere through the stack in order to determine the CO_2 concentration released. Thus, for this pollutant, a normal distribution with a μ_g of 3.10×105 kg/TJ and a σ_g of 1.13 was obtained.

The profile of the LCI results for SO_2—2.12×10^2 kg/TJ (1.29)—to the same situation is similar to those for NO_x and CO due to the same order of magnitude in the σ_g for the incinerator's emissions. In Figure 9.10, the LCI results obtained by MC simulation for the particulate matter (particles) in the former situation (Scenario 1) are also illustrated (μ_g 1.50×10^2 kg/TJ, σ_g 1.93). If the results of the former situation and the current situation are compared, a clear change can be seen in the probability distribution from a log-normal to a rather normal one after the installation of the advanced AGTS.

The mean values with confidence intervals for the former situation and the current situation related to all studied pollutants are presented in Figure 9.11. Heavy metals were only considered as a summed parameter. For the PCDD/Fs, HMs, SO_2 and HCl, a clear reduction can be observed with the installation of the advanced AGTS, especially for the first one. On the other hand, for CO_2, CO, PM and NO_x, no variation in the life-cycle emissions per TJ of electricity produced is found. For these cases, changes were smaller than the given confidence intervals. Here, it is evident that the detected uncertainty and variability interfere in the results and influence their interpretation (Figure 9.11).

9.8.1.3   Determining the Most Relevant Parameters in LCI by Sensitivity Analysis

In order to determine the most relevant parameters in the case of the MSWI of Tarragona, a sensitivity analysis of the LCI results was carried out. The results of the sensitivity analysis for the PCDD/Fs are presented for the former situation and the current situation in Figures 9.12 and 9.13, respectively. In both situations, the PCDD/Fs emitted by the incinerator process were the most important parameter, with contributions to the variance of 99.9% and 99.6%, respectively. The same results were obtained for the other pollutants: percentages over 95% with the exception of the particulate matter. Thus, this contaminant will be discussed in more detail.

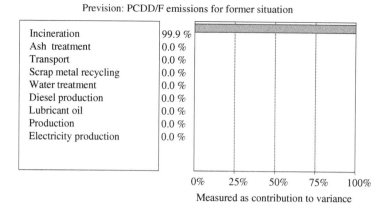

Prevision: PCDD/F emissions for former situation

FIGURE 9.12 Sensitivity analysis of the LCI results for PCDD/Fs in the former situation. (Reprinted from *J. Cleaner Prod.*, 11, Sonnemann, G.W. et al., pp. 279–292, Copyright 2002b with permission from Elsevier.)

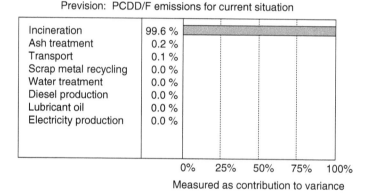

Prevision: PCDD/F emissions for current situation

FIGURE 9.13 Sensitivity analysis of the LCI results for PCDD/Fs in the current situation. (Reprinted from *J. Cleaner Prod.*, 11, Sonnemann, G.W. et al., pp. 279–292, Copyright 2002b with permission from Elsevier.)

In Scenario 1, which is defined by the former situation, the major emission of PM was due to the process of incineration. The effect of the other steps of the system under study was practically negligible, as can be seen in Figure 9.14.

On the other hand, in the current situation (Scenario 1), the process of production of the lime used in the advanced AGTS has been added to the life-cycle, especially considering that this process generates a huge amount of dust. Thus, as shown in Figure 9.15, it contributes with 83.6% to the variance of the PM from the global system. Finally, the particles emitted during the incineration concur with only 15.6%, and all the other processes embraced by the boundaries sum less than 1%.

As a conclusion, the advanced AGTS reduces the concentration of heavy metals and PCDD/Fs, PM, SO_2, and HCl in the gas flow emitted to the atmosphere from the incinerator. The concentrations of other pollutants, such as NO_x and CO emissions, are kept constant by their turn. Consequently, from the point of view of environmental risk assessment, the risk of causing hazardous effects to human health and the environment in the area surrounding the incineration plant is clearly reduced. Also, on the basis of an LCA, it was possible to observe a reduction of atmospheric emissions from the whole system per TJ of electricity produced and for the pollutant heavy metals, PCDD/Fs, HCl and SO_2. The same cannot be observed in the case of particulate

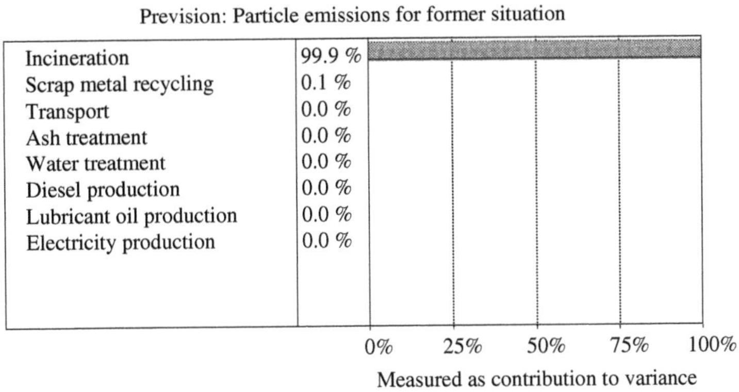

Prevision: Particle emissions for former situation

Incineration	99.9 %	
Scrap metal recycling	0.1 %	
Transport	0.0 %	
Ash treatment	0.0 %	
Water treatment	0.0 %	
Diesel production	0.0 %	
Lubricant oil production	0.0 %	
Electricity production	0.0 %	

0%   25%   50%   75%   100%

Measured as contribution to variance

FIGURE 9.14 Sensitivity analysis of the LCI results for particles in the former situation. (Reprinted from *J. Cleaner Prod.*, 11, Sonnemann, G.W. et al., pp. 279–292, Copyright 2002b with permission from Elsevier.)

Prevision: Particle emissions for current situation

CaO production	83.6 %	
Incineration	15.6 %	
Transport	0.5 %	
Scrap metal recycling	0.3 %	
Ash treatment	0.0 %	
Electricity production	0.0 %	
Water treatment	0.0 %	
Diesel production	0.0%	
Lubricant oil production	0.0 %	

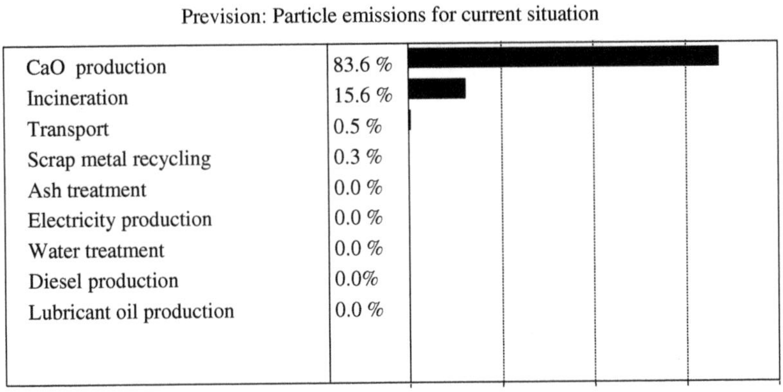

FIGURE 9.15 Sensitivity analysis of the LCI results for particles in the current situation. (Reprinted from *J. Cleaner Prod.*, 11, Sonnemann, G.W. et al., pp. 279–292, Copyright 2002b with permission from Elsevier.)

matter. Moreover, the absolute values show a very slight, though insignificant, increase of life-cycle emissions per TJ of electricity produced for CO_2, CO and NO_x. These emissions are higher in comparison with those measured in Scenario 1 due to the lime production process added to the system, the longer transportation distances resulting from higher inflows and outflows, and a lower efficiency in production of electricity (former situation = 158.56 TJ/yr.; current situation = 149.55 TJ/yr).

9.8.2 APPLICATION OF FRAMEWORK TO IMPACT PATHWAY ANALYSIS OF WASTE INCINERATOR EMISSIONS ON A LOCAL SCALE

Following the same procedure used for the LCI results, the framework for uncertainty assessment in IPA was also applied to the case study of local human health impacts due to the emissions of the MSWI in Tarragona. As in the approach used previously, some goals were proposed:

1. Assigning probability distributions to the parameters considered in the study
2. Assessing the uncertainties and the variation in application
3. Determining the most relevant parameters in such an IPA by sensitivity analysis

9.8.2.1   Assigning Probability Distributions to Considered Parameters

As explained by Rabl and Spadaro (1999), the probability distributions mainly used in environmental damage estimations are the normal distribution and the log-normal probability distribution. As mentioned earlier, all normal distributions are symmetric and have bell-shaped density curves with a single peak. The log-normal distribution, in turn, is calculated assuming that the logarithm of the variable has a normal distribution. As in the previous case of uncertainty assessment in LCI, the proper determination of the probability distribution is only possible if measured data are extensively available, as in the case of atmospheric emissions, electricity production, working hours and flow gas volume. If the parameters are based on little proper information, literature values must be applied to determine the probability distribution, e.g., in the case of PCDD/Fs emissions, dispersion modeling results, dose–response and exposure–response functions, population data and monetary valuation.

The necessary information for determining the probability distributions from measured data was taken from the LCA study of the Servei de Tecnologìa Química, STQ (1998), and from information given by the director of Tarragona's MSWI (Nadal, 1999). Probability distributions published in the literature were available from the ExternE project (EC, 1995, 2000) and, in particular, from the publication about uncertainty analysis of environmental damages and costs by Rabl and Spadaro (1999). The estimation of the uncertainty for the dispersion model was taken from McKone and Ryan (1989). Further local information was obtained from the Public Health Plan of the Tarragona region (GenCat, 1997) and from a diagnosis on the socioeconomic development of the Tarragona province by Soler (1999).

The probability distributions used in the study are summarized from Tables 9.11 through 9.13. "Variable mean" stands for an enormous number of values that are not constant, but differ depending on the grid and pollutant considered.

TABLE 9.11
Technology and Modeling Data (Emission Values for Former Situation 1)

Parameter	Units	Distribution	Mean	Dev.	Reference
Electricity production	MW	Normal	5.02	(σ) 0.23	Nadal (1999)
Working hours per year	h	Normal	8,280	CV 0.05	Nadal (1999)
Flue gas volume	Nm^3/h	Normal	90,000	CV 0.05	Nadal (1999)
SO_2 (emissions)	mg/Nm^3	Log-normal	81.13	(σ_g) 1.5	STQ (1998)
NO_x (emissions)	mg/Nm^3	Log-normal	191	(σ_g) 1.5	STQ (1998)
PM (emissions)	$mgNm^3$	Log-normal	28.57	(σ_g) 2.1	STQ (1998)
CO (emissions)	mg/Nm^3	Log-normal	40	(σ_g) 1.5	STQ (1998)
As (emissions)	$\mu g/Nm^3$	Log-normal	15.1	(σ_g) 3.4	STQ (1998)
Cd (emissions)	$\mu g/Nm^3$	Log-normal	19.9	(σ_g) 1.7	STQ (1998)
Ni (emissions)	$\mu g/Nm^3$	Log-normal	33.27	(σ_g) 2.2	STQ (1998)
PCDD/F (emissions)	ng/Nm^3	Log-normal	2	(σ_g) 2	STQ (1998); Rabl and Spadaro (1999)
Flue gas temperature	K	Point estimate	503	—	Nadal (1999)
Stack height	m	Point estimate	50	—	Nadal (1999)
Stack diameter	m^2	Point estimate	1.98	—	Nadal (1999)
Anemometer height	m	Point estimate	10	—	Nadal (1999)
Geographical latitude	°	Point estimate	41.19	—	Nadal (1999)
Geographical longitude	°	Point estimate	1.21	—	Nadal (1999)
Elevation at site	m	Point estimate	90	—	Nadal (1999)
Incremental emission concentration	mg/Nm^3	Log-normal	Variable	(σ_g) 2	McKone and Ryan (1989)

Note: CV = coefficient of variation; σ = normal standard deviation; σ_g = geometric standard deviation; dev. = deviation.

TABLE 9.12

Impact Human Health Data

Parameter	Unit	Distribution	Mean	Dev.	Reference
		Dose–Response and Exposure–Response Functions			
Chronic YOLL		Log-normal	0.00072	(σ_g) 2.1	IER (1998); Rabl and Spadaro (1999)
Acute YOLL		Log-normal	Variable	(σ_g) 2.1	Rabl and Spadaro (1999)
Cancer		Log-normal	Variable	(σ_g) 3	Rabl and Spadaro (1999)
Others		Log-normal	Variable	(σ_g) 2.1	Rabl and Spadaro (1999)
Damage factors					
Chronic YOLL		Log-normal	1	(σ_g) 1.5	Rabl and Spadaro (1999)
Acute YOLL		Log-normal	1	(σ_g) 4	Rabl and Spadaro (1999)
Cancer		Log-normal	1	(σ_g) 1.6	Rabl and Spadaro (1999)
Others		Log-normal	1	(σ_g) 1.2	Rabl and Spadaro (1999)
% pop. above 65 years	%	Point estimate	13	—	IER (1998); GenCat (1997)
% pop. adults	%	Point estimate	57	—	IER (1998); GenCat (1997)
% pop. children	%	Point estimate	24	—	IER, 1998; GenCat, 1997
% pop. asthma adults	%	Normal	4	CV 0.1	IER (1998); GenCat (1997)
% pop. asthma children	%	Normal	2	CV 0.1	IER (1998); GenCat (1997)
% pop. baseline mortality	%	Normal	0.864	CV 0.1	IER (1998); GenCat (1997)
Population	# inhab	Normal	Variable	CV 0.01	Soler (1999)

Note: CV = coefficient of variation; σ_g = geometric standard deviation; dev. = deviation.

TABLE 9.13

Monetary Valuation Data

Parameter	Units	Distribution	Mean	Dev.	Reference
Chronic YOLL	Euro	Log-normal	84,330	(σ_g) 2.1	Rabl and Spadaro (1999)
Acute YOLL	Euro	Log-normal	155,000	(σ_g) 2.1	Rabl and Spadaro (1999)
Cancer	Euro	Log-normal	1,500,000	(σ_g) 2.1	Rabl and Spadaro (1999)
Others	Euro	Log-normal	Variable	(σ_g) 1.2	Rabl and Spadaro (1999)

Note: σ_g = geometric standard deviation; dev. = deviation.

In Table 9.11 the technology and modeling parameters are presented together with their respective probability distributions and characteristics. The technology parameters in Table 9.11 consist, respectively, of electricity production, working hours and specific characteristics of the incinerator (stack dimensions, geographical situation, etc.) and the parameters properly related to the emissions (concentration of pollutants, total volume, temperature, etc.). The increment of the emission concentration is considered a modeling parameter. According to the probability distributions obtained using Crystal Ball, the variations of electricity production, working hours and flow gas volume have a normal distribution, and the emissions behave like cadmium in a log-normal way. The electricity production has a normal standard deviation of 0.23; working hours and flue gas volume have, respectively, a coefficient of variation of 0.05. The variations of the emissions are characterized by geometric standard deviations ranging from 1.5 to 3.4. The parameters corresponding to the stack dimensions and the geographical situation are point estimates and were provided by the director of Tarragona's MSWI (Nadal, 1999). The incremental emission concentration has log-normal distribution with a geometric standard deviation of 2, according to the uncertainty estimates for the dispersion model by McKone and Ryan (1989).

Table 9.12 presents the human health parameters. As can be seen in Table 9.12, uncertainty and variation are parts of the public health data (e.g., population, percentage of children, adults and elders, percentage of asthmatics or baseline mortality). The dose–response and exposure–response functions are characterized by the log-normal probability distribution provided by Rabl and Spadaro (1999). The mean value for chronic years of life lost (YOLL) due to particulate matter was provided by IER (1998) and the other means vary in function of the respective pollutants. The definition and calculation of the damage factors (e.g., chronic YOLL, acute YOLL and cancer) involve another factor of uncertainty. The probability distributions for these factors, which are used only for aggregation and further multiplication (therefore equal to 1), have been taken from Rabl and Spadaro (1999), who supposed them to have log-normal distribution with a geometric standard deviation between 1.2 and 4. The description of the population properties has been identified as a point estimate, or with normal distribution, according to the values provided once more by IER (1998) and by GenCat (1997). Finally, the possible variation of the number of inhabitants in each grid can be described by a normal distribution based on the study made by Soler (1999).

Table 9.13 shows the monetary valuation parameters. All probability distributions are log-normal and were taken from Rabl and Spadaro (1999).

9.8.2.2 Assessing Uncertainties and Variation in the Calculation of Impact Pathway Analysis

Following the IPA framework, a final result expressed in environmental damage costs due to the air emissions per kWh of electricity produced has been calculated for both situations considered (Table 9.14). By using the obtained probability distributions for the essential parameters in an MC simulation, the result for environmental damage costs has been transformed from a concrete value into a probability distribution around a mean value. Because the distributions of most parameters in Tables 9.11 through 9.13 are log-normal and not normal, the final distribution of each result has a log-normal distribution too. In the same way as in the application to LCI, each simulation has been made in one separate run

TABLE 9.14
Statistical Parameter Describing Results of the Environmental Damage Estimation in Scenarios 1 and 2 as External Costs[a]

Parameter	Scenario 1[b]	Scenario 2[c]
Normal mean	3.73	0.87
Normal standard deviation	5.16	1.08
Geometric mean	2.19	0.55
Geometric standard deviation	2.81	2.62
Minimum	0.087	0.029
Maximum	221.7	74.4
Median	2.09	0.53
68% Confidence interval		
Superior	6.15	1.44
Inferior	0.78	0.21

[a] mU.S.$ per kWh (1E-3 U.S.$/kWh).
[b] Without filters.
[c] With filters.

per scenario. Because of the inherent variability of the Monte Carlo model, it is not possible to affirm that the set of values for the input variables used in the run of Scenario 2 (current situation) will be the same as those used in Scenario 1 (former situation). The same was done with the calculation of LCI uncertainties and, due to the generation of random numbers, every run occurs in a different way. In order to verify the importance of this variability on the final outcome, both simulations have also been made in one run; it could be checked that the results are the same because the variations are negligible.

Figure 9.16 presents the first result, which is the case of the incineration process supported by the advanced AGTS. In the x-axis, it is possible to observe the environmental damage cost per energy output. The y-axis shows the probability of each cost value. The mean of the environmental damage cost in Scenario 2 is 0.87 mU.S.$ per kWh (with $m = 10^{-3}$). The total number of iterations carried out with the software Crystal Ball is 10,000. A summary of all the results generated can be found in Table 9.14. The geometric standard deviation from them is 2.62.

The second case occurs in time before the first one, when the incinerator did not have an advanced AGTS. The emissions of pollutants were more important and, consequently, the environmental damage cost is much higher, with 3.73 mU.S.$ per kWh. The probability distribution of this result can be found in Figure 9.17. This is explained because the only important change corresponds to the mean values of 10 parameters, including pollutants and electricity production, which are lower with an advanced AGTS installed.

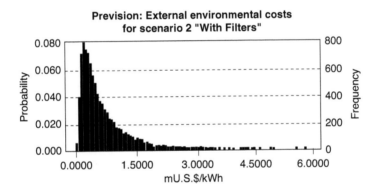

FIGURE 9.16 MC simulation results for IPA of MSWI emissions in the current situation. (Reprinted from *Environ. Int.*, 28, Sonnemann, G.W. et al., pp. 9–18, Copyright 2002a with permission from Elsevier.)

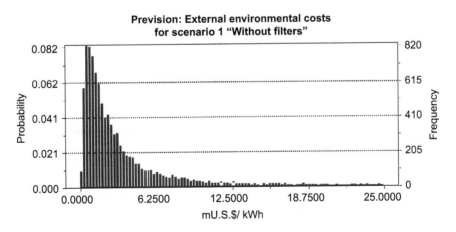

FIGURE 9.17 MC simulation results for IPA of MSWI in the former situation. (Reprinted from *Environ. Int.*, 28, Sonnemann, G.W. et al., pp. 9–18, Copyright 2002a with permission from Elsevier.)

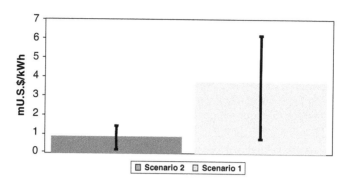

FIGURE 9.18 Comparison of the MC simulation results for the former situation (Scenario 1) without advanced gas cleaning system and the current situation (Scenario 2) with such an installation. (Reprinted from *Environ. Int.*, 28, Sonnemann, G.W. et al., pp. 9–18, Copyright 2002a with permission from Elsevier.)

The results in Table 9.14 show that the uncertainty and variability calculated using MC simulation are less than those calculated by analytical methods due to the dynamic characteristics of this stochastic model. The presented results have a geometric standard deviation of less than 3, whereas the geometric standard deviation obtained by analytical methods is higher than 4, according to Rabl and Spadaro (1999).

Figure 9.18 illustrates the differences between the means obtained in Scenarios 1 and 2, both with confidence intervals of 68%. It is possible to see a clear reduction of the damage cost. As can also be seen in Table 9.14, the 68% confidence interval for the scenario with advanced AGTS embraces a range from 0.21 to 1.44 mU.S.$ per kWh, while for the other the same confidence interval has a range between 0.78 and 6.15 mU.S.$ per kWh. The inferior bound of the confidence interval is the maximum range of possible errors; for the result of the former situation, the inferior bound is in the same order of magnitude as the mean for the current situation. Thus, according to the results, there are important uncertainties. However, a clear reduction in terms of the damage cost can be foreseen within a confidence interval of 68% when comparing the two different operation scenarios (Figure 9.18).

9.8.2.3 Determining the Most Relevant Parameters in Impact Pathway Analysis by Sensitivity Analysis

The results of the sensitivity analysis for the pollutants are presented in Figure 9.19 for Scenario 1 without advanced AGTS. Some interesting results can be extracted. The graph shows all the pollutants and their contribution to the final result. Obviously, the emission of particulate matter is the most important parameter, with 92.1% of the total damage. The NO_x seems to be the second important pollutant, while the rest produce negligible damage. Figure 9.20 for Scenario 2, i.e., with advanced AGTS, is very similar; the particulate emissions contribute more than 99.6% to the total environmental damage cost. Because of the major emissions of particulate matter, in the former situation (Scenario 1), the mentioned percentage of NO_x is practically negligible. Taking into account that the MSWI is the emission source and the major public concern of producing dioxins, the result of the IPA for the case study shows that little of the total human health damage is contributed by air emissions. Figures 9.21 and 9.22 present the results of the sensitivity analysis on the sources of health impacts to total damage costs. Most damage is caused by the loss of life expectancy, expressed as YOLL. If the damage appears in the near term, it is called acute, and if it appears in the long term, it is called a chronic impact. The chronic YOLL and the acute YOLL account together for more than 99% of the total environmental damage costs. Other parameters like hospital admission or cancer are less important, by far. Figure 9.22 (the results for Scenario 1, without advanced AGTS) is a little bit different from Figure 9.21 because of the major concentration of particulate matter. This pollutant matter has more influence on the chronic YOLL and an increase in its concentration produces an increase in the importance of the chronic YOLL.

Prevision: Total environmental cost due to human health damages

PM10 (mg/Nm³)	92.1%	
NOx (mg/Nm³)	7.7%	
SO2 (mg/Nm³)	0.2%	
CO (μg/Nm³)	0.0%	
As (μg/Nm³)	0.0%	
Dioxins (pg/Nm³)	0.0%	
Cd (μg/Nm³)	0.0%	
Ni (μg/Nm³)	0.0%	

Measured as contribution to variance

FIGURE 9.19 Sensitivity analysis for pollutants of the current situation. (Reprinted from *Environ. Int.*, 28, Sonnemann, G.W. et al., pp. 9–18, Copyright 2002a with permission from Elsevier.)

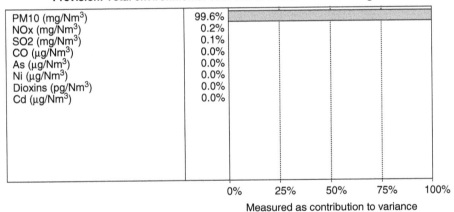

Prevision: Total environmental cost due to human health damages

PM10 (mg/Nm³)	99.6%
NOx (mg/Nm³)	0.2%
SO2 (mg/Nm³)	0.1%
CO (μg/Nm³)	0.0%
As (μg/Nm³)	0.0%
Ni (μg/Nm³)	0.0%
Dioxins (pg/Nm³)	0.0%
Cd (μg/Nm³)	0.0%

Measured as contribution to variance

FIGURE 9.20 Sensitivity analysis for pollutants of the former situation. (Reprinted from *Environ. Int.*, 28, Sonnemann, G.W. et al., pp. 9–18, Copyright 2002a with permission from Elsevier.)

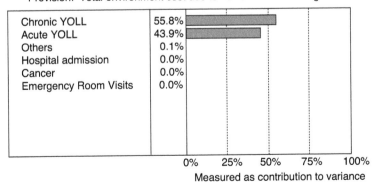

Prevision: Total environment cost due to human health damages

Chronic YOLL	55.8%
Acute YOLL	43.9%
Others	0.1%
Hospital admission	0.0%
Cancer	0.0%
Emergency Room Visits	0.0%

Measured as contribution to variance

FIGURE 9.21 Sensitivity analysis for health impacts of the current situation. (Reprinted from *Environ. Int.*, 28, Sonnemann, G.W. et al., pp. 9–18, Copyright 2002a with permission from Elsevier.)

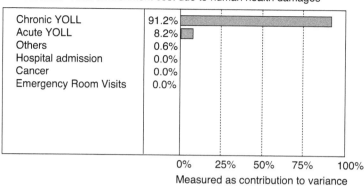

FIGURE 9.22   Sensitivity analysis for health impacts of the former situation. (Reprinted from *Environ. Int.*, 28, Sonnemann, G.W. et al., pp. 9–18, Copyright 2002a with permission from Elsevier.)

9.8.3   COMPARISON OF UNCERTAINTIES IN LIFE-CYCLE INVENTORY AND IMPACT PATHWAY ANALYSIS

As a general conclusion, it can be said that the damage estimations carried out in the IPA contain more uncertainties than the LCI due to the very important uncertainties related to the dispersion models, dose–response and exposure–response functions and the weighting schemes. In detail, the results for the LCI show a geometric standard deviation between σ_g 1.13 for CO_2, 1.29 for SO_2, and 1.92 for heavy metals or 2.30 for As. That means that higher uncertainties are related to the variability in the waste input composition regarding trace elements. In comparison with the LCI results, the uncertainties in the environmental damage estimations in the form of external costs are higher and sum up to a geometric standard deviation of 2.62 with filters and 2.81 without filters. These relatively high geometric standard deviations are influenced less by the emissions (except for As with a σ_g 3.4, for Ni 2.2 and for the PM 2.1) than by the impact of human health data, especially dose–response functions for cancer (σ_g 3) and damage factor of acute YOLL (σ_g 4). Moreover, monetary valuation for YOLL and cancer is an important source of uncertainties with σ_g of 2.1 in the same way as the dispersion model with σ_g of 2. To overcome incomparability related to the dispersion models, using an internationally accepted reference model is proposed. Using homogeneous dose–response and exposure–response functions approved by the World Health Organization (WHO) could have the same effect in the future.

9.9   APPLICATION OF MONTE CARLO SIMULATIONS TO THE HUMAN HEALTH RISK ASSESSMENT OF OCCUPATIONAL EXPOSURE TO NANOPARTICLES

The Monte Carlo simulation presented here builds upon the case study presented in Chapter 4 on the occupational exposure and impacts relating to manufacturing and handling nano-TiO_2. In the prior assessment, the human health risk assessment (HHRA) was carried out using a deterministic approach. Consequently, the dose–response assessment, exposure assessment and risk characterization were all estimated using single fixed-point values.

As was mentioned in prior chapters about emerging technologies such as engineered nanomaterial or nanoparticle (ENM) there is a general lack of relevant and available data describing these product systems. This introduces several sources of uncertainty into the HHRA process, including both the dose–response assessment and exposure assessment. ENM specifically have their own material-specific sources of uncertainty due to lack of relevant toxicological data, lack of known emissions across the life-cycle, lack of measured exposure or lack of appropriate models for making these calculations.

Thus, there are important sources of uncertainty to consider and which need to be communicated when calculating and reporting the potential risk involved with exposure to ENM. It is curious that the definition of risk necessitates the quantification of likelihood, however historical use of ERA and HHRA has demonstrated the tendency to utilize deterministic approaches. While this may better serve risk assessors and regulatory authorities, a probabilistic approach provides more information to the decision-maker and, thus, might better serve designers and developers of a technology.

The following sections describe how the parameters and values used in the HHRA were transformed to accommodate a probabilistic approach. It further explains the implication of evaluating the risk from a probabilistic versus deterministic point of view for the purposes of determining potential human health impacts when working with ENMs such as nano-TiO_2.

9.9.1 Assigning Probability Distributions to Considered Parameters

9.9.1.1 Dose–Response Assessment

The BMC_a initially calculated in Chapter 4 was converted to a distribution using PROAST's parametric bootstrap option (i.e., sampling with replacement) over 10,000 simulations. Because of this, the lower bounds of those values (e.g., 90% lower-bound) were not calculated as is done in the deterministic RA procedure, since this would have introduced a redundant level of precaution in the POD (Crump, 1995; Slob and Pieters, 1998). Additionally, the extrapolation factors used to convert the animal-based BMC to a human-based one were defined using log-normal distributions using similar approaches presented by Slob et al. (2014). Population-based differences, either between species or within species, in response to exposure of a noxious substance (e.g., survival times after disease onset) are generally asymmetrical, as opposed to being described by normal "bell-shape" curves (Limpert et al., 2001). In deterministic RA procedures, it is typical to use factors of 10 to extrapolate from animal-based toxicity thresholds to acceptable human equivalent doses. Although extrapolation factors of 10 are supposed to be conservative protective values, it is understood that this does not always provide protection. However, for transparency, demonstration of the approach and ease of interpretation, the log-normal distributions were defined such that a value of 10 was one-order of magnitude greater than the mean and lies at the 99th-percentile. In this regard, the selection of 10 as a conservative extrapolation factor remains (Slob and Pieters, 1998). The geometric mean (GM) and geometric standard deviation (GSD) were 1 and 2.7, respectively (Tsang et al., 2017). Ten thousand Monte-Carlo (MC) simulations were conducted using Crystal Ball (Oracle Corporation) to complete the extrapolation of BMC_a to BMC_h distributions.

After applying EF_{inter} and EF_{intra} over 10,000 Monte-Carlo simulations, the results showed that the best-fit describing the BMC_h was a log-normal distribution with a GM of 0.91*/4.72 mg/m^3. Approximately 59% and 40% of the distribution was driven by uncertainty in the values of EF_{inter} and EF_{intra}, respectively. Approximately 59% and 40% of the distribution was explained by uncertainty in the values of EF_{inter} and EF_{intra}, respectively, while little variation was a consequence of the dose-response data used in the analysis (Figure 9.23).

9.9.1.2 Exposure Assessment

Realistically, there are many sources of uncertainty and variability within the exposure model used in Chapter 4. These include such things as proximity of workers to the ENM dust plumes, ventilation rates in the workplace and sizes of the particles emitted into the air. In the current assessment, a standard uncertainty associated with DI_{resp} is employed for defining the probability distribution of the exposure. This is because the substance emission potential, related to particle characteristics, is found to be the most sensitive and strongest determinant input parameter in the model (Liguori et al., 2016). This uncertainty is given by Equation 9.1 ($R^2 = 65.8$; $p < 0.001$):

$$\log(\sigma_{DI}) = 0.871 \cdot \log\left(DI_{resp}\right) - 0.625 \qquad (9.1)$$

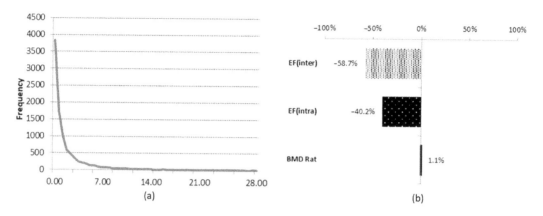

FIGURE 9.23 Results of (a) the 10,000 Monte-Carlo simulations used to estimate the benchmark concentration for humans (BMC$_h$) and (b) the contribution of each parameter used to estimate the benchmark concentrations for humans.

σ_{DI} is the standard deviation of the DI_{resp}. This equation was derived based on statistical analysis (Minitab® version 17.2.1) of the experimental standard deviation related to the dustiness indices of 59 powder samples made available from additional testing at the National Research Centre for the Working Environment (Tsang et al., 2017). The equation shows that the standard deviation (i.e., uncertainty) increases with increasing DI_{resp}. This should be expected given that a higher DI_{resp} generally results in a greater absolute standard deviation resulting from the higher release of particulates and, thus, uncertainty in fate and transport of this material. The DI_{resp} for the TiO$_2$ as considered in this study was 15 mg/kg, resulting in a standard deviation of 2.5 mg/kg. The standard deviation was then applied to ER_i for quantifying the exposure distribution of each ES.

9.9.1.3   Risk Characterization

Risk characterization ratio (RCR) values were calculated in the same manner as was described in Equation 9 of Chapter 4; however, this was done by randomly sampling the BMC$_h$ and exposure values over 10,000 MC simulations. The MC analysis was conducted using Crystal Ball (Oracle Corporation).

The RCR values for each exposure scenario were log-normally distributed (Tsang et al., 2017). The risk characterization ratio distributions for ES1, ES2 and ES3 all contained some probability of risk to inflammation of the lung (i.e., RCR values ≥ 1). Scenario ES2 had a particularly high probability of risk compared to the other scenarios (Figure 9.24), with nearly 78% of the Monte-Carlo simulation results ≥ 1 (i.e., 22% of the results resulted in no risk to the exposed workers). ES1 and ES3 resulted in 10% and 9% of their RCR ≥ 1 (Tsang et al., 2017). For all of the exposure scenarios, roughly 75% of the variation in the RCR distributions was influenced by uncertainty in the dose–response analysis (i.e., EF$_{inter}$ and EF$_{intra}$), while the remaining 25% was the result of uncertainty in the exposure estimations (Tsang et al., 2017). In accordance with the lower airborne nano-TiO$_2$ concentrations found in the FF, the probability of individual FF RCR values ≥ 1 were much lower than compared with the NF. In total, approximately 56% of the results for FF ES2 were ≥ 1 (Figure 9.24). Only 3% and 1% of ES1 and ES2 FF scenarios were likely to result in a RCR ≥ 1 (Tsang et al., 2017).

9.9.2   IMPLICATIONS OF EMPLOYING A PROBABILISTIC APPROACH IN THE HUMAN HEALTH RISK ASSESSMENT OF ENGINEERED NANOMATERIAL OR NANOPARTICLE

The utility of the probabilistic approach is demonstrated upon comparison with deterministically derived values. In 2011, the US National Institute for Occupational Safety and Health (NIOSH) reported on the human health hazards to TiO$_2$ and recommended both carcinogenic and noncarcinogenic recommended exposure limit (REL). The carcinogenic REL was estimated as

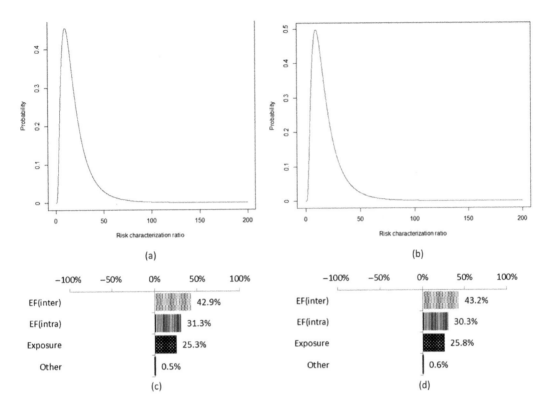

FIGURE 9.24 Results of 10,000 Monte-Carlo risk characterization ratio simulations for exposure scenario 2 in the (a) near-field and (b) far-field. Note that right-end tails of the distribution are artificially truncated for presentation. Contributions to the uncertainty and variation are displayed in (c) for the near-field and in (d) for the far-field.

0.3 mg/m³ using the lung tumor data from two bulk- (Lee et al., 1985; Muhle et al., 1991) and one nano-TiO₂ (Heinrich et al., 1995) chronic inhalation studies. The noncarcinogenic REL of 0.004 mg/m³ was estimated from four subchronic inhalation studies, three of which analyzed inflammation (i.e., response in PMN count) in the lung in response to bulk-TiO₂ exposure (Bermudez et al., 2002; Cullen et al., 2002; Tran et al., 1999) and the other upon nano-TiO₂ exposure (Bermudez et al., 2004). In context of the BMC_h distribution defined in this paper, the noncarcinogenic NIOSH value represents a REL with 0.02% frequency of likely occurrence, where 99.9% of the BMC_h values (Figure 9.3) are greater than the NIOSH value. Although we are not making any formal opinion on NIOSH's recommended value, it is worth pointing out the utility of being able to interpret the likelihood of occurrence of a proposed regulatory value. Furthermore, using the same dose–response data used to calculate our BMC_h, we derived a no observed adverse effect level (NOAEL) value of 2 mg/m³ and compared it to the BMC distributions for rats. The NOAEL value resulted in an individual probability of occurrence of 0.7%, where 99% of the BMC_a values are greater than this NOAEL. It is worth noting that, NOAELs ignore the shape of a dose–response curve, meaning that the value may not actually reflect a dose at which no or the lowest effect occurs (Allen et al., 1994; Barlow et al., 2009; Crump, 1984, 1995; Davis et al., 2011; Jager et al., 2001). The magnitude of these indicators also un-intuitively decrease (i.e., become less conservative) as the certainty of a study increases. For example, studies with more animals compared with ones with less animals per dosing group have greater chances of showing a statistically significant difference in response versus the control, potentially eliminating the possibility of establishing a NOAEL. Although the study with less animals per dosing group will have a lower chance of showing statistical significance between the dosing and control groups, leading to the establishment of an NOAEL (Crump, 1984). However, a BMD or BMC will increase when greater observational power and response data is available (e.g., more animals per study) (Crump, 1984). Applying default EF_{inter} and EF_{intra} of 10 to the estimated NOAEL value from Bermudez et al. results in a BMC_h of 0.02 mg/m^3. Compared with our BMC_h distribution, this deterministic value has a frequency of 10^{-3}, with 99.9% of all BMC_h values being greater than it. The REL should represent a safe limit and usually a precautionary (conservative) one.

9.10   QUESTIONS AND EXERCISES

1. Explain the main differences between uncertainty and variability in environmental systems analysis.
2. In which categories are the sources of uncertainty classified in tools?
3. Distinguish which of the following related situations would be sources of uncertainties and which would be sources of variability in the tools presented in Chapter 9 of a coal power plant:
   - Location of the factory
   - Time of the study
   - Data related to the prevalence of the winds
   - Percentages of age, gender and diseases of population
   - Extrapolation from animal studies to humans
   - Sample sizes for animal and human studies

4. Associate the following situations in the same tool to the types of uncertainties and variability listed in the column on the right side:

Use of data corresponding to emissions associated with a former situation	Uncertainty/choices
• Selection of kJ per kg of product as functional unit	• Variability among sources and objects
• Use of data belonging to a system originally from a geographical area different from that subject to study	• Parameter uncertainty
• Human preferences involved in the system under study	• Model uncertainty
• Use of simplified models for the calculation of factors	• Spatial/temporal variability
• Presence of multi-waste processes	• Uncertainty choices

5. What are the main sources of parameter uncertainty and how are they reflected in the outcomes of an LCA?
6. Which implications for an LCIA have the non-consideration of the model spatial and temporal characteristics?
7. Would the use of a parameter uncertainty analysis in an LCA be suitable when large model uncertainties have been detected?
8. What are the general strategies for the assessment of uncertainties in LCA studies?
9. List the main steps of an MC simulation.
10. Explain the relative importance of the different parameters for the final result when estimated by MC simulation.
11. In the framework of uncertainty assessment in IPA, a large amount of damage function data is compiled. Why can't all of them be considered? Give an explanation.
12. In an assessment study by MC simulation, data uncertainties appear as a percentage of the standard deviation, and a log-normal distribution is used. Discuss the suitability of this choice.

13. Explain the main sources of uncertainties in IPA and LCI, respectively.
14. What is the importance of determining the probability distributions of ERA data? Describe which kind of probability distribution can be associated with the following events:
   - The results of a weight study for a group of people: 155.6 lb; 141.7 lb; 158.6 lb; 174.2 lb; 172.1 lb; 164 lb; 168.9 lb; 139.2 lb; 147 lb; 156.2 lb; 171.7 lb
   - The concentrations in ppm of a toxin in samples taken at different points of a river: 0.123; 0.095; 0.154; 0.298; 0.365; 0.612; 0.389; 0.474; 0.299; 0.494; 0.341; 0.612; 0.511; 0.744; 0.519; 0.654; 0.476; 0.437; 0.365; 0.26; 0.166; 0.198; 0.108; 0.165
   - The results of a study about relative abundance (%) of endangered species in an ecosystem: 43; 26; 12.1; 7.4; 3.6; 2.5; 1.4; 1.2; 1; 0.6; 0.3; 0.2; 0.1; 0.08; 0.07; 0.06; 0.05; 0.04

   Do any of these distributions correspond to normal distribution and to log-normal?
15. Using the example of the LCA of a chair given in Chapter 2:
   - Assuming experimental data related to inputs and wastes of a similar process in Europe, discuss the probability distributions of the parameters considered in an eventual study in your country.
   - Identify the possible uncertainties and variability if an MC simulation is performed.

REFERENCES

Allen, B.C., Robert, J.K., Carole, A.K., and Elaine M.F. (1994). Dose-response assessment for developmental toxicity: II. Comparison of generic benchmark dose estimates with no observed adverse effect levels. *Fundamental and Applied Toxicology* 23(4): 487–495.

Arija, V., Salas, J., Fernández-Ballart, J., Cucó, G., and Martí-Henneberg, C. (1996). Consumo, hábitos alimentarios y estado nutricional de la población de Reus (IX). Evolución del consumo de alimentos, de su participación en la ingestión de energía y nutrientes y de su relación con el nivel socioeconómico y cultural entre 1983 y 1993. *Medicina Clinica* 106: 174–179.

Barlow, S. et al. (2009). Guidance of the scientific committee on a request from EFSA on the use of the benchmark dose approach in risk assessment. *The EFSA Journal* 1150: 1–72.

Bermudez, E. et al. (2002). Long-term pulmonary responses of three laboratory rodent species to subchronic inhalation of pigmentary titanium dioxide particles. *Toxicological Sciences: An Official Journal of the Society of Toxicology* 70(1): 86–97.

Bermudez, E. et al. (2004). Pulmonary responses of mice, rats, and hamsters to subchronic inhalation of ultra-fine titanium dioxide particles. *Toxicological Sciences* 77(2): 347–357.

Crump, K.S. (1984). A new method for determining allowable daily intakes. *Fundamental and Applied Toxicology* 871(5): 854–871.

Crump, K.S. (1995). Calculation of benchmark doses from continuous data. *Risk Analysis* 15(1): 79–89.

Cullen, R.T. et al. (2002). *Toxicity of Volcanic Ash from Montserrat*. Institute of Occupational Medicine, Edinburgh, Scotland.

Davis, J.A., Jeffrey, S.G., and Zhao, Q.J. (2011). Introduction to benchmark dose methods and U.S. EPA's benchmark dose software (BMDS) version 2.1.1. *Toxicology and Applied Pharmacology* 254(2): 181–191.

Decisioneering. (1996). *Crystal Ball Version 4.0—Guide de l'utilisateur*. Editions MEV, Versailles, France.

EC (European Commission). (1995). *ExternE—Externalities of Energy*, 6 vols., EUR 16520–162525. DG XII Science, Research and Development, Brussels, Belgium.

EC (European Commission). (2000). Externalities of fuel cycles, vol. 7: Methodology update, report of the JOULE research project ExternE (Externalities of Energy), EUR 19083. DG XII Science, Research and Development, Brussels, Belgium.

Environmental Protection Agency (EPA). (1990). *Methodology for Assessing Health Risks Associated with Indirect Exposure to Combustor Emissions*. PB90–187055, U.S. EPA, Cincinnati, OH, 1990.

Finnveden, G. and Lindfors, L.G. (1998). Data quality of life-cycle inventory data—Rules of thumb. *The International Journal of Life Cycle Assessment* 3(2): 65–66.

Frischknecht, R. et al. (PSI Villigen). (1996). Ökoinventare von Energiesystemen—Grundlagen für den ökologischen Vergleich von Energiesystemen und den Einbezug von Energiesystemen in Ökobilanzen für die Schweiz. 3rd ed., ETH Zürich: Gruppe Energie-Stoffe-Umwelt, PSI Villigen: Sektion Ganzheitliche Systemanalysen.

GenCat (Generalitat de Catalunya). (1997). *Pla de salut de la reió sanitària Tarragona, Servei Català de la Salut.* Tarragona, Spain.

Hanssen, O.J. and Asbjornsen, A.A. (1996). Statistical properties of emission data in life-cycle assessments. *The Journal of Cleaner Production* 4(3–4): 149–157.

Hawley, J.K. (1985). Assessment of health risk from exposure to contaminated soil. *Risk Analysis* 5(4): 289–302.

Heinrich, U. et al. (1995). Chronic inhalation exposure of wistar rats and two different strains of mice to diesel engine exhaust, carbon black, and titanium dioxide. *Inhalation Toxicology* 7(4): 533.

Hofstetter, P. (1998). *Perspectives in Life-Cycle Impact Assessment—A Structured Approach to Combine Models of the Technosphere, Ecosphere and Valuesphere.* Kluwer Academic Publishers, London, UK.

Huijbregts, M.A.J. (1998). Application of uncertainty and variability in LCA (part I)—A general framework for the analysis of uncertainty and variability in life-cycle assessment. *The International Journal of Life Cycle Assessment* 3(5): 273–280.

Huijbregts, M.A.J. and Seppälä, J. (2000). Towards region-specific, European fate factors for airborne nitrogen compounds causing aquatic eutrophication. *The International Journal of Life Cycle Assessment* 3(5): 65–67.

Huijbregts, M.A.J. et al. (2000). Framework for modeling data uncertainty in life-cycle inventories. *The International Journal of Life Cycle Assessment* 6(3): 127–132.

IER (Institut für Energiewirtschaft und Rationale Energieanwendung). (1998). *EcoSense 2.1, Software,* University of Stuttgart, Stuttgart, Germany.

Jager, T., Theo, G.V., Rikken, M.G.L., and der Poel, P.V. (2001). Opportunities for a probabilistic risk assessment of chemicals in the European Union. *Chemosphere* 43(2): 257–264.

Katsumata, P.T. and Kastenberg, W.E. (1997). On the assessment of health risks at superfund sites using Monte Carlo simulations. *Journal of Environmental Science and Health* A32 (9 and10): 2697–2731.

LaGrega, M.D., Buckingham, P.L., and Evans, J.C. (1994). *Hazardous Waste Management,* Mc Graw-Hill, New York.

Lee, K.P., Trochimowicz, H.J., and Reinhardt, C.F. (1985). Pulmonary response of rats exposed to titanium by inhalation for two years dioxide (TiO2). *Toxicology and Applied Pharmacology* 79: 179–192.

Liguori, B. et al. (2016). *Sensitivity Analysis of NanoSafer Control Banding Web-Tool Version 1.1: Ranking of Determining Parameters and Uncertainty.* Lyngby, Denmark, Technical University of Denmark, DTU Environment.

Limpert, E., Stahel, A.W., and Abbt, M. (2001). Log-normal distributions across the sciences: Keys and clues. *BioScience* 51(5): 341.

Maurice, B., Frischknecht, R., Coehlo-Schwirtz, V., and Hungerbühler, K. (2000). Uncertainty analysis in life-cycle inventory—Application to the production of electricity with French coal power plants. *The Journal of Cleaner Production* 8(2): 95–108.

McKone, T.E. and Ryan, P.B. (1989). Human exposures to chemicals through food chains—An uncertainty analysis. *Environmental Science & Technology* 23: 1154–1163.

Meier, M. (1997). *Eco-Efficiency Evaluation of Waste Gas Purification Systems in the Chemical Industry, LCA Documents.* Vol. 2, Ecomed Publishers, Landsberg, Germany.

Metzger, J.N., Fjeld, R.A., Hammonds, J.S., and Hoffman, F.O. (1998). Evaluation of software of propagating uncertainty through risk assessment models. *Human and Ecological Risk Assessment* 4(2): 263–290.

Muhle, H. et al. (1991). Pulmonary response to toner upon chronic inhalation exposure in rats. *Fundamental and Applied Toxicology* 17(2): 280–299.

Nadal, R. (1999). Planta incineradora de residuos sólidos urbanos de Tarragona, report for Master en Enginyeria i Gestió Ambiental (MEGA). Universitat Rovira i Virgili, Tarragona, Spain.

National Research Council. (1994). *Science and Judgement in Risk Assessment, Committee on Risk Assessment of Hazardous Air Pollutants.* Board on Environmental Studies and Toxicology. Commission on Life Sciences, National Academy Press, Washington, DC.

Nessel, S.G., Butler, J., Post, G.B., Held, J.L., Gochfeld, M., and Gallo, M.A. (1991). Evaluation of the relative contribution of exposure routes in a health risk assessment of dioxin emissions from a municipal waste incinerator. *Journal of Exposure Analysis and Environmental Epidemiology* 1(3): 283–307.

Pesonen, H. et al. (2000). Framework for scenario development in LCA. *The International Journal of Life Cycle Assessment* 5(1): 21–30.

Rabl, A. and Spadaro, J.V. (1999). Environmental damages and costs—An analysis of uncertainties. *Environment International* 25(1): 29–46.

Rolaf van Leeuwen, F.X. and Younes, M. (1998). WHO revises the tolerable daily intake (TDI) for dioxins. *Organohalogen Compounds* 38: 295–298.

Ros, M. (1998). Unsicherheit und Fuzziness in ökologischen Bewertungen—Orientierungen zu einer robusten Praxis der Ökobilanzierung, Ph. D. Thesis, ETH Zürich, Switzerland.

Roseberry, A.M. and Burmaster, D.E. (1991). A note: Estimating exposure concentrations of lipophilic organic chemicals to human via finfish. *Journal of Exposure Analysis and Environmental Epidemiology* 1: 513–521.

Schuhmacher, M., Domingo, J.L., Llobet, M., Süinderhauf, W., and Müller, L. (1998b). Temporal variation of PCDD/F concentrations in vegetation samples collected in the vicinity of a municipal waste incinerator (1996–1997). *Science of the Total Environment* 218: 175–183.

Schuhmacher, M., Granero, S., Xifró, A., Domingo, J.L., Rivera, J., and Eljarrat, E. (1998a). Levels of PCDD/Fs in soil samples in the vicinity of a municipal solid waste incinerator. *Chemosphere* 37: 2127–2137.

Shin, D., Lee, J., Yang, J., and Yu, Y. (1998). Estimation of air emission for dioxin using a mathematical model in two large cities of Korea. *Organohalogen Compounds* 36: 449–453.

Slob, W. and Pieters, M.N. (1998). A probabilistic approach for deriving acceptable human intake limits and human health risks from toxicological studies: General framework. *Risk analysis: An Official Publication of the Society for Risk Analysis* 18(6): 787–798.

Slob, W., Bakker, M.I., Te Biesebeek, J.D., and Bokkers, B.G.H. (2014). Exploring the uncertainties in cancer risk assessment using the integrated probabilistic risk assessment (IPRA) approach. *Risk Analysis: An Official Publication of the Society for Risk Analysis* 34(8): 1401–1422.

Soler, S. (1999). Diagnosi socioeconòmica i estratègies de desenvolupament de la provincia de Tarragona, report of the Unitat de Promició i Desenvolupament, Universitat Rovira i Virgili, Tarragona, Spain.

Sonnemann, G.W., Pla, Y., Schuhmacher, M., and Castells, F. (2002a). Framework for the uncertainty assessment in the impact pathway analysis with an applicaiton on a local scale in Spain. *Environment International* 28: 9–18.

Sonnemann, G.W., Schuhmacher, M., and Castells, F. (2002b). Uncertainty assessment by Monte Carlo simulation in a life-cycle inventory of electricity produced by a waste incinerator. *The Journal of Cleaner Production* 11: 279–292.

STQ (Servei de Tecnologia Química). (1998). Análisis del ciclo de vida de la electricidad producida por la planta de incineración de residuos urbanos de Tarragona, technical report. Universitat Rovira i Virgili, Tarragona, Spain.

Tran, C.L. et al. (1999). *Investigation and Prediction of Pulmonary Responses to Dust*. Sulfolk, UK.

Tsang, M.P. et al. (2017). Probabilistic risk assessment of emerging materials: Case study of titanium dioxide nanoparticles. *Nanotoxicology* 11(4): 558–568.

U.S. Environmental Protection Agency. (1992). Guidelines for exposure assessment; notice. *Federal Register* 57(104): 22888–22938.

Weidema, B. (1998). Multi-user test of the data quality matrix for product life-cycle inventory data. *The International Journal of Life Cycle Assessment* 3(5): 259–265.

Weidema, B. and Wesnaes, M.S. (1996). Data quality management for life-cycle inventories—An example of using data quality indicators. *The Journal of Cleaner Production* 4(3–4): 167–174.

10 Environmental Damage Estimations for Industrial Process Chains[1]

Guido Sonnemann, Marta Schuhmacher, and Francesc Castells

CONTENTS

---

[1] Extracts of this chapter referring to the mathematical foundation and the case study are reprinted from *J. Hazardous Mater.*, 77, Sonnemann, G.W. et al., pp. 91–106. Copyright 2000 with permission from Elsevier.

10.1   INTRODUCTION TO AN INTEGRATED APPROACH

Looking at the picture of life-cycle and risk assessment methods that has been presented; it seems necessary to come to a spatial differentiation of life cycles in order to facilitate a more integrated way of calculating environmental damage estimations in a chain perspective. In this way, the poor accordance between impact potentials and actual impacts can be overcome in life-cycle impact assessment (LCIA) and the results can become more consistent with the risk assessment approach. Thus, another approach is needed that differentiates life cycles according to the number of processes considered. This means using different levels of sophistication for different applications that are defined by their chain length.

It seems to be unfeasible to estimate environmental damages for each process of a full life-cycle assessment (LCA), i.e., of a complex product system with a huge number of industrial processes (e.g., computers), because all the local or regional information is not accessible and each process is contributing only marginally to the total environmental impact. Such a life cycle is illustrated in Figure 10.1.

However, if the LCA methodology is applied to industrial process chains, i.e., chains with a small number of industrial processes (e.g., <100 processes)—waste treatment process chains, for example—the localization of the processes is often known. Moreover, in general, only a small number of processes is responsible for the main part of the environmental impact, as can be seen in the example of Figure 10.2. Therefore, for such applications of LCA, the main individual processes can be assessed in their corresponding surroundings.

This differentiation of the life-cycle type according to chain length is crucial for estimating environmental damages in the most accurate way possible. This work will focus on the methodology development for damage estimations in industrial process chains, defined here as life cycles with a relatively small number of processes involved, in contrast to product systems, i.e., process chains with a high number of different sites to consider. This chapter presents a comprehensive methodology for such life-cycle types. It is evident in these cases that different levels of detail in the impact assessment can be used.

Based on these considerations, a methodology has been developed that allows estimating environmental damages for industrial process chains. These chains are understood here as chains of industrial processes with less than 100 processes. A framework is needed that allows evaluating environmental damages as accurately as possible, because today's damage assessment methods generate results different from those of the evaluation of potential impacts. This is demonstrated, for instance, in the study undertaken by Spirinckx and Nocker (1999)

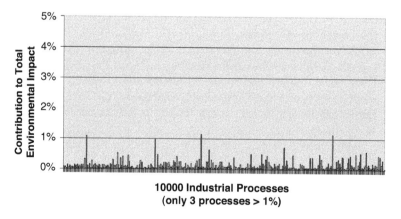

FIGURE 10.1   Full LCA studies of complex products with a huge number of industrial processes, e.g., a computer. In this example, fewer than three processes contribute to more than 1% of the total environmental impact.

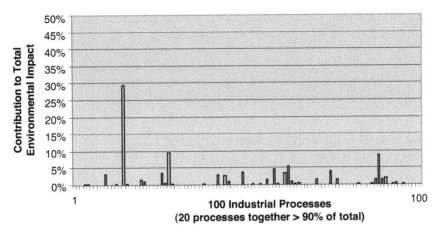

FIGURE 10.2   LCA methodology applied to industrial process chains with a small number of industrial processes, e.g., waste treatment. In this example, 20 processes together contribute to more than 90% of the total environmental impact.

that compares both approaches. This methodology is useful for certain life-cycle management (LCM) applications such as end-of-life strategies and supply chain management. Possible applications are further discussed in Section 10.6 of this chapter.

10.2   CHALLENGES AND STRATEGY FOR A COMBINED FRAMEWORK OF LIFE-CYCLE AND RISK ASSESSMENT

The bases of life-cycle inventories are the emissions of pollutants and the consumption of resources. In this methodology, the focus is on pollutant emissions and the damages that they may cause. After their emission, pollutants are transported through the environment and cause a concentration increase. On the pathway, they then can affect sensible receptors, such as humans, and may produce damages. The receptor density clearly depends on local or regional geographic characteristics for non-global impact categories. These environmental damages can be evaluated and aggregated according to socioeconomic evaluation patterns as indicators or as external costs. The methodology takes a step out of the LCA framework and integrates other environmental tools, according to the idea of CHAINET (1998). Such a methodology is confronted with the following special challenges:

1. Consider each process, or at least the main ones.
2. Find a compromise between accuracy and practicability.
3. Apply the damage functions as far as possible to the emissions in their respective continent, region or location.
4. Aggregate the damages by economic evaluation or other forms of weighting to a small number of indicators.
5. Show transparency; analyze uncertainties and sensitivity.

First, a general strategy is necessary with regard to the environmental damage estimations for industrial processes. This strategy includes an approach to make the methodology more practicable. Starting with a conventional life-cycle inventory (LCI), such a strategy can be described as:

1. Creating an algorithm to consider site-specific aspects
2. Calculating the potential impact score

3. Estimating global damages by the best available midpoint indicators
4. Determining main media, pollutants and processes
5. Using fate models to obtain the concentration increment in the respective regions
6. Relating increments with dose– and exposure–response functions and receptors
7. Disposing of methods for aggregation by accepted weighting schemes
8. Relating to other environmental management tools

10.3 COMPARISON OF ENVIRONMENTAL RISK ASSESSMENT AND LIFE-CYCLE ASSESSMENT

Before presenting the methodology, in this section, a comparison of LCA and environmental risk assessment (ERA), the two environmental tools that are further integrated, is outlined in Table 10.1. The comparison is illustrated by the example of electricity generated from coal and produced in the same way, but in two different regions, in which the combustion of coal is obviously an important part of the life cycle:

- Case 1: In a very populated and acidification-sensitive area next to the mining site
- Case 2: In a purely populated and no acidification-sensitive area far from the mining site

According to Sonnemann et al. (1999), the LCA will probably state the minimal total emissions and energy demand for Case 1 due to the importance of the additional transport and the negligence of the specific region. By contrast, the ERA will state the minimal risk to the environment for Case 2, because the focus is put only on the main process within the life cycle, but the extra transport is not considered. This example shows, in a simple way, the significance of the difference highlighted in Table 10.1. It also clearly demonstrates the need for a more integrated approach that does not so easily allow two environmental impact analysis tools to provide such contradictory and inconsistent results.

Olsen et al. (2001) emphasize the feature of LCA as a relative assessment due to the use of a functional unit, while ERA is an absolute assessment that requires very detailed information, e.g., on exposure conditions. It is concluded that the conceptual background and the purpose of the tools are different, but that overlaps in which they may benefit from each other occur.

TABLE 10.1
Comparison of Environmental Risk Assessment and Life-Cycle Assessment

Criteria	Environmental Risk Assessment	Life-Cycle Assessment
Object	Industrial process or activity	Functional unit, i.e., product or service, with its life cycle
Spatial scale	Site-specific	Global/site-generic
Temporal scale	Dependent on activity	Product life
Objective	Environmental optimization by risk minimization	Environmental optimization by reduction of potential emissions and resource use
Principle	Comparison of intensity of disturbance with sensitivity of environment	Environmental impact potential of substances
Input data	Specific emission data and environmental properties	General input and output of industrial processes
Dimension	Concentration and dose	Quantity of emissions
Reference	Exposure potential to threshold	Characterization factor
Result	Probability of hazard	Environmental effect score

10.4   MATHEMATICAL FOUNDATION

Because LCA and ERA belong to the environmental system analysis and are based on models of the real world, it is evident that the first step is to base the general strategy described, so far, on a mathematical framework that enables carrying out further steps. It includes, in particular, a procedure that allows knowing the quantity, situation and moment of the generation of the specific environmental interventions in the LCI. As introduced in Chapter 2, Castells et al. (1995) have presented an algorithm that uses an eco-vector. The different types of environmental loads, like chemical substances, can be classified in categories according to their environmental impacts; then different impact potentials can be calculated by the characterization factors presented in Heijungs et al. (1992). For example, CO_2, CH_4, CFC-11 and others can be aggregated in the global warming potential (GWP) and be expressed as CO_2 equivalents. The different chemicals are characterized by a specific weighting factor, i.e., the eco-vector e_v is converted into a weighted eco-vector $\tilde{e}_v$, as shown in Equation 10.1, where e_v^i is the specific [EL (environmental load)/kg] of the EL_i, $\tilde{e}_v^i$ is the specific weighted [ELeq/kg] of the EL_i, and λ^i is the specific weighting factor of the EL_i.

$$\tilde{e}_v = \tilde{e}_v^i = e_v^i \lambda^i \tag{10.1}$$

According to the goal of the general strategy, the eco-vector algorithm must be changed in order to make possible the assessment of the actual impacts caused by a specific process in a particular environment. Considering the example of a process chain with three processes, each process (PR) consumes raw material (RM) or an intermediate product (IP) and generates emissions and/or waste (SO_2, PM_{10}, Cd, etc.) per functional unit to obtain the functional unit, the product. When the LCI analysis is applied to the three processes, an eco-vector with the environmental interventions for all three processes is obtained, as illustrated in Figure 10.3. It is evident that the following LCIA generates only potential impacts because all site-specific information is lost when the LCIA is carried out.

Another approach to performing the environmental assessment of a functional unit consists in analyzing the actual impacts of each process according to a site-specific damage estimation concept (Figure 10.4). Consequently, the environmental loads of each process are also accounted for, but the evaluation is carried out for each process in its specific region. Each assessment contains the three consecutive procedures of fate analysis, application of impact factors and weighting across the impact categories. The results of each process assessment can be summed up if they are expressed in monetary units or by the same indicators. A damage profile is provided.

In order to express this different method in an algorithm based on the same principles as introduced before, the eco-vector must be transformed into an eco-technology matrix E_M. In this matrix, which is similar to the technology matrix mentioned by Heijungs (1998), there are M columns for M

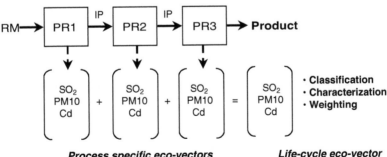

Process specific eco-vectors   Life-cycle eco-vector

FIGURE 10.3   Life-cycle inventory analysis according to the eco-vector principle (IP = intermediate product, PR = process, RM = raw material). (Reprinted from *J. Hazardous Mater.*, 77, Sonnemann, G.W. et al., pp. 91–106, Copyright 2000, with permission from Elsevier.)

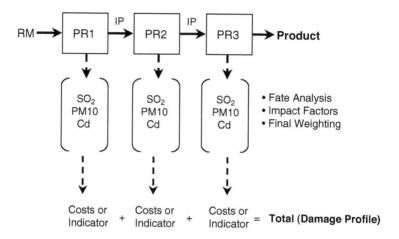

FIGURE 10.4 Determination of damage estimations by site-specific assessment (IP = intermediate product, PR = process, RM = raw material). (Reprinted from *J. Hazardous Mater.*, 77, Sonnemann, G.W. et al., pp. 91–106, Copyright 2000, with permission from Elsevier.)

linear processes, and N rows for the N environmental loads as those for the eco-vector. The example has three columns for the three processes and three rows for the three environmental loads, SO_2, PM_{10} and Cd. (See Equation 10.2 for an illustration.) It is evident that the environmental loads must be in the same order for all the processes and that a certain environmental intervention must be expressed always in the same unit.

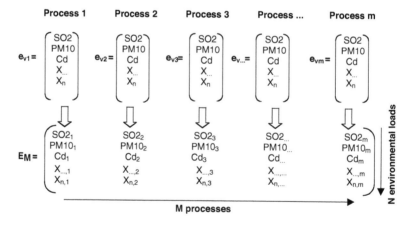

By analogy to the multiplication of the eco-vector by a weighting vector, the eco-matrix can be multiplied by another matrix, the weighting or damage-assigning matrix $\mathbf{W_M}$, which contains the fate analysis, impact factors and final weighting. This matrix assigns the damage cost or another damage indicator caused by one specific EL_i to each process. In $\mathbf{W_M}$ there are N columns for the N environmental loads and M rows for the M linear processes. In this way an M × M matrix is obtained, the weighted eco-technology matrix $\mathbf{WE_M}$, as shown in the following equation:

$$\mathbf{W_M}.\mathbf{E_M} = \mathbf{WE_M} \tag{10.3}$$

The main magnitude will be the trace D of $\mathbf{WE_M}$ (Equation 10.4), where E_M^{ij} is the specific [EL/kg] of the EL_i in the process j, W_m^{ji} is the specific damage factor for the EL_i in the process j, and D^j is the environmental cost or another indicator of the process j. D expresses the total environmental damage cost of the life cycle, if $\mathbf{W_M}$ represents a weighting by costs, or the damage indicator for the life cycle, if $\mathbf{W_M}$ represents a weighting by an indicator.

$$D = tr(\mathbf{WE_M}) = \sum_{ji}^{(W_M{}^{ji} \cdot E_M{}^{ij})} = \sum_{j}^{M}\left[\sum_{i}^{N}\left[{}^{(W_M{}^{ji} \cdot E_M{}^{ij})} \right]\right]$$

$$= \sum_{j}^{M} D^{jj\lambda} \sum_{j}^{M} D^{j}$$

(10.4)

The algorithm is illustrated in Equation 10.5 for the example of three processes and two environmental loads SO_2 and Cd.

$$
M\begin{bmatrix} W_M{}^{1,\,SO2} & W_M{}^{1,\,Cd} \\ W_M{}^{2,\,SO2} & W_M{}^{2,\,Cd} \\ W_M{}^{3,\,SO2} & W_M{}^{3,\,Cd} \end{bmatrix} \bullet
N\begin{bmatrix} E_M{}^{SO2,\,1} & E_M{}^{SO2,\,2} & E_M{}^{SO2,\,3} \\ E_M{}^{Cd,\,1} & E_M{}^{Cd,\,2} & E_M{}^{Cd,\,3} \end{bmatrix} =
M\begin{bmatrix} D^1 & D^{12} & D^{13} \\ D^{21} & D^2 & D^{23} \\ D^{31} & D^{32} & D^3 \end{bmatrix}
$$

$$\text{with } \begin{cases} D^1 = W_M{}^{1,\,SO2} \bullet E_M{}^{SO2,\,1} + W_M{}^{1,\,Cd} \bullet E_M{}^{Cd,\,1} \\ D^2 = W_M{}^{2,\,SO2} \bullet E_M{}^{SO2,\,2} + W_M{}^{2,\,Cd} \bullet E_M{}^{Cd,\,2} \to D = D^1 + D^2 + D^3 \\ D^3 = W_M{}^{3,\,SO2} \bullet E_M{}^{SO2,\,3} + W_M{}^{3,\,Cd} \bullet E_M{}^{Cd,\,3} \end{cases}$$

The values of the weighted eco-matrix $\mathbf{WE_M}$ in the diagonals D^1, D^2 and D^3 represent the environmental damage cost or damage indicator of processes 1, 2, and 3. If the values of $\mathbf{WE_M}$ are not in the diagonal, such as D^{12}, and given the assumption of linearity, they represent the environmental damage cost or damage indicator of the corresponding process, but in a different region, e.g., the damage cost or damage indicator that the process 2 would cause in the region 1. Consequently, it is possible to compare the effects of a certain process in different regions. Each component D^{kj} of the weighted eco-matrix is obtained by Equation 10.3, and, in Equation 10.6, is given the abbreviated mathematical way of expressing the contents, where k stands for the region k and j for the process j:

$$D^{kj} = \sum_{i}^{N}\left(W_M{}^{ki} \cdot E_M{}^{ij}\right)$$

(10.6)

The weighting matrix components for the EL_i corresponding to the different processes j and k are equal if the processes are situated in the same region. Indeed, that is the case for global impacts such as for the GWP, where this simplification allows working with one eco-vector for all the processes and one weighting vector for all regions, as shown in Equation 10.7. In the case of global weighting factors, the weighting matrix has the same components for all processes.

$$W_M{}^{ji} = W_M{}^{ki} <==> \text{region } (j) = \text{region } (k)$$

(10.7)

A special topic that must be considered is the question of mobile processes. If the process is a moving one, different regions may be involved so that Equation 10.8 holds true. By choosing the size of the region, the number of regions to consider for the corresponding mobile process is determined. If there is a mobile process with sufficient transport kilometers,

$$= = > \text{exists at least } 1 \ i; \ D^{kj} \neq D^{lj} \ V \ k \neq i \qquad (10.8)$$

This mathematical framework delivers a tool for introducing site-specific aspects in the life-cycle approach. The matrix algebra provides an elegant and powerful technique for the derivation and formulation of different tools in a life-cycle perspective.

10.5 OUTLINE OF THE COMBINED FRAMEWORK

10.5.1 OVERVIEW OF THE METHODOLOGY

With the mathematical framework at our disposal, the next task is to find a way to determine the eco-technology matrix and the weighting or damage-assigning matrix. It is proposed to base the environmental damage estimations of industrial process chains on the results of a conventional LCI analysis and one or more LCIA methods to answer the environmental management problem of interest.

In this way, one or more impact scores are calculated and this information can be used for a selection process in the form of a three-fold dominance analysis. In the three phases of the dominance analysis, the main media, processes and pollutants must be identified in a combination of quantitative and qualitative evaluation. In principle, such an evaluation should be carried out for each of the selected impact scores because each considers one particular type of environmental impact or weighting scheme. Then, the relevant processes and pollutants obtained are spatially differentiated according to the site or region that should be taken into account for the environmental impact assessment. Finally, the eco-technology matrix is elaborated for the predominant processes and pollutants in the assigned sites and regions.

In a next step, the fate and exposure and consequence analysis are carried out with different levels of detail for each process identified as relevant. The results of the fate and exposure and consequence analysis are the input for the damage-assigning matrices. In the case of global indicators like GWP, it is not necessary to perform a fate and exposure and consequence analysis with different levels of detail. Thus, these indicators can be used directly in the damage-assigning matrix. In particular, the GWP and the ozone depletion potential (ODP) are considered global indicators. According to Bare et al. (2000) and Udo de Haes and Lindeijer (2001), these potentials are important for the subarea of protection life support systems, which belong to the area of protection (AoP) of the natural environment and might be seen as having intrinsic value in their own right. The life support functions concern the major regulating functions that enable life on Earth (human and non-human)—particularly, regulation of the Earth's climate, hydrological cycles, soil fertility and bio-geo–chemical cycles. In the same way, depletion of the sub-AoP natural resources (abiotic, biotic and land) can be taken into account if the decision-maker considers these potentials important.

The flowchart in Figure 10.5 gives an overview of the procedure to generate the eco-technology and the damage-assigning matrices. The multiplication of the matrices yields the damage profile of each considered alternative, as well as interesting information for the optimization of process settings. In the case of using different impact scores, the same damage endpoints considered in the fate and exposure and consequence analysis related to these scores must be summed up.

The damage profile can be divided into damages to human health (mortality, cancer and morbidity), manmade environment, natural environment and global indicators (GWP and others). The application of a weighting and aggregation scheme, determined in the goal and scope definition, avoids a multicriteria analysis for a huge number of impact parameters—for instance, emergency room visits, asthma attacks, maintenance surface for paint, or yield loss of wheat, which are results of site-specific environmental evaluations. Next, each of the steps developed in Figure 10.5 will be explained in more detail to provide the user with a better understanding.

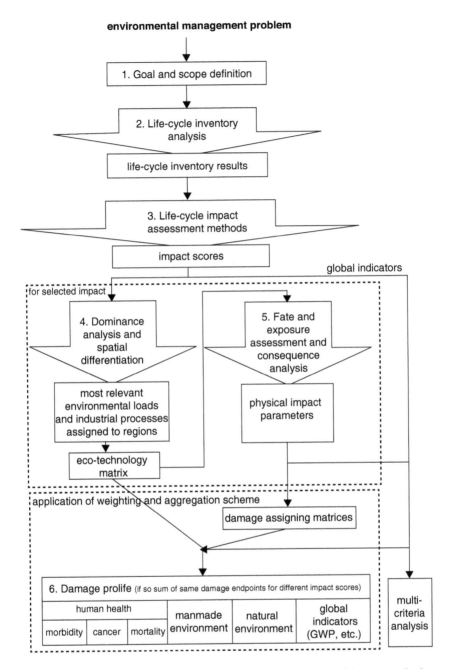

environmental management problem

1. Goal and scope definition

2. Life-cycle inventory analysis

life-cycle inventory results

3. Life-cycle impact assessment methods

impact scores

global indicators

for selected impact

4. Dominance analysis and spatial differentiation

5. Fate and exposure assessment and consequence analysis

most relevant environmental loads and industrial processes assigned to regions

physical impact parameters

eco-technology matrix

application of weighting and aggregation scheme

damage assigning matrices

6. Damage prolife (if so sum of same damage endpoints for different impact scores)

human health			manmade environment	natural environment	global indicators (GWP, etc.)
morbidity	cancer	mortality			

multi-criteria analysis

FIGURE 10.5 Overview of the procedure to generate the eco-technology and damage-assigning matrices.

10.5.2 Goal and Scope Definition

In the goal and scope definition (Figure 10.6) the decision-maker determines the cornerstones of the environmental damage estimations for industrial process chains that, in his opinion, are best fit to answer the environmental management problem of interest. Of course, he must do this while taking into account budget restrictions. In the goal definition, it must be decided which situations or scenarios will be assessed and compared. Here, situations refer to existing process chains while scenarios mean process chain options for the future (Pesonen et al., 2000).

1. Goal and scope definition

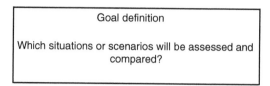

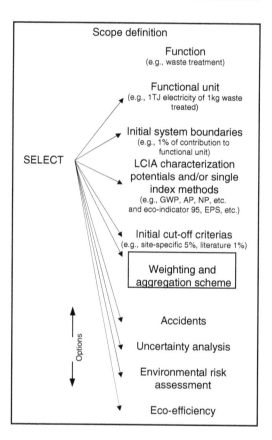

FIGURE 10.6   Goal and scope definition.

In the scope definition, the decision-maker must select the functional unit (e.g., 1 TJ electricity or 1 kg treated waste), the initial system boundaries (e.g., 1% contribution to functional unit), the LCIA characterization potentials (e.g., GWP, AP, NP, etc.) and/or single index methods or endpoint-orientated methods (e.g., eco-indicator 95 or 99, EPS, etc.). These requirements correspond to those for LCA according to the ISO 14040 series. However, in the methodology of environmental damage estimations for industrial process chains, more information is necessary to outline the study. These decision points are obligatorily the initial cut-off criteria (e.g., site-specific 5% and literature values 1%) and the weighting and aggregation scheme. Deciding if an uncertainty analysis should be included, if accidents should be considered and if the eco-efficiency of the process chain should be calculated is optional.

Initial cut-off criteria must be defined for the dominance analysis. These cut-off criteria serve to determine which media, processes and pollutants must be further studied in the fate and exposure and consequence analysis and in which way, e.g., site-specific or by literature values.

The methodology is based on the principle of transparency in the way the results are obtained. The format in which the results are desired, such as monetary values or physical impacts (e.g., cases of cancer), determines this. Therefore, the subjective elements, in particular the different parts of the weighting step, are assembled in the goal and scope definition before carrying out any analysis. For the weighting, decision-makers can follow the general decision tree presented in Figure 10.7. There are different options to evaluate the ELs, whose choice depends on the worldview of the decision-maker. Thus, the methodology avoids implicit decision-making common in endpoint-orientated LCIA methods.

For environmental loads that cause a GWP with global impact, ODP and other global indicators can be calculated. In such a case, these potentials have environmental relevance in the form of life-support functions and depletion of natural resources. First, decision-makers must select the environmental impacts they consider relevant for the environmental management problem under study;

Weighting: general decision tree

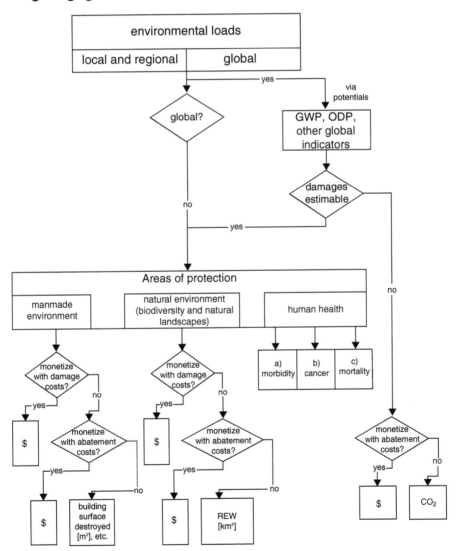

FIGURE 10.7 General decision tree for weighting.

then, they must decide, according to available knowledge, if the damages related to these potentials may be estimated. If they think that these damages cannot be estimated with acceptable reliability, then, for each global indicator, they must decide if they prefer to monetize the potential impact using abatement costs or to express it directly as a physical impact potential, e.g., CO_2 equivalent. These can be assessed in conjunction with the other environmental loads if they believe the damages to be estimable (see Figure 10.7).

Other environmental loads may cause local and regional damages, which can be divided into the AoPs published by Udo de Haes et al. (1999): manmade environment, natural environment and human health. In the cases of manmade environment and natural environment, the questions on which to decide are the same. Due to the complexity of the weighting options for the human health AoP, another decision tree has been drawn and is presented in Figure 10.8.

For the manmade environment and natural environment AoPs, the first question is whether the damage should be monetized in the form of environmental external costs, e.g., according to the ExternE approach (EC 1995). If decision-makers do not like this type of weighting, they must decide if they prefer to monetize the impacts using abatement costs or to express them directly as a physical impact parameter—in the case of manmade environment, for example, in maintenance

Weighting: decision tree for human health

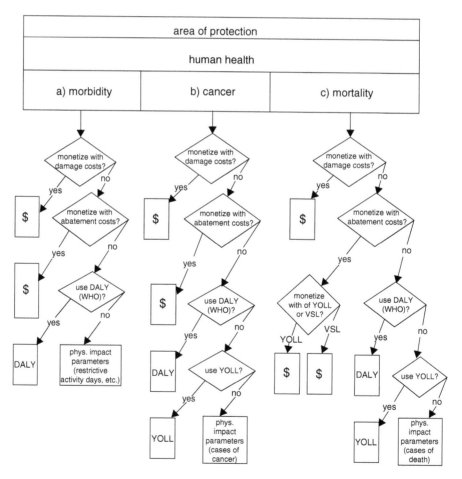

FIGURE 10.8   Decision tree for weighting for the AoP of human health.

surface for paint (m²) and yield loss of wheat (t) or for the natural environment in the REW (km²), as described in a previous chapter.

For all the damages to human health, the first questions concern the same decisions as for the other AoPs: whether to choose monetization and, if so, monetization by damage or external environmental costs, abatement costs or internal environmental costs. In the case of fatal effects, it must also be decided if the monetization of the damages should be done based on years of life lost (YOLL) or directly on value of statistical life (VSL). Additionally, due to the existence of internationally accepted damage indicators by the World Health Organization (WHO) in the form of disability-adjusted life years (DALY) (Murray and Lopez, 1996; Hofstetter, 1998) and YOLL (Meyerhofer et al., 1997), other types of weighting across the different damage endpoints are available.

Thus, in the case of damages that cause morbidity, the decision-maker must select among the assessment by DALY or physical impact parameters—for instance, emergency room visits, asthma attacks, restrictive activity days, etc. For cancer, a selection must be made among DALY, YOLL and the physical impact parameter cases of cancer. In the case of morbidity, the choice is among DALY, YOLL and the physical impact parameter cases of death. Finally, in the case of site-specific assessment, individual risk can be evaluated. These different weighting options are summarized for the decision-maker in the first table which forms part of Figure 10.9, named weighting of impacts, in which all four decision tables relevant for weighting and aggregation are presented together. Two other tables concern the discount rate for monetization and the cultural theory for DALY, and the last one concerns decisions related to the aggregation of damages. In this figure, default selections are presented as an illustration of a typical case for study of environmental damage estimations in industrial process chains.

In the table for the weighting of impacts, one entry must be made for each damage class. These damage classes are the manmade environment, natural environment and human health AoPs, as well as all the so-called global indicators such as GWP and ODP that could be related to the life support functions and resources sub-AoPs if resource depletion is considered an environmental problem (see Figure 10.9).

Human health is divided into morbidity, cancer and mortality. In principle, the monetization can be done for all the damage classes, either by damages or abatement costs. DALY can only be used for the damages to human health. Although no more than the preferred option must be selected for monetization and DALY, in the case of using physical impact parameters, the selected parameters should be mentioned here. For the monetization, a discount rate must be defined. Although in principle any rate can be chosen, here 0%, 3%, and 10% are proposed according to the standard values used in the ExternE project (EC 1995). In the case of using DALY, one cultural perspective must be selected. According to Hofstetter (1998), three archetypes represent human socioeconomic perceptions quite well: hierarchist, egalitarian and individualist.

Finally, it must be decided in which way the damage classes are aggregated. Of course, this is only possible if the classes have the same weighting unit, e.g., monetary values or DALYs. In principle, two options for aggregation exist. One option is to aggregate directly in the damage matrix, called intermediate aggregation, which is less laborious due to fewer matrix operations. The other option is to undertake a final aggregation reducing the number of components in the damage profile, which makes the steps more transparent, but risks confusion. In any case, the final result will be the same. Also, according to certain criteria, groups (for instance, AoPs), can be created to show unity.

It is generally accepted that damages to the manmade environment can be best evaluated by external environmental costs; however, in the default selections, it has been decided that this can also be done for damages to the AoP human health. The natural environment will be assessed by REW (relative exceedance weighted), and, as a global indicator, only the GWP is chosen. The default discount rate is 3%. An intermediate aggregation is selected for damages to the manmade environment and to human health.

Weighting and aggregation: decision tables

Weighting of impacts
(for each damage class you need one entry)

damage class (areas of protection, global indicator)	monetization		usage of DALY or YOLL	usage of physical impact parameters
	damages	abatement		
manmade environment	yes			
human health — morbidity	yes			
human health — cancer	yes			
human health — mortality	yes			
natural environment (biodiversity and natural landscapes)			✕	REW
GWP and other global indicator			✕	GWP

select one		
0%	3%	10%
	yes	

select one		
hierarchist	egalitarian	individualist

damage class (areas of protection, global indicator)	intermediate aggregation in the damage-assigning matrix	final aggregation in the damage profile
	create	*groups to show unity*
manmade environment	A	
human health — morbidity	A	
human health — cancer	A	
human health — mortality	A	
natural environment (biodiversity and natural landscapes)		
GWP, ODP and other global indicators		

FIGURE 10.9 Decision tables for weighting and aggregation.

Apart from the described selection of clear weighting schemes in order to obtain meaningful indicators, it must be acknowledged that, in principle, determining which dose–response and/or exposure–response functions to use, and even which dispersion model to apply, implies indirect value choices that (especially in the case of the dose–and exposure–response functions) can have very important influences on the final result. Thus, transparency on this point is recommended, as well as checking the preferences of the decision-maker; for instance, it can be said that, in general, internationally accepted standard values have a high level of reliability. All the presented criteria will be exemplified later in this chapter through a case study that has demonstrated the functionality of the weighting and aggregation scheme as one of its primary interests.

10.5.3 LIFE-CYCLE INVENTORY ANALYSIS

After the goal and scope definition, the life-cycle inventory analysis follows in the same way as in an LCA according to ISO 14040. An overview of the LCI analysis with its options is given in Figure 10.10.

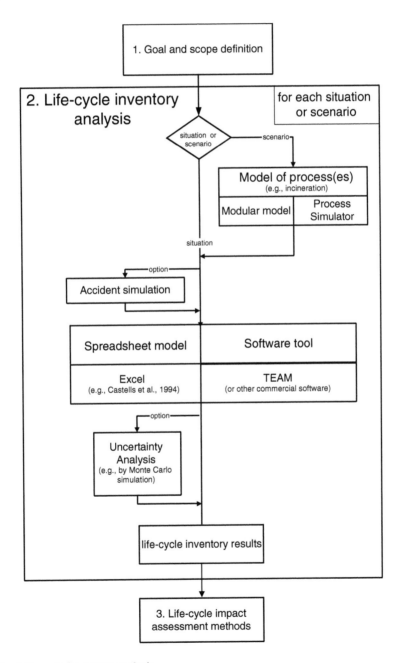

FIGURE 10.10 Life-cycle inventory analysis.

If a situation of an existing process chain is assessed, the measured data (ELs) from the core processes and those obtained from up- and downstream processes, e.g., by questionnaires, can be used to feed the LCI spreadsheet model or software tool that contains a more or less elaborated database with information for the background processes. If a future scenario for process chain options will be assessed, data can be generated by a model of core processes and linear adaptations of current data obtained from up- and downstream processes. The model can be a modular model, as described in Chapter 1, or a process simulator.

If the consideration of accidents was chosen in the goal and scope definition, potential environmental loads through the accidents must be generated by simulations with the corresponding analysis of the undesired events or accidents (AICHE, 1985; Aelion et al., 1995), as mentioned in Chapter 1. The proper LCI can be created by a spreadsheet model, e.g., Castells et al. (1995), or by a commercial software tool, e.g., TEAM, as presented in Chapter 2 and applied in Chapter 3. The incorporated database is important in the LCI model or software, especially for background processes like electricity production (e.g., Frischknecht et al. 1996). Another optional element is the uncertainty analysis, which can be carried out, for instance, by Monte Carlo (MC) simulation, as described in Chapter 9. By using probability distributions for the essential factors in an MC simulation (LaGrega, 1994), the inventory result can be transformed from a concrete value into a probability distribution around a mean value.

In principle, the proposed methodology does not need a complete LCI analysis with ISO 14041 as a basis. What it really needs is life-cycle inventory data about the process chain under study. These data can also be obtained by streamlined LCAs or simplified LCI approaches (Curran and Young, 1996). An important part of such methods is an iterative screening procedure (Fleischer and Schmidt 1997), which includes elements similar to those used in the methodology presented in this study by way of an iterative dominance analysis in order to identify the priorities. Evidently, a dominance analysis can also be applied directly in the LCI analysis during data collection (see Figure 10.10).

10.5.4 Life-Cycle Impact Assessment Methods

In the step following the ISO 14040 framework for LCA, one or more life-cycle impact assessment methods are applied to the LCI results. In the goal and scope definitions, the LCIA methods were selected. In Figure 10.11, an overview of the usage of life-cycle impact assessment methods is given; the main options are shown schematically:

- Midpoint potentials (e.g., GWP and HTP)
- Midpoint-based weighting methods (e.g., eco-indicator 95 and EDIP)
- Direct weighting methods (e.g., Tellus and EcoScarity)
- Endpoint-orientated methods (e.g., eco-indicator 99 and EPS)

More details about these methods can be found in Chapter 3.

The global indicators selected in the weighting scheme are considered separately. They are obtained in the characterization step in both options in which midpoints are calculated. Each global indicator feeds directly into the damage profile. If required, they are first monetized by abatement costs.

The midpoint-related LCIA methods allow calculating the environmental potential of the respective impact category in the characterization step. All presented LCIA methods except the midpoint potentials permit obtaining a single index to measure the environmental impact performance. The midpoint-based weighting methods require carrying out normalization and then a weighting step. Direct weighting methods omit the characterization and the normalization step. As endpoint-orientated method, eco-indicator 99 (see Chapter 3 for further details) does not contain explicitly midpoint results.

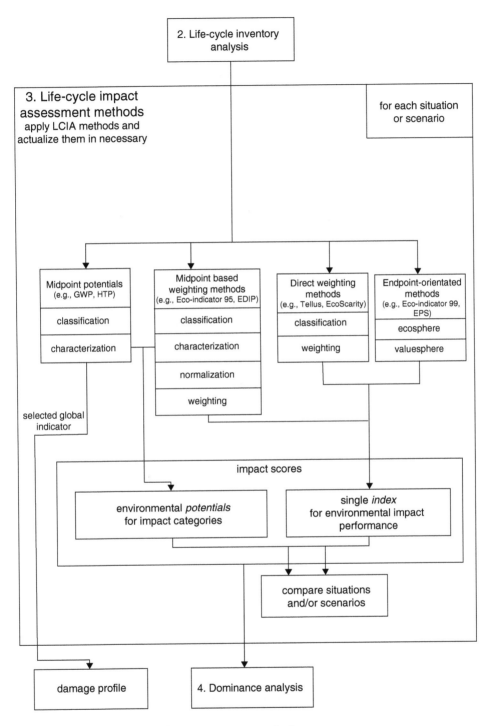

FIGURE 10.11 Usage of life-cycle impact assessment methods.

The results of the LCIA methods are called impact scores in Figure 10.11. These scores allow comparing the situations or scenarios on a midpoint level or endpoint-orientated level, but not in the most accurate way that is still feasible with regard to actual impacts and the consideration of spatial differentiation. Therefore, one or more selected impact scores are used in a dominance analysis in order to estimate in more detail the environmental damages of the main processes and pollutants in the studied process chain.

10.5.5   DOMINANCE ANALYSIS AND SPATIAL DIFFERENTIATION

Before determining the predominant processes and pollutants of the studied process chain for each selected impact score, it must be determined which media are primarily affected by the emissions of the process chain. Afterwards, the environmental loads and processes must be spatially differentiated. Therefore, in Figure 10.12, the dominance analysis for media and the spatial differentiation are presented together with a general overview of this selection procedure.

In the dominance analysis for media (i.e., air, water and soil), all emissions considered in the selected impact score are assigned to their media. Then, the contribution of each medium to the total impact score is presented graphically. Finally, a decision must be made about which of the media emitted to will be considered for further assessment. In this decision, qualitative arguments can also be used.

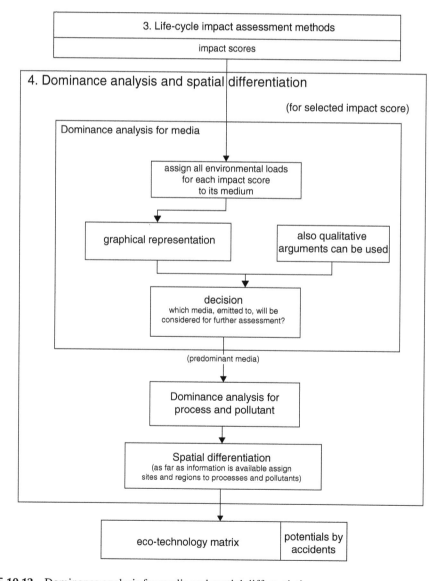

FIGURE 10.12   Dominance analysis for media and spatial differentiation.

The dominance analysis for processes and pollutants is applied to the predominant media and is structured in a similar way for processes and environmental loads (Figure 10.13). First, the percentage of the total of the selected impact score is calculated for each process and pollutant, then the defined initial cut-off criteria are applied. For processes and pollutant, respectively, the percentages and cut-off criteria are represented graphically to visualize the data. The graphical presentations

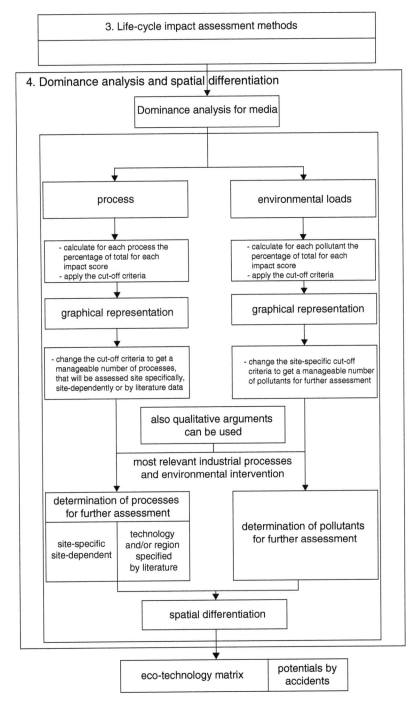

FIGURE 10.13   Dominance analysis for processes and pollutants.

may suggest redefining the respective cut-off criteria in order to obtain a manageable number of processes that can be assessed in detail. In principle, this is an iterative procedure to find the optimum. The decision-maker decides whether one or more impact scores are to be considered and in which relation to each other. This means also that dominance could be carried out with different cut-off criteria for different impact scores, for instance, 5% for the human toxicity potential and 10% for the acidification potential.

In the case of selection of processes, the processes to be assessed by site-specific and site-dependent factors (generally obtained by project-related impact assessment studies) and those to be assessed in a process-specific and/or region-dependent way (generally by values published in the literature) must be determined.

In a next step, the most relevant pollutants and industrial processes (differentiated in site-specific and site-dependent, as well as in process-specific and/or region-dependent ways) are determined. Qualitative arguments can also be used in this decision and, in a sustainable perspective, social and economic aspects are important. Therefore, it seems evident that, for instance, PCDD/Fs must be considered in a case study on waste incineration due to their relevance for discussion in society, although their percentage of the total impact score is less than the lowest selected cut-off criterion (See Figure 10.13).

Finally, the identified predominant pollutants and regions must be assigned to sites or regions. How far this is possible depends particularly on the information available about the location of a specific process. Thus, it must be taken into account that the site might be unknown; in this case, only the approximate global region can be assigned, but not the specific site. The spatial scale of the pollutants depends on their residence time in the respective medium. Many background processes whose LCI data are normally taken from databases are broadly spatially distributed. Here the question is to determine the most adequate size for a region; for example, in the case of electricity production, the LCI data for the electricity mix of a country generally are taken.

A problem that occurs in the assignment of sites is that often it is not the site that most influences the environmental damages, but the emission height. This can be concluded from the results for the site-dependent impact factors obtained in Chapter 11. Therefore, instead of regions, a differentiation is made essentially according to classes that have similar characteristics with regard to the emission situation (determined by the geographic site and the stack height). However, in this methodology, they are called regions because this term illustrates the idea behind spatial differentiation much better.

Consistent with Equation 10.7, the world is the corresponding region for pollutants that cause a global environmental impact like CO_2. In agreement with Equation 10.8, in the case of mobile processes, i.e., transports, it must be decided if the environmental loads can really be assigned to one region only or must be differentiated among two or more regions if the distance is long enough. The determined processes assigned to sites and regions are the M processes, and the chosen pollutants are the N pollutants of the eco-technology matrix. The eco-matrix may always contain a part of potential environmental loads if accidents have been considered.

10.5.6 FATE AND EXPOSURE ASSESSMENT AND CONSEQUENCE ANALYSIS

The level of detail in the fate and exposure assessment depends on the determined importance of the respective process. The few processes that contribute most to the overall environmental impact should be assessed on a site-specific basis, if possible; other important processes can be evaluated by corresponding region-dependent or technology-dependent impact assessment factors published in the literature (e.g., Krewitt et al., 2001). For airborne pollutants due to transport processes, an evaluation based on site-dependent impact assessment by statistically determined factors for generic classes seems to be most adequate. Nigge (2000) has proposed such a method; in this study it was further developed and applied to the case study explained in Chapter 11. However, it still must be considered as an approach in development.

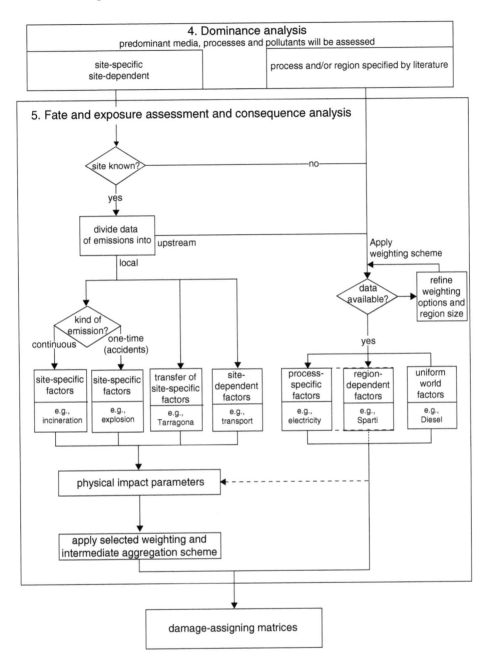

FIGURE 10.14 Fate and exposure assessment and consequence analysis.

Figure 10.14 gives an overview of the fate and exposure and consequence analysis with the different levels of detail. The results of this analysis are the basis for the damage-assigning matrices.

For the processes identified as most important, a site-specific or site-dependent assessment is carried out if the site is known. If this is not the case, the corresponding process must be treated as those processes that have been determined to be evaluated process- and/or region specified by literature values. If the site is known, the data about the emissions in the LCI must be divided into upstream-related data, which must be evaluated by literature values, and the foreground process-related data or local emissions.

Only the obtained local emission data can be further assessed. If potential emissions due to accidents are taken into account in the LCI and the eco-technology matrix, in each case the kind of emission (continuous or one time) for a site-specific assessment must be checked. An example of a site-specific assessment of continuous emissions (the ERA and the IPA) is carried out for Tarragona's municipal solid waste incinerator (MSWI) as described in Chapters 4 and 9. An example of a one-time emission is that from an explosion. Once a site-specific impact assessment has been carried out in a region, and in this way site-specific factors have been estimated, the results can be transferred to another process in the same region, using a transfer factor, if necessary, for the stack height, as proposed by Rabl et al. (1998) in the VWM (see Chapter 3). Such a transfer is the use of results of the site-specific impact assessment of Tarragona's MSWI for another process within the region, e.g., for the ash treatment operation situated in Constanti, a few kilometers away from the municipality of Tarragona.

If site-dependent impact assessment factors according to the approach outlined in Chapter 11 are available, then these factors allow estimating the environmental damages due to airborne emissions in the way of an adequate trade-off between accuracy and practicability. This holds true especially for transport (tkm) because it can be considered to be a number of industrial processes that take place (at one time after the other) in different regions. Through site-specific and site-dependent impact assessments, physical impact parameters are obtained that can be converted into indicators or environmental costs by applying the selected weighting scheme. Moreover, if laid out in the goal and scope definition, the intermediate aggregation can take place.

For processes in which literature values should be used, the first action is to check if the desired data are available in the literature. If this is not the case, the weighting options of the scope definition and/or the size of the region must be redefined. Because the assessment of site- and region-dependent damage endpoints is an issue that more or less started in the mid-1990s, not much data have been published. Therefore, it is quite possible that determined indicators or cost types are not available in the literature for the processes and pollutants of certain regions.

If data are available, then whether a classification is possible according to technology and/or region must be determined for each process. If one or both options are possible, then technology- and/or region-dependent factors from the literature are used to estimate the corresponding damage. For instance, data on external costs are available for electricity production (kWh) in regions of Spain. Another example is the mentioned region-dependent impact factors, e.g., in YOLL, published by Krewitt et al. (2001) for different European countries and some world regions. If a classification according to technology and/or region is not possible, then uniform world factors must be applied. For instance, Rabl et al. (1998) have published a uniform world model for air emissions. Diesel production and the related process chain that takes place all over the world is an example of a process difficult to classify.

Depending on the selected weighting scheme and the available data in the literature, physical impact parameters, damage indicators or environmental costs are obtained. The physical impact parameters can be summed up directly with those obtained in the site-specific and site-dependent assessment. Damage indicators and environmental costs can be gathered together according to the selected intermediate aggregation scheme. Options for site-specific impact assessment are explained in Figure 10.15 for the medium of air; in principle, these can also be applied to other media. For example, Schulze (2001) presents site-orientated impact assessments for the medium of water in relation to LCAs for detergents, using the integrated assessment model GREA-TER in an adapted version valid for products instead of chemical substances.

Based on the data of local air emissions, site-specific factors are calculated for the predominant pollutants. These factors can be expressed in the form of physical impact parameters before being weighted and aggregated according to the scheme chosen in the goal and scope definition. The fate and exposure analysis can be carried out in a generic or detailed way. The generic way uses an integrated impact assessment model, e.g., EcoSense (described in Chapter 4). Such an integrated impact assessment model consists of a Gaussian dispersion model for the pollutant transport

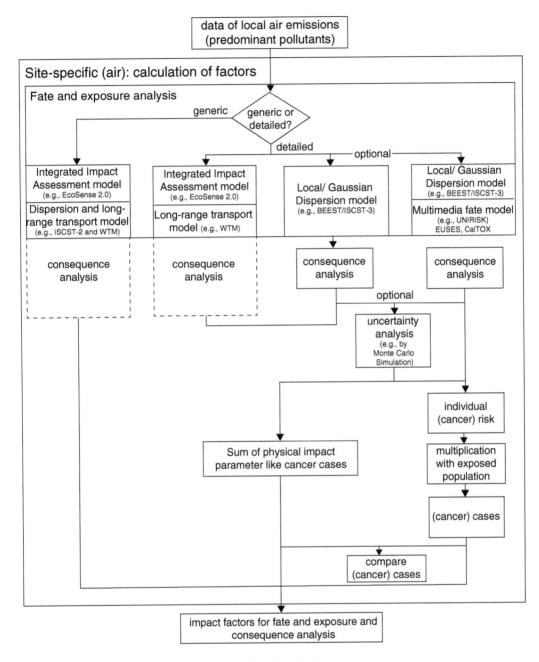

FIGURE 10.15 Site-specific impact assessment for air emissions.

near the emission point (i.e., approximately ≤100 km) and another transport model for the long-range pollutant transport (i.e., approximately >100 km). In the case of EcoSense 2.0, the models included are ISCST-2 and WTM. The integrated impact assessment model EcoSense also includes an elevated number of dose–response and exposure–response functions that can be used for the consequence analysis. The level of detail in the database of an integrated impact assessment model is limited, e.g., the resolution of population densities is not as detailed as it could be when using a geographic information system.

In the case of a more detailed assessment, only the long-range transport model of the integrated impact assessment model is used (e.g., WTM in EcoSense). An independent Gaussian dispersion model (see Chapter 4) is applied (e.g., ISCST-3 in BEEST) for the transport near the emission point and more detailed geographic data like those in ERA provided by a geographic information system are employed. Figure 10.15 proposes carrying out the consequence analysis for the damages due to long-range transport within the integrated impact assessment model and doing so separately, e.g., on a spreadsheet, for the damages generated by the emissions near the emission point. Then the physical impact parameter of both assessments can be summed up. This is a practical proposal and, in principle, the consequence analysis can also be carried out with this.

When calculating the site-specific impact factors, a lot of work is done to complete an ERA at the same time, as described in Chapter 4. The only additional step is the application of the multimedia fate model (e.g., UniRisk, EUSES, CalTOX, etc.). Another way of stating this is to say that if an ERA has been carried out (as described in Chapter 4) for one of the identified predominant processes of an industrial process chain, then it is quite easy to compute the impact factors necessary for a quite accurate environmental damage estimation of industrial process chains. The realization of an ERA is seen in Figure 10.15 as an optional element; this approach is further explored in Chapter 12.

Also, in the case of environmental risk assessment, a consequence analysis must be carried out. Due to the high level of detail in this type of study, thresholds can also be considered; thus, in the case of human health assessment, the consequence analysis is not restricted to carcinogenic effects and respiratory diseases (see Chapter 4 for further explanations) and other types of toxic effects can also be considered.

When carrying out an ERA, the individual risk must be calculated, e.g., developing cancer due to the increment of a certain pollutant in the atmosphere. Multiplication by the absolute number of population exposed would then allow obtaining an estimate of the damage in the form of physical impact parameters, e.g., cancer cases. These calculated impact parameters, like cancer cases, could also be compared with those provided by the application of the IPA on a local scale. Such a comparison can provide correction factors.

Another optional element is the uncertainty analysis that can be carried out, e.g., by MC simulation according to the framework proposed in Chapter 9. In the same way as for the LCI results, the outcomes of the fate and exposure analysis can be transformed from a concrete value into a probability distribution around a mean value.

Apart from site-specific impact assessments in this work, the focus for the fate and exposure and consequence analysis has been on site-dependent impact assessment as an adequate trade-off between accuracy and practicability. The entire method is largely explained in Chapter 11. The other options of the fate and exposure and consequence analysis (Figure 10.14) do not need further explanations because they are similar to the site-specific impact assessment or consist only in the application of published values for impact indicators.

10.5.7 DAMAGE PROFILE

Figure 10.16 presents the last part of the obligatory steps for the methodology of environmental damage estimations for industrial process chains. In principle, this flowchart consists of an illustration of the developed mathematical framework. For each selected impact score, the eco-technology matrix is multiplied with the damage-assigning matrices. The result can be another matrix or a vector for each damage-assigning matrix, or a vector for the case of global impacts. In that case, the elements of the vector need only be summed up. In the case of the matrix, a sum must be made of the elements of the main diagonal, the trace. The matrix allows checking in which location a process would have caused less damage.

The sum obtained by each matrix calculation provides a damage-endpoint-per-impact score that then forms the damage profile. If the same damage endpoints have been estimated in different impact scores, they must be summed up. The damage profile might contain potential damages in case accidents have been simulated. In principle, depending on the selected intermediate aggregation scheme, the damage profile can be broken up into damages to the human health (morbidity, cancer, mortality), manmade environment and natural environment AoPs, as well as the so-called

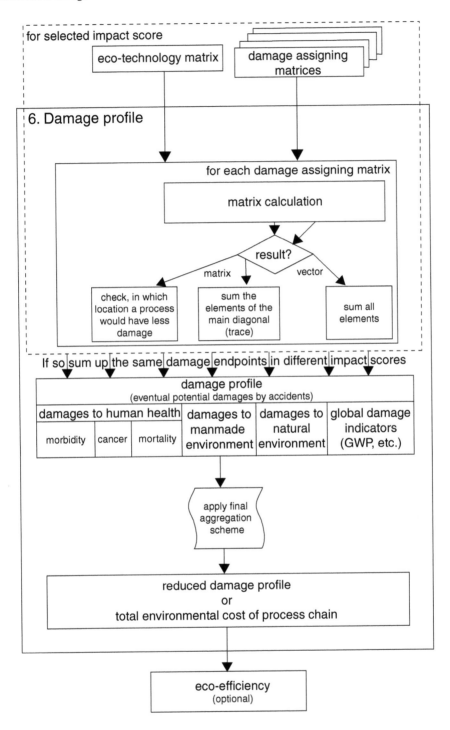

FIGURE 10.16 Damage profile.

global damage indicators, which could be related to the life support functions and resources sub-AoPs if resource depletion is considered an environmental problem.

If the damage profile consists of different damage endpoints, their number can be further reduced by applying the selected aggregation scheme. In this way, a reduced damage profile or an estimation of the total environmental cost of the process chain under study is finally obtained.

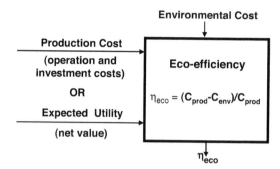

FIGURE 10.17 Eco-efficiency of an industrial process chain.

10.5.8 ECO-EFFICIENCY

A further optional element is the calculation of the eco-efficiency of the industrial process chain for which environmental damages have been estimated. The concept of eco-efficiency has been proposed as an expression of sustainability for economic activities (see Chapter 1). Eco-efficiency has been defined as the delivery of competitively priced goods and services while progressively reducing environmental impacts.

Thus, measuring eco-efficiency, η_{eco}, by the coefficient of the difference between production costs, C_{prod}, and external environmental costs, C_{env}, to the production costs, C_{prod}, has been suggested. Although production costs are easy to obtain, the environmental costs are not so visible. Nevertheless, they can be estimated by the presented methodology of environmental damage estimations for industrial process chains.

The expression of Figure 10.17 is applicable to the final result of the environmental damage estimations for industrial process chains if this is expressed in a monetary unit. This figure illustrates the procedure to calculate eco-efficiency according to this expression. Instead of production costs (operation and investment costs), the expected utility or net value could be used for determining eco-efficiency.

10.6 POSSIBLE ROLE OF THE FRAMEWORK IN ENVIRONMENTAL MANAGEMENT

Application patterns of LCA have been studied by Frankl and Rubik (1999). This section provides suggestions for possible applications of the presented methodology. The developed framework can be of interest for the following stakeholders:

- Public administration (to justify taxes and to optimize waste management strategies)
- Financial entities (to ascertain possible risks of payments)
- Companies (to contribute to a system of continuous environmental improvement, to choose between sites of a process chain from an environmental perspective, and to demonstrate reduction of environmental damages by the inversion in abatement technologies from a life-cycle viewpoint)
- Consumers and general community (to have the possibility of obtaining products and services of minor pollution and to learn about the environmental damages behind industrial process chains)

10.6.1 PUBLIC ADMINISTRATION

Table 10.2 shows in detail the possible applications for public administration, among which are the mentioned green tax by using external environmental costs, technology assessment in general, public policy planning and, in particular, environmental justice and end-of-life management. The idea behind the

TABLE 10.2

Applications for Public Administration

Application	Example	Optional Element
Technology assessment	Energy production, waste treatment, transport	None
Green tax	Electricity, transport or all other types of products and/or services	Using external environmental costs
Public policy planning	Future scenario assessment of energy production, waste treatment, transport	Possible eco-efficiency
Environmental justice	Waste incineration, landfill, cracker	None
End-of-life management	Waste incineration, landfill, integrated waste management planning (similar to Wizard by Ecobilan Group, 1999), but on a more detailed level with regard to spatial aspects (second generation)	Possible eco-efficiency

application of end-of-life management is to develop a second generation of integrated waste management software like Wisard (Ecobilan Group, 1999) that considers the problem setting also; this is quite related to environmental justice. By combining site-specific aspects with life-cycle considerations, new plans for waste management can assess transport processes and their routes in relation to waste treatment facilities and their sites and the respective environmental damages in an integrated manner.

10.6.2 FINANCIAL ENTITIES

The possible applications for financial entities (banks and insurance companies) are presented in Table 10.3. Main applications are the assessment of risk from new plants or changes in existing ones, and the respective insurance for plants, as well as the evaluation of strategic business plans. Important optional elements are accident simulations and questions of eco-efficiency.

10.6.3 PRODUCTION AND OTHER SERVICE COMPANIES

Table 10.4 represents the many possible applications for production and other service companies. Many of the applications proposed in this table could interest companies using the developed methodological framework. Especially important are the integrated end-of-life management and parallel optimization of the life cycle and problem setting from an environmental point of view. Furthermore, all the questions related to chain responsibility with the assessment of supplier and waste treatment companies and their technologies are important. The latter includes the evaluation of accident risks with regard to processes like overseas petroleum transport, for which several big companies were declared responsible in recent years.

In addition to these companies, the methodological framework is certainly relevant for the waste treatment sector that wants to document improvement of the environmental profile to the public. Moreover, these companies would probably be interested in gaining the confidence of the public concerning new plants or changes in the overall waste management plan. The developed methodology can assist greatly in this task by generating an important amount of quite objective, relevant information that should allow all interested parties to find a convincing solution.

TABLE 10.3

Applications for Financial Entities (Banks and Insurance Companies)

Application	Example	Optional Element
Planning of new plants or changes	Waste incineration, landfill, cracker	Possible accidents
New insurance of existing plant	Waste incineration, landfill, cracker	Possible accidents
Business strategic planning	Future scenario assessment	Possible eco-efficiency

TABLE 10.4
Applications for Production and Other Service Companies

Application	Example	Optional Element
Planning of new plants	Waste incineration, landfill, cracker	Possible accidents
EMAS: to show that the best solution is chosen (improvement of performance and liability by extension of EMAS)	Change of process in chemical industry	Possible eco-efficiency
Documentation of environmental profile's improvement toward the public	Advanced gas treatment system in waste incineration plant	None
Process improvement	Change of process in chemical industry, additional flue gas cleaning in waste incineration	Possible eco-efficiency
Determination of most problematic processes	Production with a huge number of complex steps	Special dominance analysis, possible accidents
Chain responsibility (assessment of supplier and waste treatment companies and their technology)	Supply chain management, avoiding use of undesired substances and occurrence of accidents	Especially dominance analysis, possibly accidents
Determination of most problematic emission and medium emitted to	Identification of points for environmental improvement; e.g., waste reduction in chemical industry plant	Especially dominance analysis
Claim for subventions	Trade association of waste management industry	Possible uncertainty analysis
Marketing strategies by communication of environmental profile to consumers	Waste incineration, chemical industry plant	None
Business strategic planning	Future scenario assessment	Possible eco-efficiency
End of life management	Waste incineration, landfill, integrated waste management planning (similar to Wizard by Ecobilan Group, 1999), but on a more detailed level with regard to spatial aspects (second generation)	Possible eco-efficiency
Optimization of setting (new plant)	Waste incineration, landfill, chemical industry	Possible ERA

10.6.4 Consumers and Society in General

The possible applications for consumers and society in general are shown in Table 10.5. The main application consists in education and communication as potential common ground for discussion about environmental damage estimations due to certain problematic industrial process chains such as the waste incineration that has been part of the public discussion on environmental aspects for the last decade.

TABLE 10.5
Applications for Consumers and Society in General

Application	Example	Optional Element
Education and communication as potential common ground for discussion	Increased information available by all considered applications	Possibly all optional elements
Eco-labeling and/or environmental product declarations	The system would be a quite accurate way to obtain relevant results about the environmental damages caused by product systems, but at the moment, it seems not to be very practicable for this purpose. Nevertheless, in the future, this information might be available due to advances in information technologies.	Possibly uncertainty analysis

Another potential application could be in future eco-labeling and environmental product declarations because the developed methodology would be a quite accurate way of obtaining relevant results about the environmental damages caused by product systems. At the moment, however, it seems to be not very practicable for this purpose. Nevertheless, in the future, this information might be available due to advances in the information technologies.

10.7   EXAMPLE: NECESSARY TECHNICAL ELEMENTS

The methodology permits various linkages with other environmental management tools and concepts as well as technical elements. In the previous chapters we have seen several of them.

Since the entire methodology is a combination of different analytical tools that, in general, have been developed for other applications, the reader is asked to identify those concepts, tools and technical elements behind the presented framework.

In principle, the LCA methodology has been developed for the environmental assessment of product systems. LCA is an important element for the LCI analysis and LCIA methods and for providing region technology-dependent impact factors. The next tool is the impact pathway analysis (IPA) that is the fruit of a project to assess the externalities of electricity production. IPA is crucial for the fate and exposure and consequence analysis, including the weighting and aggregation schemes. Furthermore, ERA has its origin in assessment of the behavior of chemical substances in the environment. It is, of course, relevant in the fate and exposure and consequence analysis and has influenced not only IPA, but also the LCIA methods. Other methods that are indirectly involved are cost-benefit analysis (CBA), accident investigation and process simulation.

Finally, several technical elements are behind the methodology and its flowchart; here, only the main technical elements are outlined. The terminology is based on Dale and English (1999). Due to the LCA part, a functional unit must be defined and allocation models must be used in the LCI analysis. In the fate and exposure analysis, fate and transport models (Gaussian, long-range transport, multimedia) are applied. In the consequence analysis, dose–response and exposure–response functions are employed. Laboratory exposure and animal tests are often the basis for dose–response functions and epidemiological studies are the basis for exposure–response functions. In order to make the weighting transparent, decision trees have been established. Socioeconomic impact assessment is conducted with the presented different methods to evaluate external costs. ERA uses individual risk or population risk based on the lifetime average dose. Accident simulation needs the help of event and fault trees. Process simulation is, in principle, an engineering model. Eco-efficiency could be calculated with the net present value of an expected utility and the uncertainty analysis carried out with MC simulation.

10.8   CASE STUDY: ENVIRONMENTAL DAMAGE ESTIMATIONS FOR THE WASTE INCINERATION PROCESS CHAIN

The methodology has successfully been applied to the case study on the process chain related to waste incineration, with interesting results. The information obtained by the developed methodology might be crucial in the future for decisions on further improvement of existing and new waste management systems. The presented algorithm is applied to the life-cycle inventory of the electricity produced by the MSWI of Tarragona, Spain.

10.8.1   GOAL AND SCOPE DEFINITION

Two operating situations of the MSWI are compared: the situation in 1996 and the current situation after the installation of an advanced gas removal system (AGRS). More data on the MSWI are presented in Chapter 1.

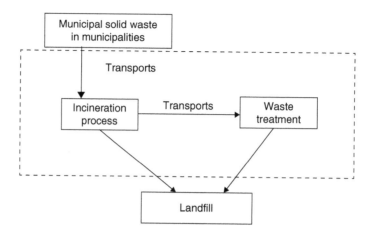

FIGURE 10.18   Boundaries of the studied system. (Reprinted from *J. Hazardous Mater.*, 77, Sonnemann, G.W. et al., pp. 91–106, Copyright 2000, with permission from Elsevier.)

The function of the MSWI is to treat the household waste of the surroundings of Tarragona. However, produced TJ of electricity is chosen as the functional unit, as was done in the existing LCA study (see Chapter 2). The study comprises in its system boundaries all the processes from municipal waste disposal in containers to the landfill of the final waste (Figure 10.18). The midpoint-based weighting method with single index, eco-indicator 95 (Goedkoop 1995) is used as LCIA method. The method is based on the characterization factors presented by Heijungs et al. (1992) and uses equal scores of distances to political targets (for more information see Chapter 3).

For reasons of resource economy, in the dominance analysis, only all processes with a contribution greater than 10% will be selected for site-specific impact assessment by a particular study. However, the remaining processes with more than 1% should be evaluated by the transfer of available site-specific damage data or the use of values published in the literature to estimate environmental damages. The selected weighting and aggregation scheme corresponds to the one presented as an example in Figure 10.9, which means that three indicators have been selected for the weighting of impacts. For the human health and manmade environment AoPs, external environmental costs (EEC), according to the European Commission (EC 1995), have been used.

A lot of criticism exists concerning monetary evaluation of environmental damages. For this reason, special attention is paid to arguments to monetize these damages. The impossibility of summing up the noneconomic impact endpoints necessarily implies a value judgment. Because most decisions must confront the reality of the marketplace, the most useful measure is the cost of the damages. This information allows society to decide how much should be done for the protection of the environment by public institutions and how much of the damage cost should be internalized so that a functional unit is consistent with the market. Further information on this topic can be found in Chapter 3 and in the huge externality studies for electricity production carried out in parallel in the EU (EC, 1995) and U.S. (ORNL/REF 1995).

No acceptable economic method exists for damage evaluation of the natural environment AoP (biodiversity and landscape). Therefore, the evaluation must be carried out through an ecological damage parameter. In the present study, the parameter applied is the REW ecosystem area in which the critical load of a pollutant is exceeded (UN-ECE, 1991); see also Chapter 3 for details.

The global damages that might occur in the future due to the emission of greenhouse gases are highly uncertain for forecasting and monetization. Therefore, the climate change has been expressed in the form of the GWP as in the LCIA (Albritton and Derwent, 1995).

The potential occurrence of accidents is not considered in this case study. Uncertainty analysis for the LCI and the site-specific environmental impact assessment of the MSWI emissions are described in Chapter 9. An environmental risk assessment for the same plant has been carried out and the results are presented in Chapters 4 and 9. Based on the results of the environmental damage estimations of the waste incineration process chain, the eco-efficiency will also be calculated.

10.8.2 Life-Cycle Inventory Analysis

The LCI analysis is described in Chapter 2. Here, the existing results are used for creating the eco-technology matrix of the environmental damage estimation for industrial process chains.

10.8.3 Life-Cycle Impact Assessment Method

The detailed results of the application of the LCIA method, eco-indicator 95, can be found in Chapter 3, in which a comparison of the results for the two situations based on the eco-indicator 95 is conducted. In the presented methodology, the impact score is further applied for the dominance analysis.

10.8.4 Dominance Analysis and Spatial Differentiation

In the current case study, the predominant medium to which the emissions are emitted is clearly air. The predominant pollutants are those that have been selected by dominance analysis for the uncertainty analysis in Chapter 9. Figure 10.19 presents the contribution of the considered processes in the LCI analysis to the total environmental impact potential measured as eco-indicator 95. It is evident that, in this case study, only the incineration process contributes with more than 10% to the total environmental impact potential. Therefore, it is the only process that will be assessed in a site-specific way by a particular study. The corresponding site is Tarragona.

The other industrial processes with more than 1% contribution will be considered in two ways: (1) the data obtained from an IPA study of the incineration process in Tarragona are considered

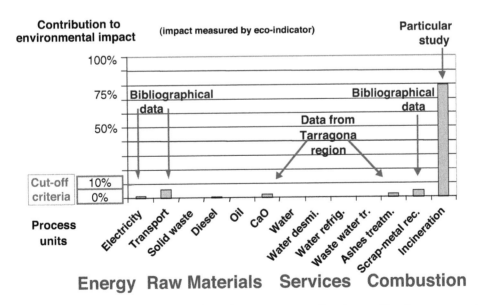

FIGURE 10.19 Dominance analysis for the relevant processes for site-specific damage assessment. (Reprinted from *J. Hazardous Mater.*, 77, Sonnemann, G.W. et al., pp. 91–106, Copyright 2000, with permission from Elsevier.)

TABLE 10.6

Eco-Matrix with Life-Cycle Inventory Data for Selected Environmental Loads, Divided into Those Assessable by Environmental Costs and Those Responsible for Ecosystem Damages and Global Warming (Current Situation)

Relevant Process	Electricity	Transport	CaO	Ash Treatment	Scrap-Metal Recycling	Incineration
Region	Spain	Tarragona/ Catalonia	Tarragona/ Catalonia	Tarragona	Madrid/ Spain	Tarragona
kWh	2.96×10^3	0.00	0.00	0.00	0.00	0.00
tkm	0.00	2.64×10^4	0.00	0.00	0.00	0.00
kg PM_{10}	2.08	8.45	1.11×10^2	4.73	6.03	2.82×10^1
kg As	3.12×10^{-4}	2.93×10^{-3}	3.74×10^{-4}	8.66×10^{-4}	2.09×10^{-3}	3.29×10^{-2}
kg Cd	6.97×10^{-4}	4.28×10^{-4}	4.74×10^{-5}	1.41×10^{-4}	3.06×10^{-4}	3.87×10^{-2}
kg Ni	1.29×10^{-4}	5.56×10^{-3}	1.31×10^{-3}	2.42×10^{-3}	3.98×10^{-3}	4.93×10^{-2}
kg VOC	4.74×10^{-1}	1.68×10^1	5.84×10^{-1}	4.86	1.21×10^1	0.00
ng Dioxins TEQ	8.05×10^1	7.32×10^2	3.83×10^2	7.57×10^2	5.22×10^2	1.17×10^4
kg CO	3.36×10^{-1}	2.92×10^1	2.46	1.16×10^1	2.09×10^1	2.35×10^2
kg SO_2	1.86×10^1	1.93×10^1	1.16×10^1	1.17×10^1	1.38×10^1	1.77×10^2
kg NO_2	3.30	9.31×10^1	5.76	3.17×10^1	6.65×10^1	1.12×10^3
kg CO_2	1.48×10^3	9.13×10^3	8.02×10^3	5.76×10^3	6.53×10^3	2.25×10^5
kg dichloro-methane	9.90×10^{-7}	3.17×10^{-6}	5.64×10^{-7}	1.11×10^{-6}	2.26×10^{-6}	0.00
kg Halon-1301	2.27×10^{-5}	1.09×10^{-3}	3.61×10^{-5}	3.21×10^{-4}	7.80×10^{-4}	0.00
kg methane	4.51	1.41×10^1	7.57	8.19	1.01×10^1	0.00
kg N_2O	1.27×10^{-2}	1.00	3.55×10^{-2}	2.75×10^{-1}	7.16×10^{-1}	0.00
kg tetrachloro-methane	3.03×10^{-7}	1.03×10^{-5}	6.41×10^{-6}	3.78×10^{-6}	7.36×10^{-6}	0.00
kg trichloro-methane	1.55×10^{-8}	1.10×10^{-6}	7.06×10^{-7}	4.03×10^{-7}	7.83×10^{-7}	0.00

valid for all processes in the Tarragona region and (2) the remaining processes must be evaluated using damage information of similar situations obtained from the literature. The processes that contribute with more than 1% and less than 10% to the total environmental impact are spatially differentiated in the following way. The "production of CaO" and "treatment of ashes" processes take place in the Tarragona region. In the LCI analysis, data for the "electricity generation" are used from the so-called Spanish mix. The environmental impacts of "transport" and "scrap-metal recycling" also depend on the Spanish region (see Figure 10.19).

The results of the inventory analysis of the current situation for the relevant processes and the selected environmental loads are presented in Table 10.6. This table includes the eco-technology matrix with submatrices: the ELs from kWh to NO_x correspond to the first matrix for the economic damage parameter; the second matrix for the ecological damage parameter consists only of SO_2 and NO_x, and the third matrix for the global damage parameter includes the other loads from CO_2 to trichloromethane (see Table 10.6).

10.8.5 FATE AND EXPOSURE AND CONSEQUENCE ANALYSIS

In the next step, the corresponding three damage-assigning matrices for the selected three indicators are established. An attempt is made to give a particular value to each environmental load for a specific region or site because the respective indicator depends on the characteristics of the respective region or site for each environmental load. If no region-dependent or process-specific damage estimates are available for a pollutant, a general model, the uniform world model (UWM; Rabl et al., 1998; see Chapter 3) is used in the case of environmental external costs.

TABLE 10.7
Damage-Assigning Matrix for External Environmental Cost (EEC) in U.S.$

Relevant Process	Region and Source	kWh	tkm	kg PM10	kg As	kg Cd	kg Ni	kg VOC	PCDD/Fs ng TEQ	kg CO	kg SO$_2$	kg NO$_2$
Electricity	Spain[a]	0.040	0.00	0	0.0	0	0.0	0.00	0.00	0.00	0	0
Transport	South Europe[b]	0.00	0.31	0	0.0	0	0.0	0.00	0.00	0.00	0	0
CaO	Tarragona, site-specific transfer	0.00	0.00	23	3.0	27	61	0.73	2.6×10^{-8}	1.03	13	11
Ash treatment	Tarragona, site-specific transfer	0.00	0.00	23	3.0	27	61	0.73	2.6×10^{-8}	1.03	13	11
Scrap-metal recycling	UWM[c]	0.00	0.00	14	156	19	2.6	0.73	1.7×10^{-5}	0.002	13	19
Incineration	Tarragona, site-specific	0.00	0.00	23	3.0	27	61	0.73	2.6×10^{-8}	1.03	13	11

[a] CIEMAT, ExternE national implementation Spain—final report, Contract JOS3-CT95-0010, Madrid, 1997.
[b] Friedrich, R. et al., *External Costs of Transport*, Forschungsbericht Band 46, IER, Universität Stuttgart, Germany, 1998.
[c] Rabl, A. et al., *Waste Manage. Res.*, 16(4), 368–388, 1998.

The largest damage-assigning matrix is elaborated for the environmental damage cost; the matrix for the current situation is presented in Table 10.7. For the electricity process, the corresponding region is Spain; the evaluation is done for the environmental load kWh (technology dependent) through the project data published by CIEMAT (1997). The transport is assessed by literature sources for truck metric ton km (tkm) in South Europe (Friedrich et al., 1998), because data for Spain are not available. The CaO supply, ash treatment and incineration processes are located in the Tarragona region. For the incineration process described in Chapter 9, the IPA obtains the weighting factors of the Tarragona region for all environmental loads that accept VOC (volatile organic carbon). The VOC values as well as the evaluation of the scrap-metal recycling are taken from the UWM (Rabl et al., 1998).

Due to lack of literature data, the matrix of the REW ecological damage indicator is not yet fully established; information is only available for the processes in the Tarragona region. Nevertheless, the matrix for the current situation is presented in Table 10.8 to illustrate the complete framework. However, in the case of the global damage indicator, the situation is different. The GWP is independent of the site where the substances like CO_2 or CH_4 are emitted; therefore a vector rather than a weighting matrix is obtained. See Table 10.9 for the current situation.

TABLE 10.8
Damage-Assigning Matrix for Relative Exceedance Weighted (REW) Area in km²

Relevant Process	Region and Source	kg NO$_2$	kg SO$_2$
Electricity	—	—	—
Transport	—	—	—
CaO	Tarragona, site-specific factor	1.31×10^{-5}	1.36×10^{-6}
Ash treatment	—	—	—
Scrap-metal recycling	Tarragona, site-specific factor	1.31×10^{-5}	1.36×10^{-6}
Incineration	Tarragona, site-specific	1.31×10^{-5}	1.36×10^{-5}

TABLE 10.9

Damage-Assigning Matrix for Global Warming Potential (GWP) in kg CO_2 Equivalents

Relevant Process	Region and Source	kg CO_2	kg dichlo-rometh.	kg Halon-1301	kg meth.	kg N_2O	kg tetrachloro-meth.	kg trichloro-meth.
All processes	World[a]	1	15	4900	11	270	1300	25

[a] Data from Albritton, D. and Derwent, R., *IPCC, Climate Change*, Cambridge University Press, Cambridge, 1995.

10.8.6 DAMAGE PROFILE

The multiplication of the damage-assigning matrices with the eco-technology matrix (the respective parts of the inventory table) yields the damage profile (see Table 10.10 for the current situation), i.e., the ensemble of the three damage parameters, per functional unit. In the weighted eco-technology matrix or damage matrix for the current situation, the external environmental cost per functional unit is estimated as 28,200 U.S.$/TJ electricity, which is in the range of other externality studies for waste incineration (CIEMAT, 1997; Rabl et al., 1998). That is the sum of the diagonal corresponding to the sum of the regions considered in the life-cycle study, here called life-cycle region. It becomes clear that the damage generated by the functional unit would be higher if all the processes were in the UWM and less if they were all in the Tarragona region (situated on the Mediterranean coast). In the case of the ecological damage parameter, only a few accurate values in the weighted eco-matrix are known. For the

TABLE 10.10

Damage Profile for the Current Situation (with Advanced Acid Gas Treatment System Scenario 2)

Regions	Electricity	Transport	CaO	Ash Treatment	Scrap-Metal Recycling	Incineration	TJ Electricity
			EEC in U.S.$				
Spain	119	0	0	0	0	0	—
South Europe	0	8.184	0	0	0	0	—
Tarragona	322	1.470	2.765	610	1.050	15.020	21.237
Tarragona	322	1.470	2.765	610	1.050	15.020	21.237
Uniform world model	328	2.130	1.827	816	1.522	23.740	30.363
Tarragona	322	1.470	2.765	610	1.050	15.020	21.237
Life-cycle region	119	8.184	2.765	610	1.522	15.020	28.220
			REW in km²				
Spain	—	—	0	0	—	0	—
South Europe	—	—	0	0	—	0	—
Tarragona	—	—	9.12×10^{-5}	4.31×10^{-4}	—	1.50×10^{-2}	1.55×10^{-2}
Tarragona	—	—	9.12×10^{-5}	4.31×10^{-4}	—	1.50×10^{-2}	1.55×10^{-2}
Uniform world model	—	—	0	0	—	0	1.55×10^{-2}
Tarragona	—	—	9.12×10^{-5}	4.31×10^{-4}	—	1.50×10^{-2}	1.55×10^{-2}
Life-cycle region	—	—	9.12×10^{-5}	4.31×10^{-4}	—	1.50×10^{-2}	1.55×10^{-2}
			GWP in kg CO_2 Equivalents				
World	1.54E+03	9.56×10^3	8.11×10^3	5.93×10^3	6.83×10^3	2.25×10^5	2.57×10^5

REW ecosystem area, the diagonal elements sum up to 0.0155 km^{2}per TJ electricity—a relative small value because the studied region is not sensitive to acidification. For the GWP, a weighted eco-vector is obtained. The sum of the vector components yields to 2.57×10^5 kg CO_2 equivalents per TJ electricity. This result is similar for electricity generation by fossil fuels (Frischknecht et al., 1996; CIEMAT, 1997).

The damage profile of the current situation, Scenario 2, can now be compared to the situation of the MSWI in 1996 without AGRS as a different process design option, Scenario 1. The damage profile for the former situation in 1996 is presented in Table 10.11. Although the eco-indicator 95 shows a reduction of 60% in the total score between the former and the current situations (Chapter 3), the environmental damage estimations show less reduction. The external environmental costs decrease 10%, but the REW increases 3% and the GWP increases 10%. These damage indicators do not show reduction due to the increased transport and the reduced energy efficiency in the MSWI process chain after the installation of the advanced gas treatment system (as explained in Chapter 2) that affects the emissions of NO_x and CO_2, which are crucial for these damage factors.

Figure 10.20 offers a comparison of the external environmental costs with the eco-indicator 95 for the plant before and after installation of the advanced gas removal system. It can be seen that the environmental external cost estimations give much more weight to the transport processes. The transport contributes approximately 25% to the external environmental costs before installation of the advanced gas treatment system and 30% afterwards. According to the eco-indicator 95, the transport adds less than 10% in the current situation and less than 5% in the former situation. In contrast, the incineration process is much more predominant in accordance with the eco-indicator 95 methodology. More than 90% of the eco-indicator 95 is attributed to the incineration in the former situation and it still accounts for 80% in the current situation. The external environmental cost

TABLE 10.11
Damage Profile for the Former Situation in 1996 (without Advanced Acid Gas Treatment System Scenario 1)

Regions	Electricity	Transport	CaO	Ash Treatment	Scrap Metal Recycling	Incineration	TJ Electricity
			EEC in U.S.$				
Spain	114	0	0	0	0	0	114
(South) Europe	0	7.812	0	0	0	0	7.812
Tarragona	308	1.400	0	100	1.001	20.890	23.699
Tarragona	308	1.400	0	100	1.001	20.890	23.699
Uniform world model	313	2.030	0	133	1.450	28.260	32.187
Tarragona	308	1.400	0	100	1.001	20.890	23.699
Life-cycle region	114	7.812	0	100	1.450	20.890	30.365
			REW in km^2				
Spain	—	—	0	0	—	0	0.0000
(South) Europe	—	—	0	0	—	0	0.0000
Tarragona	—	—	0	7.03×10^{-5}	—	0.0150	0.0151
Tarragona	—	—	0	7.03×10^{-5}	—	0.0150	0.0151
Uniform world model	—	—	0		—	0	0.0000
Tarragona	—	—	0	7.03×10^{-5}	—	0.0150	0.0151
Life-cycle region	—	—	0	7.03×10^{-5}	—	0.0150	0.0151
			GWP in CO_2 Equivalents				
World	1.47×10^3	9.11×10^3	0	9.69×10^2	6.51×10^3	2.14×10^5	2.32×10^5

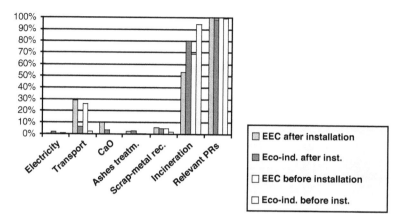

FIGURE 10.20 Comparison of external environmental cost with eco-indicator 95 for a plant before and after installation of an advanced acid gas removal system.

estimations only assign 70% to the incineration process for the situation in 1996 and even less in the current situation, namely, a little more than 50%. This different relation between the transport and incineration processes also explains why the external environmental costs have decreased less than the eco-indicator 95 score. Another important process in the current situation is CaO production. Moreover, the selected relevant processes make up nearly 100% of the total impact (Figure 10.20).

10.8.7 Eco-Efficiency

Based on the environmental cost estimations obtained in the previous sections, the eco-efficiency of the process chain can be calculated. According to the expression in Figure 10.17, production costs are needed as an additional element. According to information from SIRUSA (Nadal, 1999), the production costs are:

- Former situation: 32,000 U.S.$/TJ
- Current situation: 37,000 U.S>$/TJ

These costs are based on the yearly total operation costs of the MSWI, including the financial costs for the investments made. The production costs are higher than the market price of 0.07 U.S.$/kWh; the deficit is paid by the price for the waste treatment, which can be considered a service the MSWI is providing to society. By applying the presented formula, the following eco-efficiency for the process chain is obtained:

- Former situation: $\eta = 5\%$
- Current situation: $\eta = 25\%$

The production of electricity by waste incineration is not very eco-efficient if avoided environmental charges for the waste treatment are not considered, as is done in the existing LCA of the MSWI process chain in Tarragona (see Chapter 2 for further explanations). Nevertheless, installation of an advanced acid gas removal system increases eco-efficiency significantly.

10.9 QUESTIONS AND EXERCISES

1. Sum up the challenges and key points of a strategy for a methodology that integrates life-cycle and risk assessment.
2. Which types of impacts are global and which regional? When is spatial differentiation important?
3. What are the steps to obtain the eco-technology and damage-assigning matrices?

4. Explain which decisions must be made in the weighting and aggregations phase.
5. Give examples of LCIA methods with impact score for
   - Midpoint-based
   - Direct weighting
   - Endpoint weighting
6. Describe the fate and exposure and consequence analysis procedure.
7. Which technical elements are used in the different steps of the methodology?
8. For which reasons do you think the external cost approach gives less importance to the incineration process and more to the transports than the eco-indicator method?
9. Mention possible applications of an integrated life-cycle and risk assessment methodology.
10. The provided data correspond to the emissions of three different pollutants emitted in three different processes. By means of the correct eco-vector, express the environmental loads associated with each pollutant and process:
    1. Pretreatment process: 51.10 kg/kg feed of CO_2; 2.7 kg/kg feed of SO_2; 1kg/kg feed of NO_x
    2. Steam generation: 196 kg/kg feed of CO_2; 1.2 kg/kg feed of SO_2; 0.099 kg/kg feed of NO_x
    3. Electricity production: 293 kg/kg feed of CO_2; 1.8 kg/kg feed of SO_2; 2.3 kg/kg feed of NO_x
11. The data in Tables 10.12 through 10.14 correspond to three different scenarios of a real process. Discuss which scenario corresponds to the minimum environmental damage for the pollutants under consideration.
12. Using the following fictive data for the eco-technology matrix and the damage-assigning matrix, calculate the corresponding damage profile:
    SO_2 emission of process 1: 100 g SO_2/kg product process1
    SO_2 emission of process 2: 10 g SO_2/kg product process2
    SO_2 emission of process 3: 30 g SO_2/kg product process3
    Cd emission of process 1: 0.15 g Cd/kg product process1

TABLE 10.12
Scenario 1: Steam Production with Refining and Electricity Production

Mixture (kg emission/t C_3H_8)		Separation (kg emission/t C_3H_8)		
		Steam	Electricity	Venting
CO_2	504.49	619.94	151.59	—
SO_2	2.14	5.67	0.87	—
COD	0.055	0.0041	0.00046	—
VOC	2.73	8.51	0.77	0.00422

Source: Spanish Mix, as reported in Frischknecht et al (1996).

TABLE 10.13
Scenario 2: Steam Production with Refining and Electricity Production

Mixture (kg emission/t C_3H_8)		Separation (kg emission/t C_3H_8)		
		Steam	Electricity	Venting
CO_2	504.49	619.94	110.23	—
SO_2	2.14	5.67	0.00108	—
COD	0.055	0.0041	0.00000173	—
VOC	2.73	8.51	0.00084	0.00422

Source: Norwegian Mix, as reported in Frischknecht et al (1996).

TABLE 10.14

Scenario 3: Steam Production with Refining and Electricity Production (Natural Gas Combustion)

Mixture (kg emission/t C_3H_8)		Separation (kg emission/t C_3H_8)		
		Steam	Electricity	Venting
CO_2	504.49	619.94	573.29	—
SO_2	2.14	5.67	0.0137	—
COD	0.055	0.0041	0.00035	—
VOC	2.73	8.51	0.185	0.00422

Cd emission of process 2: 0.10 g Cd/kg product $_{process2}$
Cd emission of process 3: 0.02 g Cd/kg product $_{process3}$
Ni emission of process 1: 0.25 g Ni/kg product $_{process1}$
Ni emission of process 2: 0.15 g Ni/kg product $_{process2}$
Ni emission of process 3: 0.08 g Ni/kg product $_{process3}$
Damage due to SO_2 in region 1: 0.01 U.S.$/kg SO_2 $_{region1}$
Damage due to SO_2 in region 2: 0.006 U.S.$/kg SO_2 $_{region2}$
Damage due to SO_2 in region 3: 0.008 U.S.$/kg SO_2 $_{region3}$
Damage due to Cd in region 1: 0.005 U.S.$/kg Cd $_{region1}$
Damage due to Cd in region 2: 0.004 U.S.$/kg Cd $_{region2}$
Damage due to Cd in region 3: 0.009 U.S.$/kg Cd $_{region3}$
Damage due to Ni in region 1: 0.003 U.S.$/kg Ni $_{region1}$
Damage due to Ni in region 2: 0.002 U.S.$/kg Ni $_{region2}$
Damage due to Ni in region 3: 0.008 U.S.$/kg Ni $_{region3}$

13. Knowing the impact scores given in Figure 10.21 for different processes of a product life-cycle, discuss for this case which cut-off criteria and which further treatment method (no consideration, site specific, site dependent or technology region dependent) would be most appropriate for each of the given processes.

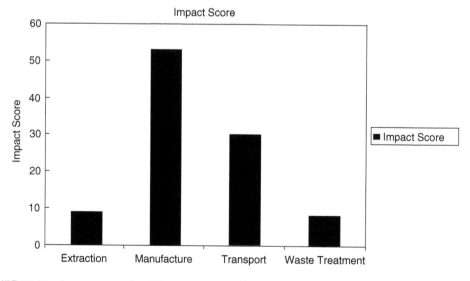

FIGURE 10.21 Impact scores for different processes of a product life cycle.

REFERENCES

Aelion, V., Castells, F., and Veroutis, A. (1995), Life-cycle inventory analysis of chemical processes, *Environ. Prog.*, 14(3), 193–200.

AICHE (American Institute of Chemical Engineers) (1985), Guidelines for hazard evaluation procedures, prepared by Batelle Columbus Division for The Center for Chemical Process Safety.

Albritton, D. and Derwent, R. (1995), Trace gas radiative forcing indices, in *IPCC, Climate Change*, Cambridge University Press, Cambridge, UK.

Bare, J.C., Hofstetter, P., Pennington, D.W., and Udo de Haes, H.A. (2000), Midpoints versus endpoints—The sacrifices and benefits, *Int. J. LCA*, 5, 319–326.

Castells, F., Aelion, V., Abeliotis, K.G., and Petrides, D.P. (1995), An algorithm for life-cycle inventory, *AICHE Symp. Series Pollut. Prev. Process Product Modifications*, 90(303), 151–160.

CHAINET (European Network on Chain Analysis for Environmental Decision Support) (1998), Definition document, CML, Leiden University, Leiden, the Netherlands.

CIEMAT (1997), ExternE national implementation Spain—final report, contract JOS3-CT95-0010. Madrid, Spain.

Curran, M.A. and Young, S. (1996), Report from the EPA conference on streamlining LCA, *Int. J. LCA*, 1(1), 57–60.

Dale, V. and English, M. (1999), *Tools to Aid Environmental Decision-Making*, Springer-Verlag, New York.

EC (European Commission) (1995), *ExternE—Externalities of Energy*, 6 vols., EUR 16520-162525. DG XII Science, Research and Development, Brussels, Belgium.

Ecobilan Group (1999), WISARD—Waste-Integrated Systems Assessment for Recovery and Disposal, Software, Paris, France.

Fleischer, G. and Schmidt, W.P. (1997), Integrative screening LCA in an ecodesign tool, *Int. J. LCA*, 2(1), 20–24.

Frankl, P. and Rubik, F. (1999), *Life-Cycle Assessment in Industry and Business—Adoption Patterns, Applications and Implications*, Springer-Verlag, Berlin, Germany.

Friedrich, R., Bickel, P., and Krewitt, W. (1998), *External Costs of Transport*, Forschungsbericht Band 46, IER, Universität Stuttgart, Germany.

Frischknecht, R., Bollens, U., Bosshart, S., Ciot, M., Ciseri, L., Doka, G., Hischier, R., and Martin, A. (ETH Zürich), Dones, R., and Gantner, U. (PSI Villigen) (1996), *Ökoinventare von Energiesystemen - Grundlagen für den ökologischen Vergleich von Energiesystemen und den Einbezug von Energiesystemen in Ökobilanzen für die Schweiz*, 3rd ed., ETH Zürich: Gruppe Energie-Stoffe-Umwelt, PSI Villigen: Sektion Ganzheitliche Systemanalysen.

Goedkoop, M.J. (1995), Eco-indicator 95—final report, NOH report 9523, Pré Consultants, Amersfoort, the Netherlands.

Heijungs, R. (1998), Economic drama and the environmental stage, PhD thesis, CML, Leiden University, Leiden, the Netherlands.

Heijungs, R., Guinée, J.B., Huppes, G., Lankreijer, R.M., Udo de Haes, H.A., and Wegener-Sleeswijk, A. (1992), Environmental life-cycle assessment of products—guide and backgrounds, technical report, CML, University of Leiden, Leiden, the Netherlands.

Hofstetter, P. (1998), *Perspectives in Life-Cycle Impact Assessment—a Structured Approach to Combine Models of the Technosphere, Ecosphere and Valuesphere*, Kluwer Academic Publishers, London, UK.

Krewitt, W., Trukenmüller, A., Bachmann, T., and Heck, T. (2001), Country-specific damage factors for air pollutants—A step towards site-dependent life-cycle impact assessment, *Int. J. LCA*, 6(4), 199–210.

LaGrega, M.D., Buckingham, P.L., and Evans, J.C. (1994), *Hazardous Waste Management*, McGraw-Hill, New York.

Meyerhofer, P., Krewitt, W., and Friedrich, R. (1997), Extension of the accounting framework, final report. ExternE Core Project, IER, Universität Stuttgart, Stuttgart, Germany.

Murray, C.J.L. and Lopez, A.D. (1996), *The Global Burden of Disease*, vol. 1 of *Global Burden of Disease and Injury Series*, WHO, Harvard School of Public Health, World Bank, Harvard University Press, Boston, MA.

Nadal, R. (1999), Planta incineradora de residuos sólidos urbanos de Tarragona, report for Master en Enginyeria i Gestió Ambiental (MEGA), Universitat Rovira i Virgili, Tarragona, Spain.

Nigge, K.M. (2000), *Life-Cycle Assessment of Natural Gas Vehicles, Development and Application of Site-Dependent Impact Indicators*, Springer-Verlag, Berlin, Germany.

Olsen, S.I., Christensen, F.M., Hauschild, M., Pedersen, F., Larsen, H.F., and Toerslov, J. (2001), Life-cycle impact assessment and risk assessment of chemicals—A methodological comparison, *Environ. Impact Assessment Rev.*, 21, 385–404.

ORNL/REF (Oak Ridge National Laboratory and Resources for the Future Oak Ridge) (1995), *U.S.—EC Fuel Cycle Study*, reports No. 1 through 8, McGraw-Hill/Utility Data Institute, Washington, DC.

Pesonen, H., Ekvall, T., Fleischer, G., Huppes, G., Jahn, C., Klos, Z.S., Rebitzer, G. et al. (2000), Framework for scenario development in LCA, *Int. J. LCA*, 5(1), 21–30.

Rabl, A., Spadaro, J.V., and McGavran, P.D. (1998), Health risks of air pollution from incinerators—A perspective, *Waste Manage. Res.*, 16(4), 368–388.

Schulze, C. (2001), Modeling and evaluating the aquatic fate of detergents, PhD thesis, Universität Osnabrück, Germany.

Sonnemann, G.W., Schuhmacher, M., and Castells, F. (1999), From LCIA via risk assessment to environmental external cost evaluation—what are the possibilities? *Presented at 9th SETAC-Europe Annual Meeting*, Leipzig, Germany.

Sonnemann, G.W., Schuhmacher, M., and Castells, F. (2000), Framework for the environmental damage assessment of an industrial process chain, *J. Hazardous Mater.*, B77, 91–106.

Spirinckx, C. and de Nocker, L. (1999), Comparative life-cycle analysis and externality analysis of biodiesel and fossil diesel fuel, *Presented at 9th SETAC-Europe Annual Meeting*, Leipzig, Germany.

Udo de Haes, H.A. and Lindeijer, E. (2001), The conceptual structure of life-cycle impact assessment, final draft for the Second Working Group on Impact Assessment of SETAC-Europe (WIA-2), in Udo de Haes et al. (2002), *Life-Cycle Impact Assessment: Striving towards Best Practice*, SETAC Order #SB02-5.

Udo de Haes, H.A., Joillet, O., Finnveden, G., Hauschild, M., Krewitt, W., and Mueller-Wenk, R. (1999), Best available practice regarding categories and category indicators in life-cycle impact assessment—background document for the Second Working Group on LCIA of SETAC-Europe, *Int. J. LCA*, 4, 66–74; 167–174.

UN-ECE (United Nations Economic Commission for Europe) (1991), Protocol to the 1979 convention on long-range transboundary air pollution concerning the control of emissions of volatile organic compounds or their transboundary fluxes, report ECE/EB AIR/30, UN-ECE, Geneva, Switzerland.

11 Site-Dependent Impact Analysis[1]

Guido Sonnemann, Ralph O. Harthan, Karl-Michael Nigge,
Marta Schuhmacher, and Francesc Castells

CONTENTS

---

[1] Selected text in Sections 11.2, 11.3, and 11.4 has been reprinted with permission from Nigge, K.M., *Life-Cycle Assessment of Naturlal Gas Vehicles: Development and Application of Site-Dependent Impact Indicators*, Copyright 2000 by Springer-Verlag, Berlin, Germany.

11.1   INTRODUCTION

In this chapter, we present a provisional approach to overcoming the disadvantages of site-generic and site-specific methods. This approach is one of the recently developed site-dependent impact assessment methods that can be considered a trade-off between exactness and feasibility. As with site-specific approaches, fate, exposure and effect information are taken into account, but indicators applicable for classes of emission sites, rather than for specific sites, are calculated. That is the trade-off between the accurate assessment of the impacts and the practicability of spatial disaggregation for impact assessments in a life-cycle perspective. We note the developing nature of these methods, but regard them as the most promising alternative for future estimations of environmental damages in industrial process chains. The approach is adapted so that it fits perfectly into the methodology presented in the previous chapter. A flowchart for site-dependent impact assessment is proposed and the algorithm of the methodology is applied using the calculated site-dependent impact indicators.

The consideration of spatial differentiation in LCIA was proposed first by Potting and Blok (1994). However, it took time until developments for site-dependent impact assessment such as those by Potting (2000) and Huijbregts and Seppälä (2000) were made in an operational way, especially for acidification and eutrophication. Moreover, several approaches have been presented for human health effects due to airborne emissions. Exemplary damage factors for a number of European countries are provided by Spadaro and Rabl (1999). Potting (2000) establishes impact indicators that take into account different release heights, population density, and substance characteristics such as atmospheric residence time and dispersion conditions. The release height is statistically linked to several industrial branches. Typical meteorological data for four zones within Europe are used, but the issue of local dispersion conditions is not addressed and no operational guidance for the determination of population densities based on sufficiently detailed data is provided.

Moriguchi and Terazono (2000) present an approach for Japan in which meteorological conditions are set to be equal for all examples. Nigge (2000) offers a method for statistically determined population exposures per mass of pollutant that considers short-range and long-range exposure separately and allows taking into account dispersion conditions and population distributions using sufficiently detailed data in a systematic way. Impact indicators are derived that depend on the settlement structure class and the stack height. However, a general framework that is also valid for other receptors is not proposed; the dispersion conditions are considered to be equal for all classes in the case study and a quite simple dispersion model is used.

In this study a general applicable framework for site-dependent impact assessment by statistically determined receptor exposures per mass of pollutant is proposed that addresses some of the mentioned shortages. Receptor density and dispersion conditions are related to form a limited number of representative generic spatial classes, as suggested by Potting and Hauschild (1997) and recommended by Udo de Haes et al. (1999). The basis for the classification is statistical reasoning, assuming no threshold (Potting 2000). For each class and receptor, incremental receptor exposures per mass of pollutant can be calculated. Finally, the incremental exposures are transformed into damage estimations.

The general framework was applied to the case of population exposures due to airborne emissions in the Mediterranean region of Catalonia, Spain. A differentiation was made with regard to dispersion conditions, stack height and atmospheric residence time; a sophisticated dispersion model was applied and a geographic information system (GIS) was used.

11.2   GENERAL FRAMEWORK FOR SITE-DEPENDENT IMPACT ANALYSIS

Udo de Haes (1996) distinguished the following dimensions of impact information that are relevant for life-cycle impact assessment (LCIA): information concerning (1) effect, (2) fate and exposure, (3) background level and (4) space. The first two dimensions directly refer to the cause–effect chain. The last two can be interpreted as additional conditions related to the processes in the considered chain.

Dimensions of impact information	Levels of sophistication
1. Information on the effect of a pollutant	standard ⟶ NOEC ⟶ slope
2. Pollutant fate information	none ⟶ some ⟶ full
3. Target information	none ⟶ some ⟶ full

FIGURE 11.1 Dimensions of impact information and levels of sophistication in life-cycle impact assessment. (From Wenzel, H. et al., *Environmental Assessment of Products. Volume 1–Methodology, Tools and Case Studies in Product Development*, Chapman & Hall, London, UK, 1997. With permission.)

Based on Udo de Haes' (1996) proposal, Wenzel et al. (1997) suggested the relation of dimensions of impact information and levels of sophistication presented in Figure 11.1. The first dimension in this figure is in accordance with the proposal of Udo de Haes (1996); only the exposure information has been removed. In this way, the second dimension covers all information connected to the source (emission, distribution/dispersion, and concentration increase). The third dimension comprises the third dimension from Udo de Haes (1996) and, additionally, all other types of information about the receiving environment and/or target system (background concentration, exposure increase, sensitivity of the target system, etc.)

With each of the three dimensions, the characterization modeling addresses different levels of sophistication. In the effect analysis, the sophistication increases from the use of legal threshold standards via the application of no observable effect concentration (NOEC) to the integration of slope in the impact factors. These factors may include no, some or comprehensive fate information. The same holds true for the information on the target system.

Category indicators for toxicity, but also for other endpoint-oriented impacts, are generally calculated by multiplying the emitted mass M of a certain pollutant p with a fate and exposure factor F and an effect factor E, i.e., the slope of the dose–response and exposure–response functions (Nigge, 2000). In the general case in which the transfer of the pollutant across different environmental media or compartments (e.g., air, water, soil) needs to be considered, Equation 11.1 gives the category indicator of incremental damage, $\Delta D_p^{nm}/M_p^n$ (damage, such as cases of cancer, years of life lost [YOLLs] and disability-adjusted life years [DALYs] or U.S.$ for external environmental costs), characterizing the effect in the compartment m of the pollutant p emitted in the initial compartment n:

$$\Delta D_p^{nm} = E_p^m \cdot F_p^{nm} \cdot M_p^n \tag{11.1}$$

where:

M_p^n is the mass of pollutant p (kg) emitted into the initial medium n (air, water, or soil)

F_p^{nm} is the fate and exposure factor for the emission of substance p into the initial medium n and transfer into medium m in the form of $(m^2 \cdot yr/m^3)$ or $(m^3 \cdot yr/m^3)$, depending on the compartments considered and taking into account the propagation, degradation, deposition, transfer among media and food chain or bioconcentration routes

E_p^m is the effect factor $(damage/m^2 \cdot (\mu g/m^3) \cdot yr))$ representing the severity of the impact due to the substance p in medium m (air, water, soil or food chain)

The ratio $\Delta D_p^{nm}/M_p^n$ is called the damage factor (damage/kg; Hofstetter 1998).

The release and target compartments are linked by the different fate and exposure routes. For example, an emission to the air compartment can have impacts in the air (inhalation), soil (via deposition) and water (absorption) target compartments. The pollutants can be transported farther to other target compartments by other routes (soil–plant etc.). A large variety of possible routes is between the release and target compartments.

Equation 11.1 does not explicitly consider the distribution and number of receptors affected by the pollutants. Depending on the impact category, the receptor may be, among others, human population, material surface, crop yield and sensitive ecosystem area or fish population. The effective receptor density $\rho_{eff,r}^m$ is introduced into the equation in order to relate the distribution of receptors r to the distribution of the respective pollutant in the environment. Equation 11.1 then reads as follows:

$$\Delta D_{p,r}^{nm} = E_{p,r}^m \cdot \rho_{eff,r}^m \cdot F_p^{nm} \cdot M_p^n \qquad (11.2)$$

where:

$\Delta D_{p,r}^{nm}$ is the incremental damage caused (damage) due to the emission of pollutant p into the initial medium n (air, water or soil) on the receptor r in the target compartment m (air, water, soil or food chain)

$\rho_{eff,r}^m$ is the effective density of receptor r (receptors/m^2) in target medium m (air, water, soil or food chain), that is, for the receptor human population (persons/m^2) and for material surface (m^2 maintenance surface/m^2), while this means for the fish population in the compartment water (fishes/m^3), i.e., we do not need to consider a density, but a concentration

$E_{p,r}^m$ is the effect factor (damage/receptors·(μg/m^3)·yr) representing the severity of the impact due to the substance p in medium m (air, water, soil or food chain) on receptor r

In the context of this framework, the incremental receptor exposure (ΔRE) is then defined as the product of the number of receptors exposed to a certain concentration during a certain period of time, as shown in the following equation:

$$\Delta RE_{p,r}^{nm} = \Delta D_{p,r}^{nm} / E_{p,r}^m = \rho_{eff,r}^m \cdot F_p^{nm} \cdot M_p^n \qquad (11.3)$$

where:

$\Delta RE_{p,r}^{nm}$ is the incremental receptor exposure during a certain period of time (receptors·(μg/m^3)·yr) due to the emission of pollutant p into the initial medium n (air, water or soil) on the receptor r in the target compartment m (air, water, soil or food chain)—for example, for airborne emissions with the receptor human population (persons·μg/m^3 air·yr) and for the receptor material surface (maintenance surface·μg/m^3 air·yr) and for water with the receptor fish population (fishes·mg/m^3 water·yr).

If the dose–response or exposure–response function is linear or the emission source contributes only marginally to the background concentration, the incremental damage becomes independent of the time pattern of the emission and only depends on the total mass emitted. For a detailed mathematical derivation, see Nigge (2000), and for the general idea of marginality and linearity see Potting (2000). For the purpose of this study, it is assumed that the incremental damage $\Delta D_{p,r}^{nm}$ is independent of the time pattern of the emission. The incremental damage can then be calculated by Equation 11.4, using an incremental receptor exposure per mass of pollutant emitted $I_{p,r,i}^{nm} = \Delta RE_{p,r,i}^{nm} / M_{p,i}^n$. The index i refers to an emission situation rather than to an emission site only. The emission situation is determined by the emission site and by the source type such as the stack height for emissions to air or the release depth into the lake or river for emissions to water.

$$\Delta D_{p,r,i}^{nm} = E_{p,r}^m \cdot M_{p,i}^n \cdot I_{p,r,i}^{nm} \qquad (11.4)$$

where:

$\Delta D_{p,r,i}^{nm}$ is the incremental damage caused (damage) due to the emission of pollutant p into the initial medium n (air, water or soil), at the emission situation i, on the receptor r in the target compartment m (air, water, soil or food chain)

$M_{p,i}^n$ is the mass of pollutant p (kg) emitted into the initial medium n (air, water or soil) at the emission situation i

$I_{p,r,i}^{nm}$ is the incremental receptor exposure per mass of pollutant emitted (receptors·(μg/m^3)·yr/kg) due to the amount of pollutant p emitted into the initial medium n (air, water or soil) at the emission situation i on the receptor r in the target compartment m (air, water, soil or food chain)

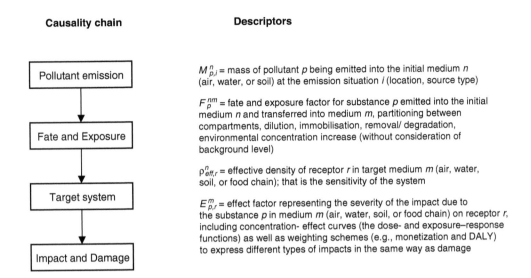

Causality chain

| Pollutant emission |

↓

| Fate and Exposure |

↓

| Target system |

↓

| Impact and Damage |

Descriptors

$M_{p,i}^{n}$ = mass of pollutant p being emitted into the initial medium n (air, water, or soil) at the emission situation i (location, source type)

F_{p}^{nm} = fate and exposure factor for substance p emitted into the initial medium n and transferred into medium m, partitioning between compartments, dilution, immobilisation, removal/ degradation, environmental concentration increase (without consideration of background level)

$\rho_{eff,r}^{n}$ = effective density of receptor r in target medium m (air, water, soil, or food chain); that is the sensitivity of the system

$E_{p,r}^{m}$ = effect factor representing the severity of the impact due to the substance p in medium m (air, water, soil, or food chain) on receptor r, including concentration- effect curves (the dose- and exposure–response functions) as well as weighting schemes (e.g., monetization and DALY) to express different types of impacts in the same way as damage

FIGURE 11.2 Relation of the factors used in the presented framework with the general cause–effect chain for the environmental impact of an emitted compound.

The mass of pollutant emitted, M, the fate and exposure factor, F, the effect factor, E, and the effective receptor density, ρ_{eff}, used in the presented framework are directly related to the causality chain illustrated in Figure 11.2. Such a comprehensive framework is the basis for using high levels of sophistication in the different dimensions of impact information corresponding to the scheme described in Figure 11.1.

This impact information should be based on a certain spatial differentiation with regard to the processes in the chain and include a minimal amount of additional data on the corresponding geographic situation. The site-dependent impact assessment can be carried out for various compartments using a multimedia fate and exposure model (see Chapter 4) or for only one release compartment n and one target compartment m by the application of a spatially explicit medium-specific model. In accordance with Potting (2000), we believe that the relevance of LCIA can be enhanced by the inclusion of a few general site parameters in the assessment procedure; we would call this site-dependent impact assessment.

For the effect analysis we propose to use the dose–response and exposure–response functions described in Chapter 4. The fate information should be obtained by using pollutant dispersion and long-range transport models and/or multimedia fate models. The target information needed corresponds to the receptor density that describes the sensitivity of the target, but we do not consider that background information is necessary, assuming that residual risk is what we want to address in LCIA and that linear dose–response and exposure–response functions exist, at least for priority pollutants. For a further discussion of this issue, see Crettaz et al. (2002), Nigge (2000) and Potting (2000).

11.3 STATISTICALLY DETERMINED GENERIC CLASSES OF AIRBORNE EMISSIONS

Considering only one pollutant, p, and one receptor, r, as well as one release compartment, n, and one target compartment, m, then $I_{p,r,i}^{nm} = I_i$ (= $\Delta RE_i/M$), the incremental receptor exposure per mass of pollutant emitted (receptors·(μg/m^3)·yr/kg), which represents the concentration increment multiplied by the receptors during a certain time period divided through the mass of pollutant.

In two-dimensional polar coordinates (r, φ) around the emission situation, i, within a suitable cartographic projection of the Earth's surface, this can be written as (Nigge 2000):

$$I_i = \frac{1}{Q} \int_0^R r \int_0^2 \Delta c_i(r, \varphi) \rho_i(r, \varphi) \, d\varphi \, dr \qquad (11.5)$$

where:

$Q = M_i/T$ is the constant emission rate (kg/yr) with M_i as mass of one pollutant (kg) emitted at the emission situation i and T as the duration of the emission (yr)

r is the radius (m)

R is the outer boundary of the modeling area (m)

$\Delta c_i(r, \varphi)$ is the concentration increment at a receptor point with the polar coordinates (r, φ) for the emission situation i ($\mu g/m^3$)

$\rho_i(r, \varphi)$ is the receptor density at a receptor point with the polar coordinates (r, φ) for the emission situation i (receptors/m²)

Generally speaking, the integration of Equation 11.5 should be carried out over the entire planet. Because most of the pollutants (even in the air compartment) do not disperse over the entire planet due to their residence time and dispersion characteristics in the environment, and because the calculation effort should be kept appropriate, R is chosen so that most of I_i is covered by the area. Another limitation is the spatial range of the models chosen, which often does not allow the calculations to be extended over a certain limit. The source characteristics influence the choice of R as well. The higher the stack in the case of emissions to air, the further the pollutant is transported and therefore the greater R must be chosen.

The idea of the methodology further developed and applied in this study is to define classes of emission situations i statistically, the impact of which differs significantly from class to class, but for which the deviation of impact between the emission situations covered by each class is small. The overall number of classes should be kept small to enable easy handling.

On the one hand, neglect of the spatial distribution around the emission point, I, of the receptor density, $\rho(r, \varphi)$, is the main reason for the discordance between the potential impact results and the actual impacts. To be precise, this means, for example, that in conventional LCIA potentials, the impact is the same for an air pollutant over the ocean as for one in a big city. On the other hand, the corresponding dispersion conditions in the respective medium and the resulting concentration increment, $\Delta c_i(r, \varphi)$, are relevant for the occurrence of damages. In order to relate these main factors for the estimation of environmental damages in a process-chain perspective, the present method proposes to form representative classes of receptor density and dispersion conditions. This classification must be based on statistical reasoning. For each class, receptor incremental exposures per mass of pollutant, ΔRE, must be calculated.

In a next step, the receptor incremental exposures per mass of pollutant ΔRE can be converted into damage estimates through an effect analysis based on dose–response and exposure–response functions and, if desired by the decision-maker, by the application of a weighting scheme to express different types of impacts in the way of an aggregated damage.

11.4   GENERIC CLASSES FOR HUMAN HEALTH EFFECTS

For the case of human toxicity impacts of airborne emissions, the receptors are the persons of a population and the release, as the target compartment is air. Thus, $\rho_{eff,r}^m$ corresponds to the population density and the receptor incremental exposures, $\Delta RE_{p,r,i}^{nm}$, are then called population incremental exposures to airborne pollutants ΔPE, expressed in units of (persons $\mu g/m^3 \cdot yr$). PE is also called pressure on human health (Nigge 2000).

Many laws and regulations with respect to emission limitation and pollution prevention exist for pollutants emitted into and solely transported by the air, so this study is confined to air as the only release and target compartment. Considering only human beings as receptors, for one pollutant, Equation 11.4 can then be simplified to:

$$\Delta D_i = E \cdot M_i \cdot I_i \tag{11.6}$$

where:

ΔD_i is the incremental damage caused (damage) due to the exposure to one pollutant being emitted at the emission situation i

E is the effect factor (damage/persons·(µg/m³)·yr) representing the severity of the impact due to one pollutant

M is the mass of one pollutant (kg) being emitted at the emission situation i

I_i is the incremental population exposure per mass of one pollutant emitted at the emission situation i (persons·(µg/m³)·yr/kg)

Equation 11.5 can be divided into two integrals accounting for the short-range dispersion and the long-range transport to the outer boundary of the modeling area R:

$I_{i,near}$ is the short-range contribution to the incremental receptor exposure per mass of one pollutant emitted at the emission situation i (persons·(µg/m³)·yr/kg).

$I_{i,far}$ is the long-range contribution to the incremental receptor exposure per mass of one pollutant emitted at the emission situation i (persons·(µg/m³)·yr/kg).

The reason for this procedure is that the concentration increment is usually highest within the first kilometers around the stack. Therefore, the impact indicator is very sensitive to the receptor density close to the stack. The population density can vary strongly within only a few kilometers. Consider, for example, that the population density often rapidly decreases from a big city to the countryside; therefore, the population exposure is subject to drastic changes within a few kilometers as well. The long-range contribution, however, only depends on the average receptor density of the region to which the pollutants are transported and is not particularly subject to changes on a local scale; the concentration increment is small due to dilution on transport, and the concentration does not change very much with the distance. Long-range contributions are known and have been well studied for pollutants like SO_2 and NO_x (due to their importance for acidification in regions far away from the emission source). Strong differences for the long-range exposure are likely to appear between densely inhabited areas, such as western and middle Europe, and scarcely inhabited regions such as Scandinavia or, in the U.S., the East Coast and the less populated Rocky Mountains. Therefore, as a good approximation, country averages for $I_{i,far}$ seem to be appropriate.

A major problem with deriving $Ii_{,near}$ is the fact that $\Delta c_i(r, \varphi)$ depends very much on the meteorological conditions, especially the wind direction, which can vary significantly within a few kilometers for the different emission sites. It is therefore desirable to eliminate φ in order to simplify Equation 11.5. Nigge (2000) assumes that $\Delta c_i(r, \varphi)$ and $\rho_i(r, \varphi)$ are not correlated and that the population density is independent of the angle φ if the emission sites considered in each class are spread over a large area. In this way, no direction is preferable for the spatial variation of the population density. Moreover, a simplification (Equation 11.7) can be introduced into Equation 11.5 (Nigge 2000):

$$\Delta c_i(r) \equiv \frac{1}{2}\pi \int_0^{2\pi} \Delta c_i(r, \varphi)\, d\varphi \tag{11.7}$$

According to Nigge (2000), Equation 11.5 then reads:

$$I_{C,\text{near}} = \frac{1}{Q} \int\limits_{0}^{100\text{km}} r \int\limits_{0}^{2\pi} \Delta c_C(r)\rho_C(r)2\pi dr \tag{11.8}$$

In Equation 11.8, 100 km is a value for orientation; it is proposed as the limit between the short- and long-range contribution to I, the incremental receptor exposure per mass of one pollutant emitted. The index C indicates that Equation 11.8 does not refer to a single emission situation, i, but to a generic class of emission situations, statistically correlated with respect to dispersion conditions and population density and with the same source characteristics. A further mathematical analysis is given in Nigge (2000). Remembering that $I_{i,\text{far}}$ is calculated as a country or regional average, the overall impact indicator for each class then reads:

$$I_C \equiv I_{C,\text{near}} + I_{\text{far}} \tag{11.9}$$

where:

I_C is the incremental population exposure per mass of one pollutant emitted (persons·(μg/m³)·yr/kg) at the generic class of emission situations that are statistically correlated with respect to dispersion conditions and receptor density and have the same source characteristics.

In order to compute the impact indicator for each class, the following elements of Equation 11.8 must be calculated:

$\rho_C(r)$ is the radial receptor density (receptors/m²) for each class (in our case, population density; persons/m²).

$\Delta c_C(r)$ is the radial concentration increment profile for each class (μg/m³).

The definition of classes of meteorological conditions and the derivation of generic meteorological data files to calculate the radial concentration increment profile, $\Delta c_C(r)$, are questions of fate analysis. The definition of classes of population densities and the calculation of the radial population density, $\rho_C(r)$, belong to the exposure analysis.

11.4.1   Fate Analysis to Characterize Dispersion Conditions

This section discusses the transport of airborne pollutants from the emission source to the receptors. For the purpose of short-range dispersion modeling (dispersion up to the radius of 100 km around the stack), the program ISCST-3 (US EPA 1995), incorporated in the software BEEST (Beeline, 1998), can be used. The calculations for the long-range transport can be carried out using the program EcoSense (IER, 1998). Both programs are mentioned in Chapter 4 and will be applied to more cases in Chapter 12.

For modeling the short-range exposure, only primary pollutants are considered due to the long formation time of secondary pollutants. In order to calculate the radial concentration, $\Delta c_C(r)$, to derive I_{near}, a statistical set of meteorological data must be used. Because emissions occurring in a life cycle usually cannot be assigned to the calendar time when they happen, only mean average pollutant concentrations on an annual basis are calculated. The BEEST program requires input data as presented in Table 11.1, where the data used in the example of this chapter are also indicated. Test runs have shown that the concentration increment results are not sensitive to the ambient air temperature. The mixing height is calculated as a function of the stability class (VDI, 1992). An equal distribution of wind directions is assumed. Therefore, as derived by Nigge (2000), the combined frequency distribution of wind speed and stability class is independent of the wind direction.

TABLE 11.1

Values Required by BEEST and Respective Data Used in This Study

Values Required by BEEST	Data Used in This Study
Hourly Values	
Wind speed	Weibull parameters, average annual wind speed to generate hourly wind speed values
Wind direction	Random values in intervals of 15°
Ambient air temperature	Annual average: 287 K
Stability class	Derived from a combined frequency distribution of wind speed and stability class, using hourly wind speed data
Rural and urban mixing height	Set to be equal for rural and urban areas
Friction velocity	Function of wind speed
Monin-Obukhov length	Function of the stability class
Surface roughness	Rural character of Catalonia: 0.3 m
Precipitation	Not taken into account: 0 mm
Fixed Values	
Elevation of modeling area	Entirely flat
Release heights	5, 100 and 200 m
Exit temperature of stack	423.15 K
Stack diameter	5 m
Volume flow	$\dot{V} = 26862 \cdot 10^{0.0196 \cdot h_{stack}}$ (Nm³/h) (11.10)
Exit velocity	Function of volume flow and stack diameter
Emission mass flow	$\dot{M}_p = \dot{V} \cdot C_{threshold,p}$ (kg/h) (11.11) where $C_{threshold,p}$ is legal threshold concentration of pollutant p

The remaining task is to determine a statistical distribution of wind speed, u, and stability class, s, for each class of meteorological conditions in the region under study. If the distribution parameters of the Weibull distribution are known, an hourly wind speed file can be generated from the average annual wind speed. Manier (1971) states that the distribution of stability class and that of wind speed classes are correlated. As a consequence, the only parameter required as additional input for using the impact indicator is the mean annual wind speed of the considered district.

Mass flow, volume flow and exit velocity determine the outcome of the concentration calculations of ISCST-3. However, unlike in EcoSense, the concentration increment calculated by BEEST does not change linearly by changing certain parameters of source characteristics. Therefore, statistical values for these parameters (volume flow and mass flow) must be defined.

A set of nine different industrial processes with stack heights ranging from 10 to 250 m were evaluated for the example with respect to their volume flows. The correlation between volume flow and stack height has a trend line that can be described with the potential approach in Equation 11.10, where $\dot{V}$ is the volume flow in (Nm³/h) and h_{stack} the stack height in (m), and the regression coefficient r^2 equals 0.799. Equation 11.10 is only a rough approximation to calculate the volume flow. Nine processes is not at all a representative statistical number that allows making general conclusions.

The mass flow of each pollutant is obtained from the volume flow and the respective threshold of each pollutant according to Equation 11.11, where $\dot{M}_p$ is the mass flow of pollutant, p, and cthreshold, p is the legal threshold concentration of pollutant p in flue gas. The threshold values for municipal waste incinerators are taken from the regulations valid for the region under study (in the example, the Catalan District 323/1994 that includes the European Guideline 89/369/EEC). In order to apply the correct threshold for

TABLE 11.2

Effective Stack Heights Depending on Wind Speed and Actual Stack Height

h_{stack} (α)	h_{eff} (m)		
	0 to 2 m/s	2 to 3 m/s	3 to 4 m/s
5	133	60	43
100	297	195	167
200	676	435	379

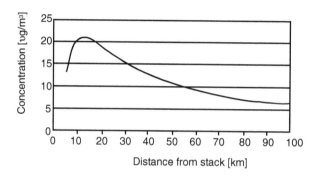

FIGURE 11.3 Concentration curve for PM_{10}, 100 km around the stack of 5 m height (average annual wind speed of 2.5 m/s).

the organic substances considered, the share of total organic carbon (TOC) of every pollutant is calculated, and, in this way, the threshold of TOC is adapted to each single organic substance considered.

The use of a threshold at this stage is probably not the best solution; in further works, basing the mass flow also on statistical reasoning according to industry types should be attempted. An alternative would be to use the mean average emission value of the respective industry.

As a matter of fact, for dispersion, the decisive parameter relating to the release height is not the stack height, but the effective stack height, h_{eff}, which also takes into account the momentum rise and the buoyancy rise of the plume and is automatically calculated by BEEST. In order to make the results of this work comparable to other studies that relate the impact indicators to the h_{eff} (e.g., Nigge 2000), the effective stack height is calculated for the indicators derived in this study (Table 11.2). The calculation of h_{eff} is carried out according to Israël (1994). The comparison of the results from different studies must be done with care, thoroughly checking the congruence of the applied algorithms for h_{eff}.

Figure 11.3 shows an example of a concentration curve for PM_{10} and the wind speed class 2 to 3 m/s, 100 km around the stack of 5 m height. The resolution of the grid is higher close to the stack in order to represent the sharp decrease of concentration there.

11.4.2 EXPOSURE ANALYSIS TO DETERMINE THE POPULATION DENSITY AROUND THE EMISSION SOURCE

In order to make results comparable, the basic classification of regions and districts according to population density is taken from Nigge (2000), who calls the combination of region and districts settlement structure classes.

For the calculation of the radial population density, the radius of 100 km around each municipality in the region under study is considered, which corresponds to the modeling area of the short-range transport covered by the Gaussian dispersion model used. Annuli are formed in intervals of

10 km that lead to 10 annuli around each municipality. Each municipality will be counted to the respective annulus if its center lies within the considered annulus. The interval of 10 km is chosen assuming that the linear extension of the municipalities in the region under study is in the range of up to 10 km of both sides of the center of the municipality. Assuming that every municipality in the region under study has the shape of a circle, the maximum area allowed for one municipality lying in the center of all annuli (origin) is 314 km². If the biggest municipality in the region comprises less than 314 km², this assumption is valid.

The calculation of the radial population density has two problems: (1) the radial population density in districts close to the sea is not independent of the direction in which one looks (the population density at the coast falls down abruptly to $\rho = 0$ persons/km²) and (2) no data might be available for the municipalities lying 100 km outside the regional borders of the region under study.

The first problem implies that the total population of all municipalities lying in the considered annulus is divided by the area of the annulus. Thus, the fact that a considerable area around the municipality has $\rho = 0$ persons/km² (municipalities close to the sea) is respected. Due to the absence of population living in the sea, the overall population density of the annulus is reduced. In order to solve the second problem for the calculation of the population density in the municipalities of the area adjacent to the area of study for population, data on a district level are used rather than the population data for the municipalities.

Because most districts are bigger than municipalities, uncertainties are introduced. As discussed earlier, the area of every municipality is assumed not to exceed a circle with a radius of 10 km if the municipality lies in the center of the circle. The average area of districts definitely exceeds this value. This means that the population density of the annuli is often determined by the entire population of one district, even though this district extends over more than one annulus and therefore should "assign" its population to more than one annulus. It is assumed, however, that the uncertainties introduced are not too big, because the average exceeding the center circle is within a tolerable range. Moreover, this procedure is assumed to be valid because it is only chosen to include the area adjacent to the region under study, while the region is dealt with using a higher resolution; so, the overall uncertainties related to the radial population density are considered to be quite low.

11.4.3 Effect Analysis to Transform Incremental Exposures into Damage Estimations

The effect analysis links the results of the fate and exposure analysis to the damage due to the emitted pollutant. This analysis is independent of the fate and exposure analysis and is based on epidemiological and toxicological studies, as well as on socioeconomic evaluation. See Chapters 3 and 4 for further details.

The effect factor represents the number of health incidences (like asthma or cancer cases or restricted activity days) per person—exposure time concentration. The dose–response and exposure–response functions used in this study are taken from IER (1998) and Hofstetter (1998). In this study, only carcinogenesis and respiratory health effects are taken into account, because they are considered to be the main contributors to the overall human health effects due to environmental pollution (Krewitt et al., 1998).

In order to aggregate different health effects into a single indicator, the DALY concept developed by Murray and Lopez (1996) is implemented. However, an economic valuation by external costs could also be applied easily, for example, the scheme used by the European Commission (1995). The DALY value not only depends on the pollutant and the type of disease, but also on the socioeconomic perspective of each person. Thompson et al. (1990) introduced the concept of the cultural theory for different perspectives; Hofstetter (1998) distinguished individualist, egalitarian and hierarchist cultural perspectives to represent the archetypes of socioeconomic behavior and related this to the DALY concept. Corresponding to age weighting for the different cultural perspectives, the economic evaluation discount rates of 0% and 3% can be chosen. An example of different factors for the conversion of population exposure values into damage estimates is given in Table 11.3.

TABLE 11.3

Example of Conversion Factors between Impact Indicators and DALY and External Costs

Pollutant	DALY I	DALY E	DALY H	Costs[a]	Costs[b]
Acetaldehyde	2.7×10^{-7}	4.1×10^{-7}	4.1×10^{-7}	0.05	0.05
As	2.2×10^{-4}	3.5×10^{-4}	3.5×10^{-4}	43.1	31.5
BaP	1.3×10^{-2}	2.0×10^{-2}	2.0×10^{-2}	2,497	1,825
1,3-Butadiene	3.4×10^{-5}	5.2×10^{-5}	5.2×10^{-5}	6.6	6.6
Cd	2.7×10^{-4}	4.1×10^{-4}	4.1×10^{-4}	51.7	37.8
Ni	5.0×10^{-5}	7.8×10^{-5}	7.8×10^{-5}	10.3	7.3
NO_x	—	2.5×10^{-6}	—	0.27	0.43
SO_2	—	5.4×10^{-6}	5.4×10^{-6}	0.57	0.89
$PM_{2.5}$	2.8×10^{-4}	2.9×10^{-4}	2.9×10^{-4}	136.5	135.6
PM_{10}	1.7×10^{-4}	1.8×10^{-4}	1.8×10^{-4}	79.7	79.1
Nitrate	9.0×10^{-5}	1.8×10^{-4}	1.8×10^{-4}	79.7	79.1
Sulfate	2.8×10^{-4}	2.9×10^{-4}	2.9×10^{-4}	132.7	131.8

Notes: DALY: a/(persons μg/m³); external costs: U.S.\$/(persons μg/m³). I: individualist (age-weighting (0,1)); E: egalitarian (no age-weighting (0,0)); H: hierarchist (no age-weighting (0,0)).

[a] Discount rate 0%.
[b] Discount rate 3%.

11.5   SITE-DEPENDENT IMPACT ASSESSMENT IN THE METHODOLOGY OF ENVIRONMENTAL DAMAGE ESTIMATIONS FOR INDUSTRIAL PROCESS CHAINS

On the basis of the calculation of site-dependent impact assessment factors described beforehand, in this section, the site-dependent method is introduced as a further element into the methodology of environmental damage estimations for industrial process chains outlined in the previous chapter. Site-dependent impact factors can be perfectly used in the mathematical framework developed and are particularly recommended for transport processes. The main task remaining is to establish a flowchart of the site-dependent impact assessment by statistically determined generic classes that can be included in the overall methodology. Figure 11.4 gives such an overview of the different working steps for this site-dependent method.

For the calculation of the impact factors, first the considered region must be divided into classes. In this chapter, this has been done for the receptor human population. However, in principle, this can be done also for other receptors, as proposed in the general framework described in Section 11.2 of this chapter.

In the fate and exposure analysis, each class is divided into the near ($\leq$100 km) and the far ($>$100 km) contribution. For the near contribution, the radial concentration increment for each pollutant and the radial receptor density are calculated. The Gaussian dispersion model ISCST-3, as incorporated in BEEST, has been used in the application to Catalonia as transport model on the local scale. The multiplication of both results obtained yields the receptor exposure per class and pollutant $I_{near, pollutant, class}$. In a similar way for the far contribution, first, the average concentration increment on a continental grid is calculated for each pollutant. Then, this result is multiplied by the average receptor density in each grid. Finally, the receptor exposure per class and pollutant $I_{far, pollutant, region}$ is obtained for the region under study. In a next step, I_{far} and I_{near} are added, and so, finally, the overall receptor exposure per pollutant for each class and specific region $I_{total, pollutant, class\&region}$ is determined.

Site-dependent (air) (calculation of factors)

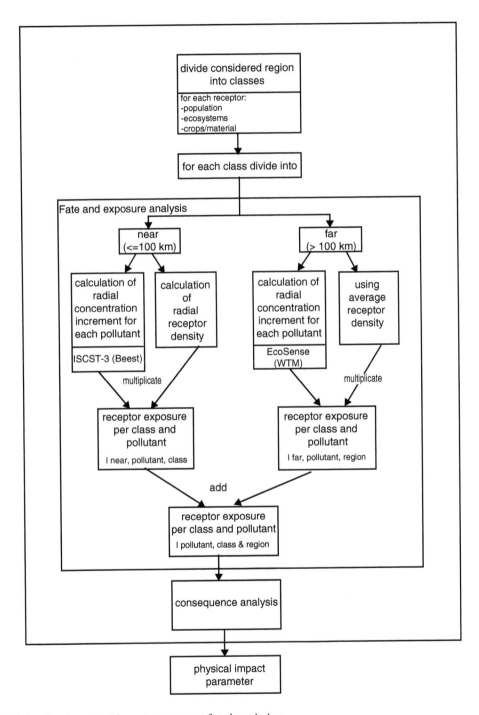

FIGURE 11.4 Site-dependent impact assessment for air emission.

By carrying out the consequence analysis factors of physical impact, parameters are obtained. Additionally, when the selected weighting and aggregation scheme is applied, the factors can also be expressed directly in the form of environmental costs or damage endpoint indicators. These two steps have been summed up as effect analysis in Section 11.4.3 of this chapter.

11.6 EXAMPLE: CALCULATION OF SITE-DEPENDENT IMPACT FACTORS FOR CATALONIA, SPAIN

In this study, the fate and exposure analysis is carried out for the Mediterranean region of Catalonia next to the sea, which significantly influences how to apply the presented method.

11.6.1 WIND SPEED CLASSES

Harthan (2001) shows the derivation of the Weibull parameters for Catalonia required to generate hourly wind speed data from a mean annual wind speed value and describes the analysis of the combined frequency distribution of wind speed and stability class. The variance coefficient of the average annual wind speed data for Catalonia is high (47.4%; Cunillera, 2000), with an average for 1996 to 1999 for all 14 stations considered. Thus, taking into account only the mean annual average value to represent all average annual wind speed data considered for Catalonia does not suffice to describe the area's situation as a whole. The formation of wind speed classes on the district level is required (see the overview of districts in Catalonia in Figure 11.5). The districts are assigned the following codes for the classes of wind speed: A: 0 to 2 m/s, B: 2 to 3 m/s, and C: 3 to 4 m/s. Because only wind speed data from 14 stations were provided in a detailed manner, average annual wind speed data for 48 stations in 1997 and 1998 are taken from http://www.gencat.es/servmet. The overall distribution of wind speed classes among the districts in Catalonia can be seen in Figure 11.5 and the portions of districts belonging to each class of meteorological conditions are shown in Table 11.4.

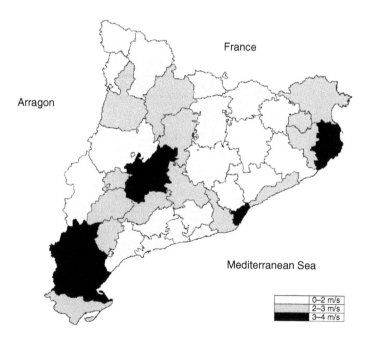

FIGURE 11.5 Classes of average wind speed for all districts in Catalonia (data of 1997 and 1998).

TABLE 11.4
Number and Share of Districts Belonging to Each Class of Wind Speed

Average Wind Speed Class (m/s)	Districts	Share (%)
0 to 2	20	48.8
2 to 3	14	34.1
3 to 4	7	17.1
0 to 4	31	100

11.6.2 SETTLEMENT STRUCTURE CLASSES

As a region, in the context of the example, the province level in Catalonia is considered. Furthermore, the districts are taken into account. In this way, four settlement structure classes have been identified to represent statistically the Catalonian situation sufficiently (Figure 11.6); this is characterized by the rapid decrease of the population from the city boundaries toward the countryside. For the purpose of this study, population data for Catalonia are taken from the MiraMon GIS (Pons and Masó 2000).

To calculate the radial population density for every municipality, the distance between each municipality must be calculated to assign the municipalities to the respective annuli. These data are also taken from MiraMon, which provides the coordinates in UTM (universal transverse mercator grid system) units expressed in meters, as well as the respective population of each municipality. The coordinates describe the outer limits of each municipality. Assuming circular areas, the coordinates of the center of the municipalities are calculated forming the average of all coordinates describing

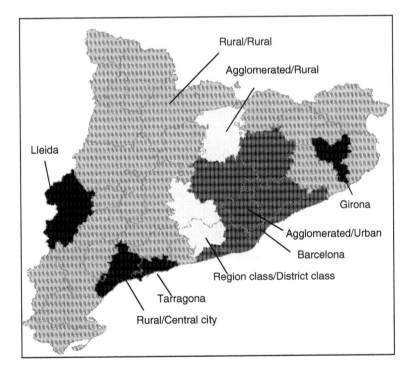

FIGURE 11.6 Settlement classes in Catalonia.

the outer limits. The latitudes and longitudes of the centers for the district outside Catalonia are converted into UTM coordinates using one of the converters available from the Internet.

After the distances between the municipalities are calculated, it is known which municipality lies in which annulus around a considered municipality. When data about the population of each municipality are used, the population of each annulus around a certain municipality and the population density for each annulus and each municipality can be calculated. The radial population density is then calculated for every generic class subsuming the population density for each annulus of each municipality belonging to the respective class and dividing the sum by the number of municipalities considered in this class. Figure 11.7 shows the radial population density for each class graphically. Each generic class is assigned a number from I to IV: I = agglomerated—urban; II = agglomerated—rural; III = rural—central city; and IV = rural—rural.

The radial population density for the agglomerated region was calculated for the province of Barcelona, Catalonia. For the urban districts of the province, to which the city center of Barcelona belongs, the radial population density is decreasing rapidly from the center toward the outer radius of 100 km. The main reason for this behavior is the concentration of population within greater Barcelona, whereas outside the province districts are scarcely populated.

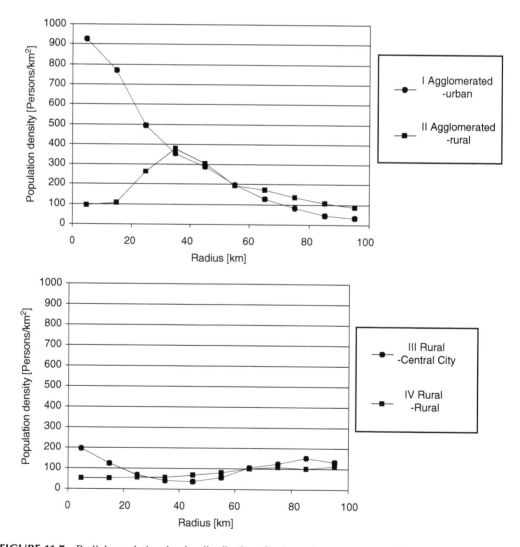

FIGURE 11.7   Radial population density distributions for the settlement structure classes in Catalonia.

The curve of the rural districts within the province of Barcelona shows the typical behavior of a rural region close to a big city. Within the first kilometers, the population density is rather low. A peak of population density occurs between 30 and 40 km from a fictive stack in this district. This can be explained by the proximity of Barcelona City, which influences the radial population density for rural districts in the province of Barcelona also.

Within the rural regions of Girona, Lleida and Tarragona are districts classified as "central city" and "rural." For central cities to which the districts of Tarragonès, Baix Camp, Segrià and Gironès belong, the population density also decreases from the center toward the countryside. However, the population density is not as high in the center as for the respective districts in the agglomerated region. In the particular situation of Catalonia, after a minimum of between 40 and 50 kα, the population density reaches a local maximum between 80 and 90 km. A look at Figure 11.6 explains this behavior: all rural regions are located around the agglomerated region—especially the central cities in the rural regions, which are within a distance of 100 km of the agglomerated region.

For the rural districts of the rural regions, the population density is much below the average of Catalonia ($\rho = 190$ persons/km^2 at the radius $r = 0$), which is self-explanatory in rural districts. Toward $r = 100$ km, the population density increases; this explains the settlement structure of Catalonia. Statistically, a big city (central city) is within a distance of 100 km of each rural district in a rural area. Another look at Figure 11.6 underlines this reasoning.

From the curve for each class, it can be seen that the population density reaches a similar value for all classes at a radius of 100 km. Therefore, it is statistically justified to treat the long-range transport in an averaged manner (regional average) for all considered districts and regions and to set the limit between short-range and long-range modeling at this distance.

11.6.3 Short- and Long-Range Exposure

The classes derived in the fate and exposure analysis are given combined class codes in order to enable easy discovery of the desired impact indicator for the combined classes. For example, the impact indicator for class III B describes the impact for a central city in a rural region with annual wind speed conditions between 2 and 3 m/s. Moreover, the stack height is mentioned and the effect analysis method chosen.

For the short-range exposure, calculations are carried out for every combination of pollutant, stack height and generic meteorological file. Concentration curves such as those in Figure 11.7 are overlapped with the curves of radial population density (Figure 11.7) for each generic class of population density and integrated over the modeling area (see Equation 11.8). The integration over the modeling area is done by summing up the products of concentration increment, population density and area for all values lying in the chosen grid.

For the long-range exposure, the concentration increment is calculated for eight grid cells of fictitious emission sites covering Catalonia, using the EcoSense software (IER 1998). The runs are carried out for stack heights of 5, 100 and 200 m. As a first step, the background population exposure is calculated (zero emission run). For this purpose a "no emission" run is carried out for each of the eight grid cells. The concentrations obtained are the background values for the respective grid cells. Multiplying the background concentration and the population of each grid cell leads to the background population exposure per grid cell. The results are then summed up over all grid cells, which leads to the overall background population exposure for each emission cell considered.

The next step is to carry out the calculations for all considered emission cells and for all stack heights. The population exposure is calculated in the same way as for the background population exposure. In order to obtain the long-range contribution due to an incremental emission of a pollutant, first, the background population exposure must be subtracted from the calculated population exposure. Because the short-range population exposure is calculated using the BEEST software, the population exposure within the radius of 100 km from the stack must be subtracted as well. For this purpose, the population exposure of the emission cell as well as the population exposure of part of the adjacent cells is subtracted.

TABLE 11.5

Diameter and Atmospheric Residence Time τ of Several Particles

Pollutant (diameter)	τ (h)
$PM_{2.5}$	135
PM_{10}	22
PM (d = 4 to 10 mm)	15
PM (d = 10 to 20 mm)	11
PM (d > 20 mm)	3

Source: Nigge, K.M., *Life-Cycle Assessment of Natural Gas Vehicles, Development and Application of Site-Dependent Impact Indicators*, Springer-Verlag, Berlin, Germany, 2000.

After summing up the population exposure of each grid cell and subtracting the background and the short-range exposure, the sum is divided by the emitted mass. The result is the population exposure per mass of emitted pollutant for the long-range transport I_{far} for each pollutant, emission cell and stack height considered. Because the modeling area is comparatively small (which explains the existence of only eight emission cells), the long-range exposure per mass of emitted pollutant does not differ very much for the considered emission cells. The variation coefficient for the different emission cells lies between 11.75% and 23.49%. Therefore, an average Catalonian value is calculated for every pollutant and every stack height and applied as I_{far} for the whole of Catalonia.

I_{far} for pollutants such as acetaldehyde and 1,3-butadiene that are not included in the EcoSense software is approximated by their atmospheric residence time. For that purpose, I_{far} is calculated for a set of particles with different diameters and atmospheric residence times (Table 11.5). A linear regression is carried out and an approximation formula is calculated:

$$I_{far} = \exp(-0.2043 \cdot \ln \tau)^2 + 2.3916 \cdot \ln(-7.3174) \qquad (11.12)$$

where:

I_{far} is the long-range contribution to the population exposure per mass of emitted pollutant (persons·µg/m³·yr/kg)

τ is the atmospheric residence time (h)

11.6.4 OVERALL IMPACT INDICATORS

The overall impact indicator, I_{total}, is the sum of I_{near} and I_{far}. Table 11.6 shows the results for several pollutants, including the values for I_{far}, in order to show the long-range contribution to I_{total}. The impact indicators for the secondary pollutants nitrate and sulfate refer to the mass of primary pollutant emitted, i.e., to NO_x and SO_2, respectively, and are represented by the results for the long-range exposure that show only very slight variations for different stack heights. The average values for all stack heights are 0.25 persons·µg/m³·yr/kg for nitrate and 0.13 persons·µg/m³·yr/kg for sulfate, respectively.

Generally speaking, it can be said that the population exposure is smaller the lower the population density. Another general correlation is that an increasing stack height leads to a decreasing impact indicator. Because Catalonia is quite populated in comparison with other regions in Spain,

TABLE 11.6

Site-Dependent Human Health Impact Indicators for Several Pollutants and Stack Heights[a]

Pollutant		Acetaldehyde			1,3-Butadiene			NO$_x$			SO$_2$			PM 2.5			PM 10		
Stack Height		5 m	100 m	200 m	5 m	100 m	200 m	5 m	100 m	200 m	5 m	100 m	200 m	5 m	100 m	200 m	5 m	100 m	200 m
Class																			
District Wind	A, B, C																		
Cat	A, B, C	0.05	0.05	0.05	0.06	0.06	0.06	0.06	0.06	0.06	0.17	0.18	0.18	0.60	0.60	0.60	0.06	0.06	0.06

I_{far} = long range contribution to the incremental receptor exposure per mass of pollutant

I_{total} = total incremental receptor exposure per mass of pollutant

District	Wind	Acet. 5 m	100 m	200 m	But. 5 m	100 m	200 m	NO$_x$ 5 m	100 m	200 m	SO$_2$ 5 m	100 m	200 m	PM2.5 5 m	100 m	200 m	PM10 5 m	100 m	200 m
I	A	3.29	0.18	0.06	3.67	0.22	0.07	4.72	0.26	0.08	5.93	0.43	0.19	6.52	0.91	0.61	5.61	0.43	0.09
I	B	2.47	0.29	0.09	2.69	0.32	0.10	3.27	0.36	0.11	3.97	0.51	0.23	4.44	0.95	0.65	3.64	0.42	0.11
I	C	2.14	0.30	0.09	2.31	0.33	0.11	2.76	0.36	0.12	3.30	0.51	0.24	3.72	0.93	0.65	2.97	0.40	0.11
II	A	1.24	0.16	0.05	1.50	0.19	0.07	2.32	0.24	0.07	3.42	0.42	0.19	3.87	0.92	0.61	3.06	0.41	0.09
II	B	0.96	0.20	0.08	1.13	0.23	0.09	1.59	0.27	0.10	2.22	0.43	0.22	2.64	0.87	0.64	1.93	0.34	0.10
II	C	0.84	0.20	0.08	0.97	0.23	0.10	1.32	0.26	0.11	1.82	0.41	0.23	2.22	0.84	0.64	1.54	0.31	0.10
III	A	0.71	0.10	0.05	0.83	0.12	0.07	1.23	0.15	0.07	1.88	0.31	0.19	2.29	0.79	0.61	1.67	0.24	0.08
III	B	0.57	0.12	0.06	0.65	0.15	0.08	0.88	0.17	0.08	1.28	0.31	0.20	1.69	0.75	0.62	1.08	0.21	0.08
III	C	0.50	0.12	0.06	0.57	0.15	0.08	0.75	0.17	0.08	1.07	0.30	0.21	1.48	0.73	0.62	0.88	0.19	0.08
IV	A	0.45	0.09	0.05	0.56	0.12	0.07	0.90	0.15	0.07	1.49	0.30	0.19	1.90	0.76	0.60	1.27	0.22	0.08
IV	B	0.37	0.11	0.06	0.44	0.13	0.08	0.64	0.15	0.08	1.01	0.29	0.20	1.42	0.73	0.62	0.82	0.18	0.08
IV	C	0.33	0.11	0.06	0.39	0.13	0.08	0.55	0.15	0.08	0.84	0.28	0.20	1.25	0.71	0.62	0.66	0.17	0.08
Cat	A	1.22	0.12	0.05	1.41	0.15	0.07	1.97	0.18	0.07	2.76	0.34	0.19	3.22	0.82	0.61	2.51	0.29	0.08
Cat	B	0.94	0.16	0.07	1.06	0.19	0.08	1.38	0.21	0.09	1.85	0.36	0.21	2.28	0.80	0.63	1.62	0.26	0.09
Cat	C	0.82	0.17	0.07	0.92	0.19	0.09	1.16	0.21	0.09	1.54	0.35	0.21	1.95	0.78	0.63	1.31	0.24	0.09

Note: District population densities: I: agglomerated – urban; II: agglomerated – rural; III: rural – central city; IV: rural – rural; Cat: Catalonian average. Wind speeds: A: 0 to 2 m/s; B: 2 to 3 m/s; C: 3 to 4 m/s.

[a] Persons * µg/m^3 * year/kg.

because every pollutant deposited on the Mediterranean does not lead to a human health effect via inhalation, and because the modeling area of EcoSense is limited to Europe (therefore neglecting harmful effects of Spanish emissions to North Africa, for instance), every molecule or particle going into the long-range transport favors the decrease of the overall population exposure.

Another obvious effect is that I_{total} decreases with higher wind speeds. Moreover, the influence of atmospheric residence time and decay can be derived. If one compares the impact indicators for PM_{10}, $PM_{2.5}$, NO_x and SO_2, it can be seen that the values decrease from $PM_{2.5}$ over SO_2 and PM_{10} to NO_x, according to their atmospheric residence time and decay rate. The span between the highest and the lowest value of I_{total} for each pollutant ranges from a factor of 10 for $PM_{2.5}$ to a factor of 70 for PM_{10}. This can be explained with the fact that $PM_{2.5}$ accounts for a much greater long-range contribution to the population exposure than PM_{10} due to its long atmospheric residence time. Therefore, the lowest value of $PM_{2.5}$ is determined by the comparatively high long-range contribution, which leads to the comparatively small span between highest and lowest value.

Using the dose–response and exposure–response functions, physical impacts (e.g., cases of chronic bronchitis) per mass of pollutant can be calculated. Applying these functions and the respective unit values, it is possible to convert the impact indicators into DALY and external costs per kilogram of pollutant using the conversion factors in Table 11.3.

11.6.5   ESTIMATES FOR ADJACENT REGIONS AND OTHER STACK HEIGHTS

Process chains often comprise processes outside Catalonia, so an approximation formula for other regions in Spain is presented in the following equation. It is supposed that the long-range exposure for other regions in Spain does not vary significantly from the values in Catalonia. Therefore, it holds that:

$$I_{far,\ other\ regions} = I_{Catalonia}.\qquad\qquad(11.13)$$

With respect to the short-range exposure, it is supposed that significant variations are due to the population density. Nigge (2000) presents an approach to approximate indicators according to meteorological conditions as well. Due to a lack of data and time, only the population density is considered here. For this purpose, the population density for all other provinces (the administration level between region and district) is calculated. The impact indicators are calculated using the average impact indicator for Catalonia according to Equation 11.14. Because no distinction is made with respect to meteorological conditions, the respective values of $I_{near,\ Catalonia}$ for the different classes of meteorological conditions are weighted with the share of occurrence in the districts of Catalonia (Table 11.4).

$$I_{near,\ other\ province} = I_{near,\ Catalonia}.\left(\frac{\rho_{other\ province}}{\rho_{Catalonia}}\right)\qquad\qquad(11.14)$$

Finally, it should also be mentioned that an interpolation can be carried out for emission heights different from the given ones.

11.7   CASE STUDY: SITE-DEPENDENT IMPACT INDICATORS USED FOR THE WASTE INCINERATION PROCESS CHAIN

In this section, the impact indicators derived in Section 11.6.4 are applied to the industrial process chain of the case study, the municipal solid waste incinerator (MSWI) plant of Tarragona

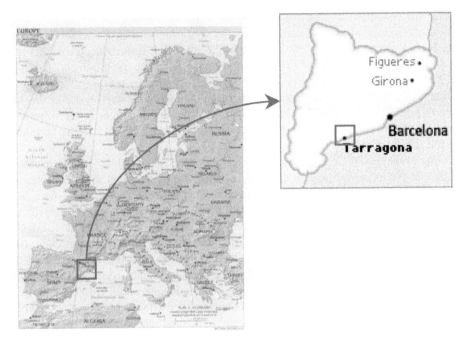

FIGURE 11.8 Physical map of Europe and regional map of Catalonia.

(SIRUSA), and the related transport processes. Chapter 1 describes the data of the MSWI as well as the related transport processes. A comparison between the results for the applied impact indicators and the results for a midpoint indicator is provided and discussed. Figure 11.8 shows a regional map of Catalonia together with a physical map of Europe, so that the localization of the study can be better visualized.

11.7.1 GOAL AND SCOPE DEFINITION

Two situations and one scenario of the MSWI process chain are compared to estimate avoided environmental damages on human health.

Scenario 1 is the basis for all comparisons. It describes the SIRUSA incinerator in 1996, including the treatment of scrap in Madrid, ash treatment at the company TRISA in Constantí, and the disposal of the treated ash and the slag in Pierola and Reus, respectively. In the ash treatment plant, cement is mixed with the ash. The cement is transported from Sta Margarida i els Monjos.

Scenario 2 is the situation after the installation of an advanced acid gas removal system (AGRS) in 1997. In addition to the features in Scenario 1, there is an additional transport of CaO for the acid gas treatment. CaO is transported from Sta Margarida i els Monjos to the waste incinerator in Tarragona. The data for Scenarios 1 and 2 are taken from the existing LCA study (STQ, 1998) (see Chapter 2).

As a third alternative, a future scenario has been created based on the modular model for the waste incineration process described briefly in Chapter 1. In addition to the features in Scenario 2, a fictitious DeNox system is installed to eliminate NO_x emissions. Ammonia for this purpose is transported from the industrial area of Tarragona to the waste incinerator. It should be stressed that, although this is a scenario with a non-validated MSWI model, it is introduced into this case study in order to discuss possible results of further improvement in the flue gas treatment. This alternative is called Scenario 3.

The function of the MSWI is its service of treating the municipal solid waste of the Tarragona region. In this particular case study, the functional unit presented here is 1 year of treated waste (which equals 153,467 t). However, 1 t of treated waste as well as 1 TJ and 1 kWh of produced electricity could also be chosen.

According to the lessons learned in the application of the methodology of environmental damage estimations for industrial process chains in Chapter 10, the focus in this study was restricted to the transport and incineration processes. Therefore, it is not necessary to carry out a dominance analysis and no cut-off criteria need to be defined. However, to compare the results with midpoint indicators, the two LCIA methods with the human toxicity potential (HTP) explained in Chapter 3 are used: Institute of Environmental Sciences (CML), Leiden University (Heijungs et al., 1992), and Environmental Design of Industrial Products (EDIP) (Hausschild and Wenzel, 1998).

For the weighting of the human health impacts, the physical impact parameter "population exposure" and the DALY concept with the egalitarian perspective are applied. Because only the AoP (area of protection) of human health is taken into account here, no further aggregation needs to be carried out. The options of uncertainty analysis, environmental risk assessment, accidents and eco-efficiency are not considered.

11.7.2 LIFE-CYCLE INVENTORY ANALYSIS

In order to obtain the life-cycle inventory (LCI) data of this abbreviated waste incinerator process chain, the volume flow (Nm^3/a) of the waste incineration plant is multiplied by the unit emission files for the waste incinerator in the different cases, and the overall transport (tkm) is multiplied with the unit emission files for the transport. These unit emission files provide the emission of different air pollutants per Nm^3 and per tkm, respectively. The previously mentioned data and the unit emission files for SIRUSA can be found in the case study sections of previous chapters; the unit emission files for the transport processes are presented in Table 11.7. The table only lists the pollutants for which

TABLE 11.7
Unit Emissions for Transport Processes

Pollutant	Emission (kg/tkm)
Benzo[a]pyrene	4.00×10^{-10}
Cd	2.10×10^{-9}
NMVOC	4.40×10^{-4}
NO_x	3.12×10^{-3}
PM_{10}	2.00×10^{-4}
SO_2	2.08×10^{-4}

Source: Kern, M., Environmental damage estimations for the process chain of the ash treatment plant TRISA, Master's thesis, Technical University of Berlin, Berlin, Germany, 2000.

impact indicators are available that are considered crucial for environmental problems related to transport. The unit emission data for the MSWI for Scenarios 1 and 2 stem from the existing LCA study (STQ, 1998; see Chapter 2) and, for Scenario 3, from Hagelueken (2001; see Chapter 1).

The most obvious result is that the overall tkm increases significantly from Scenario 1 over Scenarios 2 and 3. Thus, the additional transport processes somehow compensate for the decrease of emissions due to the improvement of flue gas treatment. To what extent transport compensates for the damages avoided through more powerful gas treatment needs to be shown by environmental damage estimations.

11.7.3 LIFE-CYCLE IMPACT ASSESSMENT METHOD

The HTP is calculated according to the CML method (Heijungs et al., 1992) and the EDIP method (Hauschild and Wenzel, 1998; the nomenclature of the latter is changed). Further explanations of the HTP can be found in Chapter 3. Table 11.8 presents the HTP results for the three alternatives.

11.7.4 DOMINANCE ANALYSIS AND SPATIAL DIFFERENTIATION

No dominance analysis is necessary for this application. The medium is air because the impact factors have been calculated for that medium. The pollutants are those for which site-dependent impact factors have been determined in Section 11.6.4. They correspond largely to the pollutants identified as predominant in Chapter 9. The processes studied are those that have resulted as the most important in Chapter 10: the incineration and transport processes.

In addition to assigning the corresponding region to each process, in this application, the transport process must be differentiated with respect to the districts crossed. This is an example of what is meant by Equation 10.8 on mobile processes in the mathematical foundation of the methodology in Chapter 10. The size of the region (here the district level) determines the number of regions by which a mobile process, i.e., transport, is represented in the eco-technology matrix.

Table 11.9 describes the transport routes for every process and the districts crossed. Table 11.10 summarizes the transport related to each process taking place in each class of population density

TABLE 11.8
HTP Results for the Three Alternatives

Process	CML (–/a)	%	EDIP (m³/a)	%
Scenario 1				
MSWI	912,885	97.9	2.85×10^{12}	98.3
Transport	19,365	2.1	4.88×10^{10}	1.7
Total	932,251	100.0	2.90×10^{12}	100.0
Scenario 2				
MSWI	160,492	88.0	8.95×10^{11}	94.2
Transport	21,885	12.0	5.51×10^{10}	5.8
Total	182,378	100.0	9.50×10^{11}	100.0
Scenario 3				
MSWI	43,361	64.5	2.31×10^{11}	79.3
Transport	23,912	35.5	6.02×10^{10}	20.7
Total	67,272	100.0	2.91×10^{11}	100.0

TABLE 11.9
Transport Routes and Districts Crossed for the Considered Processes

Purpose	Route	Districts Crossed (Class)
Municipal waste	Tarragonès–Constantí	Tarragonès (III A)
Slag	Constantí–Reus	Tarragonès (III A), Baix Camp (III A)
CaO	Els Monjos–Constantí	Alt Penedès (II A), Baix Penedès (IV A), Tarragonès (IIIA)
Ammonia	Tarragona–Constantí	Tarragonès (III A)
Scrap treatment	Constantí–Madrid	Tarragonès (III A), Alt Camp (IV A), Conca de Barberà (IV B), Garrigues (IV B), Segrià (III A), Huesca (HU), Zaragoza (ZZ), Soria (SO), Guadalajara (GU), Madrid (M)
Ash treatment	Constantí	Tarragonès (III A)
Ash disposal	Constantí–Pierola	Tarragonès (III A), Baix Penedès (IV A), Alt Penedès (II A), Baix Llobregat (I B), Anoia (II B)
Cement	Els Monjos–Constantí	Alt Penedès (II A), Baix Penedès (IV A), Tarragonès (III A)

Note: District population densities: I: agglomerated – urban; II: agglomerated – rural; III: rural – central city; IV: rural – rural; Cat: Catalonian average.
Wind speeds: A: 0 to 2 m/s; B: 2 to 3 m/s; C: 3 to 4 m/s.

TABLE 11.10
Transport Distances in Each Class

Purpose	Distance in Combined Wind Speed and Population Density Class (km)											
	IB	IIA	IIB	IIIA	IVA	IVB	HU	ZZ	SO	GU	M	Sum
Municipal waste	—	—	—	9	—	—	—	—	—	—	—	9
Slag	—	—	—	16	—	—	—	—	—	—	—	16
CaO	—	9	—	24	17	—	—	—	—	—	—	50
Ammonia	—	—	6	5	—	—	—	—	—	—	—	5
Scrap treatment	—	—	—	41	12	53	49	206	36	97	40	534
Ash treatment	—	—	—	4	—	—	—	—	—	—	—	4
Ash disposal	21	32	11	24	17	—	—	—	—	—	—	105
Cement	—	9	6	24	17	—	—	—	—	—	—	50

Note: See Table 11.9 for abbreviations of classes.

and meteorology class, as well as the distance in the provinces outside Catalonia. (For the abbreviation of classes, see Section 11.6.) The emission height for the municipal waste incinerator is 50 m and for the transport processes is 5 m. Using the transport distances in each class from Table 11.11 and the tkm presented in Chapter 1, it can be calculated how many tkm refer to each class of population density and meteorological conditions. Multiplication by the unit emission file of transport leads to the spatially differentiated LCI, the eco-technology matrix.

11.7.5   Fate and Exposure and Consequence Analysis

Table 11.12 shows the damage-assigning matrix for DALY (egalitarian). It should be noted that this matrix only represents an example of the many possibilities according to the decision

TABLE 11.11

Eco-Technology Matrix for the Former Situation 1[a,b]

Pollutant	MSWI	Transport[b]										
	IIIA[g]	IB	IIA	IIB	IIIA	IVA	IVB	HU	ZZ	SO	GU	M
As	15	0	0	0	0	0	0	0	0	0	0	0
BaP	0	1×10^{-5}	2×10^{-5}	7×10^{-6}	2×10^{-3}	4×10^{-5}	1×10^{-4}	1×10^{-4}	5×10^{-4}	8×10^{-5}	2×10^{-4}	9×10^{-5}
Cd	15	7×10^{-5}	1×10^{-4}	4×10^{-5}	9×10^{-3}	2×10^{-4}	6×10^{-4}	6×10^{-4}	2×10^{-3}	4×10^{-4}	1×10^{-3}	5×10^{-4}
PM_{10}	20,411	6	10	3	877	19	58	54	226	39	106	44
Ni	22	0	0	0	0	0	0	0	0	0	0	0
NO_x	142,333	101	158	53	13,677	296	906	838	3522	616	1658	684
SO_2	602,643	7	11	4	912	20	60	56	235	41	111	46
O_3^c	0	14	22	7	1,929	42	128	118	497	87	234	96
O_3^d	142,333	101	158	53	13,677	296	906	838	3522	616	1658	684
Nitrate[e]	142,333	101	158	53	13,677	296	906	838	3522	616	1658	684
Sulfate[f]	602,643	7	11	4	912	20	60	56	235	41	111	46

[a] kg/a.
[b] h_{source} = 5 m.
[c] as NMVOC.
[d] as NO_x.
[e] as NO_x.
[f] as SO_2.
[g] h_{stack} = 50 m.

TABLE 11.12
Damage-Assigning Matrix for DALY (Egalitarian)[a]

Class	As	BaP	Cd	PM 10	Ni	Pollutant NO$_x$	SO$_2$	O$_3$[c]	O$_3$[d]	Nitrate[e]	Sulfate[f]
III A[g]	5.4×10^{-4}	3.2×10^{-2}	6.5×10^{-4}	1.7×10^{-4}	1.2×10^{-4}	1.8×10^{-6}	6.1×10^{-6}	1.0×10^{-6}	4.1×10^{-7}	4.5×10^{-5}	3.7×10^{-5}
I B[b]	1.5×10^{-3}	8.9×10^{-2}	1.8×10^{-4}	6.4×10^{-4}	3.4×10^{-4}	8.3×10^{-6}	2.1×10^{-5}	1.0×10^{-6}	4.1×10^{-7}	4.5×10^{-5}	3.6×10^{-5}
II A[b]	1.3×10^{-3}	7.7×10^{-2}	1.6×10^{-4}	5.4×10^{-4}	3.0×10^{-4}	5.9×10^{-6}	1.8×10^{-5}	1.0×10^{-6}	4.1×10^{-7}	4.5×10^{-5}	3.6×10^{-5}
II B[b]	9.1×10^{-4}	5.3×10^{-2}	1.1×10^{-4}	3.4×10^{-4}	2.1×10^{-4}	4.0×10^{-6}	1.2×10^{-5}	1.0×10^{-6}	4.1×10^{-7}	4.5×10^{-5}	3.6×10^{-5}
III A[b]	7.9×10^{-4}	4.6×10^{-2}	9.5×10^{-4}	2.9×10^{-4}	1.8×10^{-4}	3.1×10^{-6}	1.0×10^{-5}	1.0×10^{-6}	4.1×10^{-7}	4.5×10^{-5}	3.6×10^{-5}
IV A[b]	6.6×10^{-4}	3.8×10^{-2}	7.9×10^{-4}	2.2×10^{-4}	1.5×10^{-4}	2.3×10^{-6}	8.0×10^{-6}	1.0×10^{-6}	4.1×10^{-7}	4.5×10^{-5}	3.6×10^{-5}
IV B[b]	4.9×10^{-4}	2.8×10^{-2}	5.9×10^{-4}	1.4×10^{-4}	1.1×10^{-4}	1.6×10^{-6}	5.4×10^{-6}	1.0×10^{-6}	4.1×10^{-7}	4.5×10^{-5}	3.6×10^{-5}
HU[b]	2.6×10^{-4}	1.5×10^{-2}	3.1×10^{-4}	3.4×10^{-5}	5.8×10^{-4}	4.4×10^{-7}	1.7×10^{-6}	1.0×10^{-6}	4.1×10^{-7}	4.5×10^{-5}	3.6×10^{-5}
ZZ[b]	3.9×10^{-4}	2.2×10^{-2}	4.6×10^{-4}	9.6×10^{-5}	8.7×10^{-4}	1.2×10^{-6}	3.7×10^{-6}	1.0×10^{-6}	4.1×10^{-7}	4.5×10^{-5}	3.6×10^{-5}
SO[b]	2.4×10^{-4}	1.4×10^{-2}	2.9×10^{-4}	2.7×10^{-5}	5.5×10^{-4}	3.6×10^{-7}	1.5×10^{-6}	1.0×10^{-6}	4.1×10^{-7}	4.5×10^{-5}	3.6×10^{-5}
GU[b]	2.6×10^{-4}	1.5×10^{-2}	3.1×10^{-4}	3.4×10^{-5}	5.8×10^{-4}	4.4×10^{-7}	1.7×10^{-6}	1.0×10^{-6}	4.1×10^{-7}	4.5×10^{-5}	3.6×10^{-5}
M[b]	2.6×10^{-3}	1.5×10^{-1}	3.1×10^{-3}	1.1×10^{-3}	5.8×10^{-4}	1.3×10^{-5}	3.8×10^{-5}	1.0×10^{-6}	4.1×10^{-7}	4.5×10^{-5}	3.6×10^{-5}

[a] year × year/kg.
[b] $h_{source} = 5$ m.
[c] as NMVOC.
[d] as NO$_x$.
[e] as NO$_x$.
[f] as SO$_2$.
[g] $h_{stack} = 50$ m.

table described in Chapter 10. The matrix is based on the results for the site-dependent impact indicator, expressed as population exposure per mass of emitted pollutant, calculated in Section 11.6. Another damage-assigning matrix has also been established with the values of the impact indicator expressed with the unit (persons·$\mu g/m^3$·yr/kg). It should be mentioned that the impact indicators for the heavy metals As, Cd and Ni, as well as for benzo[a]pyrene (BaP), are supposed to be the same as for $PM_{2.5}$. It is assumed that these substances are adsorbed on particles of $PM_{2.5}$ and therefore behave in the same terms of fate and exposure. However, the DALY value is specific for each substance because the dose–response and exposure–response functions are substance specific as well.

The ozone value is taken as country average from Krewitt et al. (2001). The values for nitrate and sulfate do not differ because these are secondary pollutants for which country averages have been calculated. Only the value for the sulfate in the first row differs slightly, due to the stack height of the MSWI (50 m), which differs from the one of the transport processes (5 m stack height). The highest values for the primary pollutants appear in Madrid and the lowest in the small town of Huesca. These values are calculated according to Equations 11.13 and 11.14 described in Section 11.6.5 for the transfer of impact factors to other regions. In comparison to Catalonia, Madrid is densely populated, and Huesca is scarcely populated, so these results are obvious.

11.7.6 Damage Profile

Table 11.13 shows the damage matrix for Scenario 1 (former situation) resulting from the multiplication of the eco-technology matrix (Table 11.11) and the damage-assigning matrix (Table 11.12). If the first column is considered, the first box (the first element of the diagonal) represents the damage for the MSWI. The second element in the second column represents the damage for the transport in district class I B and so on. If one leaves the diagonal, the impact of the processes at (fictitious) other locations is shown. For example, the last element of the first column shows the damage of the waste incinerator if it were to be located in Madrid. Of course, it must be admitted that the impact indicator applied to this cell refers to a stack height of 5 m (transport) rather than to 50 m (MSWI),

TABLE 11.13
Damage Matrix with DALY (Egalitarian) for Scenario 1[a]

	MSWI	Transport									
III A[c]	36.12	0.01	0.01	0.00	0.83	0.02	0.06	0.05	0.21	0.04	0.10	0.04
I B[b]	55.17	0.01	0.02	0.01	1.34	0.03	0.09	0.08	0.35	0.06	0.16	0.07
II A[b]	50.97	0.01	0.01	0.00	1.22	0.03	0.08	0.07	0.31	0.05	0.15	0.06
II B[b]	42.73	0.01	0.01	0.00	1.01	0.02	0.07	0.06	0.26	0.05	0.12	0.05
III A[b]	40.54	0.01	0.01	0.00	0.96	0.02	0.06	0.06	0.25	0.04	0.12	0.05
IV A[b]	37.77	0.01	0.01	0.00	0.88	0.02	0.06	0.05	0.23	0.04	0.11	0.04
IV B[b]	34.46	0.01	0.01	0.00	0.80	0.02	0.05	0.05	0.21	0.04	0.10	0.04
HU[b]	29.81	0.01	0.01	0.00	0.69	0.01	0.05	0.04	0.18	0.03	0.08	0.03
ZZ[b]	32.38	0.01	0.01	0.00	0.75	0.02	0.05	0.05	0.19	0.03	0.09	0.04
SO[b]	29.53	0.00	0.01	0.00	0.68	0.01	0.05	0.04	0.18	0.03	0.08	0.03
GU[b]	29.81	0.01	0.01	0.00	0.69	0.01	0.05	0.04	0.18	0.03	0.08	0.03
M[b]	76.05	0.01	0.02	0.01	1.87	0.04	0.12	0.11	0.48	0.08	0.23	0.09

[a] year/(kg/year).
[b] $h_{source} = 5$ m.
[c] $h_{stack} = 50$ m.

in which the damage would be twice as high as it is currently. Nevertheless, this provides an impressive demonstration of the importance of spatial differentiation.

Next, the diagonal elements of the damage matrix are added to obtain the damage profile, then the part corresponding to transport is compared with the value of the waste incinerator. In this way, the damages due to the waste incinerator and the transport can be compared.

The most obvious result of this environmental damage estimation study for industrial process chains is that the contribution of transport to the overall damage increases significantly from Scenario 1 over Scenarios 2 and 3. On the one hand, this is due to the sharp decrease of the overall damage and, on the other, the decrease of damage due to the improvement of flue gas treatment is partly compensated for by the additional transport processes.

From the results of all scenarios, it can be seen that the ratio between the damage due to transport and the overall damage is in the same magnitude for all chosen indicators in this study (PE and DALY). Whether the contribution of transport to the overall result is more significant for the population exposure or DALY in the case of each pollutant depends on the relationship between the toxicity and the mass of pollutants emitted. The more toxic a substance is, the higher is the increase in DALY.

From the results using the endpoint indicators derived in this study on one hand, and the HTP on the other hand, it seems that the HTP concept underestimates the environmental importance of transport. Although the share of transport in Scenario 1 is 2.1/1.7% for HTP (CML/EDIP), it is between 3.2% and 4.2% for the endpoint indicators derived in this study. Although the share of transport for the endpoint indicators reaches between 17.1% and 19.6% in Scenario 2, the share for HTP is still quite small (12.0/5.8%). However, the CML approach is closer to the results obtained by the endpoint approach than the EDIP approach. The gap widens even more in Scenario 3: 35.5/20.7% for HTP and 44.1% to 49.2% for the endpoint indicators, respectively.

The reason for the differences between the results for the HTP and the endpoint indicators is clear. The studied HTP methods consider the fate of the substances, but do not include exposure information. The environmental impact of transport is highly dependent on the location where it takes place, so the deviation from the HTP results is obvious. The results using the impact indicators of this study, however, show the limits of the ecological benefits of a further technical improvement of the flue gas cleaning. Scenario 3 indicates a clear overall reduction; however, it shows also that nearly half of the overall environmental impact is due to transport. Therefore, further technical improvement at the waste incinerator should only be carried out if transport does not increase significantly, which would worsen the overall environmental efficiency of the process chain.

It has been found that the chosen HTP methods underestimate transport in this case study and, therefore, do not identify very well the differences in the environmental impact for the two different processes considered. Figure 11.9 shows the results for all three cases—differentiated in waste incinerator and transport—for the population exposure, DALY (egalitarian) and HTP (EDIP).

The HTP indicators may be misleading in the comparison of the absolute environmental burden as well. For instance, if significant reductions only happen in regions with a low population density and high wind speeds (the factors that account for a low population exposure), the reduction for the endpoint indicators derived in this study will be rather small, while the HTP indicators will identify significant reductions. It can therefore be concluded that the use of endpoint indicators as derived in this study is beneficial with respect to a gain of information for both purposes: the comparison of different scenarios and the comparison of different processes within one scenario.

11.7.7 Uncertainties in the Applied Framework

Essential for the validity of the presented application of site-dependent impact assessment is the question of whether the uncertainties introduced are justified by the gain of information in comparison to the traditional impact potential used in LCIA. In this context, it must be highlighted that the methodology described in this chapter should be seen as a balance between the uncertainties

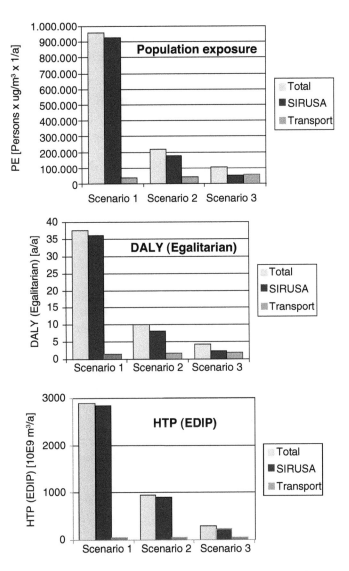

FIGURE 11.9 Results of the case study for all three scenarios for the population exposure—PE, DALY (egalitarian) and HTP (EDIP), where SIRUSA stands for the MSWI process.

introduced on the one hand and the handiness and feasibility of the applied method on the other. In general, it can be said that the larger the amount of data available, the lower the uncertainty behind the damage estimates is.

11.7.7.1 Goal and Scope Definition and Inventory Analysis

In the goal and scope definition, subjective elements exist in the form of the selection of the functional unit, which implies whether benefits are (or are not) given for avoided environmental loads. Moreover, the selection of the cut-off criteria for the system boundaries and the dominance analysis can influence the outcome. It is only possible to address these uncertainties and influences by sensitivity analysis and the use of scenarios for the different options.

In the inventory analysis, the main types of uncertainties and variability analyzed by the Monte Carlo simulation in Chapter 9 are those referring to parameters, including frequency of sampling, method of measurement (continuous on-line, or from time to time by more or less sophisticated analytical methods) and homogeneity of fuels. For the exactness of the modular process model, as for all process simulations, the adequate estimation of physical properties that are not well established is crucial. The estimations are taken from the publication of Kremer et al. (1998), so the quality depends on the values proposed by these authors.

11.7.7.2   Dominance Analysis and Spatial Differentiation

The dominance analysis adds a level of uncertainty by reducing the media, environmental loads and processes considered. However, it improves the relevance of the remaining information for further assessment.

With respect to the spatial differentiation, it should be said that the area under study, Catalonia, is rather small (compared to other administration units like Spain, European Union or the U.S.). Therefore, a statistical reasoning is per se limited due to the restricted area. Therefore, the determination of class limits with respect to the administrational units and settlement structure, and according to meteorological conditions, must be done with special care.

The problem with choosing the outer boundaries of the classes (in this case provinces and districts) is that information on dispersion conditions (wind speed) as well as on the settlement structure (population density) is not always bound to administrational units. Meteorological conditions are particularly subject to geographic situations such as topography or latitude, while administrational units are generally linked to settlement structures. For instance, the district of Barcelonès comprises the city of Barcelona with a few other municipalities and represents an urban district in an agglomerated region.

However, not all district boundaries delimit the settlement structure that clearly. For example, the Tarragonès district (central city in rural region) comprises the city of Tarragona as well as rural municipalities, which contradicts the definition of each class. Administrational units also depend on history and political decisions, because determination of the limits of the municipalities usually is not a recent decision. Thus, the choice of administrational units is problematic in terms of how representative settlement and dispersion conditions are. However, these limits are chosen because data are usually available for these administrational units; population data are available for municipalities, districts and provinces.

11.7.7.3   Fate and Exposure Assessment

Comparing the parameter uncertainty and variability involved in the LCI in relation to the impact pathway analysis (IPA; see Chapter 9) shows that uncertainties in the fate and exposure assessment as well as in the consequence/effect analysis are more important than those in the LCI analysis. This is due to the dose–response and exposure–response functions, weighting schemes and fate models. Examples of fate models used in this study include dispersion software, multimedia models and long-range transport programs (as included in integrated impact assessment models).

The program BEEST, as well as EcoSense, was primarily developed for the calculation of concentration increments due to power plant emissions. This work, however, also applies these programs for lower stack heights and volume flows (in the case of BEEST) in order to make the method applicable to a wider range of industrial process chains. In this case particularly, in order to include transport processes, several assumptions are needed. Unfortunately, the uncertainties cannot be quantified.

In general, the uncertainties related to Gaussian dispersion models are important; they are rather simple descriptions of quite complicated natural processes. Because substance data influence the dispersion of pollutants, uncertainties in these data are directly introduced into the results of the damage estimations. Another major source of uncertainty related to the fate models used is the choice of the modeling area and the grid size of the dispersion programs. In the long-range transport program WTM, the overall modeling area is confined to Europe (Eurogrid as implemented in EcoSense); a contribution to the population exposure from outside is neglected. An outside contribution is related to North Africa, which lies close to Spain, but which is not included in the Eurogrid. The outside contribution cannot be quantified, however. Because the long-range exposure does not depend very much on local variations, the resolution of 100×100 km seems to be appropriate for the use of Gaussian dispersion models. The grid for BEEST calculations is chosen so that the proximity of the stack (where the concentration increment is very sensitive to the distance from the stack) is described quite well without increasing the uncertainties for greater distances from the stack up to 100 km.

In the case of the site-dependent impact assessment study, with respect to the BEEST program, it has been seen that the outcome for I_{near} is highly sensitive to the volume flow and the mass flow chosen. Although EcoSense allows the introduction of any mass or volume flow and still leads to the same results for I_{far}, the volume flow and the mass flow must be determined carefully for BEEST calculations. The volume flow is derived as a function of the stack height from the evaluation of several industrial processes. The mass flow is derived from the volume flow and the thresholds for emissions of waste incinerators that apply in Catalonia. It must be stated, though, that the assumption for the volume flow is based only on a small number of industrial processes and that the used threshold is a political value.

As performed by Nigge (2000), the calculation of the effective emission height h_{eff}, which includes the physical stack height as well as information about stack temperature and volume flow, may resolve these problems of uncertainties. For this reason, h_{eff} is provided in this study as well. If the impact indicators are given for h_{eff}, the results may be applicable to all kinds of industries under the condition that the stack height, volume flow and temperature are known. Moreover, it should be stated that a large variety of algorithms can calculate the effective emission height for different purposes. Generally speaking, it seems to be difficult to find one procedure of resolving these problems in defining source characteristics. This requires further research.

Another important source of uncertainty for the numerical results of the site-dependent human health impact factor for Catalonia is the fact that the models ISCST-3 in BEEST (short-range) and WTM in EcoSense (long-range) may be in poor accordance in terms of dispersion. As stated earlier, the dispersion results for BEEST are highly sensitive to the source characteristics. Moreover, these characteristics must be chosen so that the results at the outer boundary comply with the results calculated by the EcoSense modeling area that begins there. This is especially difficult because the EcoSense program calculates concentration increments for the grid cells as a whole, i.e., the actual concentration at the modeling boundary with the BEEST program cannot be determined. This means that, if the source characteristics for BEEST are chosen "wrong," the overall outcome of I_{total} would be erroneous because the EcoSense model is not adapted to the source characteristics introduced in BEEST.

A related problem is that higher wind speeds usually lead to lower values of I_{near}. One may suppose that at least parts of the substances are therefore going into long-range transport covered by the Eurogrid of EcoSense. Thus, I_{far} should increase, but EcoSense is not able to take this problem into account because it provides a coherent set of meteorological data for every grid cell in Europe and is usually not applied in combination with other models. This leads to lower values of I_{total} for higher wind speeds, which is accepted because one may suppose that the population density outside Catalonia is smaller. For this reason, an enhanced long-range transport due to higher wind speed leads to lower values of I_{total}, because the increase in long-range exposure is smaller than the decrease in short-range exposure. However, this is only a theoretical reasoning because no change in I_{far} can be observed.

The same problem applies to the stack height. Although the height has a great influence on the outcome for the short-range modeling (because it is assumed that the higher the stack, the more pollutants go into the long-range transport), the results for long-range modeling are quite insensitive to stack height. The longer the atmospheric residence time of the pollutant is, the greater the uncertainties to which problems related to wind speed and stack height lead. A pollutant with a short atmospheric residence time is deposited and decays mostly in the short-range modeling area, while a pollutant with a longer atmospheric residence also accounts for a significant long-range exposure and is therefore subject to greater uncertainties with respect to wind speed and stack height.

Moreover, the local population exposure subtracted from the population exposure in the EcoSense area cannot be compared to the results of I_{near} because I_{near} is much more differentiated with respect to meteorological conditions and population density than the EcoSense results for each grid cell. Therefore, compliance between BEEST and EcoSense is not achievable by saying that the population exposure subtracted by EcoSense must be equal to the I_{near} results.

In multimedia models for ERA, additional sources of uncertainty must be considered—in particular, site characteristics and transfer factors for transport among media. The site characteristics include human lifestyles, because different diets and media composition show important variations from one site to another.

The evaluation of the dispersion conditions of pollutants in air has been carried out using meteorological data of a limited number of years (site-specific assessment) and measurement stations (site-dependent assessment). The meteorological conditions are strongly determined locally and, therefore, a reduction of data always increases the uncertainties. Also, a smaller number of years and stations makes the results less representative for the site or Catalonia as a whole. However, the derivation of the meteorological data files in this study has been carried out on this basis because no more data were available. Of course, the statistical evidence would increase if more data were available; however, for pragmatic reasons, the limited number of data was accepted.

If one considers the derivation of the statistical meteorological data files by Harthan (2001) mentioned in Section 11.4.1, it must be said that the formation of classes of wind speed (0 to 2 m/s, 2 to 3 m/s and 3 to 4 m/s) has not been undertaken according to statistical reasoning. The limits of these classes are chosen according to the limits defined by the Deutscher Wetterdienst (DWD) and constitute a good compromise between the concept of classes and the meaningfulness of the classes. If broader class limits were chosen, the handiness of the results would increase because the overall number of classes would be reduced. However, this would lead to a large variation of actually occurring wind speeds within each class, i.e., the statistical determination of the classes—describing the wind speed of all locations lying in this class with a reasonable standard deviation—would no longer be well founded. If narrower class limits were chosen, this would lead to a smaller standard deviation within each class and would therefore decrease the uncertainties. Nevertheless, the number of classes would increase and the number of districts lying in each class would decrease and a statistical reasoning combining several districts in one class would no longer be possible. The class limits chosen here seem to be appropriate because they allow a minimum differentiation of wind speed (into three classes), but still with a reasonable number of districts per class.

Neglect of terrain elevations and precipitation is necessary in the site-dependent impact assessment due to the absence of statistical reasoning for this parameter; nevertheless, this leads to uncertainties. In particular, the concentration increment of particles calculated is overestimated because wet deposition is not considered. An evaluation of different temperatures has led to the conclusion that the results for the concentration increment are not sensitive to temperature. Therefore, it is valid to choose the country average for all data sets and throughout the whole year. The neglect of wind direction leads to uncertainties, especially on a local level, because wind direction strongly influences dispersion on that level. However, in order to form class averages accounting for several wind directions, the neglect of wind directions is assumed to be valid. The height of the mixing layer is calculated according to VDI 3782/1, which is a German guideline on dispersion modeling. The uncertainties introduced are therefore considered to be small. The surface roughness is chosen as one value for the whole of Catalonia as a rural value, which is assumed to be a good estimate for the general settlement structure of Catalonia.

Classes according to the population density are formed. It is argued that the statistical basis is good enough to calculate the radial population density and that it is valid to use an interval of 10 km for the annuli considered because the biggest municipality in Catalonia has a smaller surface. However, reducing the resolution outside Catalonia to the district level increases the uncertainties. It is assumed that this is a reasonable procedure because it only concerns a limited number of adjacent districts to Catalonia and, with respect to working loads, this is a feasible way.

11.7.7.4 Consequence and Effect Analysis

One of the most important sources of uncertainties relates to the dose–response and exposure–response functions of pollutants further described in Chapter 4. These functions determine the consequence and effect analysis. Therefore, uncertainties due to these functions directly apply to the endpoint-related indicators or damage estimates (physical impacts such as cancer cases, as well as YLD, YOLL, DALY and external environmental costs). If one wants to avoid these uncertainties, the impact indicators can be applied as "pressure on human health." However, in order to take into account the differences in the toxicity of the pollutants and sensitivity to human health, dose–response and exposure–response must be considered. For instance, the EcoSense database offers a variety of dose–response functions that can be chosen according to the value preferences of the user and which show huge relative differences. More functions can be obtained from other public health or environmental authorities.

YLD, YOLL, DALY and external environmental costs are determined by subjective judgment that directly influences the outcome. In order to increase the transparency of and reduce the subjective influence by the methodology developer, this work offers several options. For uncertainties about YLD, YOLL and DALY values for several pollutants, see Hofstetter (1998). For more information on uncertainties in the evaluation of external environmental costs, see the EC (2000).

11.8 QUESTIONS AND EXERCISES

1. Why is a differentiation between near (≤ 100 km) and far (>100 km) necessary for the impact assessment in the case of air emissions?
2. How is the fate analysis carried out for the presented site-dependent approach?
3. Explain how the exposure analysis is performed for the presented site-dependent approach.
4. What are the advantages and disadvantages of the presented effect analysis approaches?
5. Describe the procedure on how to apply site-dependent impact assessment in the framework of the overall methodology introduced in Chapter 10.
6. Discuss whether there are limits for the usefulness of end-of-pipe technologies in light of the results in the case study on the installation of different levels of advanced gas treatment systems.
7. What are considered to be the main uncertainties in environmental damage estimations for industrial process chains? What are the next steps ahead to reduce them?
8. Calculate the incremental damage caused in air by the emissions to air of 200 kg of a pollutant with a fate and exposure factor $F_p^{air, air} = 2.5 \times 10^{-5}$ and an effect factor $E_p^{air} = 1.2 \times 10^{-6}$.
9. Calculate the DALY for the egalitarian case and external cost of discount rate = 3% per person for a PM_{10} emission increment of 2×10^3 µg/m³.
10. The incremental exposure per mass of pollutant for the population of an urban area of Massachusetts with a population density $\rho_2 = 550$ persons/km², is $I_{Massachusetts} = 0.18$. Calculate the incremental receptor exposure per mass of pollutant in a farming area of New Hampshire with a population density $\rho_1 = 90$ persons/km².

REFERENCES

Beeline Software, Inc. (1998), Beest for Windows. User's Manual. Beeline Software, Asheville, NC.
Crettaz, P., Pennington, D., Brand, K., Rhomberg, L., and Jolliet, O. (2002), Assessing human health response in life-cycle assessment using ED10s and DALYs—Part 1, cancer effects, *Risk Anal.*, 22(5), 929–944.
Cunillera, J. (2000), Algorithm for meteorological data obtained from the Meteorological Service of Catalonia, technical paper, Meteorological Service of Catalonia, Barcelona, Spain.
European Commission (EC) (1995), ExternE – Externalities of Energy, 6 vols., EUR 16520-162525. DG XII Science, Research and Development, Brussels.

EC (European Commission) (2000), *Externalities of Fuel Cycles*, vol. 7, methodology update, report of the JOULE research project ExternE (externalities of energy), EUR 19083. DG XII Science, Research and Development, Brussels.

Hagelueken, M. (2001), Effects of different models of municipal solid waste incinerators on the results of life-cycle assessments, student research project, cooperation of TU Berlin and Universitat Rovira i Virgili, Tarragona, Spain.

Harthan, R. (2001), Development of site-dependent impact indictors for human health effects of airborne pollutants in Catalonia, Spain, master thesism cooperation of TU Berlin and Universitat Rovira i Virgili, Tarragona, Spain.

Hauschild, M. and Wenzel, H. (1998), *Environmental Assessment of Products—Scientific Background*, (vol. 2), Chapman & Hall, London, UK.

Heijungs, R., Guinée, J.B., Huppes, G., Lankreijer, R.M., Udo de Haes, H.A., and Wegener-Sleeswijk, A. (1992), Environmental life-cycle assessment of products—Guide and backgrounds, Technical report, CML, University of Leiden, the Netherlands.

Hofstetter, P. (1998), *Perspectives in Life-Cycle Impact Assessment—a Structured Approach to Combine Models of the Technosphere, Ecosphere and Valuesphere*, Kluwer Academic Publishers, London, UK.

Huijbregts, M.A.J. and Seppälä, J. (2000), Towards region-specific, European fate factors for airborne nitrogen compounds causing aquatic eutrophication, *Int. J. LCA*, 3(5), 65–67.

IER (Institut für Energiewirtschaft und Rationale Energieanwendung) (1998), EcoSense 2.1. Software, University of Stuttgart, Germany.

Israël, G. (1994), Transmission von Luftschadstoffen, educative report, Environmental Engineering Institute, TU Berlin, Germany.

Kern, M. (2000) Environmental damage estimations for the process chain of the ash treatment plant TRISA. Master's thesis, Technical University of Berlin, Berlin, Germany.

Kremer, M., Goldhan, G., and Heyde, M. (1998), Waste treatment in product-specific life-cycle inventories, *Int. J. LCA*, 3(1), 47–55.

Krewitt, W., Mayerhofer, P., Trukenmüller, A., and Friedrich, R. (1998), Application of the impact pathway analysis in the context of LCA, *Int. J. LCA*, 3, 86–94.

Krewitt, W., Trukenmueller, A., Bachmann, T., and Heck, T. (2001), Country-specific damage factors for air pollutants—A step towards site dependent LCIA, *Int. J. LCA*, 6(4), 199–210.

Manier, G. (1971), Untersuchungen über meteorologische Einflüsse auf die Ausbreitung von Schadgasen. Berichte des Deutschen Wetterdienstes Nr. 124, German Meteorological Service (DWD), Offenbach, Germany.

Moriguchi, Y. and Terazono, A. (2000), A simplified model for spatially differentiated impact assessment of air emissions, *Int. J. LCA*, 5, 281–286.

Murray, C.J.L. and Lopez, A.D. (1996), *The Global Burden of Disease*, vol. 1 of *Global Burden of Disease and Injury Series*, WHO, Harvard School of Public Health, World Bank. Harvard University Press, Boston, MA.

Nigge, K.M. (2000), *Life-Cycle Assessment of Natural Gas Vehicles, Development and Application of Site-Dependent Impact Indicators*, Springer-Verlag, Berlin, Germany.

Pons, X. and Masó, J. (2000), The MiraMon map-reader—A new tool for the distribution and exploration of geographical information through internet or on CD, *Presented at 19th ISPRS Congress and Exhibition*, Amsterdam, the Netherlands.

Potting, J. (2000), Spatial differentiation in life-cycle impact assessment, PhD thesis, University of Utrecht, Utrecht, the Netherlands.

Potting, J. and Blok, K. (1994), Spatial aspects of life-cycle assessment, in: SETAC-Europe, Eds., *Integrating Impact Assessment into LCA*, SETAC publication, Brussels.

Potting, J. and Hauschild, M. (1997), Predicted environmental impact and expected occurrence of actual environmental impact (part 2)—spatial differentiation in life-cycle assessment via the site-dependent characterization of environmental impact from emissions, *Int. J. LCA*, 2, 209–216.

Spadaro, J.V. and Rabl, A. (1999), Estimates of real damage from air pollution–Site dependence and simple impact indices for LCA, *Int. J. LCA*, 4, 229–243.

STQ (Servei de Tecnologia Química) (1998), Análisis del ciclo de vida de la electricidad producida por la planta de incineración de residuos urbanos de Tarragona, technical report, Universitat Rovira i Virgili, Tarragona, Spain.

Thompson, M., Ellis, R., and Wildavsky, A. (1990), *Cultural Theory*, Westview Print, Boulder, CO.

Udo de Haes, H.A. (1996), *Towards a Methodology for Life-Cycle Impact Assessment*, SETAC-Europe publication, Brussels.

Udo de Haes, H.A., Joillet, O., Finnveden, G., Hauschild, M., Krewitt, W., and Mueller-Wenk, R. (1999), Best available practice regarding categories and category indicators in life-cycle impact assessment, background document for the Second Working Group on LCIA of SETAC-Europe, *Int. J.LCA*, 4, 66–74; 167–174.

US EPA (1995), Users' guide for the industrial source complex (isc3)—dispersion models, volumes I and II. EPA-454/B-95-003a, EPA-454/B-95-003b. Triangle Park, NC.

VDI (Verein Deutscher Ingenieure) (1992), *Dispersion of Pollutants in the Atmosphere, Gaussian Dispersion Model for Air Quality Management—Part 1, VDI*, Verlag, Düsseldorf, Germany.

Wenzel, H., Hauschild, M., and Alting, L. (1997), *Environmental Assessment of Products. Volume 1 – Methodology, Tools and Case Studies in Product Development*, Chapman & Hall, London, UK.

12 Applications of Environmental Impact Analysis in Industrial Process Chains

Francisco Sánchez-Soberón, Montserrat Mari, Vikas Kumar, Julio Rodrigo, Montserrat Meneses, Haydée Yrigoyen, Francesc Castells, and Marta Schuhmacher

CONTENTS

12.1 INTRODUCTION

In this chapter, the methodologies explained previously will be applied in three other cases. These examples are pilot studies to further show the potential for integration of the tools for the life-cycle and site-specific impact assessment described in the book. We note the academic character of the projects, but we wanted to include them to demonstrate the strategy outlined in Chapter 10 for cases other than waste incineration.

In the first example, life-cycle assessment (LCA) and risk assessment are applied to the case of the landfill activity of mixed household solid waste. The use of the WISARD (waste integrated systems assessment for recovery and disposal) software model (Ecobalance, 1999) and database made it possible to perform the inventory analysis and impact assessment phases of the LCA. In the risk assessment, the ISCST-3 Gaussian dispersion model (U.S. EPA, 1995) was used for the calculation of the considered substance concentrations in the region; the CalTOX™ multimedia exposure model (DTSC, 2002) was used as well to evaluate the human health risk.

The second case shows the use of complementary tools to evaluate particulate matter exposure. A comprehensive calculation is undertaken to assess particulate matter (PM) exposure of three population groups (kids, adults, and retired people) living in an area influenced by a cement plant and a couple of highways. To do so, three fractions of environmental PM10, PM2.5 and PM1 (those smaller than 10, 2.5, and 1 μm respectively) were sampled. Subsequently, these outdoor levels were used in simulation software (IAQX) to calculate PM concentrations in different indoor environments. Then, these PM levels were combined with biological and time activity data to study the deposition pattern of PM inside the respiratory tract of the three population groups using a dosimetry model (MPPD). Finally, deposition patterns were combined with PM concentrations in different environments to calculate deposited masses along the breathing system.

In the third case, an environmental impact analysis of an industrial separation process including life-cycle assessment, environmental risk assessment and impact pathway analysis is carried out. In this case, TEAM™ (tool for environmental analysis and management), an LCA software (Ecobalance, 1997), was applied to the potential impact assessment for the inventory analysis and impact assessment, while CalTOX and the integrated assessment model Ecosense (IER, 1997) were the tools used for the site-specific assessment.

12.2 EXAMPLE 1: LANDFILLING OF MIXED HOUSEHOLD SOLID WASTE

12.2.1 INTRODUCTION

Landfilling of mixed household solid waste (MHSW) was chosen as the first example for its similarity to the case study "Waste Incineration." The first applications example stems from the same life-cycle stage as the case study of the previous chapters: "Recycle—Waste Management."

A further description of the typical features of that stage for a life-cycle assessment study can be found in Ciambrone (1997).

Landfill, traditionally, has been the most widely used method of waste treatment. However, the practice of landfilling has shown that the disposal of wastes that have not been pre-treated causes emissions corresponding to those of a bioreactor. These emissions are considered high risk, and landfills are ranked as the worst option in the waste hierarchy according to the pollution-preventing principle described in Chapter 1. In modern landfills, the emissions are collected and treated as much as possible by biogas combustion and leaching effluents purifications.

In this example, an LCA is performed for an average Spanish landfill; then, a risk assessment is carried out for the substance that has the highest potential danger according to the human toxicity indicator used in the life-cycle impact assessment (LCIA). A fictitious site has been designated for this example, which can be understood as a fully developed exercise for the sequence life-cycle assessment, dominance analysis for human toxicity potential and environmental risk assessment.

12.2.2 LIFE-CYCLE ASSESSMENT FOR LANDFILLING OF MIXED HOUSEHOLD SOLID WASTE

12.2.2.1 Introduction

A case study of an LCA concerning the treatment of 50,000 t of mixed household solid waste (MHSW) in a medium-size Spanish landfill is performed. The data of the inventory include the consumption of raw materials and energy through the use of containers, collection and transport of wastes and the management of the landfill, and the corresponding emissions to air, water and soil. The following nine environmental impact categories have been considered in the impact assessment phase of the LCA: water eutrophication; depletion of nonrenewable resources; air acidification; greenhouse effect; aquatic ecotoxicity; human toxicity; terrestrial ecotoxicity; depletion of the ozone layer; and photochemical oxidant formation. The software model and database WISARD®[1] (waste integrated systems assessment for recovery and disposal) of Ecobalance Price Waterhouse & Coopers has been used in the inventory analysis and impact assessment phases of the LCA (Figure 12.1).

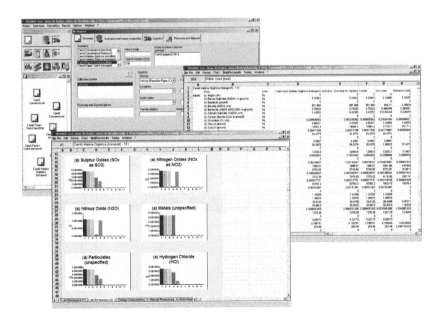

FIGURE 12.1   The software model and database WISARD.

12.2.2.2 Goal and Scope Definition

In this case study, the LCA methodology is applied to the treatment of MHSW in a typical medium-size Spanish landfill in order to evaluate the environmental impact of landfilling through the entire life cycle associated with this waste management activity. Wastes have been considered the main input to the system; 50,000 t of MHSW produced and landfilled annually in the studied area are considered the functional unit of the LCA. Figure 12.2 shows the flow diagram of the system studied with the corresponding main inputs and outputs considered.

12.2.2.3 Inventory Analysis

The considered landfill is situated in Spain. To carry out the inventory analysis, real and bibliographic Spanish data have been considered. For the collection data, the following main steps have been taken into account: use and maintenance of waste containers, collection and transport of wastes from the point of generation to the landfill, and management of the wastes in the landfill with partial (50%) collection and flaring of the generated biogas, and with partial (80%) collection and biological treatment of the produced leachate.

The following section and Tables 12.1 through 12.4 summarize the main input data introduced into the software for characterizing, from an environmental point of view, all the elements directly and indirectly, implicated in the landfilling activity (containers, vehicles, landfill and MHSW).

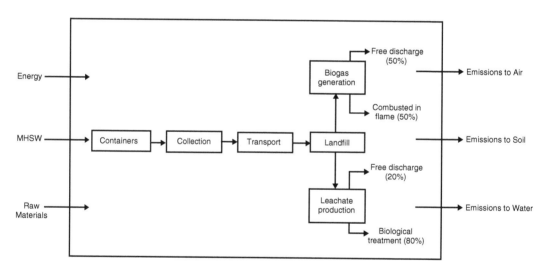

FIGURE 12.2 Flow diagram of landfilling activity.

TABLE 12.1
Characterization and Composition of MHSW

	Weight (%)	Density (kg/m³)	Moisture (%)	Biogas (kg/kg)
Organics	37.62	600	77.50	0.63
Paper/cardboard	21.01	183	22.32	0.29
Glass	8.21	300	1.84	0
Plastics	15.81	20	22.50	0
Metals	5.31	83	7.70	0
Others	12.04	94	20.93	0.18

TABLE 12.2
Vehicle Emission Factors during Collection and Transport

Flow	During Collection	During Transport
Ammonia (NH_3)	1.2	0.5
Carbon dioxide (CO_2)	306,240	132,000
Carbon monoxide (CO)	1,983	855
Methane (CH_4)	19	8
Nitrogen oxides (NO_x)	4,420	1,905
Nitrous oxide (N_2O)	131	57
Non methane hydrocarbons	726	313
Particulates	415	179
Sulfur oxides (SO_x)	88	38

Note: Grams per 100 km.

TABLE 12.3
Main Biogas Air Emissions of MHSW

Flow	Biogas Free Discharge	Biogas Combusted in Flare
Carbon dioxide (CO_2, biomass)	126,977	199,150
Carbon monoxide (CO)	73.8	89.4
Hydrocarbons (CxHy)	176.3	97.0
Methane (CH_4)	39,332	—
Nitrogen oxides (NO_x)	—	10.6
Sulfur oxides (SO_x)	—	3.7

Note: Grams per tonne.

TABLE 12.4
Main Leachate Composition of MHSW

Flow	Leachate Free Discharge
Ammonia (as N)	280.4
BOD_5 (biochemical oxygen demand)	1224.6
Chlorides (Cl^-)	1010.4
Metals	149.6
Nitrate (NO^{3-})	1.4
Nitrite (NO^{2-})	0.3
Phosphates (as P)	0.3
Potassium (K^+)	345.1
Sodium (Na^+)	618.7
Sulfates (SO_4^{2-})	117.2

Note: Grams per tonne.

Containers:
  General—capacity: 1.1 m³; lifetime: 6 years; and container load rate: 75%
  Composition—HDPE: 59 kg; steel: 4 kg; rubber: 1.5 kg; and aluminum: 0.5 kg
  Cleaning—frequency: 1 per week; water: 30 L; and cleaning product: 0.07 kg

Collection and transport vehicles:
  General—load: 21.5 m³; compacting ratio: 6.5%; lifetime: 185,000 km; collection: daily
  Vehicle body composition—steel: 13 t; PP: 500 kg; aluminum: 250 kg; glass: 50 kg
  Cleaning—frequency: 15 times/1000 km; water: 300 L; cleaning product: 5 kg
  Diesel consumption—collection: 116 L/100 km; transport: 50 L/100 km
  Average distances—collection: 9 km; transport: 18 km
  Oil consumption—motor oil: 108 L/10,000 km; hydraulic oil: 165 L/10,000 km

Landfill:
  General—tonnage stored: 4167 t/month; total storage capacity: 2,000,000 m³
  Construction—clay: 19,000 t; sand and gravel: 12,000 t; HDPE: 150 t; diesel: 30,000 L
  Operation (per month)—sand: 350 t, electricity: 2250 kWh; diesel: 4500 L; oil 100 L
  Biogas—free discharge: 50%; combusted in flare: 50%
  Leachate—total production: 55 L/t; free discharge: 20%; biologically treated: 80%
  Final capping—clay: 13,500 t; top soil: 6000 t; diesel: 13,500 L

After introducing the previously presented input data in the software model (WISARD), the eco-balance or inventory of the landfilling activity studied (50,000 t of MHSW) has been automatically calculated. The complete inventory was integrated by 307 environmental loads (inputs and outputs): energy and raw materials consumed and emissions to air, water and soil. The main inputs and outputs of the system that contribute more than 5% in any of the environmental impact indicators subsequently considered are shown in Table 12.5.

12.2.2.4 Impact Assessment

To carry out this phase of the LCA case study, the following nine impact categories have been considered: water eutrophication (WE), depletion of nonrenewable resources (DNRR), air acidification (AA), greenhouse effect (GE), aquatic ecotoxicity (AE), human toxicity (HT), terrestrial ecotoxicity (TE), depletion of the ozone layer (DOL) and photochemical oxidant formation (POF). The specific environmental impact indicators used in each of the environmental impact categories mentioned are shown in Table 12.6.

The environmental loads (inputs and outputs) previously inventoried have been classified under their corresponding environmental impact indicator, following the classification criteria specified in these indicators. Characterization factors have then been used to prioritize the environmental loads or, in other words, to quantify the potential contribution of each environmental load in the different impact indicators. These characterization factors are pre-established for each impact indicator. Finally, the corresponding potential contributions have been determined by multiplying the mass of the environmental loads in the inventory by these factors (for example, 54,533 g trichloroethane × 1200 g equiv. of 1-4-dichlorobenzene/g trichloroethane = 65,439,600 g equiv. of 1-4-dichlorobenzene).

The inventoried environmental loads are classified under their corresponding impact indicators with their respective characterization factors in Table 12.7. The total potential contribution of each environmental load in each impact category, as well as its corresponding contribution percentage for the four stages of the landfilling activity (containers, collection, transport and landfilling), are presented.

12.2.2.5 Interpretation

As can be seen in Table 12.8 and Figure 12.3, landfilling is the stage that contributes more to WE, AA, GE, AE, HT, DOL and POF, indicators being the transport of wastes to the landfill, the main contributor to DNRR and TE indicators.

TABLE 12.5
Inventory of Landfilling Activity[a]

Flow	Units	Total	Containers	Collection	Transport	Landfilling
Inputs						
(r) Natural gas (in ground)	kg	26,749	9,526	3,896	5,024	8,303
(r) Oil (in ground)	kg	194,463	8,935	64,523	58,819	62,186
(r) Phosphate rock (in ground)	kg	16,148	0	1	1	16,147
(r) Zinc (Zn, ore)	kg	289	18	90	180	0
Outputs						
(a) Aromatic hydrocarbons (unspecified)	g	783,231	5	37	74	783,115
(a) Cadmium (Cd)	g	24	1	7	12	4
(a) CFC 12 (CCl_2F_2)	g	46,624	0	0	0	46,624
(a) Ethylene (C_2H_4)	g	1,727,757	687	1,249	1,316	1,724,505
(a) Hydrocarbons (except methane)	g	2,512,440	6,211	899,164	801,240	805,825
(a) Lead (Pb)	g	635	34	193	380	27
(a) Mercury (Hg)	g	10	2	3	5	1
(a) Methane (CH_4)	g	989,662,857	105,419	2,198,955	2,063,482	985,295,000
(a) Nitrogen oxides (NO_x as NO_2)	g	8,504,624	122,639	2,944,386	2,629,864	2,807,735
(a) Sulfur oxides (SO_x as SO_2)	g	1,209,893	83,348	251,866	333,398	541,280
(a) Trichloroethane (1,1,1-CH_3CCl_3)	g	54,533	0	0	0	54,533
(a) Zinc (Zn)	g	2,899	184	898	1,791	26
(w) Ammonia (NH_4^+, NH_3, as N)	g	2,834,934	166	5,243	4,590	2,824,935
(w) Cadmium (Cd^{++})	g	171	10	5	5	151
(w) Chromium (Cr III)	g	1,074	1	3	5	1,065
(w) Mercury (Hg^+, Hg^{++})	g	65	62	0	1	2

Note:　(r): natural resources; (a): emission to air; (w): emission to water.
[a]　50,000 t MHSW.

TABLE 12.6
Environmental Impact Categories and Indicators Used in WISARD

Acronym	Category	Indicator—Source (Year)
WE	Water eutrophication	CML (water)—CML Leiden University (1992)
DNRR	Depletion of nonrenewable resources	EB (R*Y)—Ecobalance USA (1998)
AA	Air acidification	ETH (air acidification)—Swiss Federal Institute of Technology, Zurich (1995)
GE	Greenhouse effect	IPCC (direct, 20 years)—International Panel on Climate Change (1998)
AE	Aquatic ecotoxicity	USES 1.0—CML Leiden University (1996)
HT	Human toxicity	USES 1.0—CML Leiden University (1996)
TE	Terrestrial ecotoxicity	USES 1.0—CML Leiden University (1996)
DOL	Depletion of the ozone layer	WMO (average)—World Meteorological Organization (1998)
POF	Photochemical oxidant formation	WMO (average)—United Nations Economic Commission for Europe (1991)

TABLE 12.7

Impact Assessment of Landfilling Activity[a]

Impacts	Characterization Factors	Total	Containers (%)	Collection (%)	Transport (%)	Landfilling (%)
Water eutrophication (g eq. PO_4)	*	1,206,626	0.2	0.5	0.7	98.6
(w) Ammonia (NH_4^+, NH_3, as N)	0.42	1,190,672	0.0	0.2	0.2	99.6
Depletion of nonrenewable resources (yr^{-1})	*	28,995	8.6	27.9	40.8	22.7
(r) Zinc (Zn, ore)	40.29	11,626	6.4	31.2	62.4	0.0
(r) Oil (in ground)	0.0557	10,832	4.6	33.2	30.2	32.0
(r) Natural gas (in ground)	0.117	3,130	35.6	14.6	18.8	31.0
(r) Phosphate rock (in ground)	0.115	1,857	0.0	0.0	0.0	100.0
Air acidification (g equiv. H^+)	*	228,251	2.3	31.6	29.9	36.2
(a) Nitrogen oxides (NOx as NO_2)	0.0217	184,883	1.4	34.6	30.9	33.0
(a) Sulfur oxides (SOx as SO_2)	0.0313	37,809	6.9	20.8	27.6	44.7
Greenhouse effect (direct, 20 years) (g equiv. CO_2)	*	64,802,966,673	0.1	0.6	0.6	98.7
(a) Methane (CH_4)	64	63,338,422,828	0.0	0.2	0.2	99.6
Aquatic ecotoxicity (g equiv. 1-4-dichlorobenzene)	*	1,157,851	7.8	7.7	12.4	72.1
(w) Cadmium (Cd^{++})	4,500	769,211	6.1	2.8	2.7	88.4
(a) Mercury (Hg)	16,000	161,391	17.6	25.1	48.5	8.7
(w) Chromium (Cr III)	84	90,228	0.1	0.3	0.5	99.2
Human toxicity (g equiv. 1-4-dichlorobenzene)	*	173,132,057	2.1	9.3	16.8	71.8
(a) Trichloroethane (1,1,1-CH_3CCl_3)	1,200	65,439,600	0.0	0.0	0.0	100.0
(w) Ammonia (NH_4^+, NH_3, as N)	17	48,193,881	0.0	0.2	0.2	99.6
(a) Lead (Pb)	67000	42,548,314	5.4	30.4	59.9	4.3
Terrestrial ecotoxicity (g eq. 1-4-dichlorobenzene)	—	6,271,415,948	13.3	27.0	46.7	13.1
(a) Cadmium (Cd)	130,000,000	3,159,589,505	3.9	30.3	49.8	16.0
(a) Zinc (Zn)	660,000	1,913,290,536	6.4	31.0	61.8	0.9
(w) Mercury (Hg^+, Hg^{++})	8,200,000	532,756,761	95.5	0.5	1.0	3.0
Depletion of the ozone layer (average) (g equiv. CFC-11)	*	40,031	0.0	0.4	0.3	99.3
(a) CFC 12 (CCl_2F_2)	0.82	38,231	0.0	0.0	0.0	100.0
Photochemical oxidant formation (average) (g equiv. ethylene)	*	11,687,352	0.8	3.6	3.2	92.5
(a) Methane (CH_4)	0.007	6,927,640	0.0	0.2	0.2	99.6
(a) Ethylene (C_2H_4)	1	1,727,757	0.0	0.1	0.1	99.8
(a) Hydrocarbons (except methane)	0.416	1,045,175	0.2	35.8	31.9	32.1
(a) Aromatic hydrocarbons (unspecified)	0.761	596,039	0.0	0.0	0.0	100.0

Note: (r): natural resources; (a): emission to air; (w): emission to water.

[a] 50,000 t MHSW.

TABLE 12.8
Stages' Contribution (%) in Each Impact Indicator

Impacts	Containers	Collection	Transport	Landfilling
WE	0.2	0.5	0.7	98.6
DNRR	8.6	27.9	40.8	22.7
AA	2.3	31.6	29.9	36.2
GE	0.1	0.6	0.6	98.7
AE	7.8	7.7	12.4	72.1
HT	2.1	9.3	16.8	71.8
TE	13.3	27.0	46.7	13.1
DOL	0.0	0.4	0.3	99.3
POF	0.8	3.6	3.2	92.5

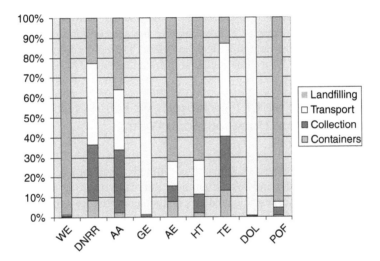

FIGURE 12.3 Stages' contribution (%) in each impact indicator.

Direct and indirect consumption of scarce natural nonrenewable resources contributes significantly to DNRR indicator. During the construction of vehicle bodies for collecting and transporting wastes to the landfill, these scarce raw materials are directly and indirectly consumed (e.g., zinc, oil, natural gas, etc.), transport and collection of wastes being the main contributors to this indicator.

Free discharge of biogas, estimated at 50% of the total generated in the landfill, and its associated methane air emissions is the main contaminant contributor to the GE indicator. Biogas is also composed by other contaminants such as CFC 12, which is the main contributor to the DOL indicator, and ethylene, hydrocarbons and aromatic hydrocarbons are the main contributors to the POF indicator. Also, free discharge of biogas (estimated at 50%) and leachate (estimated at 20%) have other associated atmospheric and water emissions; the presence of trichloroethane in biogas and ammonia in leachate are the main contaminant contributors to the HT indicator, and cadmium and chromium in leachate are the main contaminant contributors to the AE indicator.

The combustion in a flare of the biogas collected (estimated at 50% of the total generated) and the consequent air emissions of nitrogen and sulfur oxides are the main contributors to the AA indicator. Also, significant nitrogen and sulfur emissions are produced during collection and transport of wastes to the landfill.

The potential risk of leachate-free discharges into soil (estimated at 20% of the total generated) significantly increases the risk of WE; ammonia is the main contaminant contributor to this indicator. In the TE indicator, the main contributors are cadmium and zinc air emissions generated primarily during the construction of vehicle bodies, but also during diesel oil production and its combustion during collection and transport of wastes. Mercury in water, generated during the container manufacturing process, is also an important contaminant contributor to this indicator.

As a final comment, it should be mentioned that impact assessment interpretation is a particularly difficult task in a landfilling activity, mainly because of the temporary dependence of its environmental consequences. In landfilling, the chemical life of wastes can be estimated approximately 30 years before being considered an inert waste, which implies that biogas and leachate emissions will vary in quantity and composition during this period. In this case study, any emission to air, water and soil produced along the chemical life of wastes (30 years) has been directly assigned to the functional unit considered (50,000 t of MHSW).

12.2.3   SELECTION OF POLLUTANT FOR SITE-SPECIFIC IMPACT ASSESSMENT IN LANDFILLING EXAMPLE

As pointed out in the introduction to this example, a dominance analysis is to be carried out for the human toxicity indicator. Table 12.7 indicates that the air emissions of 1,1,1-trichloroethane $(C_2H_3Cl_2)$ contribute most to the human toxicity indicator. Therefore, the next section presents an exposure risk assessment for 1,1,1-tricholoroethane from landfilling of MHSW.

12.2.4   RISK ASSESSMENT OF THE 1,1,1-TRICHLOROETHANE EMISSIONS FROM LANDFILLING OF MIXED HOUSEHOLD SOLID WASTE

12.2.4.1   Introduction

Pollutants emitted to the atmosphere are transported through it and may subsequently impact environmental media (i.e., soil, water and vegetation) near the plant, resulting in a number of potential sources for human exposure. Because the landfill's emissions of trichloroethane were the main contaminant contribution to the human toxicity indicator according to the LCA applied in Example 1.1 (Table 12.7 and Section 12.2.3), the aim of this exercise is to calculate the incremental lifetime risk due to the 1,1,1-trichloroethane $(C_2H_3Cl_3)$ emission of the landfill of MHSW for the residents living in the surroundings of the plant. In order to obtain this, the air 1,1,1-trichloroethane concentrations in the vicinity of the landfill were quantified by application of a Gaussian dispersion model (ISCST-3). Then, human health risks due to 1,1,1-trichloroethane emissions from the landfill were calculated by application of a multimedia exposure model (CalTOX).

12.2.4.2   Air Dispersion Models

Pollutants emitted to the air are dispersed depending on the meteorological conditions, e.g., wind speed and solar radiation, and the characteristics of the region, e.g., elevations of the terrain and land use. Accordingly, they occur in the atmosphere in areas even farther from the emission source.

The atmosphere is the starting medium of the environmental fate and transport of pollutants of the landfill air emissions. The pollutants are dispersed in the air depending on the meteorological and topographic conditions in the location of the emission source and can be transported in the atmosphere over large distances. However, a portion of the pollutants is deposited in the surrounding area of the emission source and accumulates in other environmental media such as soil, surface water or vegetation. If air concentrations of pollutants cannot be determined with measurements, they can be calculated using air dispersion models, which simulate the atmospheric dispersion using meteorological and topographic information of the considered region. In the current exercise, the air dispersion of emitted 1,1,1-trichloroethane was modeled for the surroundings of the landfilling of MHSW. In this risk assessment, the ISCST3 model, described in Chapter 4, was used to estimate air concentration dispersion of the 1,1,1-trichloroethane emissions.

12.2.4.3   Data for the ISCST3 Model

ISCST3 is based on a Gaussian plume model (see Chapter 4). It is most common to compute ambient air concentrations and surface deposition fluxes at specific receptors near a steady-state emission source. The model is capable of simulating air dispersion of pollutants from point, area, volume and line sources. A full description of the ISCST3 model and its algorithms can be found in the ISCST3 User's Guide (U.S. EPA, 1995).

The results of the air dispersion model rely on four basic data sets: (1) meteorological conditions; (2) facility characteristics; (3) location of buildings near the emission sources; and (4) location of receptors (distance to the emission source and elevation on the terrain). To calculate the air dispersion of contaminants, the ISCST3 model requires hourly meteorological data. They include values of (1) wind speed and flow vector; (2) ambient air temperature; (3) atmospheric stability class; and (4) rural and urban mixing height.

In order to calculate the dry and wet deposition fluxes to the ground, additional information is needed: (1) friction velocity; (2) Monin–Obukhov length; (3) surface roughness length; and (4) precipitation code.

The meteorological data used in this exercise contained hourly values of wind speed and wind direction, ambient temperature, precipitation and solar radiation. All further parameters can be calculated using this information. Figure 12.4 shows the wind rose (distribution of the flow vector) corresponding to the meteorological data used in this case study. It can be observed that wind blowing from the north is the most frequent, and that wind blowing from the east is strongest. The studied landfill of MSHW is situated in a zone with a high percentage of calm hours. About 27% of all hourly wind speed values did not exceed 0.1 m/s and the average wind speed was 2.75 m/s.

The studied area was defined with an extension of 10 × 10 km, locating the plant in the middle. A set of topographic data for the studied area, including the elevations of the terrain, was used. Figure 12.5 shows the topography of the study area. It can be observed that in the eastern direction, 3 km away from the plant, the terrain presents the most considerable elevation in the area. In the southern direction, 1 km away from the plant, the terrain presents an elevation that, because it is closer to the plant, can be suspected of affecting the air dispersion of the pollutant emissions.

A network of 2602 Cartesian receptors (10 × 10 grid) was established to model the dispersion of pollutants in the entire studied area. The measuring points were set 200 m apart, representing a total area of 100 km². Each receptor was assigned an elevation based on the topographic map of the region.

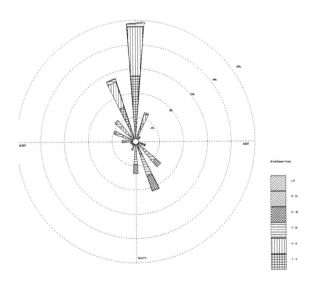

FIGURE 12.4   Wind rose (wind speed and flow vector).

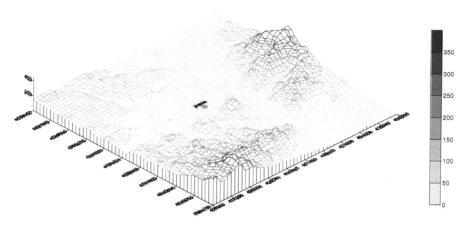

FIGURE 12.5   Topographic map of the study area.

The characteristics of the emission source have a high importance to the resulting air disper-sion concentrations of a pollutant. In addition to the dimensions of the sources—location, surface, height—and the physical characteristics of the emission flow, the concentrations of pollutants in the emission flow are required by the ISCST3 model. The landfill is located in the middle of the topographic map; its dimensions are 173 × 173 × 70 m. The emissions of 1,1,1-trichloroethane, as explained in the application of the LCA to the landfill of MHSW, were 54,533 g per year.

12.2.4.4   Air Dispersion Results of 1,1,1-Trichloroethane

This section of the study presents the results of the air dispersion modeling for the emission of 1,1,1-trichloroethane. The atmospheric distribution of the 1,1,1,-trichloroethane emissions is illus-trated in Figure 12.6, which shows the average concentration of 1,1,1-trichloroethane at ground level in the vicinity of the landfill.

The main wind in the study area is blowing from the north (most frequent; see the wind rose in Figure 12.4. Therefore, it was expected that the highest air concentration of the pollutants would occur south of the landfill. This prediction was confirmed by the modeling results. The annual aver-age concentrations southward were substantially higher than in other parts of the region. However, it can be observed that air dispersion in the south was influenced by topography, more concretely

1,1,1-TRICHLOROETHANE

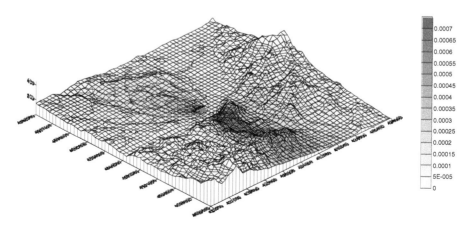

FIGURE 12.6   Air 1,1,1-trichloroethane concentration (μg/m^3).

by the elevation present 1 km away in the south. On the other hand, it can be observed that the wind blowing from the southeast influenced the air dispersion to the northwest direction.

In order to assess the air concentrations of 1,1,1-trichloroethane, medium concentrations for the entire study area were estimated. The medium air concentration of the pollutant was 9.69×10^{-2} ng/m^3; it was estimated with the annual averages in all 2584 Cartesian receptors representing the study area.

12.2.4.5 Exposure Calculation Model

For the evaluation of the exposure of the population living in the area, a multiple pathway exposure, transport and transformation model (CalTOX) was used. In Chapter 4, this model is compared with the other multimedia model EUSES. This model includes modules for the distribution of substances in the environment, exposure scenario models for humans and the environment, human risk estimation and efforts to quantify and reduce uncertainty in multimedia. It has been designed to assist in assessing human exposure, define soil clean-up goals at (uncontrolled) hazardous waste sites and improve the quality of risk assessment information, especially as required for regulatory decisions. The models and data sets have been compiled as Microsoft Excel 4.0 (or higher) spreadsheets. They are available together with information and documentation via the Internet through various websites.

Each pathway can be included or excluded separately in the calculations, depending on the scope of the study. The exposure model defines air, drinking water, food and soil as the four main sources for human exposure to a substance via different pathways such as inhalation, ingestion, or dermal contact.

Figure 12.7 shows the eight compartments implemented in CalTOX: air, surface water, groundwater, sediment, surface soil, root-zone soil, vadose-zone soil and plants. Usually, a chemical equilibrium between the phases is assumed. Unidirectional transport is assumed only from soil to water and from upper to lower soil zones, mainly because of the much higher diffusion speed in these directions compared to the other direction. The equations used in CalTOX to estimate exposure and risk are taken from the U.S. Environmental Protection Agency Risk Assessment Guidance for Superfund (U.S. EPA, 1989) and from the California Department of Toxic Substances Control (DTSC, 1992a, 1992b). They are based on conservation of mass.

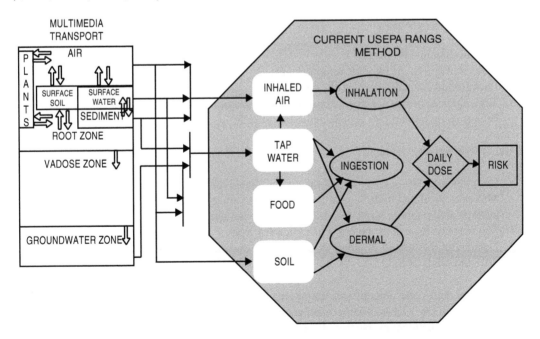

FIGURE 12.7   The structure of the CalTOX model with the multimedia transport, the intermedia transfer, and its exposure pathways. (From DTSC, CalTOX version 2.3 (www-pages), Department of Toxic Substances Control of the California EPA, State of California, http://www.dtsc.ca.gov/ScienceTechnology/caltox.html, July 2002.)

The release of a substance can be continuous (to air, water and surface soil) or a batch process with an initial concentration within deeper soil zones. The exposure model of CalTOX calculates daily human doses for various pathways (e.g., inhalation of air, ingestion of soil, milk, meat, etc.) based on daily intake rates and predicted concentrations in the respective exposure medium. Finally, risk values based on the doses are calculated.

12.2.4.6   Spatial Scale, Time Scale, and Assessable Substances

There is only one spatial scale because CalTOX is intended to be used site specifically (i.e., to assess a specific existing site rather than a big area such as an entire country). Although it is rather flexible, the assessed area should not be smaller than 1000 m² with a maximum water fraction of 10% of the surface area. Theoretically, there is no upper limit for the area, but one should keep in mind that "site-specific" means specific for an actual site, but not for a country or a continent. However, this limit is set by the adjustments to the input data. The greater the area is, the more average (hence less specific) every value necessarily must be and the less site-specific is the entire assessment.

To obtain good results, the time scale should be defined as rather long, preferably from 1 year to decades. If shorter periods are assessed properly, time-averaged landscape properties must be employed. Obviously, CalTOX has not been designed to assess acute exposure at an actual site. The original idea was to have an already contaminated site and to assess the risk for human beings living or working at or near a site that led to subchronic or chronic exposure only.

The substances assessable with CalTOX, listed in descending order of reliability, are: non-ionic organic chemicals, radionuclides, fully dissociating inorganic and/organic chemicals, solid-phase metal species, and partially dissociated inorganic and organic species (the latter only if partition coefficients are well adjusted to the pH of the considered landscape). Generally, only very low concentrations that do not exceed the solubility limit in any phase can be applied. Surfactants or volatile metals cannot be assessed.

12.2.4.7   Input Data for CalTOX

Such complex models need a good range of input data in order to obtain trustworthy results:
Data describing the substance:

- Physicochemical properties
- Measured emissions into the compartments
- Background concentrations of the contaminant
- Toxicological properties consisting of cancer and noncancer potency for human beings, because only human risk is considered in CalTOX

Data characterizing the area:

- Geographical data such as size of the contaminated area
- Meteorological and hydrological data, e.g., the average depth of surface water, the annual average precipitation, wind speed or environmental temperature
- Soil properties, such as the organic carbon fractions
- Data about the human population, e.g., the average body weight or daily intake rates for different kinds of food

In the present study, the objective of the application of CalTOX is to evaluate the health of the population due to the 1,1,1-trichloroethane emissions from a landfill applied to a determined area. First, the physicochemical properties of 1,1,1-trichloroethane, as well as the risk parameters, were obtained from the database of CalTOX, and are shown in Tables 12.9 and 12.10, respectively. Then, the geographical parameters and population data from the nearest residential area within 10 km around the plant were defined (Table 12.11).

TABLE 12.9
Value for Some Physicochemical Properties of 1,1,1-Trichloroethane

Compound Name	1,1,1-Trichloroethane
Molecular weight (g/mol)	133.4
Octanol–water partition coefficient	2.7×10^2
Melting point (K)	242.75
Vapor pressure (Pa)	16,515.975
Solubility (mol/m^3)	9.89505247
Henry's law constant (Pa-m^3/mol)	1651.5975
Diffusion coefficient in pure air (m^2/d)	0.67392
Diffusion coefficient in pure water (m^2/d)	8.6386×10^{-5}
Organic carbon partition coefficient (L/kg)	110

TABLE 12.10
Value for Risk Parameters of 1,1,1-Trichloroethane

Compound Name	1,1,1-Trichloroethane
EDF substance ID	71-55-6
Inhalation dose ADI for noncarcinogenic effects	0.28
Ingestion dose ADI for noncarcinogenic effects	0.5
Dermal dose ADI for noncarcinogenic effects	0.5
Total dose ADI for noncarcinogenic effects	0

TABLE 12.11
Geographical, Meteorological, and Population Data Describing the Study Area

Property	Value
Area [km^2]	100
Inhabitants	580,245
Area fraction of water	0.01
Area fraction of natural soil	0.46
Area fraction of agricultural soil	0.46
Wind speed [m/s]	2.8
Average annual precipitation [mm/a]	455.4
Environmental temperature [°C]	15.7
Egg intake [kg/(kg*d)]	3.85^{-4}
Grain intake [kg/(kg*d)]	3.11^{-3}
Fruit and vegetable intake [kg/(kg*d)]	7.39^{-3}
Daily intake of fish [kg/d]	0.08
Daily intake of leaf crops (including fruits and cereals [kg/d]	0.674
Daily intake of meat [kg/d]	0.04
Daily intake of dairy products [kg/d]	0.069
Milk intake [kg/(kg*d)]	2.78^{-3}
Body weight [kg]	67.52

Daily Human Doses of 1,1,1-trichlorethane

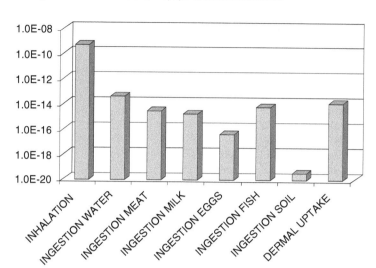

FIGURE 12.8   Daily human dose through different exposure types (mg/kg/d).

It must be taken into account that 1,1,1-trichloroethane does not occur naturally in the environment. It is found in many common products such as glue, paint, industrial degreasers and aerosol sprays. Since 1996, 1,1,1-trichloroethane has not been produced in the U.S. because of its effects on the ozone layer.

An expected most important exposure pathway might be through inhaled air because the landfill is the major source. No information is available to show that 1,1,1-trichloroethane causes cancer. The International Agency for Research on Cancer (IARC) has determined that 1,1,1-trichloroethane is not classifiable as to its human carcinogenicity.

12.2.4.8   Exposure Results of 1,1,1-Trichloroethane

Figure 12.8 represents the daily human doses through eight types of exposure that were calculated based on the simulated air concentration due to the 1,1,1-trichloroethane emissions of the landfill of MSHW. As expected, the main exposure pathway is air inhalation. Depending on the exposure level, adverse health effects other than cancer can be associated with all chemical substances. The hazard ratio result of the actual human exposure due to the emission of the 1,1,1-tricholoroethane is 1.7×10^{-10}; thus human exposure does not exceed the defined exposed level.

12.2.5   CONCLUSIONS FROM THE LANDFILLING EXAMPLE OF MIXED HOUSEHOLD SOLID WASTE

A comprehensive LCA in the context of integrated solid waste management software was carried out. It is evident that in the case of landfilling, the site-specific component is relevant to evaluate if potential impacts really correspond to actual impacts. To further study this question in the example, the main pollutant contributing to the human health indicator was chosen to carry out an exposure risk assessment.

It could be shown that the potential impact does not correspond to any unacceptable risk for the neighborhood. Nevertheless, the value of LCA clearly lies in its highlighting of the existing emissions that must all be considered according to the precautionary principle and in its holistic approach by assessing several pollutants and the related impact categories at once along the studied life cycle. Further work is necessary and should include performing similar studies for other media, such as water and soil, with adequate models and taking into account the accident risks.

12.3   EXAMPLE 2: USE OF COMPLEMENTARY TOOLS TO EVALUATE PARTICULATE MATTER EXPOSURE

12.3.1   INTRODUCTION

Inhalable particulate matter (PM) is a mixture of solid and liquid droplets ubiquitous in the atmosphere with sizes ranging from 10 µm to the nanometer scale (Perez et al., 2009). They have a variable composition, the presence of toxicants being usual, such as heavy metals or polycyclic aromatic hydrocarbons (PAHs) (Ercan and Dinçer, 2015). Consequently, PM is recognized nowadays as the most harmful air pollutant, provoking different diseases in the cardiovascular and pulmonary systems (US EPA, 2015; WHO, 2014).

In the present example, a comprehensive calculation is undertaken to assess PM exposure of three population groups (kids, adults, and retired people) living in an area influenced by a cement plant and a couple of highways. To do so, sampled outdoor levels of PM were used in simulation software (IAQX) to calculate PM concentrations in different indoor environments. Then, these PM levels were combined with biological and time activity data to study the deposition pattern of PM inside the respiratory tract of the three population groups using a dosimetry model (MPPD). Finally, deposition patterns were combined with PM concentrations in different environments to calculate deposited masses along the breathing system. The exercise here explained is based in the methodology published by Sánchez-Soberón et al. (2015a).

12.3.2   METHODOLOGY

As said in the induction, PM comprises several sizes of airborne materials. In the present study, we focused on three different ranges of PM: between 10 and 2.5 µm ($PM_{10-2.5}$), between 2.5 and 1 µm ($PM_{2.5-1}$), and smaller than 1 µm (PM_1). PMs were collected in an urban-industrial neighborhood located in the outskirts of Barcelona (Spain) using high volume samplers. Apart from domestic sources of PM, such as heating systems, atmosphere in the area is influenced by a couple of highways and a cement plant. PM sampling was developed during two seasons: winter and summer (Sánchez-Soberón et al., 2015b). Levels of PM are depicted in Table 12.12.

TABLE 12.12
Input Parameters in the IAQX Simulations

Parameter	Value			Reference
Building				
Number of air zone	1			
Volume (m³)	90 m² × 2.5 m = 225 m³			
Deposition considered?	Yes			
Ventilation				
Home air exchange rate (h⁻¹)	Winter: 0.44; Summer: 1.30			Wallace et al. (2002)
Workplace/school (h⁻¹)	Winter and Summer: 1.00			Orosa and Baaliña (2008)
PM Properties				
PM Fraction	PM_1	$PM_{2.5-1}$	$PM_{10-2.5}$	
Deposition rate (h⁻¹)	0.8	1	2.5	He et al. (2005)
Infiltration factor (unit-less)	0.8	0.6	0.35	Chen and Zhao (2011)
Winter outdoor concentration (µg/m³)	31	1	19	Sánchez-Soberón et al. (2015b)
Summer outdoor concentration (µg/m³)	13	7	1	Sánchez-Soberón et al. (2015b)

Once the outdoor levels were had, the simulation of indoor concentrations of PM was proceeded. For this purpose, Indoor PM module, from the software IAQX v1.1 (US EPA, 2000) was used. In this example, two different indoor microenvironments were considered: workplace/school and home. Both were simulated as a single air room of 90 square meters and a ceiling 2.25 meters high. The only difference between them is the ventilation rate: while at home it is variable between winter and summer (i.e., higher window ventilation in summer), it has a constant value in the workplace. Since, in this example we are focused on evaluating exposure from atmospheric particulates, we did not consider any indoor PM source. The rest of the parameters used as inputs, as well as their values, can be consulted in Table 12.12. For any other parameter demanded by the software, default values were used.

The next step to evaluate the exposition properly is to identify how much air is breathed in every microenvironment (outdoors, workplace/school, and home). Therefore, it is needed to know not only the time spent in every microenvironment, but also the activities performed within them. Since the use of time is dependent on the age, it was decided to study the exposure for three different population groups: kids, adults, and retired people. Adapting time activity studies previously published (Cohen Hubal et al., 2000; IEC, 2012; INSEE, 2010) we divided the standard day into five activities: Sleeping, Sitting, Light Intensity activities indoor, Light Intensity activities outdoor, and Heavy Intensity activities. Sleeping and Sitting were considered as fully indoor activities, and Heavy Intensity as exclusive outdoor action. Daily time spent in every one of the activities is shown in Table 12.13.

To simulate the PM deposition in the different regions of the respiratory tract, MPPD v 2.11 model was used (ARA, 2014). This model can calculate the fraction of PM deposited (referred as Deposition Fraction, or DF) in three different regions of the human respiratory tract: head/throat (Head), pulmonary/tracheobronchial (TB), and pulmonary/alveolar (P). To run this model it is needed to introduce different data to define the airway morphometry, PM properties, and exposure conditions. Regarding the airway morphometry, we used the default Yeh and Schum symmetric model. This model requires the input of two physiological parameters, Functional Residual Capacity (FRC) and Upper Respiratory Tract volume (URT). FRC is defined as "volume of the lung at end of a normal expiration," while URT is the "volume of the respiratory tract from the nostril or mouth down to the pharynx." It is possible to use the default values of these parameters provided by the software, but, in this example, we used UTR values provided by Brown et al. (2013), and calculated FRC values following the equations provided by Stocks and Quanjer (1995):

$$FRC_{kids}(mL) = 0.125 \times 10^{-3} \times Height^{3.298} (cm)$$

$$FRC_{adults\ and\ retired}\ (L) = 2.34 \times Height\ (m) + 0.01 \times Age\ (years) - 1.09$$

Concerning PM properties, we considered the diameter of the PM as single instead of multiple, since this last option does not allow to calculate depositions under variable exposures. Exposure conditions were considered variable. Time-activity patterns were introduced here, as well as the Breathing Frequency (i.e.: breaths per minute; BF) and Tidal Volume (i.e.: single breath volume; TV). Inspiratory Fraction was considered as 0.5, and Pause Fraction was 0. In this case, we considered only nasal breathing. Finally, in order to evaluate worst case scenario, not clearance was taken into account for our simulation. All input data used in the simulation is summarized in Table 12.13. The rest of the parameters needed to run the software were remained as "by default."

TABLE 12.13

Daily Activity Patterns and Physiological/Morphological Parameters for Kids, Adults, and Retired

	Age (years)	Activity	Time (hours/day)	Height (cm)	BW (kg)	BF (breaths/min) [f]	TV (mL) [g]	FRC (mL) [h]	UTR (mL) [i]
Kid	10	Sleeping	9.6[a]	139.5[d]	36[d]	17	304	1478	25
		Sitting	11.5[a]			19	333		
		Light indoor	0.4[a]			32	583		
		Light outdoor	0.5[a]			32	583		
		Heavy	2.0[a]			45	752		
Adult	45	Sleeping	8.1[b]	175[e]	69.7[e]	12	625	3455	50
		Sitting	10.4[b]			12	750		
		Light indoor	2.8[b]			20	1250		
		Light outdoor	2.3[b]			20	1250		
		Heavy	0.4[b]			26	1920		
Retired	75	Sleeping	8.6[c]	175[e]	69.2[e]	9	625	3755	50
		Sitting	7.8[c]			9	750		
		Light indoor	3.7[c]			22	1250		
		Light outdoor	3.7[c]			22	1250		
		Heavy	0.2[c]			25	1920		

Source: [a]Cohen Hubal, E.A. et al., *Environ. Health Perspect.*, 108, 475–486, 2000; [b]IEC, Enquesta de l'ús del temps 2010–2011? Principals resultats. – (Estadística Social). Institut d'Estadística de Catalunya, 2012; [c]INSEE, Conditions de vie-Société–Depuis 11 ans, moins de tâches ménagères, plus d'Internet. Insee. Available at: http://www.insee.fr/fr/themes/document.asp?ref_id=ip1377, 2010; [d]Carrascosa, A. et al., *An. Pediatría*, 68, 552–569, 2008; [e]INE, Índice de masa corporal por grupos de edad, 2011–2012, Inst. Nac. Estadística, Available http//www.ine.es/jaxi/tabla.do?type=pcaxis&path=/t00/mujeres_hombres/tablas_1/l0/&file=d06001.px, 2012; [f]US EPA, Exposure Factors Handbook, 2011 edition, National Centre of Environmental Assessment, U.S. Environmental Protection Agency, Washington, D.C., September 2011; [g]ICRP, *Ann ICRP*, 24, 1–482, 1994; [h]Stocks, J. and Quanjer, P.H., *Eur. Respir. J.*, 8, 492–506, 1995; [i]Brown, J.S. et al., *Part. Fibre Toxicol.*, 10, 12, 2013.

Note: BW: body weight; BF: breathing frequency; TV: tidal volume; FRC: Functional Residual Capacity; UTR: upper respiratory tract volume.

Finally, to calculate the deposited mass in respiratory region j, for activity i, developed in microenvironment k, the next equation was applied (Yeh and Schum, 1980):

$$\text{Deposited Mass}_{i,j,k} \ (\mu g) = \text{DF}_{i,j} \times \text{PM}_k \left(\frac{\mu g}{m^3} \right) \times \text{TV}_i \left(\frac{m^3}{\text{breath}} \right) \times \text{BF}_i \left(\frac{\text{breath}}{\text{min.}} \right) \times T_i (\text{min.})$$

where:

$\text{DF}_{i,j}$ is the deposition fraction in region j for activity i

PM_k is the PM concentration in microenvironment k

TV_i is the tidal volume for activity i

BF_i is the breathing rate for activity i

T_i is the time spent daily in activity i

12.3.3   RESULTS

12.3.3.1   PM Concentrations in Microenvironments

Outdoor concentrations of PM, as well as indoor simulated concentrations are presented in Figure 12.9. As depicted in the figure, outdoor concentrations were higher than those experienced indoors (workplace/school and home). This phenomenon is understandable, since in this example we wanted to study the exposure to outdoor PM, and we did not include any indoor sources. However, as recognized in previous studies, indoor sources, such as tobacco smoke, dust resuspension, or stove combustion products can be prevalent in some indoor environments, and should be taken into account in holistic assessments (Morawska et al., 2013). As a consequence of outdoor PM levels, winter indoor concentrations of PM were higher than in summer. In this example, it is also possible to see the influence of ventilation rates. Home PM levels in summer are higher than at workplace/school, while in winter, results are the opposite, thus confirming that the greater the ventilation rate, the higher the concentration of indoor PM from outdoor origin. It is also interesting to note the increased contributions of $PM_{2.5-1}$ and PM_1 to total PM indoors, which is explained by the higher infiltration rate and slower deposition of these two PM fractions.

12.3.3.2   Deposition Fractions

Daily Deposition Fractions (DF) in the different respiratory regions can be seen in Table 12.14. There were no differences between summer and winter in DF values, indicating no effect of PM levels on this parameter. Close DF values were obtained for adults and retired, since these two groups show similar physiological characteristics. In most cases, Head was the region experiencing highest DF regardless of PM fraction, population group, and activity. Only a couple of values were found different in kids: deposition of $PM_{2.5-1}$ during Sleeping and Sitting were higher in the pulmonary region. Impaction and sedimentation processed taking place in this region explain these results (Behera et al., 2015). TB region is characterized by higher average deposition rate of $PM_{10-2.5}$ and $PM_{2.5-1}$ in kids. This difference is especially significant when children are developing Light and Heavy intensity activities. This is a consequence of the smaller TB system in kids, implying a higher surface/volume ratio, and, therefore, higher particle impaction. Regardless activity and PM fractions, kids experienced the highest DF in lungs, due to their greater ventilation ratio per body mass. $PM_{2.5-1}$ was the prevalent fraction in this region, where $PM_{10-2.5}$ is almost absent.

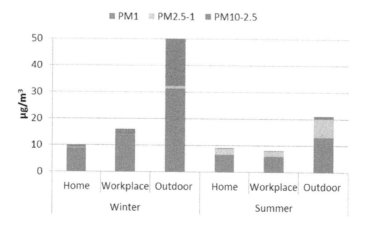

FIGURE 12.9   Levels of PM in the different microenvironments.

TABLE 12.14

Deposition Fractions for the Different PM Size Fractions for Child, Adults, and Retired in the Nose/Throat (Head), Pulmonary/Tracheobronchial (TB), and Pulmonary/Alveolar (P) Regions

	Kids			Adult			Retired		
	Head	TB	P	Head	TB	P	Head	TB	P
PM₁									
Sleeping	0.26	0.04	0.17	0.14	0.05	0.09	0.11	0.06	0.10
Sitting	0.27	0.04	0.16	0.16	0.05	0.10	0.13	0.06	0.12
Light indoor	0.32	0.05	0.11	0.35	0.04	0.08	0.37	0.04	0.07
Light outdoor	0.32	0.05	0.11	0.35	0.04	0.08	0.37	0.04	0.07
Heavy	0.34	0.07	0.08	0.54	0.04	0.05	0.53	0.04	0.05
Average	**0.27**	**0.04**	**0.16**	**0.20**	**0.05**	**0.09**	**0.20**	**0.05**	**0.10**
PM₂.₅₋₁									
Sleeping	0.28	0.06	0.37	0.47	0.09	0.17	0.39	0.11	0.19
Sitting	0.29	0.06	0.36	0.51	0.08	0.19	0.43	0.10	0.22
Light indoor	0.35	0.13	0.23	0.77	0.03	0.10	0.79	0.03	0.09
Light outdoor	0.35	0.13	0.23	0.77	0.03	0.10	0.79	0.03	0.09
Heavy	0.36	0.34	0.09	0.89	0.02	0.05	0.89	0.02	0.05
Average	**0.27**	**0.08**	**0.33**	**0.56**	**0.07**	**0.16**	**0.53**	**0.08**	**0.17**
PM₁₀₋₂.₅									
Sleeping	0.64	0.27	0.04	0.91	0.03	0.01	0.90	0.05	0.01
Sitting	0.68	0.27	0.02	0.92	0.03	0.01	0.91	0.04	0.01
Light indoor	0.80	0.19	0.00	0.95	0.01	0.01	0.95	0.01	0.00
Light outdoor	0.80	0.19	0.00	0.95	0.01	0.01	0.95	0.01	0.00
Heavy	0.86	0.14	0.00	0.96	0.00	0.00	0.96	0.00	0.00
Average	**0.68**	**0.26**	**0.03**	**0.93**	**0.03**	**0.01**	**0.92**	**0.03**	**0.00**

Note: Average values are calculated taking into account the time spent in each activity on a daily basis.

12.3.3.3 Deposited Masses

Deposited masses in the different regions of the respiratory tract for the three population groups can be seen in Figure 12.10. In line with outdoor concentrations, greater deposited masses are experienced in winter. The influence of outdoor concentrations is also patent in the contribution from different PM fractions to total deposited mass. Thus, for instance, $PM_{10-2.5}$ deposited share in winter is higher than in summer, since outdoor contributions of this fraction are higher in the winter.

Head is the region registering highest values of deposited masses, as a consequence of the high DF values obtained for this region. Consequently, irritation and inflammation in the upper respiratory tract are the diseases most likely to happen. Retired people experience the highest deposition in this region for both seasons. This result is explained because this population group is the one spending more time outdoors, and, consequently, exposed to greater PM levels.

Tracheobronchial region is characterized by being the region with lowest overall deposited mass. In this region, contributions from $PM_{10-2.5}$ are poor, except in the case of kids in winter. This result is directly related with the greater levels of $PM_{10-2.5}$ experienced in winter and the higher DF of kids in TB, as explained before.

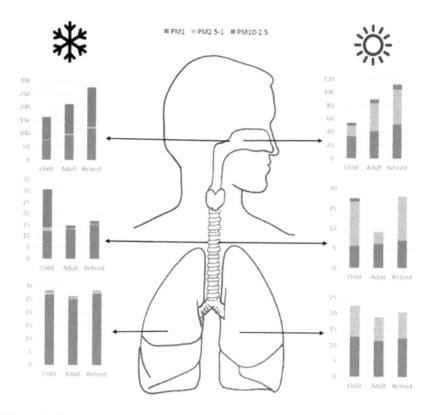

FIGURE 12.10 Daily deposited masses (µg) per region, fraction, season, and population group.

As in TB, deposition in lungs is almost absent of $PM_{10-2.5}$. Deposition among the three groups is very similar, which is especially harmful for kids. Since body weight in children is close to half the body weight in adults and retired, deposited dose in this group is close to twice the dose in grown-up groups. This result is a great concern, since immunologic system in kids is still not fully developed.

12.3.4 CONCLUSIONS AND LIMITATIONS

The exercise here presented depicts a way of calculating human exposure to environmental PM by using two different software models. Particulate infiltration simulation shows the high influence of outdoor PM levels on indoor concentrations and the importance of ventilation rates to control them. Dosimetry simulation presents how different activities, performed by diverse population groups could lead to important changes in PM deposition within the human respiratory tract. The combination of both results has proved to be useful in the study of exposure to PM, which highly involved in the assessment of human risks.

The example shown here is one of the simplest cases. Constant values have been used for most of the inputs, and indoor sources were considered nulls. By adjusting these parameters to ranges instead of fixed values, simulations will be more realistic. However, the approach here described shows higher accuracy than conventional calculation methods, being lower time and resource-consuming and easy to undertake.

12.4 EXAMPLE 3: ENVIRONMENTAL IMPACT ANALYSIS OF AN INDUSTRIAL SEPARATION PROCESS, APPLYING LCA AND ERA

12.4.1 INTRODUCTION

An environmental impact analysis of the isopentane separation process from a naphtha stream was developed. The system consists of a distillation column, which requires electricity and steam. The evaluation covered the distillation process and units of utility production. In this academic case study, the environmental assessment was carried out in two ways: potential impact and site-specific impact assessments. By this, the most important aspects of LCA, impact pathway analysis (IPA) and environmental risk assessment (ERA) approaches, were analyzed.

This example is in the heart of chemical engineering. The LCA study carried out is a cradle-to-gate study. This means that one industrial process, situated in the petrochemical complex of the Tarragona region (Spain), with its associated environmental loads due to raw material and energy consumption is considered. Two related sub-processes, electricity generation and steam production, both by on-site small fossil-based thermal plants, are simulated more in detail. The data for the other materials are taken from the LCA software database. Based on the LCA results, one process is to be further assessed by site-specific impact assessment. In this example, we will try to carry out a more generic IPA first, and then a more detailed ERA to deal with open questions not solvable with the IPA software Ecosense.

12.4.2 GOAL AND SCOPE DEFINITION

In this case study, the environmental impact analysis is applied to the separation process from a naphtha stream in order to evaluate the environmental effects of all process stages, including its requirements of electric power and steam. The system separates isopentane from naphtha about 16 t/h production. In this sense, 16 t/h production has been considered the functional unit for the LCA. The separation plant, consisting of a distillation column, the electricity generation, and the steam production, were the three main steps taken into account. Figure 12.11 shows the flow diagram of the studied system with the inputs and outputs considered.

12.4.3 POTENTIAL IMPACT ASSESSMENT

In order to carry out the inventory analysis, the aforementioned three main stages were considered: steam production, electricity generation and, separation of isopentane. The required information in the collection data step was obtained from the real process. Nevertheless, in the case of steam generation, some values were taken from the TEAM database. TEAM is the LCA software used (see Chapter 2) to carry out the LCA for the process and to evaluate the total amount of a specific pollutant in any stage. This software calculates the environmental loads produced by a functional unit starting from given process data (Figure 12.11).

TEAM consists of an integrated group of software tools to model and analyze any system. A modeling tool is used to describe physical operations. It allows a large database to be built and the LCI to be calculated for any system (inventories conducted for a product life cycle or other consistent systems). The analysis tools for applying LCIA methods (i.e., potential impacts) based on the inventories results allow further evaluation of the system under study. The scheme created in TEAM to obtain the inventory and potential impacts is represented in Figure 12.12.

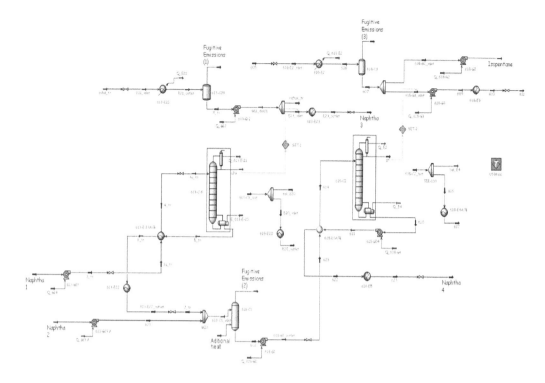

FIGURE 12.11   Flow diagram of the separation process.

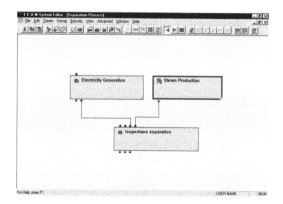

FIGURE 12.12   Diagram of the separation process obtained from TEAM.

12.4.3.1   Inventory

The main data considered and introduced in the TEAM software for creating the inventory are summarized in Table 12.15. The process has two different inputs for naphtha (corresponding to #1 and #2). Likewise, naphtha #3 and #4 represent outputs from the system. Overall energy

TABLE 12.15
Main Data (Inputs and Outputs) of the Process

Flow	Inputs	Account (kg/h)	Account (kg/kg Isopentane)	Account Total (kg/year)
	Naphtha (# 1)	2.83×10^4	1.76	2.24×10^8
	Naphtha (# 2)	7.15×10^4	4.45	5.66×10^8
Raw material	Coal	1.14×10^2	7.11×10^{-3}	9.06×10^5
	Air	2.01×10^3	1.25×10^{-1}	1.59×10^7
	Water	1.95×10^3	1.21×10^{-1}	1.54×10^7
	Outputs			
Product and by-products	Isopentane	1.61×10^4	1.00	1.27×10^8
	Naphtha (# 3)	5.10×10^2	3.17×10^{-2}	4.04×10^8
	Naphtha (# 4)	8.24×10^4	5.12	6.52×10^8
	CO_2	2.15×10^2	1.34×10^{-2}	1.70×10^6
	SO_2	1.00	6.23×10^{-5}	7.93×10^3
	NOx	4.37×10^{-1}	2.72×10^{-5}	3.46×10^3
	Particulate matter	1.67	1.04×10^{-4}	1.32×10^4
Emissions	As	1.82×10^{-1}	1.13×10^{-5}	1.44×10^3
	Hg	3.39×10^{-4}	2.11×10^{-8}	2.69
	Ni	1.06×10^{-3}	6.60×10^{-8}	8.40
	Pb	1.89×10^{-1}	1.17×10^{-5}	1.50×10^3
	Pb	1.42×10^{-1}	8.84×10^{-6}	1.13×10^3
	VOC (as fugitive emissions)	8.37×10^2	5.21×10^{-2}	6.63×10^6
Solid waste	Inorganic	6.36	3.96×10^{-4}	5.04×10^4
Water waste	COD	3.82×10^{-5}	2.37×10^{-9}	3.02×10^{-1}

requirement is 7.2 MJ/h, but the inventory was obtained taking into account that 30% of the total energy is used to produce steam.

Table 12.16 shows the inventory corresponding to the separation process. The comments in parentheses indicate the source, compounds under consideration in a specific group, properties, or specific names for the acronyms. As can be seen, volatile organic carbons (VOCs; as fugitive emissions) are the contribution for releases that can be attributed to the separation process, corresponding to 79% of the total amount of generated polluting agents (on a mass basis).

12.4.3.2 Impact Assessment

To carry out this phase of the LCA case study, the following five impact categories have been considered: AA, WE, HT, GE, and TE. The specific environmental impact indicator used in each of the environmental impact categories mentioned before is shown in Table 12.17.

The previously inventoried environmental loads are classified in different impact categories. Specific characterization factors are related to each indicator to evaluate the potential contribution of each environmental load. These factors depend on the method used; multiplying them by the environmental loads is possible to obtain the corresponding potential contributions.

Using the factors provided from TEAM, the results obtained can be seen in Table 12.18.

TABLE 12.16

Inventory of Process for Separating Isopentane from Naphtha

Flow	Units	Total	Steam Production	Isopentane Separation	Electricity Generation
		Inputs			
(r) Coal (in ground)	kg	113.64	0	0	113.64
(r) Natural gas (in ground)	kg	2.64×10^{-2}	2.64×10^{-2}	0	0
(r) Oil (in ground)	kg	2.12×10^{-2}	2.12×10^{-2}	0	0
(r) Air	kg	215.09	0	0	215.09
Naphtha (# 1)	kg	72,543.60	0	72,543.6	0
Naphtha (# 2)	kg	28,713.10	0	28,713.1	0
Water used (total)	liter	1,968.56	2.48×10^{-1}	0	1,968.32
Water: unspecified origin	liter	2.48×10^{-1}	2.48×10^{-1}	0	0
		Outputs			
(a) Carbon dioxide (CO_2, fossil)	g	218,278	140.29	0	218,138
(a) Carbon monoxide (CO)	g	4.58×10^{-2}	4.58×10^{-2}	0	0
(a) Hydrocarbons (unspecified)	g	1.93	1.93	0	0
(a) Lead (Pb)	g	144.07	0	0	144.07
(a) Mercury (Hg)	g	3.44×10^{-1}	0	0	0.34
(a) Metals (unspecified)	g	9.17×10^{-1}	9.17×10^{-4}	0	0
(a) Nitrogen oxides (NO_x as NO_2)	g	444.75	1.38	0	443.38
(a) Particulates (unspecified)	g	1.81	1.19×10^{-1}	0	1.69
(a) Sulfur oxides (SO_x as SO_2)	g	1,015.88	1.28	0	1,014.6
(a) Fugitive emissions (VOC)	g	837,000	0	837,000	0
(w) BOD5 (biochemical oxygen demand)	g	9.17×10^{-4}	9.17×10^{-4}	0	0
(w) COD (chemical oxygen demand)	g	9.17×10^{-4}	9.17×10^{-4}	0	0
(w) Hydrocarbons (unspecified)	g	9.17×10^{-3}	9.17×10^{-3}	0	0
(w) Suspended matter (unspecified)	g	9.17×10^{-4}	9.17×10^{-4}	0	0
Isopentane	kg	16,335.00	0	16,335	0
Naphtha (# 3)	kg	517.44	0	517.44	0
Naphtha (# 4)	kg	83,602.70	0	83,602.7	0
Waste (municipal and industrial)	kg	9.17×10^{-4}	9.17×10^{-4}	0	0
Waste (total)	kg	9.49×10^{-4}	9.49×10^{-4}	0	0
Waste: mineral (inert)	kg	2.75×10^{-5}	2.75×10^{-5}	0	0
Waste: slags and ash (unspecified)	kg	4.58×10^{-6}	4.58×10^{-6}	0	0

Note: (r): natural resources; (a): air emission; (w): water emission.

TABLE 12.17

Environmental Impact Categories and Indicators Used in TEAM

Acronym	Category	Indicator
AA	Air acidification	CML
WE	Water eutrophication	CML
HT	Human toxicity	CML
GE	Greenhouse effect	IPPC (direct, 100 years)
TE	Terrestrial ecotoxicity	USES 2.0

TABLE 12.18

Isopentane Separation Process Impact Assessment Obtained with TEAM™ Software

Impacts	Characterization Factors	Total
CML air acidification (g equiv. H$^+$)		4.14×10^1
(a) Nitrogen oxides (NO$_x$ as NO$_2$)	2.17×10^{-2}	9.67
(a) Sulfur oxides (SO$_x$ as SO$_2$)	3.12×10^{-2}	3.17×10^1
CML eutrophication (g equiv. PO$_4$)		5.78×10^1
(a) Nitrogen oxides (NO$_x$ as NO$_2$)	1.30×10^{-1}	5.78×10^1
(w) COD (chemical oxygen demand)	2.20×10^{-2}	2.02×10^{-5}
CML human toxicity (g)		2.47×10^4
(a) Carbon monoxide (CO)	1.20×10^{-2}	5.50×10^{-4}
(a) Lead (Pb)	1.60×10^2	2.31×10^4
(a) Mercury (Hg)	1.20×10^2	4.13×10^1
(a) Nitrogen oxides (NO$_x$ as NO$_2$)	7.80×10^{-1}	3.47×10^2
(a) Sulfur oxides (SO$_x$ as SO$_2$)	1.20	1.22×10^3
IPCC greenhouse effect (direct, 100 years) (g equiv. CO$_2$)		2.18×10^5
(a) Carbon dioxide (CO$_2$, fossil)	1.00	2.18×10^5
USES 1.0 terrestrial ecotoxicity (g equiv. 1-4-dichlorobenzene)		6.06×10^6
(a) Lead (Pb)	1.10×10^4	1.5×10^6
(a) Mercury (Hg)	1.30×10^7	4.47×10^6

Note: (a): air emission; (w): water emission.

12.4.3.3   Interpretation of the Life-Cycle Assessment

Looking at the inventory and the impact assessment results (Tables 12.16 and 12.18), we see that the important fugitive emissions (VOC) have not resulted in any potential impact in the LCIA phase. This is due to the fact that photochemical oxidant formation was not chosen as an impact category for reconsideration. Evidently, by this, the study is also an example on how important information can be lost from the LCI to the LCIA phase by subjective choices of the LCIA indicators.

12.4.4   SITE-SPECIFIC ENVIRONMENTAL IMPACT ANALYSIS

The current LCIA results presented in Table 12.18 are visibly predominated by the electricity generation sub-process. This becomes clear when looking back to the inventory in Table 12.16. Therefore, electricity generation is chosen for further site-specific environmental impact analysis. First, an impact pathway analysis is carried out with the integrated assessment model Ecosense for the priority air macro-pollutants. Next an exposure risk assessment is performed for a relevant air emission of a micropollutant that is not directly covered by the integrated assessment model used, but that could signify a risk for the neighborhood of the industrial separation process. For this particular risk assessment study, mercury—not PCDD/Fs—is chosen for further analysis (see studies for PCDD/Fs in Chapters 4 and 9). Polychlorinated dibenzo-p-dioxins (PCDDs) and polychlorinated dibenzofurans (PCDFs) are common highly toxic contaminants. PCDD/Fs are generally referred to as "dioxins" and are a group of 210 different structural congeners that are not produced intentionally but as by-products of numerous processes, in particular unwanted by-products of incineration and uncontrolled burning. Actions should be initiated as soon as possible to reduce human-generated releases of mercury due to the documented, significant adverse impacts on human health throughout the world (UNEP, 2003). The Minamata Convention adopted in 2013 is designed to limit mercury use and emissions internationally.

12.4.4.1 Impact Pathway Analysis Applied to the Energy Generation in the Industrial Separation Process

The site-specific environmental assessment has been carried out using the Ecosense model, which was developed to support the assessment of priority impacts resulting from the exposure to airborne pollutants, namely, impacts on health, crops, building materials, and ecosystems as described in Chapter 4.

To cover different pollutants and different scales, Ecosense provides two air transport models completely integrated into the system.

All input data required to run the windrose trajectory model (WTM) are provided by the Ecosense database. A set of site-specific meteorological data must be added by the user to perform the ISCST-2 model. The results of the air dispersion model rely on four basic data sets: (1) meteorological conditions; (2) facility characteristics; (3) location of buildings near the emission sources; and (4) location of receptors (distance to the emission source and elevation in the terrain). The geographic data are crucial for this type of assessment. In this case, the Tarragona region in Spain was selected as the area of study and information about elevation, population density, and the meteorological situation was provided.

The physical impacts and, as far as possible, the resulting damage costs are calculated by means of pollutant short-range and long-range transport and conversion models and the exposure–response functions for several receptors (human and ecosystems). These can be selected by the user for each individual grid cell (for the case under study, the grid cell corresponds to the Tarragona region in Spain), taking into account the information on receptor distribution and concentration levels of air pollutants from the reference environmental database (IER, 1997). Table 12.19 lists the results of the most important impacts for human health and ecosystem.

According to this table, no impacts or damages are caused by heavy metal emissions even though the inventory shows these loads in the process (Table 12.16), because the heavy metal emissions are smaller than those manageable by the software (Ecosense). Therefore, in the next section, environmental risk of the heavy metal mercury emissions will be assessed to further analyze the environmental impact of those substances.

12.4.4.2 Fate and Exposure Analysis with Risk Assessment of Mercury for the Electricity Generation in the Industrial Separation Process

An environmental risk procedure implies the inherent capacity of the substances to cause negative effects and the exposition or interaction of these substances within receptors (ecosystems or humans). These aspects are closely related to the fate analysis and the distribution in the environment.

In order to know the fate and future exposition of mercury from the electricity generation in the separation process, the software CalTOX, described in Chapter 4 and Example 1 (Section 12.2), was applied to mercury for the area around 1000 m from the emission source for 1 year of continuous emission.

The results of the fate and exposure assessment for mercury emissions are shown in Table 12.20 and in the following. According to this model, the concentrations on the compartments are constant for the considered region; however, the exposure is changing according to distance from the point of emission. (The exposure changes proportionally at the distance of the emission.) Table 12.21 shows the exposure media of mercury for the studied region. The daily human doses through several exposure types were calculated based on the modeled compartments' concentration of emissions. The main exposure pathway is air inhalation. Also important is exposure through fish ingestion.

Depending on the level of exposure, in principle, adverse health effects can be associated with all substances. In this sense, risk characterization is a dose–response analysis that compares the current human exposure with a defined level of exposure. On the other hand, hazard ratio expresses noncarcinogenic effects as a proportion of an exposure intake rate and a reference dose related to the selected exposure pathways and chronic exposure duration (the hazard ratio for mercury is 8.25×10^{-5}).

TABLE 12.19
Impacts and Damage Assessment for Energy Generation in Process

Receptor	Impacts		Damages	
	Unit	Value	Unit	Value
Human Health				
Impact: chronic YOLL				
Dose–response functions				
Pollutant: TSP	Years	6.69	mUS$	5.62×10^5
Pollutant: nitrates	Years	3.71×10^{-2}	mUS$	3.71×10^4
Pollutant: sulfates	Years	8.82×10^{-2}	mUS$	7.47×10^3
Impact: acute mortality				
Dose–response functions				
Pollutant: TSP	Cases	6.41×10^{-2}	mUS$	2.00×10^5
Pollutant: nitrates	Cases	3.57×10^{-4}	mUS$	3.54×10^5
Pollutant: sulfates	Cases	1.16×10^{-3}	mUS$	2.69×10^3
Pollutant: NO_x	Cases	5.34×10^{-2}	mUS$	1.66×10^5
Pollutant: SO_2	Cases	1.14×10^{-1}	mUS$	2.64×10^2
Ecosystem				
Dose–response functions from: UN-ECE 1997				
Pollutant: SO_2				
Impact: SO_2 exceedance area	km² exceedance area	0.00	mUS$	NA
Impact: REW SO_2 exceedance area	km² exceedance area	9.44×10^{-5}	mUS$	NA
Impact: RCW SO_2 ecosystem area	km² exceedance area	1.25×10^{-3}	mUS$	NA
Pollutant: NO_x				
Impact: NO_x exceedance area	km² exceedance area	0.00	mUS$	NA
Impact: REW NO_x exceedance area	km² exceedance area	0.00	mUS$	NA
Impact: RCW NO_x ecosystem area	km² exceedance area	4.97×10^{-4}	mUS$	NA

Source: IER, Ecosense, Version 2.0, Software, University of Stuttgard, Germany, 1997.

Note: REW: relative exceedance weighted; RCW: relative concentration weighted; RDW: relative deposition weighted; NA: not available; $m = 10^{-3}$.

TABLE 12.20
Time Average Concentration in On-Site Environmental Media

Compartment	Unit	Mercury
Air	mg/m³	3.13×10^{-10}
Total leaf	mg/kg(total)	1.92×10^{-12}
Ground-surface soil	mg/kg(total)	2.96×10^{-9}
Root-zone soil	mg/kg(total)	2.13×10^{-9}
Vadose-zone soil	mg/kg(total)	1.23×10^{-9}
Groundwater	mg/L(water)	6.99×10^{-14}
Surface water	mg/L	2.39×10^{-12}
Sediment	mg/kg	2.09×10^{-9}

TABLE 12.21

Exposure Media Concentrations for Mercury

Exposure	Air (gases)	Air (dust)	Ground Soil	Root Soil	Ground water	Surface Water
Indoor air (mg/m³)	1.91×10^{-8}	1.31×10^{-14}	8.18×10^{-15}	2.02×10^{-12}	1.05×10^{-18}	5.61×10^{-16}
Bathroom air (mg/m³)	0	0	0	0	1.35×10^{-16}	7.21×10^{-14}
Outdoor air (mg/m³)	1.91×10^{-8}	1.31×10^{-14}	NA	NA	NA	NA
Tap water (mg/L)	0	0	0	0	3.50×10^{-14}	7.97×10^{-11}
Exposed produce (mg/kg)	1.92×10^{-12}	1.32×10^{-18}	9.27×10^{-10}	2.02×10^{-17}	2.49×10^{-11}	5.67×10^{-8}
Unexposed produce (mg/kg)				2.28×10^{-8}	2.41×10^{-11}	5.49×10^{-8}
Meat (mg/kg)	3.73×10^{-9}	2.55×10^{-15}	2.63×10^{-10}	1.94×10^{-18}	2.39×10^{-12}	5.44×10^{-9}
Milk (mg/kg)	1.10×10^{-9}	7.51×10^{-16}	8.83×10^{-11}	8.08×10^{-19}	9.95×10^{-13}	2.27×10^{-9}
Eggs (mg/kg)	0	0	0	0	0	0
Fish and seafood (mg/kg)	0	0	0	NA	0	3.20×10^{-6}
Household soil (mg/kg)	0	0	1.36×10^{-7}	1.12×10^{-7}	0	0
Swimming water (mg/L)	0	0	NA	0	NA	1.59×10^{-10}

Note: NA: not available.

12.4.5 INTERPRETATION OF THE ENVIRONMENTAL IMPACT ANALYSIS FOR THE INDUSTRIAL SEPARATION PROCESS

The LCI indicates the relevance of fugitive emissions (VOC) in the case of an industrial separation process of isopentane from naphtha. The LCIA in relation to the LCI allows the conclusion that the energy generation sub-process is the most relevant process of those considered. Therefore, this process is selected for further analysis if the potential environmental impacts correspond to actual impacts, i.e., damages.

The IPA shows that main damages are produced by the particles, NO_x, and the secondary pollutants nitrates and sulfates. This is in line with the results obtained for the MSWI case study (see Chapter 9). The ERA for mercury points out that there is a very low risk for human health impacts in the neighborhood due to mercury exposure based on the fictitious data used for the industrial separation process example. However, this risk is higher than 10^{-6} and is, therefore, not really acceptable according to guidelines mentioned in Chapter 4. A further reduction of the mercury emissions in electricity generation process is thus recommended.

Altogether, this comprehensive environmental impact analysis gives a much more complete picture of the environmental implications of the industrial separation process than each tool applied independently. Moreover, since the databases are common and a lot of work is needed for their collection, the subsequent application of the different analytical tools seems to be a way to get ahead in the future.

12.5 CONCLUSIONS FROM THE APPLICATIONS

Three examples have been presented in this chapter to evaluate the principle applicability of the strategy outlined in Chapter 10 to integrate LCA and ERA, where possible, in industrial processes other than the municipal solid waste incineration:

- Example 1 is a direct continuation of the MSWI case study presented through various chapters. In the same way as for waste incineration, site-oriented impact assessment makes sense for landfilling. The site-specific study is carried out as a demonstration project for one pollutant only. More air pollutants could easily be checked by the same scheme; for emissions to water and soil, other models need to be used.

- Example 2, on the other hand, shows the use of complementary tools to evaluate particulate matter (PM) exposure in line with Chapter 4. A comprehensive calculation is undertaken to assess particulate matter exposure of three population groups.
- Example 3 demonstrates the applicability of the strategy to an industrial process in the area of chemical engineering. The subsequent application of the different analytical tools gives a much more complete picture of the environmental implications of the industrial separation process than each tool applied independently.

This means the cases show the principle feasibility and the existing limitations of the integration of life-cycle and risk assessment, and clearly indicate that the same basis of data can be used. However, the examples presented here must be put into real applications within a decision-making context to be effective. Some further adaptations—especially simplifications of the links between the different assessment tools—are highly recommended in order to facilitate non-academic applications. We need to move from theory to practice in this area.

Circular Economy and the chemicals policy, which both take a life cycle or value chain approach, are currently major areas of debate in the EU and have the potential to foster the application of the integrative approach of LCA and ERA presented in this book. Looking into related policy papers, in several places, it is pointed out that interaction is needed, i.e., an integrated approach for exchanging information on the chemical/product, preventing products containing harmful chemicals, avoiding processes applying and generating hazardous substances and, consequently, avoiding emissions of chemicals.

A good starting point could also be environmental risk assessments in which the point of departure for the assessment is usually the chemical, i.e., more or less upstream in the life cycle, whereas LCA considers the functional unit, i.e., the function that the product delivers, which is further downstream. A weak point in many ERAs is the estimation of the use and disposal emissions of the chemicals. LCA methodology may potentially improve these estimates. Simultaneously, risk assessment methodology may assist in generating upstream information in life-cycle assessments—often a weak point in many LCAs. As proposed and shown in this book, data sharing for an integrated life-cycle and risk assessment seems to be the way to proceed.

REFERENCES

ARA, 2014. ARA:Products: MPPD [WWW Document]. URL http://www.ara.com/products/mppd.htm (accessed August 24, 2015).

Behera, S., Betha, R., Huang, X., Balasubramanian, R., 2015. Characterization and estimation of human airway deposition of size-resolved particulate-bound trace elements during a recent haze episode in Southeast Asia. *Environ. Sci. Pollut. Res.* 22, 4265–4280. doi:10.1007/s11356-014-3645-6.

Brown, J.S., Gordon, T., Price, O., Asgharian, B., 2013. Thoracic and respirable particle definitions for human health risk assessment. *Part. Fibre Toxicol.* 10, 12. doi:10.1186/1743-8977-10-12.

Carrascosa, A., Fernández, J.M., Fernández, C., Ferrández, A., Longás, López-Siguero, J.P., Sánchez, E., Sobradillo, B., Yeste, F., 2008. Estudio transversal español de crecimiento 2008. Parte II: valores de talla, peso e índice de masa corporal desde el nacimiento a la talla adulta. *An. Pediatría* 68, 552–569.

Chen, C., Zhao, B., 2011. Review of relationship between indoor and outdoor particles: I/O ratio, infiltration factor and penetration factor. *Atmos. Environ.* 45, 275–288. doi:10.1016/j.atmosenv.2010.09.048.

Ciambrone, D.F. (1997). *Environmental Life Cycle Analysis*, Lewis Publishers, CRC Press, Boca Raton, FL.

Cohen Hubal, E.A., Sheldon, L.S., Burke, J.M., McCurdy, T.R., Berry, M.R., Rigas, M.L., Zartarian, V.G., Freeman, N.C., 2000. Children's exposure assessment: A review of factors influencing Children's exposure, and the data available to characterize and assess that exposure. *Environ. Health Perspect.* 108, 475–486.

DTSC, 1992a. Guidance for site characterization and multimedia risk assessment for hazardous substances release sites, vol. 2, chap. 2, documentation of assumptions used in the decision to include and exclude exposure pathways, a report prepared for the State of California, Department of Toxic Substances Control, Lawrence Livermore National Laboratory, Livermore, CA, UCRL-CR-103462.

DTSC, 1992b. Guidance for site characterization and multimedia risk assessment for hazardous substances release sites, vol. 2, chap. 3, guidelines for the documentation of methodologies, justification, input, assumptions, limitations, and output for exposure models, a report prepared for the State of California, Department of Toxic Substances Control, Lawrence Livermore National Laboratory, Livermore, CA, UCRL-CR-103460.

DTSC, 2002. CalTOX version 2.3 (www-pages), Department of Toxic Substances Control of the California EPA, State of California, https://www.dtsc.ca.gov/AssessingRisk/ctox_dwn.cfm, July 2002.

Ecobalance, 1997. TEAM–Tool for Environmental Analysis and Management–Version 2, Software, Paris, France.

Ecobalance, 1998. EIME – Environmental Information and Management Explorer, Software, Paris, France.

Ecobalance, 1999. WISARD–Waste-Integrated Systems Assessment for Recovery and Disposal, Software, Paris, France.

Ercan, Ö., Dinçer, F., 2015. Atmospheric concentrations of PCDD/Fs, PAHs, and metals in the vicinity of a cement plant in Istanbul. *Air Qual. Atmos. Heal.* 9, 159–172. doi:10.1007/s11869-015-0314-y.

He, C., Morawska, L., Gilbert, D., 2005. Particle deposition rates in residential houses. *Atmos. Environ.* 39, 3891–3899. doi:10.1016/j.atmosenv.2005.03.016.

ICRP, 1994. Human respiratory tract model for radiological protection: A report of a task group of the international commission on radiological protection. *Ann ICRP* 24, 1–482.

IEC, 2012. Enquesta de l'ús del temps 2010–2011: Principals resultats. – (Estadística Social). Institut d'Estadística de Catalunya.

IER, 1997. Ecosense, Version 2.0, Software, University of Stuttgard, Germany.

INE, 2012. Índice de masa corporal por grupos de edad. 2011–2012. Inst. Nac. estadística. http//www.ine.es/jaxi/tabla.do?type=pcaxis&path=/t00/mujeres_hombres/tablas_1/l0/&file=d06001.px.

INSEE, 2010. Conditions de vie-Société–Depuis 11 ans, moins de tâches ménagères, plus d'Internet. Insee. http://www.insee.fr/fr/themes/document.asp?ref_id=ip1377.

Morawska, L., Afshari, A., Bae, G.N., Buonanno, G., Chao, C.Y.H., Hänninen, O., Hofmann, W. et al., 2013. Indoor aerosols: From personal exposure to risk assessment. *Indoor Air* 23, 462–487. doi:10.1111/ina.12044.

Orosa, J.A., Baaliña, A., 2008. Passive climate control in Spanish office buildings for long periods of time. *Build. Environ.* 43, 2005–2012. doi:10.1016/j.buildenv.2007.12.001.

Perez, L., Medina-Ramón, M., Künzli, N., Alastuey, A., Pey, J., Pérez, N., Garcia, R., Tobias, A., Querol, X., Sunyer, J., 2009. Size fractionate particulate matter, vehicle traffic, and casespecific daily mortality in Barcelona, Spain. *Environ. Sci. Technol.* 43, 4707–4714. doi:10.1021/es8031488.

Sánchez-Soberón, F., Mari, M., Kumar, V., Rovira, J., Nadal, M., Schuhmacher, M., 2015a. An approach to assess the Particulate Matter exposure for the population living around a cement plant: modelling indoor air and particle deposition in the respiratory tract. *Environ. Res.* 143, 10–18. doi:10.1016/j.envres.2015.09.008.

Sánchez-Soberón, F., Rovira, J., Mari, M., Sierra, J., Nadal, M., Domingo, J.L., Schuhmacher, M., 2015b. Main components and human health risks assessment of PM10, PM2.5, and PM1 in two areas influenced by cement plants. *Atmos. Environ.* 120, 109–116. doi:10.1016/j.atmosenv.2015.08.020.

Stocks, J., Quanjer, P.H., 1995. Reference values for residual volume, functional residual capacity and total lung capacity. ATS Workshop on Lung Volume Measurements. Official Statement of The European Respiratory Society. *Eur. Respir. J.* 8, 492–506.

UNEP, 2003. Global mercury assessment, Report, Geneva, Switzerland.

United Nations Economic Commission for Europe (UN-ECE), 1991. Protocol to the 1979 convention on long-range transboundary air pollution concerning the control of emissions of volatile organic compounds or their transboundary fluxes. Report ECE/EB AIR/30. UN-ECE, Geneva, Switzerland.

U.S. EPA, 1989. Risk assessment guidance for superfund volume in human health evaluation manual (part A), Office of Emergency and Remedial Response, EPA/540/1-89/002.

U.S. EPA, 1995. Users' Guide for the Industrial Source Complex (ISC3)—Dispersion Models Volumes I-II. EPA-454/B-95-003a, EPA-454/B-95-003b. Triangle Park, NC.

U.S. EPA, 2000. Indoor Air Quality Modeling [WWW Document]. https://www.epa.gov/air-research/simulation-tool-kit-indoor-air-quality-and-inhalation-exposure-iaqx (accessed January 1, 2017).

U.S. EPA, 2011. *Exposure Factors Handbook*, 2011 edition, National Centre of Environmental Assessment, U.S. Environmental Protection Agency, Washington, DC, September 2011.

U.S. EPA, 2015. Health|Particulate Matter|Air & Radiation [WWW Document]. http://www3.epa.gov/pm/health.html (accessed June 11, 2015).

Wallace, L.A., Emmerich, S.J., Howard-Reed, C., 2002. Continuous measurements of air change rates in an occupied house for 1 year: The effect of temperature, wind, fans, and windows. *J Expo Anal Env. Epidemiol* 12, 296–306. doi:10.1038/sj.jea.7500229.

WHO, 2014. Ambient (outdoor) air quality and health [WWW Document]. Fact sheet 313. http://www.who.int/mediacentre/factsheets/fs313/en/ (accessed June 11, 2015).

Yeh, H.-C., Schum, G.M., 1980. Models of human lung airways and their application to inhaled particle deposition. *Bull. Math. Biol.* 42, 461–480. doi:10.1007/bf02460796.

Index

Note: Page numbers in italic and bold refer to figures and tables, respectively.

Milton Keynes UK
Ingram Content Group UK Ltd.
UKHW051943071024
449327UK00026B/2142

9 780367 570880